METHODS IN MOLECULAR BIOLOGY™

Series Editor
John M. Walker
School of Life Sciences
University of Hertfordshire
Hatfield, Hertfordshire, AL10 9AB, UK

For further volumes:
http://www.springer.com/series/7651

Mass Spectrometry in Food Safety

Methods and Protocols

Edited by

Jerry Zweigenbaum

Agilent Technologies, Wilmington, DE, USA

Humana Press

Editor
Jerry Zweigenbaum, Ph.D.
Agilent Technologies
Wilmington, DE
USA
j_zweigenbaum@agilent.com

ISSN 1064-3745 e-ISSN 1940-6029
ISBN 978-1-4939-5767-5 ISBN 978-1-61779-136-9 (eBook)
DOI 10.1007/978-1-61779-136-9
Springer New York Dordrecht Heidelberg London

Softcover re-print of the Hardcover 1st edition 2011

Printed on acid-free paper

Humana Press is part of Springer Science+Business Media (www.springer.com)

Preface

Food is a complex biological material for which all life on the planet depends and is intertwined with all living things. Thus, the food chain is both a synergistic and competitive system between plants and animals. No one need be reminded that it is a key component of human survival and that we are a part of that system, albeit on the top of that chain. A safe and sufficient food supply is necessary for a healthy and productive population throughout the world. In today's world, food for human consumption is not a local commodity but is obtained through a network of supply and transportation that spans all points of the globe. Nuts from Turkey, fruits from Chili, and shrimp from Vietnam can appear at a local grocery store anywhere in the world.

Bacterial infestation is a major cause of acute toxicity from food and has brought public awareness to pathogenic testing. Where *Salmonella*, *Listeria*, *E. coli*, and other food-borne pathogens have caused sudden and serious (even sometimes fatal) outbreaks, public attention becomes highly focused on the need to assure a safe food supply. As insidious, or maybe even more so, is the possible continued exposure to chemical residues of pesticides, veterinary drugs, chemical contaminants, and naturally produced chemical toxins, such as mycotoxins. This chemical threat to the food supply usually represents chronic toxicity and does not gain the attention that acutely toxic events command. However, as in the case of melamine adulteration, where public awareness was heightened by the acute toxicity incurred, the possibility of chemical contamination of our food remains a serious threat that demands continuous attention.

Because of the competition for fruits, grains, and vegetables with insects, rodents, other small animals, and birds, the use of pesticides is a necessary supplement for farmers to obtain good yields to feed a growing population of people around the world. Through risk assessment and proper application, the use of pesticides is a safe way to assure sufficient food for the world's people. However, the possibility of exposure to elevated levels or to pesticides no longer approved for use places people at risk of chronic toxicity with implications impacting human health from cancer to possible behavior modification. For example, recent studies have implicated the possibility of a correlation with autism and attention deficit disorder. Because of the long-term effects and slow manifestation of chronic exposure, this threat to our food supply may indeed be more insidious than an acute toxic exposure.

Likewise, veterinary drugs are necessary to assure healthy animals and their products that are used for food (e.g., milk, eggs, etc.). However, there are antibiotics that have been banned because of their toxicity to humans. In addition, the overuse of approved antibiotics may cause drug-resistant bacterial strains, and exposure of veterinary drugs to humans through the food supply may directly impact human health. The use of hormones to increase yields for animal production may have deleterious effect and are banned in some parts of the world. This places even yet another dilemma for food producers; where hormones are allowed, meat and animal products may contain residual amounts, and these foods should only be imported to regions where they are not banned. With a world food supply, this is difficult and more disconcerting, in terms of a safe food supply, and it would appear that harmonized good science and practice would be in the best interest of the entire world's population.

A third area of chemical concern to the world's food supply is that of naturally produced toxins. Among these is the category of mycotoxins or toxins that are produced as secondary metabolites of fungi. Unlike bacteria that has to be a live viable organism to cause deleterious health effects, mycotoxins, once produced, are refractory small molecules that have resident times long after the fungus that produced them are gone. Among these are the aflatoxins that are known carcinogens. There are many other mycotoxins that are found in fruits, vegetables, spices, and grains and affect not only people that eat them directly but wildlife and livestock. Again these toxins represent a threat to the food supply, where the insidious effects of long-term chronic toxicity make it difficult to chart their impact on human health. However, scientists around the world are aware of their effects if not actually able to quantify them except in regions of extreme exposure.

The final area of chemical threat to our food supply is that of contaminants. This broad range of chemicals is found in the environment, in processing, and in the packaging of food. This category of residues is classified as those materials that are neither intentionally nor naturally found in our food. One cause of this chemical contamination is the migration of unwanted bi-products of packaging materials into the food. Packaging material is an important component of the safe shipping and preservation of foodstuff and is continually tested to assure that unwanted chemicals are not found in and do not migrate from the packaging material into the food. Packaging material includes plastics bags, coatings of cans, and any other containment of food and beverages. The other route of contaminants through the environment often occurs in the form of persistent organic pollutants or POPS. These compounds remain in the environment long after their use has been banished from society. An example is that of polychlorinated biphenyls which were used exclusively through the 1970s as insulators in transformers and capacitors until their ban in the end of that decade. These compounds are still found in air (dust), water, and soil and do make their way into the food supply.

It is my opinion that total elimination of all the above in the world's food supply is simply impossible. However, keeping harmful chemical residues within acceptable risk levels is not only scientifically reasonable, but also a responsibility that all societies owe each other. The only way to accomplish this is through regulation, and it is for this reason that this book begins with an overview of the regulations around the world. Few dispute that the European Union has led the world in the most up-to-date regulations following sound scientific studies of risk assessment leading to reasonable regulations to meet the goal of ensuring a safe food supply. To give a global perspective, a view of the food safety regulations of China, the USA, and Japan are also given. These four regulatory bodies have both a great influence and stake in both import and export of food throughout the world.

The only way to monitor and enforce these regulations is through extensive food testing, and that is the subject of the remainder of this book. Mass spectrometry has become the enabling technology for both identifying and quantifying low-level chemical residues in one of the most complex biological matrices: food. Even with its high degree of chemical selectivity, or its capability to distinguish one chemical from another, the need for good sample preparation remains. Thus, the next two chapters cover two powerful procedures that have become companions to the powerful techniques of tandem mass spectrometry. The preparative technique known as QuEChERS has become a routine procedure in laboratories performing complex multiresidue pesticide analysis and has found its way into many other applications, including most recently the determination of contaminants in the Gulf of Mexico's oil spill. In addition to this manual approach, automated sample preparation offers its advantages, and thus the reader is offered the opportunity to compare and contrast these important aspects of sample preparation.

The next three chapters cover the complex aspects of testing food samples for pesticide residues. Each chapter covers chromatographic techniques combined with mass spectrometry. Gas chromatography/mass spectrometry has been used for many years for pesticide residue analysis, but even these techniques have experienced rapid advances in recent years, which are covered. The approval and use of more polar pesticides combined with shipment of fresh produce around the world has contributed to the need for rapid analysis, and liquid chromatography/tandem mass spectrometry has advanced to meet that need. The complex procedures and considerations are covered using that technology. Finally, the identification of unexpected or nontargeted pesticides has become increasingly of concern, and mass spectrometry advances that address this need conclude the contributions in this book for pesticide analysis.

Mycotoxins continue to be of major concern to scientists and regulators throughout the world. Most monitoring has centered on the aflatoxins, and there are relatively selective methods for their determination in common use, mainly liquid chromatography combined with fluorescent detection. However, other mycotoxins that do not respond to this technology are finding mass spectrometry to be the analytical method of choice. Methods for some of these residues are given. In the area of testing of antibiotics, an excellent overview is given. This is followed by detailed methodology for monitoring specific antibiotics in both animal and animal products. Likewise, the need to determine hormones and the methods used are described. These chapters combined give the reader an excellent perspective of the requirements for testing veterinary drugs and how mass spectrometry meets the needs of the present day analytical food laboratory.

The final chapters of this book cover the area of chemical contaminants. The description of present day methods for evaluating packaging materials provides in-depth insight. The complex analysis of persistent organic pollutants is thoroughly reviewed. The reader will find that both the overviews and the specific methods provide a comprehensive picture of the state of chemical residue food monitoring in the 21st century. In addition, the contributors represent scientists engaged in food safety from around the world, and thus it is a world perspective. It is this editor's hope that each reader will gain both understanding and appreciation for the contribution of mass spectrometry and those who pioneer its use as it is applied to food testing and food safety.

Wilmington, DE, USA *Jerry Zweigenbaum*

Contents

Contributors

LUTZ ALDER • *Federal Institute for Risk Assessment (BfR), Berlin, Germany*
MEIKE BADEN • *TeLA GmbH, Bremerhaven, Germany*
MARCO H. BLOKLAND • *RIKILT – Instituut voor Voedselveiligheid, Wageningen UR, Wageningen, The Netherlands*
EMMA L. BRADLEY • *The Food and Environment Research Agency (Fera), Sand Hutton, York, YO41 1LZ, UK*
TOMAS CAJKA • *Department of Food Chemistry and Analysis, Faculty of Food and Biochemical Technology, Institute of Chemical Technology, Prague, Czech Republic*
LAURENCE CASTLE • *The Food and Environment Research Agency (Fera), Sand Hutton, York, YO41 1LZ, UK*
GUI-LIANG CHEN • *Shanghai Institute for Food and Drug Control, Shanghai, China*
XIAOGANG CHU • *Chinese Academy of Inspection and Quarantine, Beijing, China*
LEON COULIER • *TNO Quality of Life Utrechtseweg 48, 3704 HE, Zeist, Netherlands*
MALCOLM DRIFFIELD • *The Food and Environment Research Agency (Fera), Sand Hutton, York, YO41 1LZ, UK*
YAN-YAN FANG • *Shanghai Sunrise Pharmaceutical Co., Ltd, Shanghai, China*
FENG FENG • *Chinese Academy of Inspection and Quarantine, Beijing, China*
IMMA FERRER • *Center for Environmental Mass Spectrometry, Department of Environmental Engineering, University of Colorado, Boulder, CO, USA*
PETER FÜRST • *Chemical and Veterinary Analytical Institute Münsterland-Emscher-Lippe, Münster, Germany*
JANA HAJSLOVA • *Department of Food Chemistry and Analysis, Faculty of Food and Biochemical Technology, Institute of Chemical Technology, Prague, Czech Republic*
DOUGLAS G. HAYWARD • *Center for Food Safety and Applied Nutrition, U.S. Food and Drug Administration, College Park, MD, USA*
NORBERT HELLE • *TeLA GmbH, Bremerhaven, Germany*
STEVEN J. LEHOTAY • *United States Department of Agriculture, Agricultural Research Service, Eastern Regional Research Center, Wyndmoor, PA, USA*
KATERINA MASTOVSKA • *Greenfield Laboratories, Nutritional Chemistry and Food Safety, Covance Laboratories, Inc., Greenfield, IN, USA*
CHIN KAI-MENG • *Agilent Technologies, Wilmington, DE, USA*

XUEMEI NIE • *Chinese Academy of Inspection and Quarantine, Beijing, China*
KAJ PETERSEN • *GERSTEL GmbH & Co.KG, Mülheim an der Ruhr, Germany*
SASKIA S. STERK • *RIKILT – Instituut voor Voedselveiligheid, Wageningen UR, Wageningen, The Netherlands*
MASAHIKO TAKINO • *Agilent Technologies, Hachioji, Tokyo, Japan*
HIROKI TANAKA • *Research Center, Suntory Business Expert Limited, Osaka, Japan*
TOSHITSUGU TANAKA • *Kobe Institute of Health, Kobe, Japan*
E. MICHAEL THURMAN • *Center for Environmental Mass Spectrometry, Department of Civil, Environmental, and Architectural Engineering, University of Colorado, Boulder, CO, USA 80309*
WENZHI WANG • *Chinese Academy of Inspection and Quarantine, Beijing, China*
JON W. WONG • *Center for Food Safety and Applied Nutrition, U.S. Food and Drug Administration, College Park, MD, USA*
MILENA ZACHARIASOVA • *Department of Food Chemistry and Analysis, Faculty of Food and Biochemical Technology, Institute of Chemical Technology, Prague, Czech Republic*
FENG ZHANG • *Chinese Academy of Inspection and Quarantine, Beijing, China*
KAI ZHANG • *Center for Food Safety and Applied Nutrition, U.S. Food and Drug Administration, College Park, MD, USA*
JERRY ZWEIGENBAUM • *Agilent Technologies, Wilmington, DE, USA*

Chapter 1

European Union Regulations

Peter Fürst

Abstract

The European Union (EU) has been a leader in the development of both guidance and regulations to ensure food safety throughout the member states. Because of the free movement of food commodities among the countries that belong to the European Union, there is a great need to assure high quality monitoring of both imported food and member state products. The procedures and methods required need to be practical, state-of-the art, and harmonised. The European Commission has developed a network of laboratories and scientific studies to meet this goal. This chapter describes the current Regulations, Directives and Decisions of the European Commission that protect the food supply throughout Europe. Because imported food needs to comply with the EU requirements, and the need to have common compliance throughout the member states, the developed system could be a worldwide template for monitoring the food supply. In addition, the integral role of chromatography hyphenated to mass spectrometry is described.

Key words: European Union, Regulations, Guidance, Directives, Decisions

1. Introduction

The European Union (EU) is an economic and political union of currently 27 Member States with a total of almost 500 million citizens. Since its foundation it has developed a single market through a standardised system of laws which apply in all Member States. The single market guarantees a free movement of people, goods, services, and capital. Treaties (known as "primary" legislation) are the basis for a large body of "secondary" legislation which has a direct impact on the daily lives of EU citizens. Secondary legislation consists mainly of Regulations, Directives, Decisions, and Recommendations adopted by the EU institutions. While Regulations have direct effect and are binding in all Member States, Directives require implementation by national

Jerry Zweigenbaum (ed.), *Mass Spectrometry in Food Safety: Methods and Protocols*,
Methods in Molecular Biology, vol. 747, DOI 10.1007/978-1-61779-136-9_1, © Springer Science+Business Media, LLC 2011

legislation to be effective. In contrast, Decisions only affect those parties to whom they are addressed. The laws, along with EU policies in general, are the result of decisions taken by the institutional triangle made up of the Council which represent national governments, the European Parliament and the European Commission which is responsible for initiating legislation.

Direct free access to European Union Law is provided by EUR-Lex (http://eur-lex.europa.eu/en/index.htm) which contains the Official Journal (OJ) as well as the treaties, legislation, legislative proposals, and in addition offers extensive search facilities.

The general principles and requirements governing food and feed in general, and food and feed safety in particular, at the Community and national level are laid down in Regulation (EC) No. 178/2002 of the European Parliament and of the Council of 28 January 2002 (1). As laid down in Article 1, "this Regulation provides the basis for the assurance of a high level of protection of human health and consumers' interest in relation to food, taking into account in particular the diversity in the supply of food including traditional products, whilst ensuring the effective functioning of the internal market. It establishes common principles and responsibilities, the means to provide a strong science base, efficient organisational arrangements, and procedures to underpin decision making in matters of food and feed safety."

Through this Regulation, the European Food Safety Authority (EFSA) and the Rapid Alert System for Food and Feed (RASFF) are established. According to Article 22, EFSA "shall provide scientific advice and scientific and technical support for the Community's legislation and policies in all fields which have a direct or indirect impact on food and feed safety. It shall provide independent information on all matters within these fields and communicate on risks." Thus, EFSA is responsible for risk assessment, whereas the European Commission is in charge of risk management measures. Whereas, the purpose of the RASFF is to provide the control authorities with an effective tool for exchange of information on measures taken to ensure food safety by establishing a network for the notification of a direct or indirect risk to human health deriving from food or feed.

While the basic rules with regard to the food and feed law are laid down in Regulation (EC) No. 178/2002, a specific harmonised framework of general rules for the organization of official controls at the Community level are established by Regulation (EC) No. 882/2004 of the European Parliament and of the Council of 29 April 2004 (2). The general requirements for methods of sampling and analysis and laboratories are laid down in Articles 11 and 12.

Article 11 stipulates that sampling and analysis methods used in the context of official controls shall comply with relevant Community rules; or (a) if no such rules exist, with internationally recognised rules or protocols, for example those that the European Committee for Standardisation (CEN) has accepted or those agreed in national legislation; or(b) in the absence of the above, with other methods fit for the intended purpose or developed in accordance with scientific protocols.

Where the above does not apply, validation of methods of analysis may take place within a single laboratory according to an internationally accepted protocol. Wherever possible, methods of analysis shall be characterised by the following appropriate criteria:

- Accuracy
- Applicability (matrix and concentration range)
- Limit of detection
- Limit of determination
- Precision
- Repeatability
- Reproducibility
- Recovery
- Selectivity
- Sensitivity
- Linearity
- Measurement uncertainty
- Other criteria that may be selected as required

Article 11 also establishes that the following implementing measures may be taken by the Commission:

- Methods of sampling and analysis, including the confirmatory or reference methods to be used in the event of a dispute.
- Performance criteria, analysis parameters, measurement uncertainty, and procedures for the validation of the before mentioned methods.
- Rules on the interpretation of results.

In any case, samples must be handled and labelled in such a way as to guarantee both their legal and analytical validity.

According to Article 12 of this Regulation, the competent authority of the Member States shall designate laboratories that may carry out the analysis of samples taken during official controls. However, they may only designate laboratories that operate and are assessed and accredited in accordance with the following European standards:

- EN ISO/IEC 17025 on "General requirements for the competence of testing and calibration laboratories."

- EN ISO/IEC 17011 on "General requirements for accreditation bodies accrediting conformity assessment bodies."
- Taking into account criteria for different testing methods laid down in Community feed and food law.

The accreditation and assessment of testing laboratories may relate to individual tests or groups of tests.

In order to contribute to a high quality and uniformity of analytical results, an analytical network of European Reference Laboratories (EURL), formerly called "Community Reference Laboratories (CRL)", National Reference Laboratories (NRL), and Official National Laboratories (OFL) was designated in the past for various classes of analytes. The activities of reference laboratories cover all areas of feed and food law and animal health, in particular those areas where there is a need for precise analytical and diagnostic results.

Article 32 of Regulation (EC) No. 882/2004 lays down the following major responsibilities for EURL for food and feed:

- Providing NRL with details of analytical methods, including reference methods.
- Coordinating application by the NRL of those methods, in particular by organising comparative testing and by ensuring an appropriate follow-up of such comparative testing in accordance with internationally accepted protocols, when available.
- Coordinating, within their area of competence, practical arrangements needed to apply new analytical methods and informing NRL of advances in this field.
- Conducting initial and further training courses for the benefit of staff from NRL and of experts from developing countries.
- Providing scientific and technical assistance to the Commission, especially in cases where Member States contest the results of analyses.
- Collaborating with laboratories responsible for analysing feed and food in third countries.

According to Article 33 of Regulation (EC) No. 882/2004 Member States shall arrange for the designation of one or more NRL for each EURL. These NRL inter alia shall

- Collaborate with the EURL in their area of competence.
- Coordinate, for their area of competence, the activities of OFL responsible for the analysis of samples.
- Where appropriate, organise comparative tests between the OFL and ensure an appropriate follow-up of such comparative testing.
- Ensure the dissemination to the Competent Authority and OFL of information that the EURL supplies.

- Provide scientific and technical assistance to the competent authority for the implementation of coordinated control plans.

While the Regulations (EC) No. 178/2002 and 882/2004 contain the general principles and requirements, specific analytical requirements as well as maximum levels for a number of residues and contaminants are laid down in special legislation. Regarding analytical methods, the EU generally follows the criteria approach. This means that no fixed methods are prescribed but detailed and strict performance criteria are established by the Commission which have to be fulfilled. As long as it can be demonstrated in a traceable manor that these performance criteria are fulfilled and the method is fit for purpose the analysts can apply whatever method. The great advantage of this approach is that it does not impede the fast innovation and progress in analytical instrumentation.

2. Pesticides

Until 1 September 2008, the legislation for pesticide residues was a shared responsibility of the European Commission and the Member States. Since 1976, more than 45,000 Community maximum residue levels (MRLs) were set for various commodities for 245 pesticides on cereals (Directive 86/362/EEC), foodstuffs of animal origin (Directive 86/363/EEC), fruit and vegetables, and other plant products (Directive 76/895/EEC and Directive 90/642/EEC). For the tens of thousands of pesticide/commodity combinations for which no Community MRLs existed, Member States could set MRLs at national level to facilitate trade and to protect the health of their consumers.

As from 1 September 2008, Regulation (EC) No. 396/2005 of the European Parliament and of the Council on MRLs of pesticides in or on food and feed of plant and animal origin defines a new fully harmonised set of rules for pesticide residues (3). This Regulation simplifies the existing legislation by harmonising pesticide MRLs and making them directly applicable in all Member States. The annexes to Regulation (EC) No. 396/2005 specify the MRLs and the products to which they apply.

Annex I is the list of products to which the MRLs apply. Annex I has been established by Commission Regulation (EC) No. 178/2006. It contains 315 products, including fruits, vegetables, spices, cereals, and animal products.

Annex II is the list of EU definitive MRLs and it consolidates the existing EU legislation before 1 September 2008. It specifies MRLs for 245 pesticides.

Annex III is the list of the so-called EU temporary MRLs. It is the result of the harmonisation process as it lists pesticides for which, before 1 September 2008, MRLs were only set at national level. It specifies MRLs for 471 pesticides.

Annex IV is the list of currently 52 pesticides for which no MRLs are needed because of their low risk.

Annex V will contain the list of pesticides for which a default limit other than 0.01 mg/kg will apply. This annex has not been published yet.

Annex VI will contain the list of conversion factors of MRLs for processed commodities. This annex has not been published yet.

Annex VII contains a list of pesticides used as fumigants for which the Member States are allowed to apply special derogations before the products are placed on the market (4).

If a pesticide is not included in any of the above mentioned Annexes a default MRL of 0.01 mg/kg applies.

The new pesticide Regulation is the result of a comprehensive review programme that was launched in 1993 by the European Commission for all active substances used in plant protection products within the European Union. In this review process, each substance had to be evaluated as to whether it could be used safely with respect to human health (consumers, farmers, local residents, and passers-by) and the environment, in particular groundwater and non-target organisms, such as birds, mammals, earthworms, and bees. The review of existing pesticides has led to the removal of pesticides from the market which cannot be used safely. Of some 1,000 active substances on the market, in at least one Member State before 1993, about 250 active substances have passed the harmonised EU safety assessment. Almost 700 active substances have been eliminated because dossiers were not submitted, incomplete, or withdrawn by industry. About 70 substances failed the review and have been removed from the market, because the evaluation carried out did not show safe use with respect to human health and the environment (5).

An EU pesticides database has been created and published on the web that provides a search tool to find out which active substances are approved in Europe together with a reference to the EU legislation. Moreover, this database includes the respective relevant toxicological information and the MRLs in food and feed (6).

The method validation and quality control procedures for pesticide residues analysis in food and feed are laid down in guidance documents published by the Directorate General (DG) for Health and Consumers of the European Commission. This DG has inter alia, the task of keeping the laws on the safety of food and feed up to date. The guidance documents are reviewed and updated regularly. The currently effective requirements (implemented by 01/01/2010) are laid down in the document "SANCO/10684/2009" (7). SANCO is the abbreviation of the French term "Santé et Consommateurs" for "Health and Consumers."

The guidance in this document is intended for the monitoring of pesticide residues in the European Union. The document describes the method of validation and analytical quality control (AQC) requirements to support the validity of data used for checking compliance with maximum residue levels (MRLs), enforcement actions, or assessment of consumer exposure to pesticides.

The key objectives are to

- Provide a harmonised cost-effective quality assurance system in the EU.
- Ensure the quality and comparability of analytical results.
- Ensure that acceptable accuracy is achieved.
- Ensure that false positives or false negatives are not reported.
- Support compliance with ISO/IEC 17025.

The document is complementary and integral to the requirements in ISO/IEC 17025.

Besides detailed requirements, such as for sampling, transport, processing and storage of samples, handling of calibration standards, avoidance of contamination and interferences, performance criteria, confirmation and reporting of results, a number of detailed requirements and recommendations are also laid down concerning mass spectrometry (MS).

The following is an excerpt of SANCO/10684/2009 with respect to application of mass spectrometric techniques in official pesticide analysis:

In case of MRL exceedances or the identification of unusual residues, the use of highly specific detection systems, such as mass spectrometry is recommended.

For GC-MS procedures, the chromatographic separation should be carried out using capillary columns. For LC-MS procedures, the chromatographic separation can be performed using any suitable LC column. In either case, the minimum acceptable retention time for the analyte(s) under examination should be at least twice the retention time corresponding to the void volume of the column. The retention time (or relative retention time) of the analyte in the sample extract must match that of the calibration standard (may need to be matrix matched) within a specified window after taking into consideration the resolving power of the chromatographic system. The ratio of the chromatographic retention time of the analyte to that of a suitable internal standard, i.e. the relative retention time of the analyte, should correspond to that of the calibration solution with a tolerance of ±0.5% for GC and ±2.5% for LC.

Reference spectra for the analyte should be generated using the instruments and techniques employed for analysis of the samples.

If major differences are evident between a published spectrum and that generated within the laboratory, the latter must be shown to be valid. To avoid distortion of ion ratios, the response of the analyte ions must not overload the detector. The reference spectrum in the instrument software can originate from a previous injection without matrix present, but preferably from the same batch.

Identification relies on proper selection of diagnostic ions. The (quasi) molecular ion is a diagnostic ion that should be included in the measurement and identification procedure whenever possible. In general, and especially in single MS, high m/z ions are more diagnostic than low m/z ions (e.g. $m/z<100$). However, high m/z ions arising from loss of water or from common moieties may be of little use. Although characteristic isotopic ions, especially Cl or Br clusters, may be of particular utility, the selected diagnostic ions should not exclusively originate from the same part of the parent molecule. Choice of diagnostic ions may change depending on background interferences.

Diagnostic ion chromatograms should have peaks (exceeding S/N 3:1) of similar retention time, peak shape, and response ratio to those obtained from a calibration standard analysed at comparable concentration in the same batch. Chromatographic peaks from different diagnostic ions for the same analyte must overlap with each other. Where an ion chromatogram shows evidence of significant chromatographic interference, it must not be relied upon to quantify or identify residues. The ion that shows the best signal-to-noise ratio and no evidence of significant chromatographic interference should be used for quantification.

In case of full scan measurement careful subtraction of background spectra, either manually or automatically by deconvolution or other algorithms may be required to ensure that the resultant spectrum of the chromatographic peak is representative. Whenever background correction is applied, this must be applied uniformly throughout the batch and should be clearly indicated.

Different types and modes of mass spectrometric detectors provide different degrees of selectivity, which relates to the confidence in identification. The requirements for identification are given in Table 1. They should be regarded as guidance criteria for identification, not as absolute criteria to prove the presence or absence of a compound.

The intensities of diagnostic ions or product ions can be determined through spectra or by integrating the signals of the single mass traces (extracted ion chromatograms). The relative intensities of the detected ions, expressed as a percentage of the intensity of the most intense (abundant) ion or product ion, should correspond to those of the calibration standard at comparable concentrations and measured under the same conditions. Matrix-matched calibration solutions may need to be employed. Table 2 below indicates the maximum tolerances.

Table 1
Identification requirements for different types of mass spectrometers

MS mode	Single MS (standard mass resolution)	Single MS (high resolution/ high mass accuracy)	MS/MS
Typical systems (examples)	Quadrupole, ion trap, time-of-flight (TOF)	TOF, Orbitrap, FTMS, magnetic sector	Triple quadrupole, ion trap, hybride MS (e.g. Q-TOF, Q-trap)
Acquisition	Full scan, limited *m/z* range, selected ion monitoring (SIM)	Full scan, limited *m/z* range, SIM	Selected/multiple reaction monitoring (SRM/MRM), full scan product-ion spectra
Requirements for identification	≥3 diagnostic ions, (preferably including quasi molecular ion)	≥2 diagnostic ions (preferably including the quasi molecular ion). Mass accuracy <5 ppm. At least one fragment ion	≥2 product ions
Ion ratio(s)	According to Table 2		

Table 2
Maximum permitted tolerances for relative ion intensities using a range of spectrometric techniques

Relative intensity (% of base peak)	EI-GC-MS (relative) (%)	CI-GC-MS, GC-MS^n, LC-MS, LC-MS^n (relative) (%)
>50	±10	±20
>20–50	±15	±25
>10–20	±20	±30
≤10	±50	±50

Larger tolerances may lead to a larger percentage of false-positive results. Likewise, if the tolerances are decreased, then the likelihood of false-negatives increases. The tolerances should not be taken as absolute limits and automated data interpretation based on the criteria without complementary interpretation by an experienced analyst is not recommended.

For a higher degree of confidence in identification, further evidence may be required. This can be achieved through additional mass spectrometric information, for example evaluation of full scan spectra, additional accurate mass (fragment) ions, additional product ions (in MS/MS), or accurate mass product ions.

If the isotope ratio of the ion(s), or the chromatographic profile of isomers of the analyte, is highly characteristic it may provide sufficient evidence. Otherwise, additional evidence may be sought using a different chromatographic separation system and/or a different ionisation technique, or by any other means of providing supporting information.

3. Pharmacologically Active Substances

To guarantee a high level of consumer protection, EU Community legislation requires that the toxicity of potential residues is evaluated before the use of a medicinal substance in food producing animals is authorised. If considered necessary, MRLs are established and in some cases the use of the relevant substance is prohibited. The evaluation procedure is laid out in Regulation (EC) No. 470/2009 of the European Parliament and of the Council (8). Pharmaceutically active substances and their classification regarding MRLs are set out in the Annex of Commission Regulation (EC) No. 37/2010 (9). The Annex of this Regulation contains two tables.

Table 1 contains the allowable substances in alphabetical order along with their marker residues, MRLs, animal species, target tissues, therapeutic classification, and other specific provisions.

Table 2 lists the prohibited substances. Currently, the following compounds are included:

- *Aristolochia* spp. and preparations thereof
- Chloramphenicol
- Chloroform
- Chlorpromazine
- Colchicine
- Dapsone
- Dimetridazole
- Metronidazole
- Nitrofurans (including furazolidone)
- Ronidazole

The requirements that must be met in relation to the planning and execution of national residue control plans for live animals and products of animal origin are prescribed in Directive 96/23/EC (10). The principal objective of the legislation is to detect the illegal use of substances in animal production and the misuse of authorised veterinary medicinal products and to ensure the implementation of appropriate actions to minimise recurrence of all such residues in food of animal origin.

The annexes of this Directive establish the substances to be monitored by type of animal, their feeding stuffs, including drinking water and primary animal products. The frequency of sampling is dependent either on the number of animals slaughtered the previous year or on national annual production figures.

Detailed performance criteria for screening and confirmatory methods applied in the analysis of pharmacologically active substances and the interpretation of results are laid down in Commission Decision 2002/657/EC (11). As a general requirement, confirmatory methods for organic residues or contaminants shall provide information on the chemical structure of the analyte, especially for prohibited or non-authorised compounds. Consequently, methods based only on chromatographic analysis without the use of spectrometric detection are not suitable on their own for use as confirmatory methods. In principle, LC or GC with mass-spectrometric detection and LC or GC with infrared (IR) spectrometric detection are considered suitable as confirmatory methods for banned as well as authorised compounds. However, because of the limited detection power of IR spectrometric detectors, GC and LC with mass spectrometric detection are the techniques of choice to unequivocally confirm the analytes of interest at low trace levels.

With respect to mass spectrometry, the Annex of Decision 2002/657/EC contains a number of detailed performance criteria which have to be fulfilled. The following is an excerpt of the respective parts of this Annex:

Mass spectrometric methods are suitable for consideration as confirmatory methods only following either an on-line or an off-line chromatographic separation. For GC-MS procedures, the gas chromatographic separation shall be carried out using capillary columns. For LC-MS procedures, the chromatographic separation shall be carried out using suitable LC columns. In any case, the minimum acceptable retention time for the analyte under examination is twice the retention time corresponding to the void volume of the column. The retention time (or relative retention time) of the analyte in the test portion shall match that of the calibration standard within a specified retention time window. The retention time window shall be commensurate with the resolving power of the chromatographic system. The ratio of the chromatographic retention time of the analyte to that of the internal standard, i.e. the relative retention time of the analyte, shall correspond to that of the calibration solution at a tolerance of ±0.5% for GC and ±2.5% for LC.

Mass-spectrometric detection shall be carried out by employing MS techniques such as recording of full mass spectra (full scans) or selected ion monitoring (SIM), as well as MS-MS^n

techniques such as Selected Reaction Monitoring (SRM), or other suitable MS or MS-MS^n techniques in combination with appropriate ionisation modes. In high-resolution mass spectrometry (HRMS), the resolution shall typically be greater than 10,000 for the entire mass range at 10% valley.

Full scan: When mass spectrometric determination is performed by the recording of full scan spectra, the presence of all measured diagnostic ions (the molecular ion, characteristic adducts of the molecular ion, characteristic fragment ions, and isotope ions) with a relative intensity of more than 10% in the reference spectrum of the calibration standard is obligatory.

SIM: When mass spectrometric determination is performed by fragmentography, the molecular ion shall preferably be one of the selected diagnostic ions (the molecular ion, characteristic adducts of the molecular ion, characteristic fragment ions, and all their isotope ions). The selected diagnostic ions should not exclusively originate from the same part of the molecule. The signal-to-noise ratio for each diagnostic ion shall be ≥3:1.

Full scan and SIM: The relative intensities of the detected ions, expressed as a percentage of the intensity of the most intense ion or transition, shall correspond to those of the calibration standard, either from calibration standard solutions or from spiked samples, at comparable concentrations, measured under the same conditions, within the following tolerances:

Interpretation of mass spectral data: The relative intensities of the diagnostic ions and/or precursor/product ion pairs have to be identified by comparing spectra or by integrating the signals of the single mass traces. Whenever background correction is applied, this shall be applied uniformly throughout the batch and shall be clearly indicated.

Full scan: When full scan spectra are recorded in single mass spectrometry, a minimum of four ions shall be present with a relative intensity of ≥10% of the base peak. The molecular ion shall be included if it is present in the reference spectrum with a relative intensity of ≥10%. At least four ions shall lie within the maximum permitted tolerances for relative ion intensities. Computer-aided library searching may be used. In this case, the comparison of mass spectral data in the test samples to that of the calibration solution has to exceed a critical match factor. This factor shall be determined during the validation process for every analyte on the basis of spectra for which the criteria described below are fulfilled. Variability in the spectra caused by the sample matrix and the detector performance shall be checked.

As can be seen from Table 2 as well as from the analytical requirements above, the performance criteria for mass spectrometric methods in pesticide and veterinary drug analysis have meanwhile been harmonised for the most part. However, one substantial difference exists between these two areas as the

Directive 2002/657/EC introduces the system of "identification points" for the interpretation of results obtained with mass spectrometric methods using other techniques than full scan.

SIM: When mass fragments are measured using other than full-scan techniques, a system of ***identification points*** shall be used to interpret the data. For the confirmation of substances listed in Group A of Annex of Directive 96/23/EC (banned compounds), a minimum of four identification points shall be required. For the confirmation of substances listed in Group B of Annex I of Directive 96/23/EC (compounds with maximum residue limits), a minimum of three identification points are required. The table below shows the number of identification points that each of the basic mass spectrometric techniques can earn. However, in order to qualify for the identification points required for confirmation and the sum of identification points to be calculated:

- A minimum of at least one ion ratio shall be measured.
- All relevant measured ion ratios shall meet the criteria described above.
- A maximum of three separate techniques can be combined to achieve the minimum number of identification points (see Table 3).

Table 3
The relationship between a range of classes of mass fragments and identification points earned

MS technique	Identification points earned per ion
Low-resolution mass spectrometry (LRMS)	1.0
LRMS^n precursor ion	1.0
LRMS^n transition products	1.5
HRMS	2.0
HRMS^n precursor ion	2.0
HRMS^n transition products	2.5

1. Each ion may only be counted once
2. GC-MS using electron ionisation is regarded as being a different technique to GC-MS using chemical ionisation
3. Different analytes can be used to increase the number of identification points only if the derivatives employ different reaction chemistries
4. For substances in Group A of Annex 1 to Directive 96/23/EC, if one of the following techniques are used in the analytical procedure: HPLC coupled with full-scan diode array spectrophotometry (DAD); HPLC coupled with fluorescence detection; HPLC coupled to an immunogram; two-dimensional TLC coupled to spectrometric detection; a maximum of one identification point may be contributed, providing that the relevant criteria for these techniques are fulfilled
5. Transition products include both daughter and granddaughter products

To earn four identification points with low-resolution mass spectrometry applying GC-MS and LC-MS methods in SIM mode the determination of four diagnostic ions is required. In contrast, application of the increasingly common triple quadrupole mass spectrometry requires only one precursor ion and two transition products to earn four identification points and thus fulfil the requirements.

Besides detailed analytical performance criteria, Decision 2002/657/EC also establishes "minimum required performance limits (MRPL)" for several substances which are prohibited in food producing animals in the EU or are not authorised. MRPLs are defined as "minimum content of an analyte in a sample, which at least has to be detected and confirmed" and are the reference point for action in relation to the evaluation of consignments of food.

Currently, MRPL values are set for the compounds found in Table 4.

As these are all banned or non-authorised compounds, their confirmatory determination has to be performed using spectrometric methods. Due to the low concentration, GC and LC methods based on mass spectrometry rather than infrared spectrometry are the techniques of choice.

Table 4
Minimum required performance limits (MRPL) for banned substances

Substance and/or metabolite	Matrixes	MRPL (μg/kg)
Chloramphenicol	Meat Eggs Milk Urine Aquaculture products Honey	0.3
Medroxyprogesterone acetate	Pig kidney fat	1
Nitrofuran metabolites of Furazolidone Furaltadone Nitrofurantoin Nitrofurazone	Poultry meat Aquaculture products	1 for all
Sum of malachite green and leucomalachite green	Meat of aquaculture products	2

4. Organic Contaminants

The general Community procedures for contaminants in food are laid out in Council Regulation (EEC) No 315/93 (12). Besides definitions, this Regulation stipulates that food containing a contaminant in an amount which is unacceptable from the public health viewpoint and in particular at a toxicological level shall not be placed on the market. In addition, it has the general obligation that contaminant levels shall be kept as low as can reasonably be achieved by following good practices at all stages. In order to protect public health, the Regulation empowers to establish, where necessary, maximum levels for specific contaminants in food. These maximum levels are listed in the Annex of Commission Regulation (EC) No. 1881/2006 of 19 December 2006 setting maximum levels for certain contaminants in foodstuffs (13). According to the recitals of this Regulation, "maximum levels should be set at a strict level which is reasonably achievable by following good agricultural, fishery, and manufacturing practices and taking into account the risk related to the consumption of the food. In the case of contaminants which are considered to be genotoxic carcinogens or in cases where current exposure of the population or of vulnerable groups in the population is close to or exceeds the tolerable intake, maximum levels should be set at a level which is as low as reasonably achievable (ALARA)."

4.1. Mycotoxins

A number of maximum levels in various foodstuffs are set in Regulation (EC) No. 1881/2006 for aflatoxins, ochratoxin A, patulin, deoxynivalenol, zearalenone, fumonisins, T2 and HT-2 toxin, respectively. The maximum levels for aflatoxins are given separately for aflatoxin B1 and the sum of aflatoxins B1 + B2 + G1 + G2.

The performance criteria for methods of sampling and analysis for the official control of the levels of mycotoxins are established in Commission Regulation (EC) No. 401/2006 (14). Detailed provisions are given for the sampling procedure and especially for subdivision of lots into sublots depending on type and weight of products to be tested. As a general requirement regarding analysis of mycotoxins, the analytical methods applied shall comply with the general provisions established in Regulation (EC) No. 882/2004. In addition, some specific performance criteria are stipulated. Provided the selected method meets these criteria, analysts may select any method. The performance criteria mainly concern requirements regarding handling of blanks, recovery, and relative standard deviations calculated from results generated under repeatability and reproducibility conditions. The latter ones depend on the level of the respective mycotoxin in the food stuff. Detailed specific performance criteria for the determination of mycotoxins in food applying mass spectrometric methods are not given.

In any case, for controlling compliance with maximum levels, the analytical result corrected for recovery shall be used taking into account the expanded measurement uncertainty using a coverage factor of 2 which gives a level of confidence of approximately 95%.

4.2. Dioxins and Dioxin-Like PCBs

Maximum levels for dioxins as well as the sum of dioxins and dioxin-like polychlorinated biphenyls (dl-PCB) are stipulated in the Annex, section 5 of Regulation (EC) No. 1881/2006 for various food stuffs of animal origin. The maximum levels for dioxins [sum of polychlorinated dibenzo-para-dioxins (PCDDs) and polychlorinated dibenzofurans (PCDFs)] and the sum of dioxins and dioxin-like PCBs (sum of PCDDs, PCDFs and dl-PCBs), are both expressed as WHO toxic equivalents (WHO-TEQ) using the WHO-toxic equivalency factors (WHO-TEFs) for human risk assessment based on the conclusions of the WHO meeting in Stockholm, Sweden, 15–18 June 1997 (15). As a peculiarity all maximum levels are given as upper-bound concentrations. Upper-bound concentrations are calculated on the assumption that all the values of the different congeners below the limit of quantification are equal to the limit of quantification. The maximum levels are not based on toxicological considerations but on the frequency of occurrence in the different food categories with the aim to detach those products with the highest levels from the market. In any case, food stuffs have to comply both with the maximum levels for dioxins as well as for the sum of dioxins and dl-PCBs.

In addition to maximum levels, the EU Commission has launched separate action levels for dioxins and dioxin-like PCBs, respectively as an early warning tool through Commission Recommendation 2006/88/EC (16). The action levels apply for food and feed and are set at approximately 2/3 of the respective maximum levels. In cases where levels of dioxins and/or dioxin-like PCBs in excess of the action levels specified in this Recommendation are found, it is recommended that Member States, in co-operation with operators:

(a) Initiate investigations to identify the source of contamination.

(b) Take measures to reduce or eliminate the source of contamination.

(c) Check for the presence of non-dioxin-like PCBs.

The performance criteria that analytical methods have to fulfil are established in Commission Regulation (EC) No. 1883/2006 laying down methods of sampling and analysis for the official control of levels of dioxins and dioxin like PCBs in certain foodstuffs (17). As a basic requirement for acceptance of analytical procedures the sensitivity for PCDD, PCDF, and non-ortho-PCB must be in the picogram range. Screening methods may comprise bioassays and GC/MS methods; confirmatory methods are high-resolution gas chromatography/high-resolution mass spectrometry (HRGC/HRMS) methods. The criteria for confirmatory methods concerning trueness and precision (RSD_R) are given as −20% to +20% and <15%, respectively.

The following specific requirements for GC/MS methods to be complied with for screening or confirmatory methods are an excerpt of ANNEX II to Regulation (EC) No. 1883/2006:

- Addition of ^{13}C-labelled 2,3,7,8-chlorine substituted internal PCDD/F standards and of ^{13}C-labelled internal dioxin-like PCB standards must be carried out at the very beginning of the analytical method, for example prior to extraction in order to validate the analytical procedure. At least one congener for each of the tetra- to octachlorinated homologous groups for PCDD/F and at least one congener for each of the homologous groups for dioxin-like PCBs must be added (alternatively, at least one congener for each mass spectrometric selected ion recording function used for monitoring PCDD/F and dioxin-like PCBs). There shall be clear preference, certainly in case of confirmatory methods, of using all 17 ^{13}C-labelled 2,3,7,8-substituted internal PCDD/F standards and all 12 ^{13}C-labelled internal dioxin-like PCB standard. Relative response factors shall also be determined for those congeners for which no ^{13}C-labelled analogue is added by using appropriate calibration solutions.
- For foodstuffs of plant origin and foodstuffs of animal origin containing less than 10% fat, the addition of the internal standards is mandatory prior to extraction. For foodstuffs of animal origin containing more than 10% fat, the internal standards may be added either before extraction or after fat extraction. An appropriate validation of the extraction efficiency shall be carried out, depending on the stage at which internal standards are introduced and on whether results are reported on product or fat basis.
- Prior to GC/MS analysis, one or two recovery (surrogate) standard(s) must be added.
- Control of recovery is necessary. For confirmatory methods, the recoveries of the individual internal standards shall be in the range of 60–120%. Lower or higher recoveries for individual congeners, in particular for some hepta- and octa-chlorinated dibenzodioxins and dibenzofurans, are acceptable on the condition that their contribution to the TEQ value does not exceed 10% of the total TEQ value (based on the sum of PCDD/F and dioxin-like PCBs). For screening methods, the recoveries shall be in the range of 30–140%.
- Separation of dioxins from interfering chlorinated compounds such as non-dioxin-like PCBs and chlorinated diphenyl ethers

shall be carried out by suitable chromatographic techniques (preferably with a florisil, alumina, and/or carbon column).

- Gas chromatographic separation of isomers shall be sufficient (<25% peak to peak between 1,2,3,4,7,8-HxCDF and 1,2,3,6,7,8-HxCDF).
- Determination shall be performed according to EPA Method 1613 revision B: tetra- through octa-chlorinated dioxins and furans by isotope dilution HRGC/HRMS or another with equivalent performance criteria.
- The difference between upper-bound level and lower-bound level shall not exceed 20% for foodstuffs with a dioxin contamination of about 1 pg WHO-TEQ/g fat (based on the sum of PCDD/PCDF and dioxin-like PCBs). For foodstuffs with a low fat content, the same requirements for contamination levels of about 1 pg WHO-TEQ/g product have to be applied. For lower contamination levels, for example 0.50 pg WHO-TEQ/g product, the difference between upper-bound and lower-bound level may be in the range of 25–40%.

For the purpose of this Regulation, the accepted specific limit of quantification of an individual congener is the concentration of an analyte in the extract of a sample which produces an instrumental response at two different ions to be monitored with a S/N ratio of 3:1 for the less sensitive signal and fulfilment of the basic requirements such as, for example retention time, isotope ratio according to the determination procedure as described in EPA method 1613 revision B (http://www.epa.gov/region03/1613.pdf).

The analytical results has to be reported as $x \pm U$, whereby x is the analytical result and U is the expanded measurement uncertainty using a coverage factor of 2 which gives a level of confidence of approximately 95%. In case of a separate determination of dioxins and dioxin-like-PCBs, the sum of the estimated expanded uncertainty of the separate analytical results of dioxins and dioxin-like PCBs has to be used for the sum of dioxins and dioxin-like PCBs.

4.3. Polycyclic Aromatic Hydrocarbons

In 2002, the Scientific Committee on Food (SCF) conducted a risk assessment on polycyclic aromatic hydrocarbons (PAHs) and concluded that the following 15 PAHs, benz[*a*]anthracene, benzo[*b*]fluoranthene, benzo[*j*]fluoranthene, benzo[*k*]fluoranthene, benzo[*ghi*]perylene, benzo[*a*]pyrene, chrysene, cyclopenta[*cd*]pyrene, dibenz[*a,h*]anthracene, dibenzo[*a,e*]pyrene, dibenzo[*a,h*]pyrene, dibenzo[*a,i*]pyrene, dibenzo[*a,l*]pyrene, indeno[*1,2,3-cd*]pyrene, and 5-methylchrysene, may be regarded as potentially genotoxic and carcinogenic to humans and thus represent a priority group in the assessment of the risk of long-term adverse health effects following dietary intake of PAHs (18). As benzo[*a*]pyrene was considered a good indicator for occurrence and effect of carcinogenic PAHs in food, a number of maximum levels for various foodstuffs were established for this compound in section 6 of the Annex to Regulation (EC) No. 1881/2006. In 2008, the Panel on Contaminants in the Food Chain (CONTAM Panel) of the EFSA performed a reassessment of the SCF opinion and concluded that benzo[a]pyrene is not a suitable indicator for the occurrence of PAHs in food. Based on the currently available data relating to occurrence and toxicity, the CONTAM Panel concluded that the sum of benzo[*a*]pyrene, chrysene, benz[*a*]anthracene, and benzo[*b*]fluoranthene (PAH4) and the sum of benzo[*a*]pyrene, benz[*a*]anthracene, benzo[*b*]fluoranthene, benzo[*k*]fluoranthene, benzo[*ghi*]perylene, chrysene, dibenz[*a,h*]anthracene, and indeno[*1,2,3-cd*]pyrene (PAH8) are the most suitable indicators of PAHs in food, with PAH8 not providing much added value compared to PAH4. As a consequence of this risk assessment, the European legislation on benzo[*a*]pyrene in food is currently reviewed (19).

The performance criteria for methods of sampling and analysis for the official control of the levels of benzo[*a*]pyrene are laid down in Commission Regulation (EC) No. 333/2007 together with the respective criteria for lead, cadmium, mercury, inorganic tin, and 3-MCPD (20). Currently, the performance criteria only apply for benzo[*a*]pyrene because this is so far the only PAH which is regulated in food in the EU.

As a general requirement, laboratories must comply with the respective provisions of Regulation (EC) No. 882/2004 as mentioned above also for other contaminants. Furthermore, they shall participate in appropriate proficiency testing schemes which comply with the International Harmonised Protocol for the Proficiency Testing of (Chemical) Analytical Laboratories developed under the auspices of IUPAC/ISO/AOAC and shall be able to demonstrate that they have internal quality control procedures in place.

Specific procedures for benzo[*a*]pyrene are given as follows:

"The analyst shall ensure that samples do not become contaminated during sample preparation. Containers shall be rinsed with high purity acetone or hexane before use to minimise the risk of contamination. Wherever possible, apparatus and equipment coming into contact with the sample shall be made of inert materials, such as aluminium, glass, or polished stainless steel. Plastics such as polypropylene or PTFE shall be avoided because the analyte can adsorb onto these materials".

As no specific method for the determination of benzo[*a*]pyrene in foodstuffs is prescribed at Community level, laboratories may select any validated method of analysis provided the selected method meets the following specific performance criteria:

> The limits of determination and quantification have to be less than 0.3 μg/kg and 0.9 μg/kg, respectively. Concerning precision, it is stipulated that the $HORRAT_r$ and

$HORRAT_R$ values are less than 2. The recovery has to be between 50 and 120%. As regarding specificity, the Regulation demands that it should be free from matrix or spectral interferences and in case of positive detection verification is needed without giving further details.

Nowadays, the two main analytical techniques in the EU for determining benzo[*a*]pyrene and other PAHs in foodstuffs are high-performance liquid chromatography (HPLC) coupled to a fluorescence detector (FLD) and gas chromatography-mass spectrometry (GC-MS). Currently, GC-MS is gaining increased importance, especially because of its superior separation power and the possibility to add isotope labelled standards as ideal internal standards at the very beginning of the extraction to the samples. This addition can be done either with deuterated or ^{13}C-labelled standards of which both groups a number of PAHs are meanwhile commercially available.

In any case the analytical result shall be corrected for recovery and the level of recovery must be reported. As for other contaminants the analytical result shall be reported as $x \pm U$, whereby x is the analytical result and U is the expanded measurement uncertainty, using a coverage factor of 2 which gives a level of confidence of approximately 95% ($U = 2u$).

References

1. OJ L 31, 1.2.2002, p.1; last amended by Regulation (EC) No 596/2009, OJ L 188, 18.07.2009, p. 14
2. OJ L 165, 30.4.2004, p. 1; last amended by Regulation (EC) No 596/2009, OJ L 188, 18.07.2009, p. 14
3. OJ L 70, 16.03.2005, p. 1; last amended by Commission Regulation (EC) No 1097/2009, OJ L 301, 17.11.2009, p. 6
4. http://ec.europa.eu/food/plant/protection/pesticides/regulation_ec_396_2005_en.htm
5. http://ec.europa.eu/food/plant/protection/evaluation/docs/factsheet_pesticides_web.pdf
6. http://ec.europa.eu/sanco_pesticides/public/index.cfm
7. http://www.crl-pesticides.eu/library/docs/allcrl/AqcGuidance_Sanco_2009_10684.pdf
8. OJ L 152, 16.06.2009, p. 11
9. OJ L 15, 20.01.2010, p. 1; last amended by Commission Regulation (EC) No 914/2010, OJ L 269, 13.10.2010, p. 5
10. OJ L 125, 23.5.1996, p. 10; last amended by Regulation (EC) No 596/2009, OJ L 188, 18.07.2009, p. 14
11. OJ L 221, 17.08.2002, p. 8; last amended by Commission Decision 2004/25/EC, OJ L 6, 10.01.2004, p. 38
12. OJ L 37, 13.2.1993, p. 1; last amended by Regulation (EC) No 596/2009, OJ L 188, 18.7.2009, p. 14
13. OJ L 364, 20.12.2006, p. 5; last amended by Commission Regulation (EC) No 165/2010, OJ L 50, 27.2.2010, p. 8
14. OJ L 70, 9.3.2006, p. 12; last amended by Commission Regulation (EC) No 178/2010, OJ L 52, 3.3.2010, p. 32
15. (Van den Berg et al., (1998) Toxic Equivalency Factors (TEFs) for PCBs, PCDDs, PCDFs for Humans and for Wildlife. Environmental Health Perspectives, 106 (12), 775
16. OJ L 42, 14.2.2006, p. 26
17. OJ L 364, 20.12.2006, p. 32
18. Opinion of the Scientific Committee on Food on the risks to human health of Polycyclic Aromatic Hydrocarbons in food (expressed on 4 December 2002) http://ec.europa.eu/food/fs/sc/scf/out153_en.pdf
19. Scientific Opinion of the Panel on Contaminants in the Food Chain on a request from the European Commission on Polycyclic Aromatic Hydrocarbons in Food. The EFSA Journal (2008) 724, 1–114 (http://www.efsa.europa.eu/en/scdocs/doc/724.pdf)
20. OJ L 88, 29.3.2007, p. 29

Chapter 2

China's Food Safety Regulation and Mass Spectrometry

Xiaogang Chu, Feng Zhang, Xuemei Nie, Wenzhi Wang, and Feng Feng

Abstract

Food safety is essential to people's health and people's livelihood. To ensure that food safety is an important current strategy of the governments, both regulation and standardization are important support for implementing this strategic initiative effectively. The status and prospects of China's food laws, regulations, and standards system are introduced. China now has established a complete law regime providing a sound foundation and good environment for keeping the health of people, maintaining the order of social economy and promoting the international trade of food. At the same time, it is undoubtedly important to strengthen standardization and improve the food safety standards system. In the administration of food safety, mass spectrometry is becoming more and more important and many analytical methods developed in China are based on its application.

Key words: Food safety, Law, Regulation, Standards, Mass spectrometry, China

1. Food Laws and Regulations of China

1.1. China's Legislative system

1.1.1. Legislative System

China's current legislative system is unitary, two-class, and multi-level. "Unitary" means according to the constitution thus China is a both a unified and multi-ethnic country. Therefore, there is but one unified legislative system. The so-called "two-class" system indicates that according to the constitution, the Chinese legislation system is divided into two levels containing central and local legislation. "Multi-level" as defined by the constitution, such that both central-level legislation and local-level legislation can be divided into several categories.

The state council is empowered to enact administrative regulations, issue decisions and orders, and change or repeal inappropriate decisions and orders of the local state administrative organs at all levels, under the provision of the constitution and other related laws and regulations.

Jerry Zweigenbaum (ed.), *Mass Spectrometry in Food Safety: Methods and Protocols*,
Methods in Molecular Biology, vol. 747, DOI 10.1007/978-1-61779-136-9_2, © Springer Science+Business Media, LLC 2011

The national people's congress and its standing committee of the provinces, autonomous regions, and municipalities can enact local rules and regulations according to the administrative area of the specific conditions and actual needs, under the premise that they do not conflict with the constitution, laws, and administrative rules.

The people's congress or its provincial and municipal standing committee or special economic zones located therein, can set the regulations in the scope of special economic zones, in accordance with the authorization of the national people's congress.

The national people's congress of the autonomous regions, autonomous prefectures and autonomous counties, also has the right to develop self-government regulations and separate regulations, laws, administrative regulations, and to modify the provisions, in accordance with local political, economic, and cultural charters.

The ministries and commissions under the state council and the people's bank of China, the audit commission and the departments with administrative functions, can develop regulations within the purview of their sector, according to the administrative rules and regulations, decisions, and orders of the state council.

1.1.2. Legislative Procedure

The law-making process refers to the procedures and methods, which the state organizations with legislative power must comply with during the enactment, amendment, supplements, and repeal of regulatory legal documents. In general, China's legislative procedure comprises proposal, pondering, voting, and publishing of a law.

1.2. The Status of China's Food Laws and Regulations (2, 3)

Food laws and regulations refer to the sum of legal norms, being conducted or authorized by national mandatory authority to strengthen the supervision and management of food, ensure food hygiene, protect people's health from food contamination and health hazards, and provide adequate nutrition. The Chinese food laws and regulations system are composed of the following normative documents at different legal levels, based on its sound form and legal effect-level.

1.2.1. Laws

On February 28, 2009, the Seventh session of the China's national people's congress (NPC) standing committee passed "the Food Safety Law of the People's Republic of China," replacing the food hygiene law. This has become the highest legal effect-level normative document in food safety and the health legal system, and the basis of formulating food safety regulations, rules, and other normative documents.

The relevant laws being issued and implemented in the field of food include the Product Quality Law, the Standardization Law, the law on the quality and safety of agricultural products, the law on import and export commodity inspection, the law on

Animal and Plant Entry and Exit Quarantine, the advertising law, the consumer rights and interests protection law, the unfair competition law, and the trademark law.

1.2.2. Administrative Regulations

The administrative regulations can be developed by the state or local governments, but its legal status is second to the constitution. Administrative regulations in the food industry refer to the law normative documents drawn up by departments of the state council, in accordance with the constitution. The administrative regulations are called regulations, provisions, and measures. The regulations comprise a comprehensive and systematic foundation for the administrative work, and the provisions provide a subset of rules on the administrative work. The measures make up specific rules based on the administrative work, for example: food hygiene management for the catering industry, hygiene management for food additives, and hygiene management for health food.

Local food administrative regulations refer to the normative documents formulated by the people's congress or its standing committee of provinces, autonomous regions, and municipalities according to the constitution. The local administrative regulations are valid only in its own territory, and cannot contravene the constitution, laws, and administrative regulations, such as detailed rules for the implementation of the measures for food hygiene of Hebei province.

1.2.3. The Departmental Rules

The departmental rules include the rules and regulations stipulated by the executive department of the state council and local governments. These encompass food measures for the hygiene administration of food additives, measures for the certification of organic food, and measures for hygiene management of genetically modified food.

1.2.4. Other Normative Documents

Normative documents don't belong to laws, administrative regulations, departmental rules, or standards. These include notifications issued by the state council or individual administrative departments, the management measures of food hygiene licensing developed by the relevant administrative departments of local government, and the management measures of the procurement of food and raw materials. These kind of normative documents are indispensable, and they are an important part of the legal system. An example is "the decision of state council on further strengthening the work of food safety," "hazard analysis and critical control point (HACCP) management system of food-producing enterprises."

1.3. The Introduction to China's Food Laws and Regulations

Based on many laws such as the food safety law, product quality law, and the law on import and export commodity inspection, the main body China's food safety regulatory system is set. This in turn sets the "standardization law" and the "import and export commodity inspection law" as the basis for "food production and processing enterprise quality safety supervision and management measures," "food labeling mark rule," "food additive regulations," as well as a large number of food safety requirements related to technical standards and other regulations as the main body. In addition, these are supplemented by food safety regulations of the provincial and local governments. In addition, China's local governments have introduced a large number of local laws and regulations and local administrative regulations, which lay an important foundation for improving the level of food safety in China.

1.3.1. Food Safety Law

The Food Safety Law was implemented formally on June 1, 2009. The law provides an orderly solution to the current system of food safety and opens a new stage of China's food safety supervision.

1.3.1.1. The Background and Motivation of the Food Safety Law

- The previous food hygiene law was no longer suited to the current socioeconomic situation.

  At the inception of the food hygiene law, it played an important role in ensuring the quality of Chinese food hygiene and safety. However, the food hygiene law became unsuitable to the current food safety management with the economic development:

  1. The division work of food regulatory authorities was not clear. There was too much crossover and blind spots between the management of different departments that caused disruption in the food market.
  2. With the social and economic development, enforcement according to food hygiene law failed to exercise sufficient punishment to deter potential offenders causing the proliferation of fake and shoddy products within China. As examples, illegal ingredients such as Sudan dye and the melamine were used for food counterfeiting and forgery by unscrupulous traders.

- International trade calls for more scientific and effective laws to keep food safe.

  The development of global economic integration enables countries to produce food for the world. China is a large agricultural country and its products are so cheap and fine that they sell very well abroad. For example, chicken breasts made in China are well received by the distributors of European countries, because of their automated production, complete and clean nature. However, it is often meets limit export because veterinary drug residues do not meet EU requirements.

  Chinese foods restricted by developed countries also include vegetables and aquatic products. This phenomenon not only affects the development of food enterprises in China, but also causes significant loss of its reputation. Therefore, the development of more scientific and effective laws to guarantee food safety is not only needed in the domestic market, but also to break through technical barriers and enter the international market.

1.3.1.2. The Major Focus of the Food Safety Law

1. The establishment of a food safety commission.
   The food safety law provides that a food safety commission and its work responsibilities is established by the state council. Health administrative departments under the state council are responsible for integrated and coordinated work, risk assessment, standards establishment, and information disclosure of food safety. It also sets accreditation conditions and test specification development of the food inspection agency, the organization charged with investigating and dealing with major food safety incidents. In addition, the state council implements supervision and management of quality, industrial and commercial administration, and the state food and drug's supervision of food production, food distribution, and food service activities in accordance with the provisions of this law and the state department functions.
2. The unification of national food safety standards.
   The Food Safety Law provides that enforcement food standards of the current food quality and safety standards for agricultural products, food hygiene standards, food quality standards, and industry standards should be integrated into the national standards for food safety. At the same time they do not establish mandatory food standards other than for food safety.
3. The establishment of a monitoring and evaluation system for risk assessment.
   The Food Safety Law provides that food safety risk monitoring systems be built in China to address forborne diseases, food contamination, and harmful factors in food. Food safety risk assessment system was established in China for food additives and biological, chemical, and physical hazards.
4. The abolishment of the food exemption system.
   The Food Safety Law clearly states that food safety supervision and management departments cannot implement an exemption for food. Quality supervision, business administration, food and drug supervision, and management departments above the county level shall carry out regular or irregular food sampling tests.
5. The implementation of a recall system for problem food.
   Food safety law clearly states that: food recall system was built in China. Food producers should immediately stop

production, recall the sale food, notify the relevant production operators and consumers, and make a recording and notification of the recall if it is found that the production of food does not meet food safety standards. The law also stipulates if food business operators found the food does not meet food safety standards, it shall immediately cease operations, notify the production operators and consumers, and make a recording and notification of cessation of operations.

6. Ten times the compensation.
   The new law serves as a deterrent as the compensation standard is greatly improved and the cost to illegal operators has significantly increased. In the event of food safety incidents, the responsibility of food companies may bear civil, administrative, and criminal action. If entrepreneurial capacity is limited, civil compensation must first be met, which reflects the protection of civil rights.
7. Food additives outside of the directory are banned.
   Food additives are not only the essential raw material for production in the modern food industry, but also a double-edged sword. It will be counterproductive if they are used illegally. The approvals of food additive are being strictly managed in the present food safety law. If a food additive is not in the directory, even if there is initial proof of no hazard to the human body, it is still banned. In addition, the use of additives must be specifically labeled on the package and with how much. Legal action would be taken if the content was not labeled correctly.

1.3.2. Food Hygiene Law

The food hygiene law of People's Republic of China was adopted and promulgated on the 25th meeting of the fifth national people's congress standing committee on November 9, 1982, which marks the change of Chinese food health work from health administration to legal management. The food sanitation law of the People's Republic of China was adopted on the sixteenth meeting of the eighth national people's congress standing committee on October 30, 1995, which was promulgated as the presidential decree no. 59. However the food safety law of the People's Republic of China was adopted on the seventh meeting of the eleventh national people's congress standing committee on February 28, 2009, and the food sanitation law was revoked.

The food sanitation law was implemented by the national mandate for food hygiene in food production, supervision, and management. It was to address any state engaged in food production, business units or individuals, as well as food production, operators and consumers in the food safety and health management, and supervision of social relations especially for economic interests.

1.3.2.1. The Meaning of Developing and Implementing the Food Hygiene Law

The amendment and implementation of the food sanitation law is of great significance for promoting the healthy development of the food market economy and for food hygiene protecting the health of people. It addresses the need to protect people's lives. It establishes and improves the legal system for food hygiene. It also strengthens administrative law enforcement and increased law enforcement efforts and it is beneficial for the healthy development of China's food industry.

1.3.2.2. The Main Content of the Food Hygiene Law

The food hygiene law which consists of 9 chapters and 57 sections makes clear definitions for the applicable conditions, behavior patterns, and the legal consequences of food hygiene norms.

1. Chapter I: General Provisions
   This provides the legislative purpose and legal range, for compliance with this law if engaging in food production and operation within the territory of the People's Republic of China. All food, food additives, food containers, packaging materials and food, tools, equipment, detergents, disinfectants are applicable to this law, as well as food production and business premises, facilities and the environment. It stipulates that the health administrative department of the state council is in charge of the national food hygiene supervision and management. In addition, state council departments are responsible for food hygiene management.
2. Chapter II: Food Hygiene
   This provides for health requirements in the process of food production and operation and for production that is prohibited along with the management of food. It also provides that food should be non-toxic or harmless and nutritional and that food should be consistent with the corresponding color, smell, taste, and other sensory characteristics. In addition, it requires drugs are not added into food, while those that are both food and medicine in their raw state such as spices or traditional nutrient enhancers are accepted.
3. Chapter III: Health of Food Additives
   This provides that production, operation, and use of food additives conform to the use of hygiene standards and health management.
4. Chapter IV: Sanitation of Food Containers, Packaging Materials, Food Processing Tools, and Equipment
5. Chapter V: Development of Food Hygiene Standards and Management Practices
   This chapter provides the standards-setting bodies of national health standards, health management practices and inspection procedures, and local health standards.

6. Chapter VI: Food Hygiene Management
   The main provisions are contained in a seven step health management system.
7. Chapter VII: Food Hygiene Supervision
   This chapter contains the health monitoring system, food safety oversight responsibilities, establishment of food safety supervisors, health rights, and obligations of supervisors.
8. Chapter VIII: Legal Liability
   The administrative, civil, and criminal responsibility for violation of the food sanitation laws are contained herein. Administrative consideration, administrative litigation and administrative relief measures are also provided.
9. Chapter IX: Supplementary Provisions
   This chapter provides the interpretation of the food sanitation law. Food hygiene management for exporting is assigned by the state import and export commodities inspection department in conjunction with the state council, health administrative departments, and relevant administrative departments. Food hygiene management measures for the army special food and confession are assigned by the central military commission in accordance with this law, as well as the formal implementation date of this act.

1.3.3. Law of the People's Republic of China on Product Quality

On February 22, 1993, the seventh national people's congress standing committee 30th meeting adopted "law of the People's republic of China on product quality." This was amended according to the ninth national people's congress standing committee 16th meeting conference on July 8, 2000. Since September 1, 2000, the "law of product quality" was implemented. The "law of product quality" is the basic law and regulations of China's product quality, and important to standardize problems of product quality both comprehensively and systematically, and is the basic law containing supervision of product quality and the responsibility of product quality. The law consists of 6 chapters and 74 sections, including general principles, supervision of product quality, quality responsibilities and obligations of producers and sellers, damages, penalties provisions, and supplementary provisions.

1.3.4. China's Food-Related Laws

China's food-related laws promulgated and implemented also include the "standardization law," "measurement law," "trademark law," "patent law," "advertising law," "law of the People's Republic of China for countering unfair competition," "law of the Peoples Republic of China on the protection of consumer rights and interests," "foreign trade law," "law of the People's Republic of China on import and export commodity inspection," "law of the People's Republic of China on the entry and exit for animal and plant quarantine," "frontier health and quarantine law of the People's Republic of China," "animal epidemic prevention law of the

People's Republic of China," "radioactive pollution prevention and control law," "infectious disease prevention and control law," "agriculture law," "fisheries act," "wild animal protection law," "seed law," and the "agricultural product quality safety law."

1.3.5. Food Administrative Regulations, Departmental Rules and Regulations

1.3.5.1. Administrative Regulations

The current food administrative rules and regulations of China mainly include food distribution management regulations, the management rules of the agriculture genetically modified organisms safety, the regulations of the People's Republic of China on the import and export animal and plant quarantine, the policy of purchasing grain, regulations on pig slaughtering, edible salt monopoly regulation, regulations on edible salt iodization as a means to eliminate iodine deficiency disorders, and management regulations on veterinary drugs.

1.3.5.2. Normative Documents of the State Council

Normative documents of the state council mainly include decisions on further strengthening food safety, a circular on issuing the development of food and nutrition program, and a circular on the implementation of food and drug safety project.

1.3.5.3. Departmental Regulations

The current food regulations of China mainly include the measures of food additives health management approach, the management and supervision of the quality of food processing manufacturer, health registration regulations of the enterprises for the production of export commodities, order of the state environmental protection administration, and up to more than 50 other rules and regulations

1.3.5.4. Normative Documents of the Ministries

Normative documents of the ministries mainly include management system certification regulations for food manufacturers' hazard analyses and critical control point (HACCP), the management approach on applying for foreign registration of exporting food production enterprises, rule on import and export food labeling auditing operation and nearly more than 20 filings.

1.3.5.5. Regulations of Local Government

Regulations of local government include the management measures of Muslim food for the city of Guangzhou.

1.4. Key Issues and the Improvement of Laws and Regulations for Chinese Food Quality and Security

1.4.1. Key Issues of Laws and Regulations for Chinese Food Quality and Security

The key issues of China's current system for food safety is mainly focused on the following aspects: the food safety law is not complete; the food safety standards are not sound; there is poor internal coordination in the legal system; some of the current system is not entirely reasonable and does not operate well; there is some conflict between law regulations and the actual supervision and management system; The penalty system needs to be more strict.

1.4.2. The Improvement of Laws and Regulations for Chinese Food Quality and Security and Prospects (4)

For improvement of the laws and regulations of Chinese food quality and security, we must emphasize not only the integration of the existing system, but also the creation of a new system. The integration of the existing system would form a uniform scientific and effective regime based on massively different levels and a different emphasis simplifying the stipulations, adjusting the old measures, and solving significant conflicts. The foundation of the new system should streamline relationships while avoiding new conflicts between the laws, regulations, and rules during the time of its foundation. Meanwhile, consistent with the Chinese national situation, we should learn from our advanced foreign experience and focus on the innovations of management, including institutional innovation and system innovation. Therefore, during the reformation of present laws and regulations, we also should focus on improvement and development of the following systems.

1. Food security risk monitoring and risk assessment system.
   Food safety risk monitoring and risk assessment is a comprehensive, interdisciplinary, cross-professional technical work, reflecting food safety supervision and management based on prevention. It provides the scientific basis for food security policy making and developing food safe supervision and management. It may help to improve government decision-making of food security and the management of scientific validation.
2. Early-warning system for food safety risk.
   "Preventing accidents before they occur" is the best way to reduce the cost to society. The Food safety risk early warning system is different from what the food safety regulatory agencies have done in the past, warning after the accident, now with the intention to prevent the occurrence of adverse events. Through the early warning system, the openness and transparency of food safety management can be enhanced, while also enhancing public confidence in food safety management.
3. Special supervision and management system for new food resources.
   The basic contents of the system should include, but not be limited to the following aspects: defining the concept of new food resources; appraisal system of new food resources applying the prerequisite and the examination of the main body of new resources entering the market; the labeling requirements of new food resources, namely, it must indicate clearly its nature and any danger to the consumer; the remedial measures of the new food resources predicted risk; the related legal liability stipulation of the new food resources safety.
4. For the operator of the food production and the management system for food risk classification.

5. The management system for food production commissioned the legal relationship involving marking, the label of food production commissioned, and the specific issues of the food quality safety responsibility burden-sharing, which affects the market supervision and claims approach of the consumer. General Administration of Quality Supervision Inspection "Food Production and Processing Enterprises Quality and Safety Supervision and Management Regulations (Trial)" contains the relevant stipulations, which are still at the testing stage, but also needs further standardization and confirmation in its legal form.
6. The food production and operation management system for traceability, as some countries and regions have already established and implemented a "food status coded identification system."
7. The food recall system is implemented for unsafe or defective food in a timely manner before markets are entered to avoid the occurrence of damaging public safety, thus providing the benefit of consumer protection. In addition, a special system is needed for flawed products. Because food sale is broad and single food product values are low, a universal recall is difficult, unlike large products such as automobiles. Therefore, there is a need for special recall rules including the conditions of food recall, the main body of a recall being unsafe food, the specific ways and methods of an unsafe food recall, recall procedures of unsafe food, and finally the relevant legal responsibilities.
8. Food security for the processing system and accident prevention. There is a need to establish an effective prevention system which is fast and highly effective containing a coordinated response and defined handling mechanism to address a processing accident because human health is directly related. Therefore, it is proposed to stipulate security for accident prevention and processing by the following: 1) the prevention system for accidents involving food safety; 2) an accident report and notification system for food safety; 3) a medical and responsibility investigation system for food security; 4) administrative control measures and a medical treatment system for food safety accidents.
9. Food safety information release system.

   Through releasing of food safety information, this may encourage legitimate producers to fight against illegal trade and provide a reference for consumers that may reduce the substandard food production that causes harm.

   A food safety information system includes the following: 1) determination of the performer of an issue; 2) determination scope and content of the issue; 3) determination of the procedure for verification of the issue; 4) determination of legal liability of an omission or an unlawful act regarding the information process of the issue.

10. The quality of food production and its operation.
   In view of the spread of counterfeit and shoddy food and other unsafe food forms, as well as frequently occurring food security accidents, the responsibility system and food safety supervising and management department's legal liability system should further strengthen the food security administration legal responsibility. The civil liability for food quality standards, simultaneously, should further consummate protection rights and interests of food cost for the consumer and provide a relief system.

2. Food Standard of China

2.1. The Development Process of the Chinese Food Safety Standard System

China's food safety standard system was constructed in the 1950s and has passed through five stages, a primary stage of establishment (1960s and 1970s), a developmental stage (1980s), one of adjustment (from late 1980s to middle 1990s), consolidation and further development (from mid-and late 1990s to 2001), and now an improvement stage (2001–present).

2.1.1. The Primary Stage

A rapid development of food standardization took place in this period. The Ministry of Light Industry issued 13 standards for milk and dairy products, 8 standards for egg products, and 3 standards for salt, all of which were published as a ministerial standard. The canned food industry had a larger scale of production and higher technical level. Strengthening standardization management was first carried out in the canned food industry in order to meet the needs of export. In 1964, 153 ministerial product standards for the main varieties of canned foods were formulated and published. At the same time, the corresponding process operational elements, detection methods, and canning factory health system were established, which became a milestone in the standardization of canned food. In 1976, 42 ministerial standards for varieties of canned food were formulated and published, and some provinces and municipalities also organized companies to develop enterprise standards. In 1977, the National Standards Administration approved and issued 54 national standards of food hygiene including meat, eggs, aquatic food, beverages, wine, and food additives. These were proposed, organized, and developed by the Ministry of Health. In 1978, the National Standards Administration approved and promulgated standards for white sugar, soft white sugar, wheat, soybeans, corn, rice, wheat flour, peanuts, fruit, peanut oil, soybean oil, rapeseed oil, refined rapeseed which were presented and organized to be developed by the Department of Commerce and Ministry of Light Industry. In the same year, the Ministry of Commerce, Ministry of Health, Ministry of Light Industry and Supply and Marketing Cooperatives jointly formulated and issued food quality standards, testing methods, quality management regulations and health regulations of soy sauce, vinegar, bean paste, and flour paste.

2.1.2. The Development Stage

China's food standardization was developed in the 1980s. Much was done in this period. Not only did food standards increase greatly, but also a number of technical institutions were established, all to ensure food standardization specification and rapid organizational.

A total of 725 national standards of food were issued by The State General Bureau of Standardization, The State Bureau of Technology and the Ministry of Health successively in this 10-year period (1980–1990). These referred to food processing products, agricultural standards, food industry base and related standards, food testing method standards, food and hygiene standards for processed products, food packaging materials and container standards, and standards for food additives in which there were the "General Standard for Food Labeling." Additionally, there were 12 categories of food businesses hygienic practices, a number of important national standards including 16 infant and young children food standards issued.

At the same time, a total of nine ministerial Food Standards were issued by the Ministry of Agriculture, Animal Husbandry and Fishery. A total of 15 ministerial Food Standards were issued by Ministry of Light Industry. A total of 245 ministerial food professional standards were issued by Ministry of Commerce.

2.1.3. The Adjustment Stage

A rapid development of China's food industry took place in the late 1980s to mid-1990s. Food standards no longer met the needs of the situation at that time and thus food standardization went into the adjustment stage.

During this period, the existing standards at all levels were regulated and a total of 232 food standards were rectified as mandatory national standards while the food hygiene standards remained the main body.

After rectification, there was 108 former national standards, which were adjusted to the food industry standards; there were 198 former food products mandatory ministerial standards, professional standards and industry standards approved and released by Department of Commerce, which were all adjusted to the recommended food industry; there were 174 original food mandatory ministerial standards, professional standards, and industry standards approved and released by the Ministry of Light Industry, which were all reduced to 23 mandatory industrial standards.

Subsequently, the Ministry of Health revised a total of 140 food hygiene standards and testing methods and the relevant authorities revised a total of 28 standards for food additives.

After the comprehensive rectification of standards, China built the food standardization system with Chinese characteristics in the 1990s. Then China's food standardized management accomplished the management model that is the harmonization of food standards program, centralized review of the draft standards, and coordinated release to administrative departments. The model further promotes China's food standards into a legal and uniform track.

2.1.4. The Consolidation of the Development Stage

The consolidation and enrichment of China's food standardization and integration with the global standards (from mid and late 1990s to 2001) is an important stage in which the national standardization organization (National Standardization Management Committee) was established. In this period, a number of laws and department regulations were successively promulgated, which further strengthen the foundational position of food standardization. Meanwhile, China's food standards were further improved, mainly by amending old and out-of-date food product standards and formulating ones that the market needed. The Ministry of Health revised a total of 140 standards and testing methods for food hygiene, and the relevant authorities revised a total of 28 standards for food additives. All relevant departments also revised a number of important food standards.

2.1.5. The Improvement Stage

According to the relevant provisions of State Council for strengthening food safety management, the National Standards Committee organized people to rectify the existing agricultural standards and food standards (including national standards and industry standards) in 2001–2005. This was the second major adjustment and improvement of China's food standardization since the founding of the People's Republic of China.

A total of 867 national food standards, 347 industry standards, 268 projects of national standards and 134 industry-standard projects were canceled after collation. There were a total of 259 national food standards that continue to be valid, 49 to be modified, 282 to be amended, 122 to be revised integrally, and 125 to be repealed, 17 mandatory standards to be turned into recommended standards, and 12 to be turned into industry standards. There were a total of 68 food industry standards that continue to be valid, 109 to be amended, 124 to be revised integrally and 47 to be repealed. There were a total of 120 national standards projects to be continued, 90 to be integrated, 32 to be canceled and 26 completed. There were a total of 78 industry-standard projects to be continued, 26 to be integrated, and 30 to be canceled. The verification for local food standards were accomplished and a total of 250 standards were collated.

After China's new "Food Safety Law" was implemented, current safety standards for agricultural products and food quality, food hygiene standards, food quality standards, and industry mandatory standards for food industry were integrated by the health administrative departments of the State Council, and turned into food safety national standards.

2.2. The Status of China's Food Safety Standard System

2.2.1. The Basic Components of Chinese Food Safety Standard System

After more than 50 years of development, China's food safety standardization has made great achievements, especially remarkable in recent years.

China established a food standards system that consists of the national standards, industry standards, local standards, and enterprise standards before the implementation of new Food Safety Law.

The National Standard issued by the standardization administrative department of the State Council is a technical requirement that needed to be unified nationwide. The Industry Standard is the technical standard that has no national standard but needs to be uniform throughout industry. The Industry Standard was established by the relevant administrative departments of the State Council and then put on record by the standardization administrative department. The local standard is the uniform standard for food safety, including sanitation requirements where there are no national standards and industry standards but where it is needed at the provincial, autonomous regions, and municipalities. The Local Standard is provided by the standardization administrative departments of the provinces, autonomous regions, and municipalities and put on record by the standardization administrative department and relative administrative department of the State Council. The Enterprise Standard is the technical requirements for enterprise-wide coordination and harmonization, management and work requirements, and is set the basis of production and business activities. National Standards, Industry Standards and Local Standards are divided into mandatory and recommended ones according to the binding nature of the laws. By 2005, there were a total of 3,680 standards including all kinds of food national standards, industry standards and local standards and a large number of enterprise standards. Among national standards, industry standards , there were a total of 484 food hygiene standards, 1,298 quality standards, 170 basic standards, 1,418 methods standards and 30 management standards referring to the basics of food safety standards, toxic and harmful substances in food limited standards, food contact materials hygiene standards, food safety management and control standards, food safety inspection and testing methodology standards, the label of food safety standards, and identifying a specific food product standards. These standards initially shaped our country's food standards system, and basically met the needs of China's food production and consumption requirements.

Existing standards will be integrated into the new Food Safety Law, and the scope of Food Safety standards is limited. The food safety standards should include the following: limited requirements of pathogenic microorganisms, pesticides residues, veterinary drug residues, heavy metals, pollutants, and other substances harmful to human health in food and food-related products. It also includes the varieties, use, and consumption of food additives, nutrient requirements of food specifically for infants and young children, and other specific groups, labeling, marking and specification requirements referring to food safety and nutrition, health requirements during food production and business processes, quality requirements related food safety, food inspection methods and procedures, and other needs to develop food safety related standards.

2.2.2. The Management System of China Food Safety Standards

1. Administration
   Before implementation of the Food Safety Law, the State Standardization Administration carried out the nation's food standardization work that was authorized by the State Council. Ministries and commissions in charge of food-related industries such as National Development and Reform Commission, Ministry of Agriculture, Ministry of Commerce, Ministry of Health, State Administration of Grain manage the industry's food standardization in accordance with the work division. Local quality and technical supervision departments and relevant administrative departments carry out the food standardization management in the administrative regions with the principle of "unified management and division of labor."
2. Food Safety Standardization Organization
   Before the implementation of the Food Safety Law, the food safety standardization organization consisted of the Food Standards Technical Committee of Standardization, standardization research institutions and relevant associations. The food-related National Standardization Technical Committee includes the National Food Industry Standardization Technical Committee, the National Standardization Technical Committee on Food Additives, the National Cereals, Oils & Standardization Technical Committee, and the National Fisheries Standardization Technical Committee. The food-related industries administration established industry standardization technical committees, such as the food hygiene standards Professional Committee of Hygiene Standards, the Technical Committee of the Ministry of Health, and the tropical crops and products Standardization Technical Committee of the Ministry of Agriculture all for meeting the developing needs of food standards. Standardization research institutions and food-related trade associations play active

roles in the food standards revision, the standards promotion, monitoring the implementation of standards, and advisory services. After the implementation of the new Food Safety Law, it was authorized that the formulation and publication work of national standards for food safety would be carried out by the health administrative departments of the State Council. The standardization executive branch of the State Council is responsible for providing national standard numbers. Pesticide and veterinary drug residues limits in food and the testing methods and procedures are provided by the health administration and agriculture administration of the State Council. The inspection procedures for slaughtering livestock and poultry are provided by the relevant administration in conjunction with the health administration of the State Council. National standards of the products involved in the provisions of food safety should be consistent with National standards of food safety.

2.2.3. The Features of Chinese Food Safety Standard System

Compared with the food safety standard system of the Codex Alimentarius Commission (CAC), the International Organization for Standardization (ISO) and other international organizations as well as the USA, the European Union, Japan, Australia, Canada, and other developed countries and regions, the features of the Chinese food safety standard system are mainly embodied in the following areas:

1. Standards at all levels complement each other and form a relatively complete standard system.
   The food safety standard system of our country's compulsory standards combined with recommended standards, national standards matched with industry standards, local standards, and enterprise standards form a relatively complete set of criteria.
2. Meets the demand of food safety control and management.
   The types of Chinese food safety standards are relatively complete and cover major food species, every link of the food chain, and toxic or hazardous pollution factors, and can realize the target for hazard control of the entire food chain.
3. Consistent with international standards.
   Compared with the CAC, from a general point of view, our country's food safety standard system is basically consistent with the CAC, whether considering the standard system structure and composition the main standard index, or technical requirements.
4. Reflects the scientific principle of the WTO: SPS (World Trade Organization: Sanitary and Phytosanitary) agreement.
   The development of our country's food safety standards has fully taken into account the principle of the "SPS Agreement,"

use of or change to international standards as far as possible. The food safety standards of our country, also take into account the unique geography of Chinese environmental factors, human factors, and other special requirements in order to relegate the "appropriate level of health protection" as the goal and principle, and take "risk assessment" as the scientific basis.

2.2.4. Issues of the Food Safety Standard System

As the food industry's development is improving people's living standards, food safety standardization is facing several challenges. Challenges include the overall level of standards still need to be improved, some standards have disparity with international standards, and some standards lack coordination, and cross functionality with other standards. Finally the national standards still have some deficiencies and the status of implementation of some standards need further strengthening in their administration.

2.2.5. China's International Food Safety Standards

In 2002, China led the draft of the CAC standard for production norms for the control of aflatoxin contamination in fruit trees. This marked the first time China effectively participated in the work of international standards. Since then, China has led the preface of standards for food additives, the food classification system for food additive, and the revision of fermented product standards. In the development of the work of Codex standards such as the chloropropanol limit in soy sauce, the limit of acid hydrolyzed vegetable protein, the limit of lead in aquatic products, microbiological risk management principles and guidelines, the International Code of Hygienic Practice, and the standards of fermented bean paste and chili sauce China actively participated in the formulation group and submitted monitoring data.

In addition, agricultural products to be tested for multi-residue pesticide, pyrethrum lipids, poultry organizations testing for dichloro-dimethyl pyridinoline , and methods for veterinary drug residues, developed by the chief researchers of AQSIQ (General Administration of Quality Supervision, Inspection, and Quarantine) (1) and with the research guidance of AOAC (International Association of official analytical chemists) Professor Pang Guofang, were approved as international AOAC standards by the International Association of official analytical chemists (AOAC). This was the first AOAC international standards study that was developed by Chinese scientists, and won honor for our country in the international analytical chemistry arena.

2.3. Measures and Prospects to Speed Up Construction of China's Food Safety Standard System

To address the concern of food safety, we should first grasp the source of agricultural standardization, the whole process from farm to table carrying out standardization of management; strengthening the food safety standard system construction through the culture, production, processing and distribution of the whole process to form a complete standard system. Food safety standardization is a systematic project which should give complete attention to agriculture, health and other departments, and unified management effectively improving the food safety standards system.

Based on the situation faced by the food industry and developing trends, in order to continue building the technical base for food standardization, we should focus on the following areas.

2.3.1. Construct the Food Safety Standard System

The construction of the food safety standard system mainly includes:

1. The construction of a completed food safety standard system that covers all aspects "from farm to table." National, industrial and local standards should accurately position and coordinate supporting legislation. The enterprise standard should not only be legally effective, but aimed more to improve the level of protection.
2. We need to accelerate standard revision process. To do this effective management is needed to streamline the entire process of standards revision, and to shorten the cycle-time of revisions by using information technology and other advanced means.
3. We need to improve the level of effectiveness of the standards. This requires strengthening the theoretical research and preliminary studies needed for food safety standards, including enhancing the scientific basis of standards with the use of risk assessment. The standards should be revised according to the principle that fair, open, transparent, and broad participation will ensure their applicability. Finally, we can ensure validity of the standards and continuous improvement by frequent review and information feedback.
4. The standardization level of food enterprise should be improved. Food enterprises are not only the main body of implementation of the food safety standards, but also the main force to formulate and revise those standards. We encourage enterprises to formulate and implement enterprise standards that are more stringent than the national or industry standards, thus improving the quality of food enterprises.
5. We need to improve the expertise required food safety standards so that a high level of technical support is available for standardization and revision.
6. Finally, we need to establish emergency mechanisms of food safety standardization that can provide warning, analysis, and treatment measures when food safety accidents occur.

2.3.2. Development of Important Food Standards

Enhanced research for food standards, advanced foreign standards, and emphasis on important standards is essential for independent intellectual property. In the next few years, food national standards and their revisions will focus on:

1. Food terms and classification-based standards are the key points of the construction and management of the food safety standards system. The drafting unit should learn from international standards in the revision process. At the same time, we need to take full account of the status of development and management needs of our country's food industry, to ensure the effectiveness of the standards.
2. Limit standards for pesticide residues should be based on the investigation and study of pesticide residues and pollutants making full use of risk analysis, to supplement and perfect the relevant standards in a timely manner.
3. Standards for food additives should be revised using sanitation methods to perfect the relevant methods and standards of food additives products, and to ensure stable and effective implementation of the standards in a timely manner.
4. The relevant standards of detection methods should be a gradual rationalization of existing detection methods of food standard system. It should be the goal to improve the generality and applicability of detection standards. There should be a focus on improving the accuracy of inspection standards, and shorten testing cycles through the research of advanced detection technology.

2.3.3. Emphasize the Application of the Principles and Methods of Risk Analysis on the Development of Food Standards

Food safety risk analysis is the principle that the WTO/SPS agreement clearly stipulates that member states must comply. It is also widely recognized in the international community as a strengthening measure of food safety management and in resolving trade disputes. Food safety standards formulated by the principles and methods of risk analysis are based on scientific data to ensure the standards are both technically correct and reasonable. Therefore, the study and the application of the food safety risk analysis, is currently a cornerstone of the production of food safety standards.

2.3.4. Development of the Basic Scientific Research for Food Standardization

Basic research on food safety standards is needed to limit poisonous and harmful material in food, the migration of harmful material in packing materials, and the safety assessment and inspection methods of genetically modified food. In addition, in accordance with international and domestic food market development, the research on food labeling should be developed to regulate the normal operation of food circulation so that conditions for the safety of consumer are created. Standard measures should be actively pursued to protect domestic food markets and to break foreign technical barriers which will improve the competitiveness of the Chinese food enterprise.

2.3.5. Value the Implementation and Supervision of Food Standards

The implementation of the food standard is the key to food standardization activities, and will make it effective and achieve its intended goals. Comprehensive utilization of the means that propogate, training execution, and certification will provide market access and monitoring to ensure the effective implementation of food standards. At the same time, the establishment of a feedback mechanism for the further amendments and improve the standard system is important. That mechanism must include scientifically sound and powerful monitoring programs that provide accurate assessment of how enterprises meet the standards and where the standards do not address ever changing issues. It is chromatographic techniques combined with mass spectrometry that provides this in the area of safe residue requirements.

3. Application of Mass Spectrometry to Food Safety of China

Mass spectrometry has become a powerful tool for the analysis of complex samples such as food. During recent decades, scientists in China have made significant achievements with the application of mass spectrometry on the determination of chemical contaminants in foods. Some rapid, simple, sensitive, and accurate analysis methods based on the mass spectrometry have been applied.

Regulatory analysis of pesticides, veterinary drugs, and additives frequently encounter compounds, which cannot be readily identified by chromatography alone on two or more stationary phases. Fortunately, mass spectrometry with its powerful ability of structure identification can easily resolve this problem. In addition, mass spectrometry provides high sensitivity and good quantitation of chemical contaminants in foods.

For example, gas chromatography coupled to triple quadrupole mass spectrometry (GC-MS/MS) was applied for the first time in the determination of trace food-derived hazardous compounds (2). The sensitive method developed for 13 heterocyclic amines provided both confirmation and quantitation. The method includes a solid-phase extraction procedure with a polystyrene copolymer cartridge (LiChrolut EN), followed by a derivatization reaction with *N*,*N*-dimethylformamide di-*tert*-butylacetal. Analyses were performed by gas chromatography with triple quadrupole mass spectrometry in electron ionization mode. The MS/MS fragmentation pathway of derived heterocyclic amines was studied and the differences of fragmentation characteristics were used successively to distinguish the isomers in absence of chemical standards. The excellent selectivity and sensitivity was achieved using multiple reaction monitoring mode (MRM). The limits of quantitation of the method for these compounds ranged from 0.12 to 0.48 ng/g of sample. The method developed was applied to the analysis of Chinese cooked foods and the results demonstrated the power of GC-MS/MS for the analysis of trace food-derived hazardous compounds in complex food matrices such as meat samples.

3.1. Analysis of Pesticides Using Mass Spectrometry

Pesticides are found in plants and soil and the amount of residues and their toxic metabolites must be determined. Many methods based on mass spectrometry have been developed by Chinese scientists to determine these pesticide residues.

Wang (3) has developed a gas chromatography-mass spectrometry method for the determination of 11 herbicide residues (alachlor, acetochlor, butachlor, pretilachlor, metolachlor, dimethenamid, propachlor, napropamide, propanil, atrazine, and metribuzin) in rice and soybeans. The sample was extracted with acetone–water, defatted by liquid–liquid partition, and purified through solid-phase extraction with Florisil. Experiments on five fortification concentrations were carried out, and the limit of determination is 0.02 mg/kg. The average recoveries of soybean samples ranged from 63.3 to 96.0% and the relative standard deviations were from 2.14 to 11.2%. The average recoveries of rice samples ranged from 76.8 to 102% and the relative standard deviations are from 2.2 to 9.08%. The results indicated that the method was fast, accurate, and easy to perform. It also demonstrated that the method met the requirements of simultaneous determination of 11 herbicides in rice and soybeans. Zhang (4) developed a rapid multi-residue method for the determination of 16 herbicides in onion. The onion samples, which had been pretreated were extracted with acetonitrile followed with solid-phase extraction (SPE) clean-up. The herbicide residues in onion were detected by gas chromatography/mass spectrometry with selected ion monitoring. The limit of quantitation (LOQ) ranged from 0.003 to 0.015 mg/kg. Xu (5) developed an effective method for the trace analysis of indoxacarb residue in foodstuffs of both plant and animal origin (grapefruit, ginger, fresh soyabean, bamboo shoot, chicken, fish, and pork) using gas chromatography (GC-ECD) and liquid chromatography tandem mass spectrometry (LC-MS/MS). Samples were extracted using acetone and n-hexane mixed solvent (1:2, v/v) and then purified using SPE. The extracts were analyzed using GC-ECD and LC-MS/MS. The MS/MS used MRM with transitions of each of the precursor ions with two product ions in which one product ion for identification was m/z 529 > 293 and the another for quantification was m/z 529 > 249. The detection limits (LODs) of the methods were 0.0015 and 0.0006 mg kg^{-1}, and the quantification limits (LOQs) were 0.005 and 0.002 mg/kg for GC-ECD and LC-MS/MS, respectively. The relative standard deviations (RSDs) of recovery for indoxacarb were lower than 15% in 10 types of agro-products. Ten repetitive determinations of recovery achieved good reproducibility for indoxacarb and the recovery ranged from 72.08 to 113.74%. The proposed procedure was applied to the analysis of several real samples of different origin from Fujian Province, China, and 299 samples were screened for indoxacarb residue of which five positive samples were found.

The improvement of throughput of the analysis method is very important for food safety and many efforts on this type of method development were made in China. An efficient and sensitive method was established and validated by Yang (6) for the simultaneous determination of 118 pesticide residues in teas. This multi-residue method involved extraction with ethyl acetate-hexane, clean-up using gel permeation chromatography (GPC) and solid-phase extraction (SPE), and subsequent identification and quantification of the 118 pesticides in the extract by gas chromatography-mass spectrometry (GC-MS). For most of the target analytes, the optimized sample preparation led to no significant interference with the analysis of sample matrix for the determination of the 118 compounds was achieved in about 60 min. The limits of detection for the method were 0.00030–0.36 mg/kg, depending on each pesticide. A new approach for the extraction of nine kinds of organochlorine pesticides (OCPs) from vegetable samples coupling single-drop micro-extraction with gas chromatography-mass spectrometry was presented by Zhang (7). A novel mixed liquid of *p*-xylene and acetone was selected as an organic extraction solvent. Experimental parameters such as the extraction solvent composition, solvent volume, drop size, stirring speed, and the extraction time were optimized. The results indicated that the proposed approach is feasible for the fast determination of organochlorine pesticides in vegetable samples. Chen (8) investigated a simple and fast method for the simultaneous analysis of thiobencarb, deltamethrin and 19 organochlorine pesticide residues in fish by gas chromatography-mass spectrometry. Most of the lipids in the extract were eliminated by low-temperature cleanup (freeze out), prior to solid-phase extraction cleanup. The lipids extracted from the fish samples were easily removed without any significant losses of the pesticides. The newly developed method was demonstrated to give efficient recoveries and LODs for detecting pesticide multi-residues in fish.

In addition, the methods for determination pesticide residues in the biological samples have been developed in China. Zhou (9) developed a new method, headspace solid-phase microextraction (HS-SPME), with in situ derivatization and gas chromatography-mass spectrometry (GC-MS), which was used for the determination of trace amount of pentachlorophenol (PCP) in human plasma. The conditions for the analysis of PCP in human blood plasma using headspace SPME with situ-derivatization coupling with GC-MS was optimized with chemometrics. By using headspace SPME, the sample preparation procedure was simplified and the used quantity of organic solvents decreased substantially, meaning less associated costs and low environmental impact. The proposed method attained a detection limit (0.02 ng/mL) lower than previously described methods for detection of PCP level in human plasma samples. Moreover, only 0.5 mL of plasma sample was needed for the analysis. Therefore, the proposed method could be considered as an attractive alternative to the currently used analytical methods. Jin (10) developed a novel analytical method for the determination of 14 trace chlorophenols in clam tissues by ion chromatography (IC) coupled with atmospheric pressure chemical ionization mass spectrometry (APCI-MS) in negative mode. The method comprised a fast ultrasound-assisted extraction using a mixture of methanol/water (4:1 v/v) containing 5% triethylamine (TEA) as extraction solvent. Solid-phase extraction with an Oasis HLB cartridge was followed by ion chromatography with a gradient separation using KOH/acetonitrile at a flow rate of 1.0 mL/min on an IonPac AG11 guard column (50 mm × 4.0 mm I.D.) and an IonPac AS11 analytical column (250 mm × 4.0 mm I.D.). The deprotonated molecular ions were selected for quantification in the selected ion monitoring (SIM) mode for monochlorophenols (MCPs), dichlorophenols (DCPs), trichlorophenols (TCPs), and pentachlorophenol (PCP).

3.2. Analysis of Veterinary Medicines Using Mass Spectrometry

Jia et al. (11) reported an ultra-performance liquid chromatography coupled with tandem mass spectrometric (MS/MS) method for the simultaneous quantitation of multiclass veterinary drugs in egg. The analysis of target compounds, including seven tetracyclines and four types of quinolones, may be accomplished in total run time of 15 min. The egg was extracted with ethylenediamine tetraacetic acid–McIlvaine buffer solution and further purified using a polymer-based Oasis HLB solid-phase extraction cartridge. A C18 column was used to separate the analytes followed by MS/MS using an electrospray ion source. The overall average recoveries of the analytes based on matrix-fortified calibration ranged from 71 to 112% with acceptable relative standard deviations of <20% for six trials. For all of the target compounds, the limits of quantitation ranged between 0.02 and 4.29 μg/kg. The proposed method is sufficiently sensitive and highly selective.

Tang et al. (12) demonstrate a high-throughput screening method using LC/MS/MS with on-line extraction cartridge cleanup for measuring drug residues in animal muscle. The rapid qualitative method was developed and validated for screening 13 target veterinary drugs: four macrolides – erythromycin A, josamycin (leucomycin A3), kitasamycin (leucomycin A5), and tylosin A; six (fluoro)quinolones–ciprofloxacin, danofloxacin, enrofloxacin, flumequine, oxolinic acid, and sarafloxacin; and lincomycin, virginiamycin M1, and trimethoprim in different animal muscles. Clindamycin, norfloxacin, nalidixic acid, oleandomycin, ormetoprim, and roxithromycin were used as the internal standards. After simple deproteination and analyte extraction of muscle samples using acetonitrile, the supernatant was subjected to on-line cleanup and direct analysis by LC/MS/MS. On-line cleanup with an extraction cartridge packed with hydrophilic-hydrophobic polymer sorbent followed by fast LC using a short C18 column resulted in a total analysis cycle of 6 min for 19 drugs. This screening method considerably reduced the time and the cost for the quantitative and confirmatory analyses. The application of a control point approach was also introduced and explained.

3.3. Analysis of Additives Using Mass Spectrometry

Food additives have been defined by the Joint WHO/FAO Committee in 1955 as substances "which are added intentionally to food, generally in small quantities, to improve its appearance, flavor, texture, or storage properties" (13). This definition excludes substances, which find their way into foods, but have not been deliberately added, and are usually classed as contaminants. Here we mainly introduced the application of mass spectrometry in the analysis of original food additives. Some prevalent contaminants analyzed by mass spectrometry are briefly introduced.

There are various types of food additives used in China, including antioxidants and preservatives, colors, sweeteners, flavors, emulsifiers, bread and flour additives, and nutritional additives. Here we introduce some methods for analyzing food additives in foods by mass spectrometry.

First we present recent work done in China for quantitative or qualitative analysis of antioxidants and preservatives by mass spectrometry. Hao et al. (14) developed a method of liquid chromatography/ion trap mass spectrometry (LC/ITMS) to quantify the synthetic phenolic antioxidant, tertiary butyl hydroquinone (TBHQ), in 2005. The liner calibration curves were obtained in the concentration range of 61.8–4,542.5 μg/L (R2 = 0.9999), the detection limit was 48 μg/L, and the recoveries of TBHQ from ten samples were 81.9–109.6%. This method has been used for determination of TBHQ in edible vegetable oil and the RSD was less than 5.3% (n = 10). Zhang et al. (15) developed a method for determining TBHQ content in edible oils using ultra performance liquid chromatography-tandem triple quadrupole mass spectrometer (UPLC-QQQ) in 2009. The detection used selected reaction monitoring (SRM) mode and calibration by external standard. The detection limit of TBHQ was 0.10 mg/kg. The correlation coefficient of linear calibration curve was over 0.999 in the range of 0.05–2.5 μg/ml and the recovery rate was 83–115%.

TBHQ could also be detected by GC-MS. For example, The TBHQ in XO sauce was detected by Yue et al. (16) using gas chromatography-mass spectrometry (full scan mode) in 2004. Feng et al. (17) developed a GC-MS method for simultaneous determination of butylated hydroxyanisole (BHA), butylated hydrosytoluene (BHT) and TBHQ in foods (instant rice, instant noodles, preserved ham, apple pie, and shortening). Besides antioxidants, preservatives have also been determined by MS. For instance, Wang et al. (18) developed a method for determination of four paraben preservatives including methylparaben (MP), ethylparaben (EP), propylparaben (PP) and butylparaben (BP) in dairy products by high-performance liquid chromatography mass spectrometry (HPLC-MS).

A trend for determination of preservatives and antioxidants in food is simultaneous analysis and some marked progress has been made in this field. Li et al. (19) developed a method for determination of preservatives and antioxidants in cake by GC-MS-SIM. The method could simultaneously determine seven preservatives (sorbic acid, benzoic acid, dehydroacetic acid, methyl paraben, ethyl paraben, propyl paraben, and bulyl paraben) and three antioxidants (TBHQ, BHA, and BHT) in cakes. The detection limits of the 10 additives were 0.02–0.10 mg/kg. Zhan et al. (20) developed a method for the simultaneous determination of methylparaben, ethylparaben, propylparaben, butylparaben, BHA, BHT and TBHQ in food with complex matrices, by gas chromatography with mass chromatography (GC/MS). The detection limits were 0.05–0.5 mg/kg for all compounds based on a 5.0 g sample. Xiang et al. (21) developed a method for determination of four preservatives (methyl-*p*-hydroxybenzoate, ethyl-*p*-hydroxybenzoate, propyl-*p*-hydroxybenzoate, butyl-*p*-hydroxybenzoate) and three antioxidants (butylated hydroxy anisol, *tert*-butylhydroquinone, and 2, 6-di-*tert*-butyl-4-methylphenol) in food was developed by GC-IT/MS combined with hollow-fibre membrane liquid-phase microextraction (HF-LPME). The linear range for the method was 0.4–80 mg/kg, and the detection limits (LOD) were between 0.002 and 8.0 μg/kg.

Colorants are also a large class of food additives. Some analysis methods to determine them using mass spectrometry have also been developed. Chen et al. (22) developed a method for determination of five pigments (Tartrazine, Amaranth, Ponceau 4R, Sunset Yellow FCF, Brilliant Blue FCF) in drinks by reverse phase liquid chromatography tandem mass spectrometry under selected ion monitoring (SIM) mode via negative electrospray ionization. This method has been applied in the determination of these colorants in drinks.

Besides water-soluble colorants, fat-soluble colorants have also been detected by mass spectrometry. Sun et al. (23) developed a method for simultaneous determination of banned 10 azo-dyes (Sudan (I-IV), Sudan Orange G, Sudan Red B, Sudan Red G, Sudan Red 7B, Butter Yellow, and Para Red) in hot chili products by gel permeation chromatography-LC-MS/MS. The LOD and LOQ for the investigated dyes were in the ranges of 0.1–1.8 and 0.4–5.0 μg/kg respectively, depending on the matrices.

This method has been applied successfully for the determination of the studied ten banned dyes in hot chili products. Ning et al. (24) developed a method for determination of Para-red in food by UPLC-ESI-MS/MS. The detection limit of the method was 0.26 μg/kg ($S/N=3$), and the limited of quantitation was 0.85 μg/kg ($S/N=10$). The author also developed a method for simultaneous determination of Para-red and Sudan I–IV in food by UPLC-ESI-MS/MS with isotopic internal standard (Sudan I-D5 and Sudan IV-D6) (25).

GC-MS has also been applied in colorants detection. For instance, Huang et al. (26) developed a method for determining Sudan I–IV in food by GC-MS/SIM. The linear range of the determination Sudan I and II were 0.01–10 mg/L with a detection limit of 1 μg/kg. The linear range of the determination of Sudan III and IV was 0.1–10 mg/L with a detection limit of 5 and 10 μg/kg. The method has also been applied to the determination of Sudan I–IV in foods such as hot chili products with satisfactory results.

Ma et al. (27) developed a method of simultaneous determination of water-soluble (Tartrazine, Amaranth, Ponceau 4R, and Sunset Yellow FCF) and fat-soluble (Sudan (I–IV)) synthetic colorants in foodstuff by high-performance liquid chromatography-diode array detection-electrospray mass spectrometry. This method uses dimethylsulfoxide (DMSO) as the extraction solvent in the sample preparation process. Detection and quantitation limits of the investigated dyes were in the ranges of 0.01–4 and 0.03–11.2 ng respectively. The recoveries of the eight synthetic colorants in four matrices ranged from 93.2 to 108.3% and the RSDs were less than 8.2%. Although this method has been applied successfully in the determination of synthetic colorants in four matrices (soft drink, delicious ginger, chilli powders, and chilli spices), the mass spectral data had to be acquired under positive/negative mode requiring a separate injection for each. New-style ESI-MS systems now offer operation simultaneously under positive and negative ion mode. Therefore, the method for simultaneous determination of water-soluble and fat-soluble synthetic colorants may be a development trend for screening and quantitation of colorants in foodstuff by HPLC-ESI-MS.

Cai et al. (28) developed a method for rapid identification of betacyanins from Amaranthus tricolor, Gomphrena globosa, and Hylocereus polyrhizus, by matrix-assisted laser desorption/ionization quadrupole ion trap time-of-flight mass spectrometry (MALDI-QIT-TOF MS). Based on this method, 14 free and acylated betacyanins, belonging to amaranthin-type, betanin-type, and gomphrenin-type betacyanins, respectively, were identified. However, the related isomers could not be differentiated for the lack of LC separation.

Artificial sweeteners are also a type of food additive that needs sensitive detection for determining their identity and concentrations in foods. Yang et al. (29) developed a method of simultaneous determination of seven artificial (aspartame, saccharin, acesulfame-K, neotame, sucralose, cyclamate, and alitame) and one natural sweetener (stevioside) in foods by HPLC/ESI-MS. The correlation coefficient of the calibration curve was better than 0.998 ($n=6$). The limits of detection (LODs) were below 0.10 μg/ml, whereas the limits of quantification (LOQs) were below 0.30 μg/ml.

Huang et al. (30) developed a highly sensitive method for the determination of cyclamate in foods (such as prepared dried fruit of *Areca catechu* L and tanned fruits) by ion-pair high-performance liquid chromatography-electrospray ionization mass spectrometry. The quantitation of the target compound was completed using a selected ion monitoring (SIM) at m/z 178 obtained from ESI-mode. Tiopronin was used as internal standard for the quantitation of cyclamate. The correlation coefficient of the calibration curve was better than 0.996 in the range of 50–5,000 ng/mL. The limits of detection and quantitation were 5 ng/ml and 20 ng/ml, respectively.

Lin et al. (31) developed a method for determination of sodium cyclamate in vitamin C effervescent tablets by gas chromatography-mass spectrum with selected ion monitoring (GC-MS/SIM). Sodium cyclamate was quantified using a single ion. The linear range was 0.05–10 mg/L with a detection limit of 0.6 mg/kg. The recovery range was 90.5–96.9% with a RSD of less than 4.4%.

The merits of high sensitivity and accuracy in mass spectrometry analysis have also been applied in flavors detection. For example, Han et al. (32) developed a method for isolation and identification aromatic components in milk by GC/MS. Using this method, 17 compounds were identified. The main flavor components in milk were: piperonal (16.210%), vanillin (16.838%), ethyl vanillin (9.817%), δ-undecalactone (32.657%), τ-octalactone (6.050%), and the ethyl ester (5.163%). The author (33) also developed a method for determination of vanillin in milk by GC-MS. The result shows that this method could eliminate interferences with complex material conveniently, quickly, and accurately.

There are some other studies applying MS analysis in food additives and contaminants. For example, Zhu et al. (34) developed a method for simultaneous separation and determination of eight kinds of drugs in dietary supplements with solid phase extraction and high performance liquid chromatography-mass spectrometry. The average recoveries for the eight drugs were 60.4–93.1% with relative standard deviations of 1.13–9.82% and the limits of detection were 0.5–3.0 mg/L respectively. Xu et al. (35)

developed a method of simultaneous separation and determination of chloropropanols (1, 3-DCP, 2, 3-DCP, 3-MCPD, and 2-MCPD) in soy sauce and other flavoring with gas chromatography-mass spectrometry in negative chemical and electron ionization modes. The author used D (5)-3-MCPD and D(5)-1,3-DCP as the deuterium isotopic labeled internal standards. Shao et al. (36) developed a method of analysis of nonylphenol(NP), octylphenol(OP), and bisphenol A (BPA) in animal tissues by liquid chromatography-tandem mass spectrometry with accelerated solvent extraction. Sample concentration and purification were performed using an Oasis NH_2 solid extraction cartridge. The detection limits of the method under multiple reaction monitoring mode were 0.3, 0.05 and 0.1 μg/kg for BPA, NP, and OP respectively. Zhang et al. (37, 38) developed a LC-MS/MS method to determine the concentration of acrylamide in fried bread sticks and chicken wings. The authors found that both bamboo leaves (AOB) and extracts of green tea (EGT) have two natural antioxidants that could reduce acrylamide levels significantly in fried bread sticks. The present study indicated that both AOB and EGT could significantly reduce the acrylamide content generated in fried bread sticks while maintaining the original flavor and crispness of the fried bread sticks. This study could be regarded as an important contribution to the studies on the reduction of acrylamide by natural antioxidants.

References

1. Li, M.G. The regulation of AQSIQ.
2. Zhang, F., Chu, X. G., Sun, L., Zhao, Y. S., Ling, Y., Wang, X. J., Yong, W., Yang, M. L., Li, X. Q. (2008) Determination of trace food-derived hazardous compounds in Chinese cooked foods using solid-phase extraction and gas chromatography coupled to triple quadrupole mass spectrometry. Journal of Chromatography A. 1209. 220–229.
3. Wang, L. B., Li, C., Li, C. F., Li, X. Q., Xu, C. L. (2008) A Rapid Multi-Residue Determination Method of Herbicides in Grain by GC-MS-SIM. *Journal of Chromatographic Science*. 5. 424–429.
4. Zhang, H., Chen Z., Yang G. S., Wang, W. Z., Li X. Q., Li. R. J., Wu Y. J. (2008) Microwave pretreatment and gas chromatography-mass spectrometry determination of herbicide residues in onion. *Food Chemistry*. 1. 322–328.
5. Xu D. M., Yang F., Lu S.Y., Lan J. C., Yu K. J., Cai C. P., Liu X. J. (2008) Determination of indoxacarb residue in foodstuffs of plant and animal origin by GC-ECD and LC-MS/MS. *Agricultural Sciences in China*.10. 1671–2927.
6. Yang, X; Xu, D. C., Qiu, J. W., Zhang, H., Zhang, Y. C., Dong, A. J., Ma, Y., Wang, J. (2009) Simultaneous determination of 118 pesticide residues in Chinese teas by gas chromatography-mass spectrometry. *Chemical Papers* 1. 39–46.
7. Zhang, M. S., Huang, J. R., Wei, C. L., Yu, B. B., Yang, X. Q., Chen, X. (2008) Mixed liquids for single-drop microextraction of organochlorine pesticides in vegetables . *Talanta*.4. 599–604.
8. Chen, S. B., Yu X. J., He X. Y., Xie, D. H., Fan, Y. M., Peng J. F. (2009) Simplified Pesticide Multiresidues Analysis In Fish By Low-temperature Cleanup And Solid-phase Extraction Coupled With Gas Chromatography/mass Spectrometry. *Food Chemistry*.4. 1297–1300.
9. Zhou, Y., Jiang, Q. W., Peng, Q., Xuan, D. L., Qu, W. D. (2007) Development of a solid phase microextraction-gas chromatography-mass spectrometry method for the determination of pentachlorophenol in human plasma using experimental design. *Chemosphere*.2. 256–262.

10. Jin, M. C., Zhu, Y. (2006) Ion chromatography-atmospheric pressure chemical ionization mass spectrometry for the determination of trace chlorophenols in clam tissues. *Journal of Chromatography A* 1. 111–117.
11. Jia, X. F., Shao, B., Wu, Y. N., Yang, Y., Zhang, J. (2008) Simultaneous determination of tetracylines and quinolones antibiotics in egg by ultra-performance liquid chromatography–electrospray tandem mass spectrometry. *Journal of AOAC International.* 91, 461–468.
12. Tang, H. P., Ho, C., Lai, S. S. High-throughput screening for multi-class veterinary drug residues in animal muscle using liquid chromatography/tandem mass spectrometry with on-line solid-phase extraction. *Rapid Communications in Mass Spectrometry.* 20, 2565–2572.
13. FAO. (1956) Nutrition Report Series No. 11. Food and Agriculture Organization, Rome.
14. Hao P., Ni J., Sun W., Huang W. (2005) Quantification of TBHQ in edible vegetable oil by liquid chromatography/ion trap mass spectrometry. Journal of Chinese Mass Spectrometry Society 26, 222–227.
15. Zhang P., Zhang X., Zhang J., Wang Y. (2009) Determination of tertiary butyhydroquinone in edible oils by ultra performance liquid chromatography-tandem mass spectrometry. Food Science in Chinese 30, 143–145.
16. Yue Z., Lan F., Xie L. (2004) Determination of BHA, BHT and TBHQ in XO Sauce by GC – MS. Journal of the Chinese Cereals and Oils Association 19, 83–85.
17. Feng N., Li S., Cai H., Li J., An J., Li G.. (2008) Simultaneous determination of 3 fat-soluble antioxidants in foods by GC – MS. Chinese Journal of Health Laboratory Technology 18, 1742–1743.
18. Wang J., Fan B., Xu S. (2008) Determination of parabens in dairy products by solid phase extraction combined with high performance liquid chromatography-mass spectrometry. Food Science in Chinese 29, 328–331.
19. Li D., Li C., Li J. (2009) Determination of preservatives and antioxidants in cakes by GC-MS-SIM. Analytical Instruments in Chinese 4, 56–60.
20. Zhan J., Yu X., Huang W., Fan Y., Peng J., Xie D., Shi J., Huang S. (2008) A simple method for the simultaneous determination of 7 food additives in foods with complex matrices by GC/MS. Journal of Analytical Science in Chinese 24, 729–731.
21. Xiang J., Qi A., Mao L. (2009) Determination of preservatives and antioxidants in foods by gas chromatography - mass spectrometry combined with hollow-fibre membrane liquid phase microextraction. Journal of Instrumental Analysis 28, 560–563.
22. Chen X., Li X., Yao X. (2005) Determination of synthetical pigments in drinks by high performance liquid chromatography-mass spectrometry. Chinese Journal of Health Laboratory Technology 15, 941–942.
23. Sun H. W., Wang F. C., Ai L. F. (2007) Determination of banned 10 azo-dyes in hot chili products by gel permeation chromatography-liquid chromatography-electrospray ionization-tandem mass spectrometry. J Chromatogr A 1164, 120–128.
24. Ning X., Ge Y., Wang D., Zhang Y., Shun H. (2007) Determination of parared in food by UPLC-ESI-MS/MS. Food Science and Technology 23, 189–193.
25. Ning X., Wang D., Yu C., Lu Z., Zhang Y. (2009) UPLC-MS/MS isotopic internal standard determination of parared and Sudan I-IV in food. Journal of Chinese Mass Spectrometry Society 30, 41–46.
26. Huang X., Wu H., Huang F., Lin X., Deng X. (2005) Rapid determination of Sudan I-IV in food by GC- MS/SIM. Journal of Instrumental Analysis 24, 1–5.
27. Ma M., Luo X., Chen B., Su S., Yao S. (2006) Simultaneous determination of water-soluble and fat-soluble synthetic colorants in foodstuff by high-performance liquid chromatography-diode array detection-electrospray mass spectrometry. J Chromatogr A. 1103, 170–176.
28. Cai Y. Z., Xing J., Sun M., Corke H. (2006) Rapid identification of betacyanins from Amaranthus tricolor, Gomphrena globosa, and Hylocereus polyrhizus by matrix-assisted laser desorption/ionization quadrupole ion trap time-of-flight mass spectrometry (MALDI-QIT-TOF MS). Journal of Agricultural and Food Chemistry 54, 6520–6526.
29. Yang D. J., Chen B. (2009) Simultaneous determination of nonnutritive sweeteners in foods by HPLC/ESI-MS. J Agric Food Chem. 57, 3022–3027.
30. Huang Z., Ma J., Chen B., Zhang Y., Yao S. (2006) Determination of cyclamate in foods by high performance liquid chromatography-electrospray ionization mass spectrometry. Analytica Chimica Acta. 555, 233–237.

31. Lin Y., Hu X., Wang P., Li J., Zhao X., Fu X. (2006) Determination of sodium cyclamate in vitamin C effervescent tablets by gas chromatography-mass spectrometry. Chinese Journal of Spectroscopy Laboratory 23, 338–341.
32. Han S., Yang J., Liu N. (2007) Study on establishing chromatogram fingerprint of milk flavour by GC/MS. Food Science and Technology in Chinese 7, 200–203.
33. Han S., Yang J., Liu N. (2008) Determination of vanillin in milk by GC- MS/SIM. Chinese Journal of dairy industry 36, 53–55.
34. Zhu B., Wu X., Long C., Li M. (2007) Simultaneous determination of eight kinds of illegal adulterant in dietary supplements by high performance liquid chromatography-mass spectrometry. Journal of Analytical Science 23, 378–382.
35. Xu X., Ren Y., Wu P., Han J., Shen X. (2006) The simultaneous separation and determination of chloropropanols in soy sauce and other flavoring with gas chromatography-mass spectrometry in negative chemical and electron impact ionization modes. Food Addit Contam. 23, 110–119.
36. Shao B., Han H., Li D., Zhao R., Meng J., Ma Y. (2005) Analysis of nonylphenol, octylphenol and bisphenol A in animal tissues by liquid chromatography-tandem mass spectrometry with accelerated solvent extraction. Se Pu in Chinese 23, 362–365.
37. Zhang Y., Zhang Y. (2007) Study on reduction of acrylamide in fried bread sticks by addition of antioxidant of bamboo leaves and extract of green tea. Asia Pac. J Clin. Nutr. 16, 131–136.
38. Zhang Y., Xu W., Wu X., Zhang X., Zhang Y. (2007) Addition of antioxidant from bamboo leaves as an effective way to reduce the formation of acrylamide in fried chicken wings. Food Addit Contam. 24, 242–251.

Chapter 3

United States and Japanese Food Regulations

Jerry Zweigenbaum

Abstract

The two other major importers and exporters of the world are the USA and Japan. Each of their regulations impact how the network of food supply and delivery are viewed. Their regulations mirror that of Europe but have their own origins and focal points. It is the intent of this chapter to provide an overview of food safety through the regulatory bodies of these two world influencers. This should provide a more complete picture of the current state of implementation for keeping the food supply safe for the people of the world.

Key words: United States food law, Japanese food law, Regulations, Japanese positive list, International food safety regulations

1. Introduction to USA Food Law

Regulations for food safety in the USA rest with the US Food and Drug Administration. All regulations are found in the Code of Federal Register (CFR) and address the areas of chemical residues and contamination. However, there are two other US agencies that work together with the FDA for food regulation and safety. They are the United States Department of Agriculture (USDA) and the United States Environmental Protection Agency (US EPA). The USDA's role in food safety extends to both the Agriculture Research Service (ARS) and the Food Safety and Inspection Service (FSIS). The ARS provides research and development for many agricultural functions as well as methods of testing whereas FSIS is responsible for the inspection and testing of meat and poultry products. Finally, the US EPA is the agency responsible for setting maximum residue limits (MRLs) for pesticides in both food and feed. Although there are many other agencies, both state and federal, that are involved in protecting the

Jerry Zweigenbaum (ed.), *Mass Spectrometry in Food Safety: Methods and Protocols*,
Methods in Molecular Biology, vol. 747, DOI 10.1007/978-1-61779-136-9_3, © Springer Science+Business Media, LLC 2011

food supply, each of these agencies function as the main source of adopting and enforcing regulations for food safety as they pertain to chemical residues and contaminants.

2. US Regulations

Among the first food safety regulations in the United States was the Food and Drug Act of 1906. Also known as the 1906 Pure Food and Drug Act, it dealt primarily with mislabeling and adulterating food. The Food, Drug and Cosmetic Act of 1938 followed and, in addition to adding controls to drugs and cosmetics, for the first time brought tolerances to certain poisons in food and regulated both packaging and food quality. It was not until the 1950s that chemical residues were addressed in the law and included tolerances for pesticides and chemical food additives in the Food Additives law of 1957. In 1958, the famous "Delaney Clause" was instituted which banned any level of carcinogen in food. This was the first zero tolerance and included no risk assessment other than the identification that a chemical was a carcinogen. This zero tolerance for labeled carcinogens was later adopted by the US EPA for environmental pollutants as well. A famous example was that of saccharin, a chemical sweetener that had specific benefits for diabetics. However, a study showing carcinogenic activity by inducing bladder tumors in rats caused the sweetener to be labeled as a carcinogen (1). The FDA proposed a ban on the compound in 1977 under the Delaney clause. Later all labels and bans on this substance were removed.

In 1959, the FDA published its first list of substances in the Federal Register that was identified as generally recognized as safe (GRAS). In 1968, the FDA enacted the Animal Drug Amendment which made provision for approval of drugs for animals and in animal feed. In 1994 the Animal and Medicinal Drug Use Clarification Act (AMDUCA) was passed. In this provision, the use of "extra-label" drugs was allowed under certain conditions. This basically gave veterinarians the latitude to use drugs for which they were not "labeled" but where their prescription was warranted. This was allowed, provided they did not leave a residue in food that lead to a violation. Important to this act was a list of prohibited drugs (21 CFR Part 530). They included chloramphenicol, clenbuterol, diethylstilbestrol (DES), dimetridazole, ipronidazole and other nitroimidazoles, furazolidone, nitrofurazone, fluoroquinolones, and phenylbutazone (in female dairy cattle 20 months of age or older). The Food Quality Protection Act followed in 1996 stipulating meat inspection. In the same year the Animal Drug Availability Act (ADAA) was passed giving more flexibility to the FDAs regulation of animal drugs.

The intent of this act was to promote healthier animals and thus improve food quality.

Significant legislation was passed with Bioterrorism Act of 2002 (http://www.fda.gov/RegulatoryInformation/Legislation/ucm155769.htm). Title III of that act specifically addressed food defense and food safety. Section 301 of title III provides that: "*The President's Council on Food Safety (as established by Executive Order No. 13100) shall, in consultation with the Secretary of Transportation, the Secretary of the Treasury, other relevant Federal agencies, the food industry, consumer and producer groups, scientific organizations, and the States, develop a crisis communications and education strategy with respect to bioterrorist threats to the food supply.*" Section 302 of Title III addresses adulteration and specifically testing for adulteration: "... *For use in inspections of food under this section, the Secretary shall provide for research on the development of tests and sampling methodologies – (A) whose purpose is to test food in order to rapidly detect the adulteration of the food, with the greatest priority given to detect the intentional adulteration of food; and (B) whose results offer significant improvements over the available technology in terms of accuracy, timing, or costs.*" From this legislation the combined resources of the Center for Disease Control (CDC), the USDA, and the FDA established a Federal, State and local network of laboratories called the Food Emergency Response Network (FERN – http://www.fernlab.org/). This network is responsible for detection and identification of both chemical and biological threats to the food supply.

In recent years legislation to reform the food safety laws have been introduced but none have yet been passed as law. The National Uniformity for Food Act (H.R. 4167), passed by the US House in 2006, but was not passed by the Senate (S. 3128). This act would have harmonized the food safety laws throughout the country specifying all State laws must agree with and comply with Federal laws. Most recently, the Food Safety Enhancement Act of 2009 (HR2749) passed the house in August of 2009 and a similar bill is in the Senate (S. 510), The FDA Food Safety Modernization Act, is in the Senate. Both would give more regulatory power to the US FDA and require more extensive food tracking, inspection, and testing.

3. US Monitoring and Test Procedures

3.1. FDA Food Code of 2009

The intent of the food code is clearly stated:

"The purpose of this Code is to safeguard public health and provide to *consumers food* that is safe, *unadulterated*, and honestly presented."

That one sentence embodies all that regulations, guidelines, codes, and decisions about food the world strives to achieve. It encompasses sanitation, pathogens, additives, veterinary drugs, contaminants, natural toxins, pesticides, and labeling. The code deals with food additives and refers to 21 CFR 170 where anti-caking agents to texturizers are identified (Title 21 CFR Part 170.3). This provision includes contact materials and sets tolerance levels for materials not "generally recognized as safe (GRAS)." Part 170.39 of 21 CFR specifies that migrating contact substances must not be carcinogenic, must not represent a health or safety concern (dietary intake of the substance is less than 0.5 parts per billion), does not technically effect the food, and has no impact on the environment. In addition there are substances specifically prohibited from use as additives in food (21 CFR 189). They include calamus and its derivatives, cinnamyl anthranilate, cobaltous salts and its derivatives, coumarin, cyclamate and its derivatives, diethylpyrocarbonate (DEPC), monochloroacetic acid, nordihydroguaiaretic acid (NDGA), P–4000 (5-nitro-2-*n*-propoxyaniline), safrole, thiourea, and chlorofluorocarbon propellants. Each of these prohibited additives are analyzed by methods published in the "Official Methods of Analysis of the Association of Official Analytical Chemists," 13th Ed. (1980) with the exception of safrole and coumarin where the methods are found in the "Journal of the Association of Official Analytical Chemists" [54(4); 900–902 and 56(5); 1281–1283, respectively]; and cinnamyl anthranilate, cobaltous salts and its derivatives, DEPC, and chlorofluorocarbon propellants for which no methods are specified. The entire subject of additives is covered in 21 CFR Parts 170–189.

3.1.1. Chemical Hazards

Chemical hazards according to the code are those substances that either occur naturally or are added to food and that cause either acute or chronic toxicity. Chemical hazards that are naturally occurring include scombrotoxin found in tuna fish, mahi-mahi, blue fish, anchovies bonito, mackerel, and in cheese; ciguatoxin found in reef fin fish, barracuda, jacks, king mackerel, large groupers, and snappers; tetrodoxin found in puffer fish; mycotoxins, the most prevalent being aflatoxins found in corn and corn products, peanuts and peanut products, cottonseed, milk, and tree nuts such as Brazil nuts, pecans, pistachio nuts, and walnuts; patulin in apple juice products; paralytic shellfish poisoning (PSP) found in molluscan shellfish, mackerel, viscera of lobsters and Dungeness crabs, tanner, and red rock crabs; diarrhetic shellfish poisoning (DSP) in molluscan shellfish; neurotoxin shellfish poisoning (NSP) in molluscan; amnesic shellfish poisoning (ASP) in molluscan shellfish, viscera of Dungeness crab, tanner, red rock crabs and anchovies, pyrrolizidine alkaloids found in plants such as ragwort; and phytohemagglutinin found in raw red kidney beans.

This is not a complete list, but does cover the most common naturally occurring toxins.

Added chemicals from environmental contamination include rodenticides, fungicides, insecticides, fertilizers, antibiotics, and growth hormones. Polychlorinated biphenyls remain environmental contaminants that continue to be found in fish products. Because of that the FDA has set temporary tolerances for levels of these substances (21 CFR 509.30). Dioxins, another ubiquitous environmental contaminant that may be carcinogenic at low levels, have no tolerance. Maximum residue limits (MRLs) of pesticides that find their way into food are the responsibility of the US EPA and those levels are given in Title 40 CFR Part 180. However, these regulations must be coordinated with other laws as stated in Title 21 of the US Code, Chapter IV, part 346a. Thus the EPA tolerances must agree with those set by the Federal Insecticide, Fungicide, and Rodenticide Act (FIFRA). In addition to EPA tolerances, the Foreign Agriculture Service (FAS) of the USDA has set up a database for comparison to international tolerances. This database can be found at http://www.fas.usda.gov/htp/MRL.asp.

3.2. Mass Spectrometric Methods for Specific Contaminants

Of the banned drugs in food, 21 CFR 530, the antibiotic chloramphenicol was found to cause aplastic anemia (basically a loss of bone marrow cells) which is non-reversible and fatal. Methods for determination of this compound were first developed by FDA laboratories using gas chromatography (2, 3). However, to meet limits of detection to at least one part per billion needed for violatory samples, a method for the compound in shrimp was reported (4). Similar methodology was then reported for chloramphenicol in crab and provided criteria for LC/MS/MS confirmation (5) and another method reported for crab shortly thereafter (6). In addition, LC/UV was used for the banned antibiotic malachite green and its metabolite, leucomalachite green, found in shrimp and fish and confirmed by LC/MS (7). This method was extended to the dyes crystal violet and brilliant green with similar properties as malachite green (8). Another important group of antibiotics where there is great concern for abuse is the fluoroquinolones. A method for these residues in honey was developed again using LC/MS/MS (9). The expanded use of mass spectrometry demonstrated the need for this technology in monitoring antibiotics of concern in food commodities.

Acrylamide was found in food and thought to be migrating from packaging material. However, further investigation found that it was often a by-product of frying. Of the analytical methods that could be employed to determine this small molecule in food samples, again LC/MS/MS was selected (10). This method provided part per billion detection with high selectivity for the compound. This allowed more detailed studies to determination the origin and formation of this toxic compound.

In the area of natural contaminants, mass spectrometry has become more prominent. Different mycotoxins as described in the food code are found in various food commodities. The most common are the aflatoxins and they are known to be carcinogenic. Aflatoxin B1 is considered to be the most toxic. The action levels as defined by the FDA's "Guidance for Industry: Action Levels for Poisonous or Deleterious Substances in Human Food and Animal Feed" (http://www.fda.gov/Food/GuidanceComplianceRegulatoryInformation/GuidanceDocuments/ChemicalContaminantsandPesticides/ucm077969.htm) shows the levels that must be monitored. Table 1 is taken directly from that guidance and gives the action levels for aflatoxins in both

Table 1
Action levels for aflatoxins in animal feed and food taken from "Guidance for Industry: Action Levels for Poisonous or Deleterious Substances in Human Food and Animal Feed" prepared by the US FDA

Aflatoxin

Commodity	Action level (ppb)	Reference
Animal feeds		
Corn and peanut products intended for finishing (i.e. feedlot) beef cattle	300	CPG 683.100
Cottonseed meal intended for beef, cattle, swine, or poultry (regardless of age or breeding status)	300	CPG 683.100
Corn and peanut products intended for finishing swine of 100 pounds or greater	200	CPG 683.100
Corn and peanut products intended for breeding beef cattle, breeding swine, or mature poultry	100	CPG 683.100
Corn, peanut products, and other animal feeds and feed ingredients but excluding cottonseed meal, intended for immature animals	20	CPG 683.100
Corn, peanut products, cottonseed meal, and other animal feed ingredients intended for dairy animals, for animal species or uses not specified above, or when the intended use is not known	20	CPG 683.100
Brazil nuts	20	CPG 570.200
Foods	20	CPG 555.400
Milk	0.5 (aflatoxin M1)	CPG 527.400
Peanuts and peanut products	20	CPG 570.375
Pistachio nuts	20	CPG 570.500

The action levels must be monitored to assure the commodities do not contain these unavoidable poisons above their non- permissible concentrations animal feed and food. Methods for aflatoxins were developed using HPLC with fluorescence detection (11, 12). More often confirmation, especially for violatory action, has invoked the use of mass spectrometry. Methods for determination of aflatoxins with mass spectrometry have been developed with the realization and implementation that this methodology could be used for these and other mycotoxins in one procedure (13–15). However, HPLC with fluorescence is still being used and even a multimycotoxin method without MS has been recently reported (15). In addition, ochratoxin A can be monitored by HPLC with fluorescence (16, 17), but confirmation again relies on mass spectrometry (18, 19). This is also the case with fumonisins where HPLC (20) was initially developed but mass spectrometry is required for confirmation (21) and for deoxynivalenol (22–25). Another mycotoxin, patulin, is typically found in apple juice and has an action level of 50 ppb (CPG Sec.510.150, http://www.fda.gov/ICECI/ComplianceManuals/CompliancePolicyGuidanceManual/ucm074427.htm). That FDA guidance specifies HPLC analysis for patulin (26) with confirmation by GC/MS (27).

The last example of the use of mass spectrometry for compliance monitoring will be that of the adulterant melamine. This compound initially appeared in pet food in the US and Canada. The compound was not an approved additive, a residue from approved medicine or crop protectant, nor an environmental or packaging contaminant. Because this compound is a triamine triazine, its addition provided significant amounts of nitrogen giving the appearance of high protein content by the Kjeldahl method for nitrogen content. The procedure measures total nitrogen with the extrapolation that most nitrogen in organic matter comes from protein. Fatalities were traced to renal failure where both melamine and cyanuric acid combined to form insoluble crystals that blocked the passages of the kidney. Because the source of the melamine was identified in gluten product that was added to the pet food, a method for melamine and its analogues was developed employing GC/MS (28). Concern that the compound could find its way into fish products, an LC/MS/MS method using a triple quadrupole mass spectrometer was developed for trace analysis in catfish (29). Finally, when the compound was discovered as an adulterant in infant formula, the triple quadropole procedure was adapted for that matrix (30).

4. Japanese Regulations

Japan's food laws, like the laws of all other nations, are designed to provide safe food for their people. The Food Safety Basic Law of 2003 was the culmination of meeting this need. The basis of the current food law was the Food and Sanitation Act of 1947. It was at that time that the system of a "positive list" was introduced for additives to food. The list included only synthetic compounds added to food. However, in 1995 the positive list was amended to include most additives whether synthetic or natural.

Within the Pharmaceutical and Food Safety Bureau the Department of Food Safety is responsible, under the Food Safety Basic Law, for planning, risk communication, inspection, labeling, and inspection of food import. This includes the standards for environmental contaminants. The Bureau resides under the Ministry of Health, Labor and Welfare, which coordinates the work of the Prefecture and Municipal governments. Through these governments, the health authorities are responsible for food sanitation, meat, and poultry inspection at the local level. The policy planning division of the Department of Food Safety is responsible for quarantine stations and food sanitation inspectors at the national level. Under the Department of Food Safety is the Inspection and Safety division and under that the Office of Import Food Safety.

In 2002, it was determined that 60% of Japan's food supply came from other countries. Because of this, strong import laws were needed. In Article 27 of the Food Sanitation Law of 2002, all importers of food must submit import notification. Inspectors must then determine whether manufacturing sanitation is adequate or if sanitation problems exist. In addition, if additives meet standards and whether poisonous or hazardous substances are present in the food to be imported. In the 2003 Food Safety Basic Law, the implementation of an extensive "positive list" was mandated to cover pesticides, veterinary drugs, and feed additives. At that time there were 288 substances listed with maximum residue limits (MRLs). Of the 288 substances, 255 were pesticides and 33 were veterinary drugs. The law required that full implementation of the positive list take place three years from that date. In 2006, the positive list went into effect with 758 substances having at least provisional MRLs based on risk assessment and the best scientific data at that time. Of those substances that no MRL could be established, a level below 0.01 ppm was established as safe. At that time there were 65 substances that were identified as not posing a health risk and thus no level was set.

Methods for monitoring food for the positive list have been issued by the Japanese government and can be found at http://www.ffcr.or.jp/zaidan/FFCRHOME.nsf/pages/MRLs-p. They include a multiresidue method for agricultural Chemicals by GC/MS, a multiresidue Method I and II for agricultural chemicals by LC/MS, and a multiresidue method for agricultural chemicals by GC/MS. These methods do not cover all compounds on the positive list and there are additional substances used such as ingredients of agricultural chemicals and other chemical substances that are stipulated to be "not detected" in foods. The "not detected" list include antibiotics such as chloramphenicol, malachite green, and nitrofurans. Methods for these would generally follow those that are used elsewhere in the world. In essence the laws and stipulations for protecting Japan's food supply are specific to that country but are similar to those around the world.

5. Codex Alimentarius

Because food is a shared need in a global community, guidelines that all countries can use as a reference are important. Those guidelines, standards and recommendations were initiated in 1963 by the United Nations Food and Agriculture Organization (FOA) and the World Health Organization (WHO). The Codex Alimentarius Commission reviews and updates these guidelines and recommendations. Much of the scientific information used for risk assessment, establishment of provisional MRLs and other important decisions are drawn from the work of this organization. More information can be found at the official website of the Codex Alimentarius, http://www.codexalimentarius.net/web/index_en.jsp.

References

1. Munro, I. C.; Moodie, C. A.; Krewski, D.; Grice, H. C., A carcinogenicity study of commercial saccharin in the rat. *Toxicology and Applied Pharmacology* **1975,** *32* (3), 513–526.
2. Pfenning, A. P.; Madson, M. R.; Roybal, J. E.; Turnipseed, S. B.; Gonzales, S. A.; Hurlbut, J. A.; Salmon, G. D., Simultaneous determination of chloramphenicol, florfenicol, and thiamphenicol residues in milk by gas chromatography with electron capture detection. *J AOAC Int* **1998,** *81* (4), 714–20.
3. Pfenning, A. P.; Roybal, J. E.; Rupp, H. S.; Turnipseed, S. B.; Gonzales, S. A.; Hurlbut, J. A., Simultaneous determination of residues of chloramphenicol, florfenicol, florfenicol amine, and thiamphenicol in shrimp tissue by gas chromatography with electron capture detection. *J AOAC Int* **2000,** *83* (1), 26–30.
4. Neuhaus, B. K.; Hurlbut, J. A.; Hammack, W. LIB No. 4290 LC/MS/MS Analysis of Chloramphenicol in Shrimp *Laboratory Information Bulletin* [Online], 2002. http://www.fda.gov/Food/ScienceResearch/LaboratoryMethods/DrugChemicalResiduesMethodology/ucm113335.htm (accessed June, 2010).
5. Pfenning, A.; Turnipseed, S.; Roybal, J.; Madson, M.; Lee, R.; Storey, J. LIB No. 4294 Confirmation of Chloramphenicol Residue in Crab by Electrospray LC/MS *Laboratory Information Bulletin* [Online], 2003. http://www.fda.gov/Food/ScienceResearch/LaboratoryMethods/DrugChemicalResiduesMethodology/ucm113600.htm (accessed June, 2010).
6. Rupp, H. S.; Stuart, J. S.; Hurlbut, J. A. LIB 4302LC/MS/MSAnalysisofChloramphenicol in Crab Meat *Laboratory Information Bulletin* [Online], 2003. http://www.fda.gov/Food/ScienceResearch/LaboratoryMethods/DrugChemicalResiduesMethodology/ucm113380.htm (accessed June, 2010).
7. Andersen, W. C.; Turnipseed, S. B.; Roybal, J. E. LIB 4363 Malachite Green and Leucomalachite Green in Fish and Shrimp *Laboratory Information Bulletin* [Online], 2005. http://www.fda.gov/Food/ScienceResearch/LaboratoryMethods/DrugChemicalResiduesMethodology/ucm071511.htm (accessed June, 2010).
8. Andersen, W. C.; Turnipseed, S. B.; Karbiwny, C. M.; Lee, R. H.; Clark, S. B.; Rowe, W. D.; Madson, M. R.; Miller, K. E. LIB 4395

Quantitative and Confirmatory Analyses of Crystal Violet (Gentian Violet) and Brilliant Green in Fish *Laboratory Information Bulletin* [Online], 2007. http://www.fda.gov/Food/ScienceResearch/LaboratoryMethods/DrugChemicalResiduesMethodology/ucm071574.htm (accessed June 2010).

9. Preparation and LC/MS/MS Analysis of Honey for Fluoroquinolone Residues *Method Developed by Florida Department of Agriculture and Consumer Services* [Online], 2006. http://www.fda.gov/Food/ScienceResearch/LaboratoryMethods/DrugChemicalResiduesMethodology/ucm071495.htm (accessed June, 2010).
10. Musser, S. M. Detection and Quantitation of Acrylamide in Foods (Draft) 2003. http://www.fda.gov/Food/FoodSafety/FoodContaminantsAdulteration/ChemicalContaminants/Acrylamide/UCM053537.
11. Park, D. L.; Nesheim, S.; Trucksess, M. W.; Stack, M. E.; Newell, R. F., Liquid chromatographic method for determination of aflatoxins B1, B2, G1, and G2 in corn and peanut products: collaborative study. *J Assoc Off Anal Chem* **1990,** *73* (2), 260–6.
12. Trucksess, M. W.; Stack, M. E.; Nesheim, S.; Page, S. W.; Albert, R. H.; Hansen, T. J.; Donahue, K. F., Immunoaffinity column coupled with solution fluorometry or liquid chromatography postcolumn derivatization for determination of aflatoxins in corn, peanuts, and peanut butter: collaborative study. *J Assoc Off Anal Chem* **1991,** *74* (1), 81–8.
13. Beltran, E.; Ibanez, M.; Sancho, J. V.; Hernandez, F., Determination of mycotoxins in different food commodities by ultra-high-pressure liquid chromatography coupled to triple quadrupole mass spectrometry. *Rapid Commun Mass Spectrom* **2009,** *23* (12), 1801–9.
14. Diana Di Mavungu, J.; Monbaliu, S.; Scippo, M. L.; Maghuin-Rogister, G.; Schneider, Y. J.; Larondelle, Y.; Callebaut, A.; Robbens, J.; Van Peteghem, C.; De Saeger, S., LC-MS/MS multi-analyte method for mycotoxin determination in food supplements. *Food Addit Contam Part A Chem Anal Control Expo Risk Assess* **2009,** *26* (6), 885–95.
15. Monbaliu, S.; Van Poucke, C.; Detavernier, C.; Dumoulin, F.; Van De Velde, M.; Schoeters, E.; Van Dyck, S.; Averkieva, O.; Van Peteghem, C.; De Saeger, S., Occurrence of mycotoxins in feed as analyzed by a multi-mycotoxin LC-MS/MS method. *J Agric Food Chem* **2010,** *58* (1), 66–71.
16. Entwisle, A. C.; Williams, A. C.; Mann, P. J.; Slack, P. T.; Gilbert, J., Liquid chromatographic method with immunoaffinity column cleanup for determination of ochratoxin A in barley: collaborative study. *J AOAC Int* **2000,** *83* (6), 1377–83.
17. MacDonald, S. J.; Anderson, S.; Brereton, P.; Wood, R., Determination of ochratoxin A in currants, raisins, sultanas, mixed dried fruit, and dried figs by immunoaffinity column cleanup with liquid chromatography: interlaboratory study. *J AOAC Int* **2003,** *86* (6), 1164–71.
18. Lau, B. P. Y.; Scott, P. M.; Lewis, D. A.; Kanhere, S. R., Quantitative determination of ochratoxin A by liquid chromatography/electrospray tandem mass spectrometry. *Journal of Mass Spectrometry* **2000,** *35* (1), 23–32.
19. Lombaert, G. A.; Pellaers, P.; Chettiar, M.; Lavalee, D.; Scott, P. M.; Lau, B. P., Survey of Canadian retail coffees for ochratoxin A. *Food Addit Contam* **2002,** *19* (9), 869–77.
20. Visconti, A.; Solfrizzo, M.; De Girolamo, A., Determination of fumonisins B1 and B2 in corn and corn flakes by liquid chromatography with immunoaffinity column cleanup: collaborative study. *J AOAC Int* **2001,** *84* (6), 1828–37.
21. Senyuva, H. Z.; Ozcan, S.; Cimen, D.; Gilbert, J., Determination of fumonisins B1 and B2 in corn by liquid chromatography/mass spectrometry with immunoaffinity column cleanup: single-laboratory method validation. *J AOAC Int* **2008,** *91* (3), 598–606.
22. Biancardi, A.; Gasparini, M.; Dall'Asta, C.; Marchelli, R., A rapid multiresidual determination of type A and type B trichothecenes in wheat flour by HPLC-ESI-MS. *Food Addit Contam* **2005,** *22* (3), 251–8.
23. Trucksess, M. W.; Ready, D. W.; Pender, M. K.; Ligmond, C. A.; Wood, G. E.; Page, S. W., Determination and survey of deoxynivalenol in white flour, whole wheat flour, and bran. *J AOAC Int* **1996,** *79* (4), 883–7.
24. Razzazi-Fazeli, E.; Bohm, J.; Luf, W., Determination of nivalenol and deoxynivalenol in wheat using liquid chromatography-mass spectrometry with negative ion atmospheric pressure chemical ionisation. *J Chromatogr A* **1999,** *854* (1–2), 45–55.
25. Sasanya, J. J.; Hall, C.; Wolf-Hall, C., Analysis of deoxynivalenol, masked deoxynivalenol, and Fusarium graminearum pigment in wheat samples, using liquid chromatography-UV-mass spectrometry. *J Food Prot* **2008,** *71* (6), 1205–13.
26. Brause, A. R.; Trucksess, M. W.; Thomas, F. S.; Page, S. W., Determination of patulin in apple juice by liquid chromatography: collaborative study. *J AOAC Int* **1996,** *79* (2), 451–5.

27. Roach, J. A.; Brause, A. R.; Eisele, T. A.; Rupp, H. S., Hplc detection of patulin in apple juice with GC/MS confirmation of patulin identity. *Adv Exp Med Biol* **2002,** *504*, 135–40.
28. Litzau, J. J.; Mercer, G. E.; Mulligan, K. J. LIB No. 4423 GC-MS Screen for the Presence of Melamine, Ammeline, Ammelide, and Cyanuric Acid *Laboratory Information Bulletin* [Online], 2008. http://www.fda.gov/Food/ScienceResearch/LaboratoryMethods/DrugChemicalResiduesMethodology/ucm071759.htm (accessed June 2010).
29. Andersen, W. C.; Turnipseed, S. B.; Karbiwnyk, C. M.; Madson, M. R. LIB 4396 Determination of Melamine Residues in Catfish Tissue by Triple Quadrupole LC-MS-MS with HILIC Chromatography *Laboratory Information Bulletin* [Online], 2007. http://www.fda.gov/Food/ScienceResearch/LaboratoryMethods/DrugChemicalResiduesMethodology/ucm071796.htm (accessed June, 2010).
30. Turnipseed, S.; Casey, C.; Nochetto, C.; Heller, D. N. LIB No. 4421 Determination of Melamine and Cyanuric Acid Residues in Infant Formula using LC-MS/MS *Laboratory Information Bulletin* [Online], 2008. http://www.fda.gov/Food/ScienceResearch/LaboratoryMethods/DrugChemicalResiduesMethodology/ucm071637.htm (accessed June, 2010).

Chapter 4

QuEChERS Sample Preparation Approach for Mass Spectrometric Analysis of Pesticide Residues in Foods

Steven J. Lehotay

Abstract

This chapter describes an easy, rapid, and low-cost sample preparation approach for the determination of pesticide residues in foods using gas and/or liquid chromatographic (GC and/or LC) analytical separation and mass spectrometric (MS) detection. The approach is known as QuEChERS, which stands for "quick, easy, cheap, effective, rugged, and safe." Originally, QuEChERS was a particular "method" for pesticide residue analysis, but it is very flexible and has evolved into an "approach," which has been used in many methods, and not just for pesticide residues. Two of the QuEChERS versions using buffering have been validated in interlaboratory trials for dozens of pesticides in several food matrices, and both have successfully met performance criteria to achieve "official" status from international standard organizations (AOAC Official Method 2007.01 and CEN Standard Method EN 15662). The main aspects of the QuEChERS approach consists of extraction of a well-homogenized sample by shaking with solvent (typically acetonitrile) in a centrifuge tube, salt-out partitioning of water with salts including magnesium sulfate ($MgSO_4$), and cleanup using "dispersive solid-phase extraction" (dSPE), in which common matrix components are retained by sorbent(s) and the analytes remain in the extract. For widest analytical scope, concurrent analysis is done for hundreds of pesticides using GC-MS(/MS) and LC-MS/MS. The aim of this chapter is to review the QuEChERS sample preparation methodology and provide a summary of up-to-date information with modification options depending on the application needs.

Key words: QuEChERS, Residue analysis, Sample preparation, Gas chromatography, Liquid chromatography, Mass spectrometry, Food

1. Introduction

In 1975, Milton Luke and colleagues at the U.S. Food and Drug Administration (FDA) Office of Regulatory Affairs laboratory in Los Angeles, California introduced a new method for multiclass, multiresidue pesticide analysis of fruit, vegetable, grains, and other food samples (1). This method, which became widely

Jerry Zweigenbaum (ed.), *Mass Spectrometry in Food Safety: Methods and Protocols*,
Methods in Molecular Biology, vol. 747, DOI 10.1007/978-1-61779-136-9_4, © Springer Science+Business Media, LLC 2011

known as the "Luke method," was able to achieve high recoveries for the major types of pesticides used at that time (e.g. organochlorines, organophosphates, organonitrogens). The use of multiple selective detectors in gas chromatography (GC), such as electron-capture detection (ECD), flame photometric detection (FPD), electrolytic conductivity detection (ELCD or Hall detector), and nitrogen–phosphorus detection (NPD), allowed an expanded scope over common previous methods, which generally were effective only for single classes of pesticides, such as organochlorines using GC-ECD (2, 3). The Luke method achieved AOAC Official Method status (Method 985.22) in 1985 based on a study involving FDA labs (4). Over time, the FDA and many other labs demonstrated the recoveries of hundreds of pesticides using the Luke method (3), and it is believed to have become the most common pesticide residue method in the world.

For many years to follow, analytical technologies continued to improve and agrochemical companies registered many more pesticides from different classes. Although the registration process often required companies to first test the ability of the Luke method to recover the newly registered pesticides, fewer modern pesticides could be included in the FDA multiclass monitoring method, and this required the companies to develop single analyte methods in the registration process to be used for enforcement. However, monitoring labs had too few resources to use the typically very complicated methods for so many different pesticides, and little or no monitoring was done of those types of pesticides.

In terms of analytical technology, mass spectrometry (MS) was coupled to GC in the commercial bench-top instruments during the 1980s, and they were initially used for qualitative confirmation purposes in pesticide analyses. In the 1990s, the performance features of the instruments improved to the point that detection limits were acceptably low enough that GC-MS could be used to replace selective GC detectors for quantitative as well as qualitative analysis and reduce the need for multiple injections in GC. By the late 1990s, GC-MS had become commonplace in monitoring labs, particularly in developed countries.

Additionally, the price reduced and performance improved for high quality commercial bench-top LC-MS/MS (tandem mass spectrometry) instruments. This allowed multiclass, multiresidue analysis of many LC-type pesticides that could previously be detected only by single-analyte methods. However, the Luke method, which used acetone for extraction and partitioning from water with a combination of methylene chloride and petroleum ether (and addition of salt for more polar pesticides), was not sufficiently effective, environmentally-friendly, safe, and efficient enough for "twenty-first century" standards. A new multiclass, multiresidue sample preparation approach for pesticides was needed for the new times to take advantage of GC- and LC-MS technologies.

During the mid-1990s, Steven Lehotay and Michelangelo Anastassiades (among others) were independently investigating supercritical fluid extraction (SFE) as an environmentally-friendly, semi-automated approach to replace the Luke method (5–9). Although SFE had many advantages, its limitations in terms of a too-small sample size and too-narrow analytical scope did not meet broad analytical needs, particularly for modern more polar type of pesticides. However, in the first AOAC collaborative study of GC-MS, Lehotay used SFE for sample preparation for pesticide residues in foods, which became AOAC Official Method 2002.03 (9).

In 2000, Anastassiades joined Lehotay's group in the USDA for 2 years, and they applied the knowledge and experience they gained from SFE with the goal to develop a streamlined and effective method of analysis to replace the Luke and other less efficient methods developed prior to the wider availability of GC-MS and LC-MS/MS instruments. For one, they realized that the use of $MgSO_4$ in SFE improved recoveries for the relatively polar pesticide, methamidophos (5, 7), and they included $MgSO_4$ as one of the salts evaluated in conventional liquid–liquid partitioning between organic extraction solvents and water in the sample. Anastassiades, Lehotay, et al. conducted numerous experiments to evaluate all aspects of the sample preparation process, which are described in the first publication of their work (10). When the original version of the method was to be introduced in 2002 at the fourth European Pesticide Residue Workshop in Rome, Italy, Anastassiades and Lehotay needed a name for the method. On the basis of the method's features, they coined the acronym, QuEChERS, which stands for "quick, easy, cheap, effective, rugged, and safe."

QuEChERS lives up to its name in each aspect, and using the method for pre-homogenized samples and preweighed powders, a single analyst can prepare about 20 cleaned-up final extracts in an hour (or one sample in 10 min) using $1 of materials and generating <10 mL of acetonitrile (MeCN) waste per sample. Only a few labware items are needed in the method, which greatly reduces lab space and dishwashing requirements (a highly desired benefit among lab personnel!). Extracts are concurrently analyzed by GC-MS and LC-MS/MS to monitor hundreds of pesticides to typically achieve 90–100% recoveries, 5–10% relative standard deviations (RSDs), and detection limits (LODs) <10 ng/g (10–21). Many articles have been published on QuEChERS or its modifications demonstrating the utility, performance, and features of the concepts (10–88).

QuEChERS employs several streamlining features that can be considered novel in their combination for the first time in pesticide residue analysis, as listed below, but the most distinguishing feature of QuEChERS over previous sample preparation techniques is the use of dispersive solid-phase extraction (dSPE) for cleanup. Fillion et al. introduced an effective column/cartridge-based cleanup for MeCN pesticide extracts, which had been salted out from water, with a combination of primary secondary amine (PSA), octadecylsilyl (ODS or C_{18}), and graphitized carbon black (GCB) sorbents (89). Lehotay and Anastassiades realized that this was a "chemical filtration" approach in which certain common matrix components in foods (e.g. fatty acids, chlorophyll, sterols, anthocyanins) remained on the sorbents and the MeCN served as the elution solvent for the pesticide analytes. It is more efficient in this application to use SPE to retain matrix than to retain analytes, and they considered alternatives to the traditional SPE column approach (such as use of sorbents in pipette tips and syringes). Anastassiades had the idea to dispense an aliquot of the extract into a centrifuge tube containing loose sorbent(s), and then to take a second aliquot after shaking and centrifugation for analysis. It turns out that the inventor of chromatography, Mikhail Tswett, also had this idea in 1902 (90), but chromatography had evolved into thin-layer and column systems since the time of its origins. By using dSPE, the trappings of traditional cartridge-based SPE disappeared, such as needing a manifold, vacuum system, collection tubes, elution solvents, solvent evaporation apparatus, and reliance on limited commercial products.

Other streamlined aspects of QuEChERS, some of which had been done in different ways previously but not combined before, include: (1) reduced subsample size from a thoroughly homogenized sample of larger size to meet the needs of the application (chopping of cut frozen samples with dry ice provides the most effective homogenization/comminution process); (2) extraction by shaking of sample with solvent in a centrifuge tube in a 1/1 g/mL ratio (1/2 g/mL ratios were used previously, which entailed a post-extraction concentration step); (3) partitioning of water from the sample using $MgSO_4$ in combination with other salt(s) (NaCl or Na_2SO_4 were more commonly used previously); (4) centrifugation to separate the extract from the water and non-soluble material rather than filtration; (5) taking an aliquot of the extract rather than trying to collect the entire portion; (6) use of internal standard(s) to improve accuracy and precision of the results rather than having to make calculations of extract volume depending on water content of the sample; and (7) injection of the same extract, preferably without solvent exchange or concentration steps, in both GC-MS and LC-MS/MS analyses.

Additionally, Anastassiades independently thought to use an "analyte protectant" concept (10, 24, 71, 91, 92), originally proposed by Erney (93), in GC-MS to provide improved signals for relatively polar pesticides, such as acephate and methamidophos, which were susceptible to the matrix-induced enhancement effect (94). LC-MS/MS provides improved detection of these type of pesticides over GC-MS, which has reduced the importance of analyte protectants in combination with the QuEChERS method, but it is still an interesting innovation that can be useful in many GC applications. Analyte protecting agents not only provide lower LODs, better peak shapes, and improved precision and qualitative analysis for relatively polar GC-amenable analytes, but they can be used to avoid the use of matrix-matched standards in calibration (92).

Anastassiades returned to Germany in 2002 where he and colleagues continued to conduct QuEChERS research, and Lehotay and colleagues also continued QuEChERS investigations in the U.S. and the Netherlands. Lehotay et al. found that a few pH-dependent pesticides gave lower recoveries in the original method, in which pH was not controlled (11), and they demonstrated that an acetate buffering procedure improved the results for the small percentage of pesticides affected by pH (14). These pesticides include pymetrozine, cycloxydim, and sethoxydim. Independently, Anastassiades et al. chose to use a milder citrate buffering method using a combination of salts (12).

The two buffered versions of QuEChERS have been evaluated in interlaboratory trials, both of which were successful. In January of 2007, the acetate buffered method developed in Lehotay's lab became AOAC Official Method 2007.01 (17). In December of 2008, the citrate buffered version developed in Anastassiades' lab became CEN (European Committee for Standardization) method EN 15662 (18). Interlaboratory tested methods are the "gold standard" for analytical methods, and for QuEChERS to attain that status twice in independent studies for so many pesticide analytes and matrices is an unprecedented achievement.

Several companies have introduced QuEChERS products to save time and labor involved in the weighing of salts and powders in the lab. Due to these interactions and QuEChERS Official Method status since 2007, several companies have commercialized QuEChERS products. At the time of writing, these include (search QuEChERS on the sites): Agilent Gerstel/DPX; Macherey-Nagel; Restek; Supelco; Thermo-Fisher; Applied Separations, Bekolut, Biotage, and Waters.

A previous article on QuEChERS appears in this book series (20), and this chapter is intended to update the protocols with new information and developments. QuEChERS is not just a method, or a pair of official methods, but it is also a flexible approach to chemical sample preparation that is being used in multiple applications, such as clinical (87, 88), steroids, hormones, and veterinary drug analyses (25–27, 42, 64, 78, 80, 81), perfluorinated compounds (76, 77), ergot alkaloids (79), soil (85), and acrylamide (23, 32). Even for another streamlined Luke replacement method using LC-MS/MS and GC-MS for analysis, Mol and Banerjee used similar features of QuEChERS with 25 g subsample mixed with Na_2SO_4 extracted with 40 mL ethyl acetate and followed by dSPE with 50 mg/g sample equivalent each of PSA and GCB (36–41). In practice, small differences occur between the different modifications used in the QuEChERS approach, with minor advantages and disadvantages depending on the pesticide analytes, matrices, lab equipment, analytical instruments, and analyst preferences. In this chapter, the author will try to sort through these nuances and help the reader make informed and beneficial decisions.

2. The QuEChERS Approach for Pesticide Residues in Foods

2.1. Sample Comminution

A critical aspect of the QuEChERS approach is to minimize extracted sample size. The smallest acceptable subsample is taken for analysis, and shaking is used for extraction (blending of a larger sample is also feasible, but many of the practical advantages of QuEChERS are lost in this manner). Thus, the sample comminution process (homogenization and reduction in the sizes of the sample material pieces) is critical in obtaining valid results. *The purchase of a good food chopper on the order of thousands of dollars is more important in a monitoring lab than the purchase of GC-MS and LC-MS/MS analytical instruments costing on the order of hundreds of thousands of dollars.* Results from expensive and high performance GC-MS and LC-MS/MS will be inaccurate (but misleadingly precise) if good sample comminution is not done, independent of the quality of the procedural steps that follow.

Chopping of precut, frozen samples (e.g. 5–10 cm^3 portions) with dry ice provides the most effective homogenization process for QuEChERS (95–97), but in comparison to chopping at room temperature, it takes longer, slightly reduces worker safety, and needs dry ice and extra freezer storage space. The sample size and amount of dry ice added to a container depends on the sample, environmental conditions, and container size, *but sample volume of about one-third to two-third of the container volume typically provides an acceptable comminution process with a good chopper within a few minutes.* Chopping is conducted until a dispersed, well-homogenized flowable powder is achieved. Time must be given for the dry ice to sublime before weighing of subsamples for extraction or else a positive bias can occur in the results. It is also possible to chop frozen samples without adding dry ice. Especially in humid environments, the condensation of water into the sample from the atmosphere causes the sample weight to increase, and can create a negative bias in the results. For this reason, work quickly when using very cold conditions and/or conduct the homogenization in a dry and/or cold place.

In many labs, the sample comminution step is not done using frozen samples with dry ice. Instead, the large, representative sample (e.g. 1–10 kg) is chopped thoroughly, and then ≈200 g is taken for further homogenization in a smaller container using a probe-type blender. The larger subsample is representative of the original sample, and the use of a second subsample of 10–15 g should also be representative of the original representative sample of the commodity to be analyzed.

For less dense and more difficult-to-homogenize samples, such as broccoli, the dry ice procedure still works well, but the room temperature two-step homogenization method is more challenging. In that case, the collected sample can be chopped as usual, but accurately weigh the ≈200 g portion taken for probe-blending, and add half the sample weight of distilled water to the sample in the container. The additional water helps improve the homogenization process. *In this case, when a 15-g subsample is used in the extraction procedure, the calculations of results must take into account that this equates to 10 g of this type of commodity to which water was added.*

In a few cases of highly aqueous fruits (e.g. grapes and strawberries), the 1–10 kg original sample can be homogenized into a homogeneous liquid slurry that does not need further homogenization of a ≈200 g subsample. A 10–15 g subsample may be used if empirical experiments with the chopper demonstrate that the protocol is acceptable (97).

For dry commodities such as grains, cereals, and nuts, use a milling process to produce a fine, consistent, flowable powdery material.

For quality control (QC) purposes of the comminution process, a chemical that is not present in the sample should be fortified into the sample when it is weighed into the chopper, and then recovery (or consistency of results among multiple samples or subsamples) should be determined in the analytical sequence to demonstrate that adequate sample processing (and analysis) was conducted. One of the internal standards suggested in Subheading 2.2.2 may serve as a good QC standard for comminution.

Items needed for either comminution method:

1. Food Chopper – e.g. Stephan or Robotcoupe vertical cutters.
2. Knife and cutting board.

3. Spatula.
4. Top-loading balance.
5. Container jar(s)/beaker(s).
6. Blank sample – for control and preparing matrix-matched calibration standards.
7. QC-spiking solution – for quality control of comminution step.
8. Repeating pipette and tips (calibrated to solvent used).
9. Freezer – for sample storage and re-extraction if needed.
10. Soap, water, acetone, and sink – for cleaning materials between samples.

    Additional item for two-step comminution at room temperature:

11. Probe blender – e.g. Ultraturrax or ProScientific.
12. Distilled water for less dense samples that are difficult to homogenize (e.g. broccoli).

    Additional item for cryogenic comminution step:

13. Dry ice (or cryogenic chopper).

    Additional item for dry samples:

14. Milling device.

2.2. QuEChERS Extraction

2.2.1. Choice of Subsample Size

For high moisture fruits, vegetables, and beverages, 10 or 15 g is most common because that amount fits in a 50-mL centrifuge tube with sufficient space remaining for shaking to provide good interaction between the solvent and sample, even after the salts are added. The original and CEN versions of the QuEChERS method entails a 10-g subsample size, which is typically an acceptable size after the effective dry ice homogenization approach. The AOAC version calls for a 15-g subsample size, which gives a slightly better representative amount related to the original sample in either case, but especially if comminution is done at room temperature.

The 15-g sample size is considered the maximum about that can be used in a 50-mL volume for foods with 80–95% water content. Larger sample portions require larger containers, such as centrifuge bottles. *The QuEChERS approach can be scaled to larger sample sizes using proportionally scaled ratios of materials.* Blending is also possible if the analyst wishes, but this increases inconveniences and generates noise. When using matrix-matched standards or conducting solvent concentration at the end of the procedure, the use of a 15-g sample size gives more extract with which to work.

For dry samples, water must be added and enough time given (≥1 h of mixing) to swell the matrix and allow the solvent to gain better access into the sample and permit the partitioning process between the aqueous and organic phases when the salts are added. The amount of dry sample and water added depends on the commodity. For milled corn, flaxseeds, peanuts, and similar relatively higher fat dry matrices, weigh 1 g subsample per 4 mL each of water and MeCN extraction solvent into the 50-mL centrifuge tube (e.g. 2.5 g subsample + 10 mL water + 10 mL MeCN, or 3.75 g + 15 mL each of water and MeCN). For milled wheat, oats, almonds, doughs, and other lower fat dry matrices, add 1 g subsample per 2 mL of water and MeCN (e.g. 5 or 7.5 g + 10 or 15 mL each of water and MeCN, respectively). For milled rice, add 1 g subsample per 3 mL of water and 2 mL of MeCN (e.g. 5 g + 15 mL water + 10 mL MeCN, or 7.5 g + 22.5 mL water + 15 mL MeCN). *See* ref. (18, 19) for why these ratios were chosen. Other dry matrices can be assessed on a case by case basis depending on lipid content and amount of water that is absorbed.

Items needed for weighing of subsample:

1. Balance.
2. Funnel.
3. Spoon.
4. Centrifuge tube, 50 mL (or other size) – polycarbonate or fluorinated ethylene propylene.
5. Spiking solution(s) for QC and/or recovery studies.
6. Distilled water – for use as reagent blank (and addition to dry matrices).
7. Repeating pipette and tips (calibrated to solvent used).

Additional item needed for dry matrices:

8. Mixing apparatus – e.g. vortex mixer with platform to hold multiple tubes.

2.2.2. Choice of Extraction Solvent, Buffers, Internal Standards, and Salts

The most common and typically preferred extraction solvent in the QuEChERS approach is MeCN because of its several advantages over acetone and/or ethyl acetate (EtOAc), which include: (1) it is miscible with water, but partitions well from water when salt(s) is(are) added; (2) it is immiscible with hexane and isooctane, but miscible with toluene; (3) it does not co-extract much lipid (or proteins or salts); (4) it can serve as the injection solvent in both LC and GC; and (5) it is an excellent solvent for dSPE, which leads to cleaner final extracts.

EtOAc can be used for nonfatty foods, but should be avoided with matrices containing fats unless gel permeation chromatography (GPC) cleanup is used. EtOAc is not fully miscible with water, gives dirtier final extracts after dSPE than with MeCN, is poorly compatible in LC, and yields lower recoveries for many relatively polar pesticides, even when buffering is used. However, it is better than MeCN for GC analysis (not that MeCN is unacceptable). Certain pesticides such as chlorothalonil, captan, folpet, and captafol are more stable in EtOAc than MeCN or acetone, but certain organophosphates (e.g. fenamiphos, phorate) form sulfones and sulfoxides in EtOAc, but not in MeCN or acetone (98).

Acetone is the least preferable solvent, having each chemical disadvantage of both MeCN and EtOAc, plus it is worse in the partitioning process and dissolves more of the salts used in QuEChERS. Its only advantage is that is has a lower boiling point for faster evaporation if needed and potentially better analysis of the most volatile analytes in GC (LC-MS/MS can be used for the volatiles, though). Like EtOAc, acetone has lower cost and less toxicity than MeCN (not that MeCN is overly expensive and unsafe).

The use of buffering with 1% HOAc in the MeCN (or EtOAc or acetone) extraction solvent and NaOAc salt (as in the AOAC Official Method 2007.01) or with citrate salts (as in EN 15662) can be done to improve recoveries of pH-dependent or less stable pesticides (such as pymetrozine in acidic samples and tolylfluanid in neutral samples). However, the large majority of pesticides are readily extracted with MeCN for nearly all food matrices without the need for buffering. *It is recommended to use the original QuEChERS approach (for a 15-g sample in a 50-mL centrifuge tube) without buffering unless the analysis requires inclusion of those pesticides improved by buffering.* Buffering increases monitoring scope slightly at the cost of potentially less clean extracts and slightly higher cost and inconvenience. *If the application entails pymetrozine in citrus, then the acetate buffering method gives reproducible ≈80% recoveries, whereas the other two versions give significantly lower and more variable recoveries of pymetrozine (in all matrices). On the other hand, for cereals and grains, then the acetate buffering method does not give an adequate cleanup in dSPE even with 150 mg PSA per milliliter extract* (19). CEN Standard Method EN 15662 calls for the addition of 600 μL 5N NaOH solution per 10 g limes, lemons, and currants, and 200 μL 5N NaOH per 10 g raspberries (18).

Although results are acceptable without using an internal standard (17), an internal standard typically provides even better results due to compensation for small volume fluctuations. The internal standard solution should be added after the solvent has been mixed briefly with the sample and before the addition of the salts. The choice of internal standard is an important one, and the following considerations must be made: (1) it must not occur in the sample to begin with; (2) it must be stable; (3) it must give consistently high recoveries; (4) it must be readily available and inexpensive; (5) it must not interfere with any analytes; and

Table 1
Possible internal standards in LC-MS/MS and GC-MS(/MS)

Compound	Detector
Aldrin	GC
d_5-Atrazine	LC and GC
d_4-Carbendazim	LC
d_3-Carbofuran	LC (and GC)
d_{10}-Diazinon	LC and GC
d_6-α-HCH	GC
d_6-Malathion	LC and GC
d_6-Methoxychlor	GC
d_{10}-Parathion	GC
d_6-Parathion-methyl	GC (and LC)
d_3-Propoxur	LC and GC
Triphenylphosphate	LC and GC
Triethylphosphate	LC and GC

(6) ideally be readily detected in both GC-MS and LC-MS/MS without being affected by matrix effects in either case. Use of multiple internal standards is preferable in case of a problem with one of them in the analysis. Suggested internal standards for LC, GC, and both are listed in Table 1.

The QuEChERS approach consistently uses 0.4 g anhydrous $MgSO_4$ per milliliter extraction solvent (MeCN) in the partitioning/extraction step in combination with additional salt(s). Anastassiades et al. extensively studied the effects of different salts and their combinations in the introductory QuEChERS publication (10), and there is little reason to change this ratio for extraction. The use of 0.1 g NaCl per milliliter extraction solvent is also used in the original method to reduce polar co-extractives (sugars) without affecting recoveries. In the acetate buffered version (14), the NaCl is replaced with 0.1 g anhydrous NaOAc (or 0.167 g $NaOAc{\cdot}3H_2O$) per milliliter extraction solvent (MeCN). In the citrate buffered version, the NaCl is still included in the salt mixture, but citrate salts are also included (12).

As listed in the introduction, commercial products are available for each of the salt combinations used in the different methods, even for different sample size choices. The products come in sealed packets, 50-mL polycarbonate centrifuge tubes, or vials.

Alternatively, lab personnel can prepare the powder mixes themselves and a riffling device can be used to ease this process if desired (12, 18). For the commercial products in which the salts are provided in centrifuge tubes, it is recommended that the initial sample/solvent extract be poured from a separate tube into the tube containing the preweighed salts (static charges or hydrostatic forces on the tube walls can make it difficult to transfer the salt from one tube to the other). Otherwise the procedure can be altered using just one tube by first adding the solvent to the salts in the tube, and then adding/weighing the sample into the salt/solvent combination in the tube. *Do not add a wet sample directly to the salts or else the highly exothermic reaction between anhydrous $MgSO_4$ and water will create problems.*

Items needed for extraction of subsample (original QuEChERS):

1. Acetonitrile (MeCN) – analytical grade (or EtOAc).
2. Solvent bottle and calibrated solvent dispenser (e.g. 10 or 15 mL).
3. Anhydrous magnesium sulfate ($MgSO_4$) – powder form; purity >98%.
4. Sodium chloride (NaCl) – reagent grade.
5. 4/1 (w/w) anhydrous $MgSO_4$/NaCl (5 g for 10 mL of extraction solvent or 7.5 g for 15 mL).
6. Internal standards solution (concentration at which 100 μL spike yields 200 ng/g).
7. Repeating pipet and tips (calibrated to solvent used).

Items needed for acetate buffered QuEChERS (AOAC Official Method 2007.01):

1. Acetonitrile (MeCN) – analytical grade (or EtOAc).
2. Acetic acid (HOAc) – glacial (do not compensate for purity).
3. 1% HOAc in MeCN (v/v) – e.g. 10 mL glacial HOAc in a 1 L MeCN solution.
4. Solvent bottle and calibrated solvent dispenser (e.g. 10 or 15 mL).
5. Anhydrous magnesium sulfate ($MgSO_4$) – powder form; purity >98%.
6. Anhydrous sodium acetate (NaOAc) or trihydrate – reagent grade.
7. 4/1 (w/w) anhydrous $MgSO_4$/NaOAc (7.5 g for 15 mL extraction solvent or 5 g for 10 mL), or 4/1.67 anhydrous $MgSO_4$/NaOAc·$3H_2O$ (8.5 g for 15 mL, or 5.67 g for 10 mL extraction solvent).

8. Internal standards solution (concentration at which 100 μL spike yields 200 ng/g).
9. Repeating pipet and tips (calibrated to solvent used).

Items needed for citrate buffered QuEChERS (CEN Standard Method EN 15662):

1. Acetonitrile (MeCN) – analytical grade (or EtOAc).
2. Solvent bottle and calibrated solvent dispenser (e.g. 10 or 15 mL).
3. Anhydrous magnesium sulfate ($MgSO_4$) – powder form; purity >98%.
4. Sodium chloride (NaCl) – reagent grade.
5. Trisodium citrate dihydrate ($Na_3Cit{\cdot}2H_2O$) – CAS # 6132-04-3; reagent grade.
6. Disodium hydrogen citrate sesquihydrate ($Na_2Cit{\cdot}1.5H_2O$) – CAS # 6132-05-4; reagent grade.
7. 4/1/1/0.5 (w/w/w/w) anhydrous $MgSO_4$/NaCl/$Na_3Cit{\cdot}2H_2O$/$Na_2Cit{\cdot}1.5H_2O$ (6.5 g for 10 mL extraction solvent or 9.75 g for 15 mL).
8. 5N sodium hydroxide (NaOH) solution (for matrices with pH <3).
9. Internal standards solution (conc. at which 100 μL spike yields 200 ng/g).
10. Repeating pipet and tips (calibrated to solvent used).

2.2.3. Shake and Centrifuge

The QuEChERS approach is very flexible, and despite the modifications and options that can be used, all of the steps are easy and straightforward to perform. In the case of the extraction/partitioning step, a vortexer or mechanical shaker may be used, or even a Geno/Grinder with stainless steel balls (or not) added to the tubes, but *manual shaking for 1 min has been shown to achieve consistently high recoveries for hundreds of pesticides from numerous commodities in many labs around the globe.* QuEChERS has also consistently achieved excellent results for proficiency test samples and for incurred residues in interlaboratory trials, which demonstrates high extractability by the shaking approach.

It is important to cap the tubes well to avoid leaks, which is best accomplished by keeping salts from getting into the threads. Use gloves when shaking the tubes just in case a leak does occur. To increase sample throughput, hold multiple tubes in each hand and shake the tubes using the elbows and shoulders more so than the wrists. Personnel with big hands can achieve throughput of ten samples/minute in this way, which is actually faster than operating a mechanical shaker.

Lab personnel frequently debate about the benefits of shaking by hand or using a mechanical shaker for extraction in QuEChERS, and while some prefer the exercise, simplicity, and lower cost of shaking by hand, others complain of the exertion and worry about the inconsistency and degree of vigor in the manual process. The lab can purchase and modify a paint shaker from a hardware store to automate the shaking process, or personnel can listen to disco music to increase enjoyment while they do the 1 min "QuEChERS dance."

After shaking, it is possible to avoid the centrifugation step if enough time is given for the needed aliquot of the upper extract layer to be taken for dSPE, but centrifugation also provides a degree of cleanup. Centrifuge for 2 min at nearly the highest force that the tube and centrifuge are rated to withstand. Place equally weighted tubes across from each other in the centrifuge rotor and follow the safety precautions given in the centrifuge manual. A swinging bucket rotor gives a more convenient horizontal separation of the layers rather than the slanted layers formed with a fixed angle rotor, but the fixed angle rotor often provides greater maximal centrifugal force.

Items needed for shaking and centrifugation (all versions):

1. Centrifuge – able to achieve >2,500 relative centrifugal force (rcf).
2. Balance.
3. Rack to hold tubes.
4. Laboratory gloves.
5. Timer.

Optional item for mechanical shaking:

6. Mechanical shaker.

Optional items for manual shaking

7. Sound system and appropriate music.

2.3. QuEChERS Cleanup

2.3.1. No Cleanup, or Use of Freeze-Out or Hexane Partitioning

The choice of solvent, buffer, salt(s), and centrifugation in QuEChERS provides a degree of cleanup in the extraction/partitioning step, and ideally, no further cleanup would be needed. Indeed, the high degree of selectivity and sensitivity in LC-MS/MS often affords analysis of the initial QuEChERS extract without conducting further cleanup (but cleanup is fast, cheap, and easy enough with dSPE that it is worth the precaution). *However, 2,4-dichlorophenoxyacetic acid (2,4-D) and similarly acidic pesticides (that are retained by PSA in the dSPE cleanup step) require an aliquot of the extract prior to the addition of salts during extraction* (18, 34). In this case, the volume of the water and MeCN initial extract must be known, or an internal standard used, to compensate for MeCN–water volume changes. An aliquot of 100–200 μL taken from the ≈20–30 mL MeCN–sample extracts for LC-MS/MS analysis of acid herbicides (typically requiring electrospray negative mode) is small enough that it does not cause a bias in the results for the other analytes. In the case of EtOAc for extraction, EtOAc partially separates from water even without addition of salts and gives poor recoveries of nearly all polar pesticides that can be analyzed by LC-only. This approach is not acceptable for the analysis of phenoxy acid herbicides like 2,4-D.

When using MeCN extraction for matrices that contain lipids (oils, fats, waxes), a freeze-out procedure can be performed to reduce co-extracted lipids (18, 53, 54). In this situation, transfer the upper layer or an aliquot of the upper layer into a vial, cap the vial well, and place it into a freezer with temperature ≤−18°C for ≥2 h. If the frozen lipids/oils are known to be more dense than MeCN, such as olive and flaxseed oils, then take the upper MeCN layer for further cleanup and/or analysis. Otherwise, the floating lipids/waxes must be removed from the lower MeCN extract before further cleanup or analysis.

If the analysis only entails relatively polar analytes (typically those with solubilities in water ≥1 mg/L), then hexane (typically 0.5 mL per milliliter MeCN) may be used for removal of co-extracted lipids without decreasing analyte recoveries. In the analysis of acrylamide for example, hexane is added to the fatty samples prior to the addition of MeCN in the method (23). For more lipophilic analytes with solubility in water <1 mg/L, such as organochlorinated insecticides, partial partitioning between the MeCN extract and hexane and/or lipid phases will occur, but it will be a consistent ratio that should be empirically determined through the analysis of multiple samples of the same commodity. Different samples of the same commodity usually have quite similar lipid content (99), thus yield similar partitioning factors and recoveries, which permits the analyst to validly correct for recovery factors in the determination of the pesticide(s) in real samples of the same commodity (*incurred samples and proficiency testing should be done to verify this assertion during the method validation process*). This recovery factor is based on a consistent physicochemical property, which justifies use of the correction factor in this situation, but correction factors are generally not permitted in residue analysis.

Optional items for optional freeze-out of fatty or waxy samples:

1. Vials.
2. Freezer with temperature ≤−18°C (or container of dry ice).

Optional item for optional hexane partitioning of fatty or waxy samples:

1. Hexane – analytical grade.

2.3.2. QuEChERS Cleanup Using Dispersive SPE

Dispersive SPE entails the addition of the extract to sorbent powders contained in a centrifuge tube, followed by shaking for 30 s and centrifugation for 2 min. Again, near maximal force within safety limits should be used in the centrifugation to provide better separation of the sorbents from the QuEChERS extracts. Ideally, the interfering matrix components are adsorbed, and the analyte(s) remain in the extract, which is transferred to (an) autosampler vial(s) for analysis by GC-MS and/or LC-MS/MS. Advantages of dSPE over traditional SPE in "chemical filtration" cleanup include: (1) easier; (2) faster; (3) cheaper; (4) no vacuum needed; (5) no manifold needed (replaced by a centrifuge); (6) no flow control; (7) no pretreatment of columns; (8) no extra collection tube; (9) no elution solvent; (10) no worries about channeling or drying out the column; (11) solvent evaporation is not needed afterwards; and (12) higher analyte recoveries are often obtained.

As mentioned previously, it is not necessary to use dSPE in QuEChERS, but it is simple, inexpensive, and fast to do so, and it is usually effective. *dSPE is typically not as effective in removing matrix components as traditional cartridge-based SPE with the same sorbent(s), but it commonly provides higher recoveries and achieves high-quality cleanup of food extracts for LC-MS/MS and GC-MS(/MS) analyses.* The practical benefits of dSPE over traditional SPE make it preferable for use in the QuEChERS approach. Alternatives such as disposable pipette extraction (DPX) may further streamline the "chemical filtration" cleanup approach, but thus far, the convenience and practical benefits of dSPE are still superior. In the case of water and other large-volume sample applications, traditional SPE is superior than dSPE to concentrate the analyte(s) on the sorbent and then use a small amount of organic solvent for elution.

A disadvantage of dSPE is that ≈50% of the extract volume is lost to the sorbent(s) in the centrifuge tube, which makes it inconvenient to conduct solvent concentration of the final extract. For LC-MS/MS with 5–10 μL injections (even after 1:3 dilution with mobile phase) and GC-MS using large-volume injection (LVI) with 10 μL injections, limits of quantitation (LOQs) of 5 ng/g for most pesticides can be achieved using modern instruments from 1 g/mL equivalent final QuEChERS extracts (especially in selected ion monitoring mode in GC-MS, but also with full scan with some instruments). For dry matrices, the equivalent final extract concentrations are as low as 0.25 g/mL depending on sample sizes, and then LOQs increase concomitantly. The same relationship occurs for lipophilic pesticides in fatty samples, which may only achieve 25% recoveries prior to compensation for the lipid/MeCN partitioning process.

The Luke method was not developed with LC-MS/MS and LVI-GC in mind, and produced more concentrated final extracts of 4 g/mL sample equivalent. An option in the QuEChERS method for dry/fatty matrices and/or 1–2 μL splitless injection in GC is to scale up the dSPE step from 1 mL extract to 8 mL, and then take 4 mL and conduct a solvent exchange/concentration step to yield a 1 mL final extract (4 g/mL) in toluene. This option is integrated into AOAC Official Method 2007.01 and was shown among different labs to attain equivalent results as dSPE of 1 g/mL MeCN final extracts and 8–10 μL injection in LVI-GC-MS (17).

MeCN is the better solvent over EtOAc or acetone for use in dSPE (or SPE in general) because it provides cleaner extracts than the other two stronger solvents. MeCN extracts in QuEChERS are visibly clearer and contain measurably less and fewer matrix components than EtOAc and acetone.

Any sorbent that does not float, dissolve, or swell so much as to incorporate the entire extract volume may be used in dSPE. The most common sorbents used in QuEChERS are the same ones that are normally used in multiclass, multiresidue pesticide analysis (3, 89), which include PSA, C_{18}, and/or GCB. Another important component used in dSPE with QuEChERS is anhydrous $MgSO_4$, which reduces the water content in the extracts to improve the fatty acid retention properties of the PSA sorbent in particular. The initial QuEChERS MeCN extract is ≈16% water, and the use of 150 mg anhydrous $MgSO_4$ per milliliter extract reduces water content to ≈2% in the final MeCN extract (10). All versions of the QuEChERS approach use this ratio of $MgSO_4$ to help dry the extract in dSPE.

2.3.3. Use of PSA in Dispersive SPE

All versions of QuEChERS also employ PSA to help remove fatty acids from the food commodity extracts. PSA is a weak anion exchange sorbent on silica particles that also provide adsorption properties for some hydrocarbons akin to C_{18}. Due to the extra secondary amino group, PSA has a greater capacity than aminopropyl sorbent, thus less sorbent is able to be used to achieve equivalent performance. PSA retains 2,4-D and other acid herbicides, but does not lower recoveries of hundreds to other pesticides, except for acephate, methamidophos, and omethoate, which are only slightly affected in most cases (they can be appreciably retained by PSA in column-based SPE).

Originally without buffering, 25 mg PSA per milliliter QuEChERS extract was found to be enough to remove fatty acid and anthocyanin components from the tested fruits and vegetables. After evaluation of more matrices and introduction of acetate buffering, Lehotay et al. (14) increased the amount of PSA to 50 mg/mL extract, which compensated for the presence of acetic acid in the extract without affecting any pesticide recoveries. To account for the much higher fatty acid composition of dry cereal and grain matrices, Mastovska et al. (19) further increased PSA content in dSPE to 150 mg/mL (or equivalent to as much as 600 mg/g sample). This was done for the protocol without use of buffering, which would overwhelm even that amount of extra PSA.

Some protocols call for the checking and modification of pH to achieve pH 5–7 in the initial QuEChERS extracts prior to dSPE with PSA. This reduces the amount of acids in extracts and effectively increases the capacity of the PSA (12). *However, it is easier to actually increase the capacity of PSA by increasing the amount of PSA in dSPE*, as Lehotay, Mastovska, et al. have done (14, 19). When 50 mg/mL PSA is used in the case of orange juice for example, the dSPE provides the same degree of cleanup in final MeCN extracts independent of adjusted pH of the extract with (0.09% co-extractives) or without (0.04% co-extractives) acetate buffering (14).

In some cases, the use of acetate buffering can overwhelm the PSA with acetic acid, but recoveries of only a few pesticides are improved by buffering, and the majority of applications only need to screen for these pesticides anyway, which can be achieved without buffering (11). In regulatory applications, buffering is only truly needed for pymetrozine in citrus matrices, in which case the acetate buffering AOAC Official Method 2007.01 provides better performance.

2.3.4. Use of C_{18} in Dispersive SPE

In the original QuEChERS method (10) and in the Official versions (17, 18), only $MgSO_4$ and PSA were used in dSPE. Octadecylsilane (ODS or C_{18}) was also evaluated for use in dSPE in the original QuEChERS method, but for nonfatty matrices, it did not provide much additional if any cleanup over the use of PSA alone. Since that time, C_{18} was shown to be highly effective to reduce lipid matrix co-extractives from milk, eggs, avocado, olives, cereals, nuts, animal feeds, and other commodities (15, 16). Lehotay and Anastassiades independently found that dSPE with 50 mg/mL C_{18} is more effective at removing trace lipids from the MeCN extracts than the freeze-out step described in Subheading 2.3.1 (15, 18). Moreover, C_{18} has not been found to reduce the recoveries of any pesticides among hundreds evaluated in dSPE of MeCN extracts. Also, C_{18} is highly hydrophobic and thus has minimal effect in the reduction of recovered extract volume that occurs in dSPE. C_{18} may be ineffective in some cases, but it helps in others and cannot hurt in any known application of QuEChERS. *Therefore, 50 mg/mL C_{18} should be included in all dSPE applications in QuEChERS, and the author regrets that he did not include this sorbent in the collaborative study protocol for AOAC Official Method 2007.01.*

2.3.5. Chlorophyll Co-extractives and Use of GCB

The most complicating aspect in the QuEChERS method is the co-extraction of chlorophyll from green vegetables. There is nothing new about this issue in multiclass, multiresidue analysis of pesticides using any method (1–4, 9, 89), and trade-offs in recoveries of certain analytes vs. how much cleanup should be done to reduce the green color have been debated for many years. *One of the deceiving factors that affects decisions is that a very small amount of chlorophyll can create a dark green color in the extracts.* Even when 90% of the chlorophyll is removed from extracts of lettuce, broccoli, spinach, peas, green beans, etc., the final extracts remain quite green. Although chlorophyll makes no direct interferences in LC or GC chromatography because it is retained on the guard column in reversed-phase LC and remains in the injection liner in GC, it reduces the ruggedness of both types of analyses because replacement of the LC guard column and GC liner (and retention gap) are needed about twice more frequently than if the chlorophyll is removed. In GC, the liner needs to be changed after about 100 injections of QuEChERS extracts for many fruits and vegetables, but it should be changed after about 50 injections of spinach. In LC-MS/MS, hundreds of injections are possible without maintenance in either case, but the guard column deteriorates faster if the injected extracts contain a lot of chlorophyll.

It is very easy to remove 100% of the co-extracted chlorophyll in QuEChERS (and other methods) by using 50 mg/mL GCB in dSPE (or a traditional SPE cartridge of GCB). However, GCB also strongly retains several important pesticide analytes, such as thiabendazole, quintozene, cyprodinil, hexachlorobenzene, and other structurally planar chemicals. C_{18} helps to reduce chlorophyll content slightly, and polymer sorbents like Waters' Oasis, also retain chlorophyll well in MeCN extracts, and do not cause pesticide losses. Unfortunately, Oasis sorbent floats in dSPE, but other polymers that sink may be similarly effective. A polymeric sorbent product known as Chlorofilter may be useful in this application (it tends to swell in MeCN more than other sorbents, though).

Until the optimal solution to the chlorophyll removal problem is achieved, chemists have used GCB in different ratios and/or mixed toluene into the extracts in an attempt to increase recoveries of planar pesticides while still removing some of the chlorophyll. *This approach is a game of trade-offs and compromises because recoveries of these types of pesticides and chlorophyll removal are intertwined.* Approximately 50 mg/mL GCB in dSPE of MeCN extracts is needed to remove all the chlorophyll, which also removes nearly all of the planar pesticides. In CEN Standard Method EN 15662, Anastassiades chose to add 2.5 mg GCB per milliliter citrate buffered MeCN extract in dSPE for moderately green extracts (e.g. green beans and peas), and 7.5 mg/mL GCB for dark green extracts (e.g. spinach and broccoli). *This has moderate impact in terms of chlorophyll removal while only minimally affecting analyte recoveries* (e.g. 10% reduction in recovery of thiabendazole from spinach).

The choice to add 20–25% toluene by volume to MeCN extracts in dSPE with GCB is fine for GC applications, especially if post-cleanup and solvent exchange/concentration is to be conducted, but it is not a good idea for LC analysis. The toluene causes peak shape and solvent miscibility problems in the mobile phase. Furthermore, the small amount of toluene does not help to increase pesticide recoveries much (89), and acts to reduce chlorophyll removal. EtOAc and acetone are slightly stronger solvents that can be used with or without toluene to play the same game of trade-offs and compromises in dSPE cleanup with GCB.

Items for dSPE:

1. Anhydrous magnesium sulfate ($MgSO_4$) – powder form; purity >98%.
2. Primary secondary amine (PSA) sorbent – 40 μm particle size.
3. C_{18} sorbent – 40 μm particle size.
4. Centrifuge tubes – 2-mL mini-tubes or 15-mL tubes (glass or appropriate plastic).
5. Balance (if powders are to be weighed into tubes).
6. Centrifuge – able to achieve >2,500 relative centrifugal force (rcf).
7. Rack to hold tubes.
8. Repeating pipet and tips (calibrated to solvent used).
9. Laboratory gloves.
10. Timer.
11. Autosampler vials.

Optional items for dSPE:

12. Graphitized carbon black (GCB) – for potential use with green vegetable extracts.
13. Toluene – analytical grade (for solvent exchange/concentration in GC if needed).
14. Evaporator – e.g. Turbovap or N-Evap (for solvent exchange/concentration if needed).
15. Graduated centrifuge tubes (10–15 mL) for use in evaporator.

3. Conclusions

In this chapter, the previous QuEChERS protocol in this series (20) has been updated to include more options. A diagram of the main options appears as Fig. 1. For more specific protocols, see the Official/Standard Methods (17, 18, 20) and peruse the extensive

Step	Procedure
0.	Comminute >1 kg sample w/ vertical cutter and homogenize ≈200 g subsample w/ probe blender; Or, cut sample into pieces, freeze, and comminute sample w/ dry ice in vertical cutter.
1.	Transfer 10-15 g subsample to 50 mL cent. tube
2.	Add 10-15 mL MeCN; shake briefly
2A.	Add 10-15 mL 1%HOAc in MeCN; shake briefly
3.	Add 4-6 g anh. $MgSO_4$ + 1-1.5 g NaCl+ I.S.
3A.	Add 4-6 g anh. $MgSO_4$ +1-1.5 g anh. NaOAc+I.S.
3C.	Add 4-6 g anh. $MgSO_4$ + 1-1.5 g NaCl+ 1-1.5 g Na_3 Cit•$2H_2O$+ 0.5-0.75 g Na_2Cit•$1.5H_2O$+ I.S.
4.	Shake for 1 min; Centrifuge >2500 rcf for 2 min
5.	Transfer 1-8 mL to cent. tube w/ 150 mg anh. $MgSO_4$ + 50 mg PSA + 50 mg C_{18} per mL extract
5C.	Transfer 1-8 mL to tube w/ 150 mg anh. $MgSO_4$ + 25 mg PSA + 0-7.5 mg GCB per mL extract
6.	Shake for 30 s; Centrifuge >2500 rcf for 2 min
7.	Transfer 0.5-1 mL extract to LC vial (add QC-Std); Transfer 0.2 mL of that to GC vial w/ insert; Add 0.45-0.9 mobile phase diluent to LC vial
7b.	Transfer 0.25 mL from step 6 to LC vial; Add QC-Std and mobile phase diluent
8b.	Transfer 4 mL from step 6 to grad. cent. tube; Add QC-Std and 1 mL toluene
9b.	Evaporate at 40°C with N_2 to 0.3-0.5 mL; Add toluene to make 1 mL
10b.	Add 0.2 mL anh. $MgSO_4$ and swirl >6 mL mark; Centrifuge >2500 rcf for 1 min; Transfer 0.2 mL to GC vial w/ insert
8/11b.	Analyze by (LVI/)GC-MS(/MS) and LC-MS/MS

Fig. 1. Diagram of the main QuEChERS protocols with options without buffering or for buffering with A = acetate (AOAC Official Method 2007.01) and C = citrate (CEN Standard Method EN 15662). Option "b" gives procedure for 8 mL extract in dSPE if concentration of the extract is needed for GC.

literature referenced (10–88). Many of these papers describe novel and state-of-the-art MS applications, which further demonstrate the utility of QuEChERS concepts for sample preparation prior to MS analysis. QuEChERS is a sample preparation template that is easy to try in any application, and if it fails, the chemist has only spent a short time testing it. If QuEChERS works, then the chemist has saved a great deal of time in sample preparation method development and in routine analyses, which have been major time-consuming tasks previously.

In the author's opinion, the chemist should use the unbuffered original approach without matrix and MeCN solvent as the first choice. If that does not meet the application needs, then the acetate buffering procedure should be done if the analyte(s) is(are) pH-dependent. EtOAc (and acetone and methanol) can be evaluated as the solvent if MeCN doesn't achieve high recoveries and stability. For polar analytes that do not partition into organic solvent during the salting out step, the phase separation can be eliminated, as in the case of applications involving 2,4-D (12, 18, 34) and veterinary drugs (25–27, 80). For acidic analytes retained on PSA, the PSA can be replaced by C_{18}, which provides less cleanup (10, 25–27), but satisfactory recoveries and ruggedness in modern LC-MS techniques. This template has been used with interesting results in the even greater challenge to investigate multi-application, multi-class, multi-residue analysis for pesticides, veterinary drugs, mycotoxins, and other possible food contaminants (100).

4. Disclaimer

Mention of brand or firm name does not constitute an endorsement by the U.S. Department of Agriculture above others of a similar nature not mentioned. USDA is an equal opportunity employer.

References

1. Luke, M.A., Froberg, J.E., and Masumoto, H.T. (1975) Extraction and cleanup of organochlorine, organophosphate, organonitrogen, and hydrocarbon pesticides in produce for determination by gas-liquid chromatography. *J. Assoc. Off. Anal. Chem.* **58**, 1020–1026.
2. Mills, P.A., Onley, J.H., and Guither, R.A. (1963) Rapid method for chlorinated pesticide residues in nonfatty foods. *J. Assoc. Off. Anal. Chem.* **46**, 186–191.
3. Food and Drug Administration (1999) Pesticide Analytical Manual Volume I: Multiresidue Methods, 3rd Edition, U.S. Department of Health and Human Services, Washington, DC. http://www.cfsan.fda.gov/~frf/pami3.html
4. Sawyer, L.D. (1985) The Luke et al. method for determining multipesticide residues in fruits and vegetables: collaborative study. *J. Assoc. Off. Anal. Chem.* **68**, 64 71.
5. Eller, K.I., and Lehotay, S.J. (1997) Evaluation of hydromatrix and magnesium sulfate drying agents for supercritical fluid extraction of multiple pesticides in produce. *Analyst* **122**, 429–435.
6. Anastassiades, M., and Schwack, W. (1998) Analysis of carbendazim, benomyl, thiophanate methyl and 2,4-dichlorophenoxyacetic acid in fruits and vegetables after supercritical fluid extraction. *J. Chromatogr. A* **825**, 45–54.
7. Valverde-García, A., Fernández-Alba, A.R., Agüera, A., and Contreras, M. (1995) Extraction of methamidophos residues from vegetables with supercritical fluid carbon dioxide. *J. AOAC Int.* **78**, 867–73.
8. Lehotay, S.J. (1997) Supercritical fluid extraction of pesticides in foods. *J. Chromatogr. A* **785**, 289–312.
9. Lehotay, S.J. (2000) Determination of pesticide residues in nonfatty foods by supercritical fluid extraction and gas chromatography/mass spectrometry: collaborative study. *J. AOAC Int.* **83**, 680–697.
10. Anastassiades, M., Lehotay, S.J., Štajnbaher, D., and Schenck, F.J. (2003) Fast and easy multiresidue method employing acetonitrile

extraction/partitioning and "dispersive solid-phase extraction" for the determination of pesticide residues in produce. *J. AOAC Int.* **86**, 412–431.

11. Lehotay, S.J., Hiemstra, M., van Bodegraven, P., and de Kok, A. (2005) Validation of a fast and easy method for the determination of residues from 229 pesticides in fruits and vegetables using gas and liquid chromatography and mass spectrometric detection. *J. AOAC Int.* **88**, 595–614.
12. Anastassiades, M., Scherbaum, E., Taşdelen, B., and Štajnbaher, D. (2007) Recent developments in QuEChERS methodology for pesticide multiresidue analysis. In *Pesticide Chemistry. Crop Protection, Public Health, Environmental Safety* Ohkawa, H., Miyagawa, H., and Lee, P.W. (editors), Wiley-VCH, Weinheim, Germany, pp. 439–458.
13. Payá, P., Anastassiades, M., Mack, D., Sigalova, I., Tasdelen, B., Oliva, J., and Barba, A. (2007) Analysis of pesticide residues using the Quick Easy Cheap Effective Rugged and Safe (QuEChERS) pesticide multiresidue method in combination with gas and liquid chromatography and tandem mass spectrometric detection. *Anal. Bioanal. Chem.* **389**, 1697–1714.
14. Lehotay, S.J., Maštovská, K., and Lightfield, A.R. (2005) Use of buffering and other means to improve results of problematic pesticides in a fast and easy method for residue analysis of fruits and vegetables. *J. AOAC Int.* **88**, 615–629 & 60A.
15. Lehotay, S.J., Maštovská, K., and Yun, S.J. (2005) Evaluation of two fast and easy methods for pesticide residue analysis in fatty food matrixes. *J. AOAC Int.* **88**, 630–638.
16. Cunha, S.C., Lehotay, S.J., Mastovska, K., Fernandes, J.O., and Oliveira, B.M.P.P. (2007) Evaluation of the QuEChERS sample preparation approach for the analysis of pesticide residues in olives. *J. Sep. Sci.* **30**, 620–632.
17. Lehotay, S.J. (2007) Determination of pesticide residues in foods by acetonitrile extraction and partitioning with magnesium sulfate: collaborative study. *J. AOAC Int.* **90**, 485–520.
18. http://www.quechers.com; http://www.cen.eu
19. Mastovska, K., Dorweiler, K., Fitzpatrick, G., Wegscheid, J., Szpylka, K., and Lehotay, S. (2008) Pesticide Multiresidue Analysis in Cereal Grains using Modified QuEChERS Method. 45th Florida Pesticide Residue Workshop, St. Pete Beach, FL http://www.flworkshop.com/2008/2008%20PDFs/04-Mastovska.pdf
20. Lehotay, S.J. (2006) Quick, easy, cheap, effective, rugged, and safe approach for determining pesticide residues. In *Pesticide Protocols, Methods in Biotechnology Series* **#19**, Humana Press, Totowa, New Jersey, USA, pp. 239–261.
21. Cajka, T., Hajslova, J., Lacina, O., Mastovska, K., and Lehotay, S.J. (2008) Rapid analysis of multiple pesticide residues in fruit-based baby food using programmed temperature vaporiser injection-low-pressure gas chromatography-high-resolution time-of-flight mass spectrometry. *J. Chromatogr. A* **1186**, 281–294.
22. Aysal, P., Ambrus, A., Lehotay, S.J., and Cannavan, A. (2007) Validation of an efficient method for the determination of pesticide residues in fruits and vegetables using ethyl acetate for extraction. *J. Environ. Sci. Health B* **42**, 481–490.
23. Mastovska, K., and Lehotay, S.J. (2006) Rapid sample preparation method for LC-MS/MS or GC-MS analysis of acrylamide in various food matrices. *J. Agric. Food Chem.* **54**, 7001–7008.
24. Cajka, T., Mastovská, K., Lehotay, S.J., and Hajslová, J. (2005) Use of automated direct sample introduction with analyte protectants in the GC-MS analysis of pesticide residues. *J. Sep. Sci.* **28**, 1048–1060.
25. Fagerquist, C.K., Lightfield, A.R., and Lehotay, S.J. (2005) Confirmatory and quantitative analysis of beta-lactam antibiotics in bovine kidney tissue by dispersive solid-phase extraction and liquid chromatography-tandem mass spectrometry. *Anal. Chem.* **77**, 1473–1482.
26. Mastovska, K., and Lightfield, A.R. (2008) Streamlining methodology for the multiresidue analysis of beta-lactam antibiotics in bovine kidney using liquid chromatography-tandem mass spectrometry. *J. Chromatogr. A* **1202**, 118–123.
27. Kinsella, B., Lehotay, S.J., Mastovska, K., Lightfield, A.R., Furey, A., and Danaher, M. (2009) New method for the analysis of flukicide and other anthelmintic residues in bovine milk and liver using liquid chromatography-tandem mass spectrometry. *Anal Chim. Acta* **637**, 196–207.
28. Schenck, F.J., and Hobbs, J.E. (2004) Evaluation of the quick, easy, cheap, effective, rugged, and safe (QuEChERS) approach to pesticide residue analysis. *Bull. Environ. Contam. Toxicol.* **73**, 24–30.
29. Schenck, F.J., Brown, A.N., Podhorniak, L.V., Parker, A., Reliford, M., and Wong, J.W. (2008) A rapid multiresidue method for determination of pesticides in fruits and

vegetables by using acetonitrile extraction/ partitioning and solid-phase extraction column cleanup. *J. AOAC Int.* **91**, 422–438.

30. Zhang, K., Wong, J.W., Hayward, D.G., Sheladia, P., Krynitsky, A.J., Schenck, F.J., Webster, M.G., Ammann, J.A., and Ebeler, S.E. (2009) Multiresidue pesticide analysis of wines by dispersive solid-phase extraction and ultrahigh-performance liquid chromatography-tandem mass spectrometry. *J. Agric. Food. Chem.* **57**, 4068–4078.
31. Kovalczuk, T., Lacina, O., Jech, M., Poustka, J., and Hajslová, J. (2008) Novel approach to fast determination of multiple pesticide residues using ultra-performance liquid chromatography-tandem mass spectrometry (UPLC-MS/MS). *Food Addit. Contam. Part A Chem. Anal. Control Expo. Risk Assess.* **25**, 444–457.
32. Dunovská, L., Cajka, T., Hajslová, J., and Holadová, K. (2006) Direct determination of acrylamide in food by gas chromatography-high-resolution time-of-flight mass spectrometry. *Anal. Chim. Acta* **578**, 234–240.
33. Cunha, S.C., Fernandes, J.O., Alves, A., and Oliveira, M.B. (2009) Fast low-pressure gas chromatography-mass spectrometry method for the determination of multiple pesticides in grapes, musts and wines. *J. Chromatogr. A* **1216**, 119–126.
34. Koesukwiwat, U., Sanguankaew, K., and Leepipatpiboon, N. (2008) Rapid determination of phenoxy acid residues in rice by modified QuEChERS extraction and liquid chromatography-tandem mass spectrometry. *Anal. Chim. Acta* **626**, 10–20.
35. Díez, C., Traag, W.A., Zommer, P., Marinero, P., and Atienza, J. (2006) Comparison of an acetonitrile extraction/partitioning and "dispersive solid-phase extraction" method with classical multi-residue methods for the extraction of herbicide residues in barley samples. *J. Chromatogr. A* **1131**, 11–23.
36. Mol, H.G., Rooseboom, A., van Dam, R., Roding, M., Arondeus, K., and Sunarto, S. (2007) Modification and re-validation of the ethyl acetate-based multi-residue method for pesticides in produce. *Anal. Bioanal. Chem.* **389**, 1715–1754.
37. van der Lee, M.K., van der Weg, G., Traag, W.A., and Mol, H.G. (2008) Qualitative screening and quantitative determination of pesticides and contaminants in animal feed using comprehensive two-dimensional gas chromatography with time-of-flight mass spectrometry. *J. Chromatogr. A* **1186**, 325–339.
38. Banerjee, K., Oulkar, D.P., Dasgupta, S., Patil, S.B., Patil, S.H., Savant, R., and Adsule, P.G. (2007) Validation and uncertainty analysis of a multi-residue method for pesticides in grapes using ethyl acetate extraction and liquid chromatography-tandem mass spectrometry. *J. Chromatogr. A* **1173**, 98–109.
39. Banerjee, K., Oulkar, D.P., Patil, S.B., Jadhav, M.R., Dasgupta, S., Patil, S.H., Bal, S., and Adsule, P.G. (2009) Multiresidue Determination and uncertainty analysis of 87 pesticides in mango by liquid chromatography-tandem mass spectrometry. *J. Agric. Food Chem.* **57**, 4068–4078.
40. Patil, S.H., Banerjee, K., Dasgupta, S., Oulkar, D.P., Patil, S.B., Jadhav, M.R., Savant, R.H., Adsule, P.G., and Deshmukh, M.B. (2009) Multiresidue analysis of 83 pesticides and 12 dioxin-like polychlorinated biphenyls in wine by gas chromatography-time-of-flight mass spectrometry. *J. Chromatogr. A* **1216**, 2307–2319.
41. Banerjee, K., Oulkar, D.P., Patil, S.B., Patil, S.H., Dasgupta, S., Savant, R., and Adsule, P.G. (2008) Single-laboratory validation and uncertainty analysis of 82 pesticides determined in pomegranate, apple, and orange by ethyl acetate extraction and liquid chromatography/tandem mass spectrometry. *J. AOAC Int.* **91**, 1435–1445.
42. Stepan, R., Cuhra, P., and Barsova, S. (2008) Comprehensive two-dimensional gas chromatography with time-of-flight mass spectrometric detection for the determination of anabolic steroids and related compounds in nutritional supplements. *Food Addit. Contam. Part A Chem. Anal. Control Expo. Risk Assess.* **25**, 557–565.
43. Leandro, C.C., Hancock, P., Fussell, R.J., and Keely, B.J. (2007) Quantification and screening of pesticide residues in food by gas chromatography-exact mass time-of-flight mass spectrometry. *J. Chromatogr. A* **1166**, 152–162.
44. Leandro, C.C., Hancock, P., Fussell, R.J., and Keely, B.J. (2007) Ultra-performance liquid chromatography for the determination of pesticide residues in foods by tandem quadrupole mass spectrometry with polarity switching. *J. Chromatogr. A* **1144**, 161–169.
45. Leandro, C.C., Fussell, R.J., and Keely, B.J. (2005) Determination of priority pesticides in baby foods by gas chromatography tandem quadrupole mass spectrometry. *J. Chromatogr. A* **1085**, 207–212.
46. Leandro, C.C., Hancock, P., Fussell, R.J., and Keely, B.J. (2006) Comparison of ultra-performance liquid chromatography and high-performance liquid chromatography for the determination of priority pesticides in baby foods by tandem quadrupole mass spectrometry. *J. Chromatogr. A* **1103**, 94–101.

47. Walorczyk, S. (2007) Development of a multi-residue screening method for the determination of pesticides in cereals and dry animal feed using gas chromatography-triple quadrupole tandem mass spectrometry. *J. Chromatogr. A* **1165**, 200–212.
48. Walorczyk, S. (2008) Development of a multi-residue method for the determination of pesticides in cereals and dry animal feed using gas chromatography-tandem quadrupole mass spectrometry II. Improvement and extension to new analytes. *J. Chromatogr. A* **1208**, 202–214.
49. Walorczyk, S. (2008) Application of gas chromatography/tandem quadrupole mass spectrometry to the multi-residue analysis of pesticides in green leafy vegetables. *Rapid Commun. Mass Spectrom.* **22**, 3791–3801.
50. Nguyen, T.D., Lee, B.S., Lee, B.R., Lee, D.M., and Lee, G.H. (2007) A multiresidue method for the determination of 109 pesticides in rice using the Quick Easy Cheap Effective Rugged and Safe (QuEChERS) sample preparation method and gas chromatography/mass spectrometry with temperature control and vacuum concentration. *Rapid Commun. Mass Spectrom.* **21**, 3115–3122.
51. Nguyen, T.D., Han, E.M., Seo, M.S., Kim, S.R., Yun, M.Y., Lee, D.M., and Lee, G.H. (2008) A multi-residue method for the determination of 203 pesticides in rice paddies using gas chromatography/mass spectrometry. *Anal. Chim. Acta* **619**, 67–74.
52. Ji, F., Zhao, L., Yan, W., Feng, Q., and Lin, J.M. (2008) Determination of triazine herbicides in fruits and vegetables using dispersive solid-phase extraction coupled with LC-MS. *J. Sep. Sci.* **31**, 961–968.
53. Li, L., Xu, Y., Pan, C., Zhou, Z., Jianc, S., and Liu, F. (2007) Simplified pesticide multiresidue analysis of soybean oil by low-temperature cleanup and dispersive solid-phase extraction coupled with gas chromatography/mass spectrometry. *J. AOAC Int.* **90**, 1387–1394.
54. Li, L., Zhang, H., Pan, C., Zhou, Z., Jiang, S., and Liu, F. (2007) Multiresidue analytical method of pesticides in peanut oil using low-temperature cleanup and dispersive solid phase extraction by GC-MS. *J. Sep. Sci.* **30**, 2097–2104.
55. Lee, J.M., Park, J.W., Jang, G.C., and Hwang, K.J. (2008) Comparative study of pesticide multi-residue extraction in tobacco for gas chromatography-triple quadrupole mass spectrometry. *J. Chromatogr. A* **1187**, 25–33.
56. Hu, J., Sheng, Y., Fang, Z., Jin, J., and He, K. (2007) [Determination of forchlorfenuron in Chinese gooseberry with dispersive solid phase extraction and HPLC] *Se Pu* **25**, 441–442.
57. Tsai, W.H., Huang, T.C., Huang, J.J., Hsue, Y.H., and Chuang, H.Y. (2009) Dispersive solid-phase microextraction method for sample extraction in the analysis of four tetracyclines in water and milk samples by high-performance liquid chromatography with diode-array detection. *J. Chromatogr. A* **1216**, 2263–2269.
58. Alcudia-León, M.C., Lucena, R., Cárdenas, S., and Valcárcel, M. (2009) Dispersive solid phase extraction for in-sorbent surface attenuated total reflection infrared detection. *Anal. Chem.* **81**, 1184–1190.
59. Shen, W., Lin, H., Zhao, Z., Ding, T., Xu, J., and Shen, C. (2008) [Determination of triadimenol residue in foods with dispersive solid phase extraction cleanup by gas chromatography-negative chemical ionization mass spectrometry] *Se Pu* **26**, 515–518.
60. Hayama, T., and Takada, M. (2008) Simple and rapid method for the determination of ethylenebisdithiocarbamate fungicides in fruits and vegetables using liquid chromatography with tandem mass spectrometry. *Anal. Bioanal. Chem.* **392**, 969–976.
61. Galera, M.M., García, M.D., and Valverde, R.S. (2008) Determination of photoirradiated high polar benzoylureas in tomato by HPLC with luminol chemiluminescence detection. *Talanta* **76**, 815–823.
62. Dagnac, T., Garcia-Chao, M., Pulleiro, P., Garcia-Jares, C., and Llompart, M. (2009) Dispersive solid-phase extraction followed by liquid chromatography-tandem mass spectrometry for the multi-residue analysis of pesticides in raw bovine milk. *J. Chromatogr. A* **1216**, 3702–3709.
63. Romero-González, R., Garrido Frenich, A., and Martínez Vidal, J.L. (2008) Multiresidue method for fast determination of pesticides in fruit juices by ultra performance liquid chromatography coupled to tandem mass spectrometry. *Talanta* **76**, 211–225.
64. Aguilera-Luiz, M.M., Vidal, J.L., Romero-González, R., and Frenich, A.G. (2008) Multi-residue determination of veterinary drugs in milk by ultra-high-pressure liquid chromatography-tandem mass spectrometry. *J. Chromatogr. A* **1205**, 10–16.
65. Garrido Frenich, A., Martínez Vidal, J.L., Pastor-Montoro, E., and Romero-González, R. (2008) High-throughput determination of pesticide residues in food commodities by use

of ultra-performance liquid chromatography-tandem mass spectrometry. *Anal. Bioanal. Chem.* **390**, 947–959.

66. Fernández Moreno, J.L., Garrido Frenich, A., Plaza Bolaños, P., and Martínez Vidal, J.L. (2008) Multiresidue method for the analysis of more than 140 pesticide residues in fruits and vegetables by gas chromatography coupled to triple quadrupole mass spectrometry. *J. Mass Spectrom.* **43**, 1235–1254.
67. Bolaños, P.P., Moreno, J.L., Shtereva, D.D., Frenich, A.G., and Vidal, J.L. (2007) Development and validation of a multiresidue method for the analysis of 151 pesticide residues in strawberry by gas chromatography coupled to a triple quadrupole mass analyzer. *Rapid Commun. Mass Spectrom.* **21**, 2282–2294.
68. Gilbert-López, B., García-Reyes, J.F., Ortega-Barrales, P., Molina-Díaz, A., and Fernández-Alba, A.R. (2007) Analyses of pesticide residues in fruit-based baby food by liquid chromatography/electrospray ionization time-of-flight mass spectrometry. *Rapid Commun. Mass Spectrom.* **21**, 2059–2071.
69. Mezcua, M., Ferrer, C., García-Reyes, J.F., Martínez-Bueno, M.J., Albarracín, M., Claret, M., and Fernández-Alba, A.R. (2008) Determination of selected non-authorized insecticides in peppers by liquid chromatography time-of-flight mass spectrometry and tandem mass spectrometry. *Rapid Commun. Mass Spectrom.* **22**, 1384–1392.
70. Kmellár, B., Fodor, P., Pareja, L., Ferrer, C., Martínez-Uroz, M.A., Valverde, A., and Fernandez-Alba, A.R. (2008) Validation and uncertainty study of a comprehensive list of 160 pesticide residues in multi-class vegetables by liquid chromatography-tandem mass spectrometry. *J. Chromatogr. A* **1215**, 37–50.
71. Kirchner, M., Húsková, R., Matisová, E., and Mocák, J. (2008) Fast gas chromatography for pesticide residues analysis using analyte protectants. *J. Chromatogr. A* **1186**, 271–280.
72. Hercegová, A., Dömötörová, M., Kruzlicová, D., and Matisová, E. (2006) Comparison of sample preparation methods combined with fast gas chromatography-mass spectrometry for ultratrace analysis of pesticide residues in baby food. *J. Sep. Sci.* **29**, 1102–1109.
73. Hercegová, A., Dömötörová, M., and Matisová, E. (2007) Sample preparation methods in the analysis of pesticide residues in baby food with subsequent chromatographic determination. *J. Chromatogr. A* **1153**, 54–73.
74. Liu, M., Hashi, Y., Song, Y., and Lin, J.M. (2005) Simultaneous determination of carbamate and organophosphorus pesticides in fruits and vegetables by liquid chromatography-mass spectrometry. *J. Chromatogr. A* **1097**, 183–187.
75. Kawata, K., Asada, T., and Oikawa, K. (2005) Determination of pesticides in compost by pressurized liquid extraction and gas chromatography-mass spectrometry. *J. Chromatogr. A* **1090**, 10–15.
76. Powley, C.R., George, S.W., Ryan, T.W., and Buck, R.C. (2005) Matrix effect-free analytical methods for determination of perfluorinated carboxylic acids in environmental matrixes. *Anal. Chem.* **77**, 6353–6358.
77. van Leeuwen, S.P., and de Boer, J. (2007) Extraction and clean-up strategies for the analysis of poly- and perfluoroalkyl substances in environmental and human matrices. *J. Chromatogr. A* **1153**, 172–185.
78. Posyniak, A., Zmudzki, J., and Mitrowska, K. (2005) Dispersive solid-phase extraction for the determination of sulfonamides in chicken muscle by liquid chromatography. *J. Chromatogr. A* **1087**, 259–264.
79. Krska, R., Stubbings, G., Macarthur, R., and Crews, C. (2008) Simultaneous determination of six major ergot alkaloids and their epimers in cereals and foodstuffs by LC-MS-MS. *Anal. Bioanal. Chem.* **391**, 563–576
80. Stubbings, G., and Bigwood, T. (2009) The development and validation of a multiclass liquid chromatography tandem mass spectrometry (LC-MS/MS) procedure for the determination of veterinary drug residues in animal tissue using a QuEChERS (QUick, Easy, CHeap, Effective, Rugged and Safe) approach. *Anal. Chim. Acta* **637**, 68–78.
81. Costain, R.M., Fesser, A.C., McKenzie, D., Mizuno, M., and Macneil, J.D. (2008) Identification of hormone esters in injection site in muscle tissues by LC/MS/MS. *Food Addit. Contam.* **21**, 1–10.
82. Kruve, A., Künnapas, A., Herodes, K., and Leito, I. (2008) Matrix effects in pesticide multi-residue analysis by liquid chromatography-mass spectrometry. *J. Chromatogr. A* **1187**, 58–66.
83. Wang, J., and Leung, D. (2009) Determination of 142 pesticides in fruit- and vegetable-based infant foods by liquid chromatography/electrospray ionization-tandem mass spectrometry and estimation of measurement uncertainty. *J. AOAC Int.* **92**, 279–301.

84. Wang, J., and Leung, D. (2009) Applications of ultra-performance liquid chromatography electrospray ionization quadrupole time-of-flight mass spectrometry on analysis of 138 pesticides in fruit- and vegetable-based infant foods. *J. Agric. Food Chem.* **57**, 2162–2173.

85. Lesueur, C., Gartner, M., Mentler, A., and Fuerhacker, M. (2008) Comparison of four extraction methods for the analysis of 24 pesticides in soil samples with gas chromatography-mass spectrometry and liquid chromatography-ion trap-mass spectrometry. *Talanta* **75**, 284–293.

86. Walz, I., and Schwack, W. (2007) Multienzyme inhibition assay for residue analysis of insecticidal organophosphates and carbamates. *J. Agric. Food Chem.* **55**, 10563–10571.

87. Plössl, F., Giera, M., and Bracher, F. (2006) Multiresidue analytical method using dispersive solid-phase extraction and gas chromatography/ion trap mass spectrometry to determine pharmaceuticals in whole blood. *J. Chromatogr. A* **1135**, 19–26.

88. Giera, M., Plössl, F., and Bracher, F. (2007) Fast and easy in vitro screening assay for cholesterol biosynthesis inhibitors in the post-squalene pathway. *Steroids* **72**, 633–642.

89. Fillion, J., Sauvé, F., and Selwyn, J. (2000) Multiresidue method for the determination of residues of 251 pesticides in fruits and vegetables by gas chromatography/mass spectrometry and liquid chromatography with fluorescence detection. *J. AOAC Int.* 83, 698–713.

90. Ettre, L.S. (2003) M.S. Tswett and the invention of chromatography. LCGC North America **21**(5), 458–467.

91. Anastassiades, M., Maštovská, K., and Lehotay, S.J. (2003) Evaluation of analyte protectants to improve gas chromatographic analysis of pesticides. *J. Chromatogr. A* **1015**, 163–184.

92. Maštovská, K., and Lehotay, S.J. (2005) Combination of analyte protectants to overcome matrix effects in routine GC analysis of pesticide residues in food matrixes. *Anal. Chem.* 77, 8129–8137.

93. Erney, D.R., and Poole, C.F. (1993) A study of single compound additives to minimize the matrix induced chromatographic response enhancement observed in the gas chromatography of pesticide residues. *J. High Resolut. Chromatogr.* **16**, 501–503.

94. Erney, D.R., Gillespie, A.M., Gilvydis, D.M., and Poole, C.F. (1993) Explanation of the matrix-induced chromatographic enhancement of organophosphorus pesticides during open tubular column gas chromatography with splitless or hot on-column injection and flame photometric detection. *J. Chromatogr.* **638**, 57–63.

95. Fussell, R.J. Jackson-Addie, K., Reynolds, S.L., and Wilson, M.F., (2002) Assessment of the stability of pesticides during cryogenic sample processing. 1. Apples. *J. Agric. Food Chem.* **50**, 441–448.

96. Young, S.J., Parfitt, C.H., Jr., Newell, R.F., and Spittler, T.D. (1996) Homogeneity of fruits and vegetables comminuted in a vertical cutter mixer. *J. AOAC Int.* **79**, 976–980.

97. Maestroni, B., Ghods, A., El-Bidaoui, M., Rathor, N., Ton, T., Jarju, O.P., Phakaeiw, Y., and Ambrus, Á. (2000) Testing the efficiency and uncertainty of sample processing. In *Principles and Practices of Method Validation*, Fajgelj, A., and Ambrus, Á. (editors), Royal Society of Chemistry, Cambridge, UK, pp 49–88.

98. Maštovská, K., and Lehotay, S.J. (2004) Evaluation of common organic solvents for gas chromatographic analysis and stability of multiclass pesticide residues. *J. Chromatogr. A* **1040**, 259–272.

99. U.S. Department of Agriculture, Agricultural Research Service (2008) USDA National Nutrient Database for Standard Reference, Release 21. Nutrient Data Laboratory Home Page, http://www.ars.usda.gov/ba/bhnrc/ndl

100. Mol, H.G., Plaza-Bolaños, P., Zomer, P., de Rijk, T.C., Stolker, A.A., and Mulder, P.P. (2008) Toward a generic extraction method for simultaneous determination of pesticides, mycotoxins, plant toxins, and veterinary drugs in feed and food matrixes. *Anal. Chem.* **80**, 9450–9459.

Chapter 5

Automated Solid Phase Extraction

Norbert Helle, Meike Baden, and Kaj Petersen

Abstract

An overview is given in this chapter of the main potential benefits of using automated Solid phase Extraction (SPE) in the preparation of food samples for LC-MS analysis, both in terms of quality of results and in terms of performance and productivity. Automated SPE instrumentation is described and a range of application examples are given. The foods used in these applications range from non-fatty vegetables, to more complex vegetables, fish, prawn meat, and water, a vital raw product for the food and beverage industries. In most applications previously reported, the SPE technique was mainly used for sample cleanup prior to analysis. Additional examples are given here in which automated SPE is combined with analyte concentration, analyte derivatization, or addition of liquids such as internal standards to further improve limits of quantitation, selectivity, stability and quality of the analysis.

Key words: Solid phase extraction, SPE, SPE automation, Automated SPE, Food analysis, Sample clean-up, Derivatization, LC-MS, LC-MS/MS

1. Introduction

The development of LC-MS technology and instrumentation initially lead analytical chemists to believe that there would be less need for sample preparation or even for chromatographic separation due to the high selectivity of the LC-MS/MS technique. With increasing use of LC-MS for routine analysis of foods, however, it has become clear that especially the API-ES technique, despite its high selectivity and sensitivity, encounters problems in providing the required accuracy of quantitation of analytes in complex matrices. Techniques like spiking and matrix matched calibration only partially compensate for matrix effects.

Solid phase extraction (SPE) is a powerful and fast technique for separating analytes of interest from matrix and cleaning up a sample for analysis. In order to get results of good quality, however,

Jerry Zweigenbaum (ed.), *Mass Spectrometry in Food Safety: Methods and Protocols*,
Methods in Molecular Biology, vol. 747, DOI 10.1007/978-1-61779-136-9_5, © Springer Science+Business Media, LLC 2011

manual SPE methods largely depend on the experience of the user and on how meticulously each step is performed. Recovery and reproducibility can be subject to extreme deviations. If the SPE process and all associated liquid handling steps can be automated, the entire process becomes more reliable and efficient. Automating the SPE process eliminates extensive and tedious manual sample preparation steps that are known sources of errors, for example when cartridges run dry or when sample matrix restricts the flow of liquids through the cartridge. The benefits of automation are manifold: more uniformity in the handling of samples, higher recovery, improved reproducibility and repeatability, higher sample throughput and reduced exposure of laboratory staff to potentially hazardous solvents. Automated SPE can be performed using two different setups. The SPE system can be coupled directly to the analytical instrumentation, enabling direct sample introduction of the extracts and fully automated operation, for example, from SPE to LC-MS/MS. Alternatively, the SPE system can be operated as a sample preparation workstation separate from the analysis instruments. The WorkStation set-up provides the flexibility of choosing between different techniques or different instruments for the sample analysis in order to meet individual laboratory requirements. Different types of automated sample preparation systems are available in the market. Sequentially operating SPE systems are mainly designed for integrated operation with direct introduction of the extract to the analytical instrument. Systems that perform parallel processing of multiple samples are designed for high throughput off-line sample preparation. In this chapter, some examples of automated SPE combined with other sample preparation steps are presented for the analysis of foods, specifically the determination of some trace contaminants that have recently received increased attention. The examples given in this chapter further demonstrate that excellent analytical results can be achieved when SPE is automated and combined with further automated sample preparation steps such as liquid additions in the form of internal standards or derivatization reagents and with sample introduction to various LC-MS systems. Optimal conditions for the SPE process can be combined with optimal conditions for the HPLC separation and MS ionization by exchanging or adding solvents at various stages in the sample preparation. This provides significant improvements both in selectivity and in limits of determination. The results obtained using automated systems are typically better and more reliable than those obtained when performing the operations manually. This is especially the case for many routine laboratories, where sample preparation is performed by more than one person. Moreover, the risk of errors during transfer of samples from the sample preparation laboratory to the analysis instrument is eliminated when using integrated SPE systems. Additional sample preparation steps such as centrifugation and ultrasonication can be combined with the SPE technique. This enables the automation of even more complex sample preparation processes. The modular nature of automated sample preparation instrumentation means that systems can be tailored to individual laboratory requirements.

2. Materials

2.1. SPE Cartridges and Packing Materials

Leading producers of SPE cartridges have over the past decades developed cartridge packing materials for hundreds of food related applications covering a wide range of analytes that have mainly been determined by HPLC, by GC, or by GC-MS. Many of these are available both for manual SPE processes and for automated systems. LC-MS has only in recent years been used more widely for routine food analysis. SPE clean-up on commercially available silica based SPE cartridges has become the method of choice in food analysis laboratories. This is due to the speed, reliability, and ruggedness of the SPE process. Batch to batch reproducibility of SPE cartridges has improved vastly over past years and the technique has gained wider acceptance. In addition to numerous silica-based materials, other SPE materials have become available, such as polymer-based materials that offer improved ruggedness and in many cases do not require conditioning, saving both time and costly solvents. Further, analyte specific materials are available such as mixed-bed cartridges, for example specifically for the extraction of acrylamide (see Note 6), and immunoaffinity cartridges with immobilized antibodies in SPE cartridges that offer extremely selective cleanup and extremely low limits of determination. Cartridges for immunoassay work do place higher demands on the automation equipment used since they often need to be stored at sub-ambient temperatures until shortly prior to the start of the SPE process. The cartridges must be kept filled with aqueous phase and they lose their activity when exposed to high pressure, which means they can only be used in low-pressure SPE processes. Suitable automation equipment for immunoaffinity SPE is readily available.

Manual SPE is typically performed using standard 1-mL, 3-mL or 6-mL cartridges that contain from 0.1 to 1 g of solid packing material. A manifold under vacuum is used to hold the cartridges, vacuum is applied at the cartridge outlet and liquids are introduced manually into the cartridge from above. Systems for automated operation are available that use standard SPE cartridges directly or in slightly modified form. Yet others rely on specially developed cartridges (see Note 1). The automated SPE system used for the application examples shown in this chapter

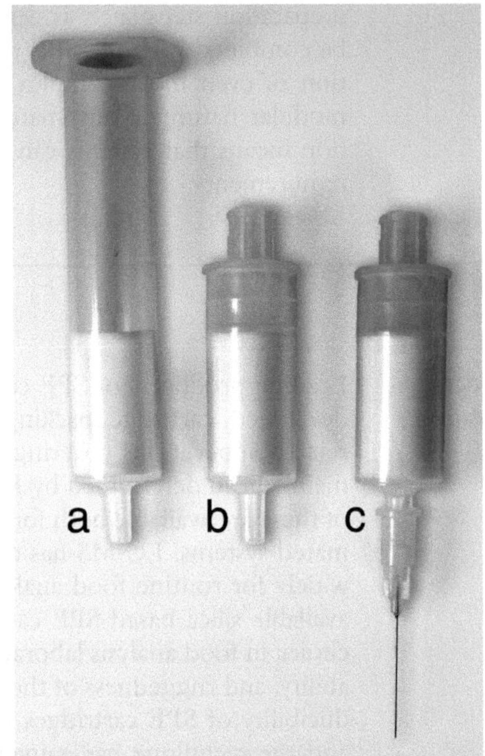

Fig. 1. Cartridges for manual SPE (**a**) and automated SPE (**b**) and (**c**). The cartridge (**c**) is fitted with a syringe needle for liquid transfer into a sealed vial.

operates with slightly modified standard cartridges. These are cut shorter at the top and equipped with a transport adapter that enables the autosampler to move the individual cartridge to an SPE station and also serves as a connector for the autosampler syringe for liquid transfer. Cartridges for manual SPE as well as for automated SPE are shown in Fig. 1. The height of the packing including frit typically ranges from 4.5 to 23 mm with a particle size range from 20 to 200 μm. Liquid transfer through SPE cartridges with relatively large beds of adsorbent material requires only limited pressure differential. An SPE process using such larger packed bed cartridges can be classified as low-pressure SPE. All applications shown in this chapter were performed using this technique with standard packing materials. Other automated SPE systems are available that are based on SPE cartridges or columns typically with smaller ID, particle size, volume and capacity. These cartridges are inserted into a flow system using an HPLC pump to deliver liquid at more elevated pressure. Systems that deliver the SPE eluate directly to the HPLC system are typically referred to as "on-line" SPE systems. Two basic types of high-pressure on-line SPE systems are used. (1) A fixed SPE column or scavenger column connected to the LC injection valve. The column is used to separate matrix from analytes and it is subsequently back-flushed and cleaned between injections. All samples are cleaned using the same fixed column. (2) Exchangeable SPE columns of e.g. 10 mm length with 2 mm ID, designed for on-line elution at pressures up to 300 bar directly to the HPLC column. One such system is produced by Spark Holland B.V., Emmen, The Netherlands. Cartridge dimensions are designed to provide sufficient extraction capacity combined with small elution volumes. The high-pressure on-line systems have limited capacity compared with standard SPE cartridges and are less useful for most food samples especially those with significant matrix load. Also, such systems cannot be used for immunoaffinity SPE since the packing material cannot withstand high pressures without losing its activity and it needs to be immersed in aqueous phase at all times. If relatively clean samples are analyzed, on-line systems can provide better sensitivity than standard SPE systems since the smaller volume of eluent used enables a higher concentration factor and quantitative introduction of the eluate to the LC-MS system. On-line systems, however, and especially those with a single fixed column for all samples, do introduce the risk of sample to sample carry over since all samples are transported through the same connecting tubing and, in the case of the fixed column version, through the same column during cleanup. Additionally, a fixed column system gives less flexibility to handle different samples in a single automated sequence and availability of packing materials is more limited compared with standard SPE cartridges.

2.2. The SPE Process

In order to complete the SPE process, various liquids must flow through the SPE cartridge. These include solvents for conditioning, sample or extract, wash solvent(s), and elution solvent(s). An overview of the SPE steps is shown in Fig. 2. Liquid flow can be achieved by various means: By applying a fixed pressure; by direct positive displacement as provided by a syringe or by a piston pump; or by applying vacuum at the outlet of the cartridge. Positive displacement, in our experience, is best suited to ensure constant, controlled and reproducible flow independent of cartridge to cartridge variations and independent of matrix-induced changes in flow restriction across the cartridge.

Following sample introduction and cleanup, analytes must be eluted from the SPE cartridge using sufficient solvent volume for adequate recovery. The concentration factor, assuming close to quantitative analyte recovery, is normally calculated as the sample volume to eluate volume ratio. If analytes must be determined at very low concentration levels, the eluate can be concentrated further in order to reach the required limits of quantitation. The most commonly used concentration technique is evaporative concentration, which is performed in the following way: The sample is

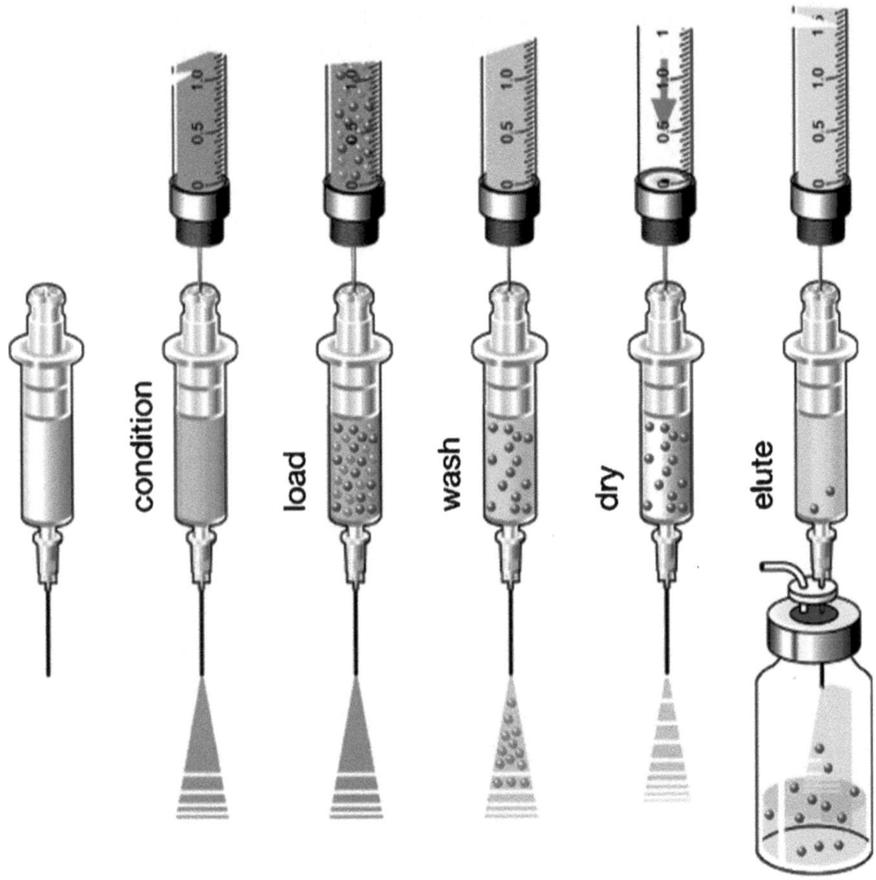

Fig. 2. Diagram showing the SPE cartridge used for automated operation (*left*) and individual steps in the SPE process.

heated to a preset temperature level, which depends on the volatility of the analytes and of the solvent that needs to be evaporated. A flow of inert gas is passed across the sample inside the vial, effectively purging the headspace and removing volatile solvent in the process. A graphical representation of the evaporation step and solvent change-over is shown in Fig. 3. One challenge when using this approach is that the eluate can normally not be allowed to be evaporated to dryness since this could lead to loss of analyte. A commonly used technique to avoid loss of analyte during the evaporation process is to add a solvent, which has a higher boiling point than the original solvent and therefore keeps the analytes in solution. Such a solvent is referred to as a keeper solvent. Nevertheless, an automated system should control critical parameters such as evaporation temperature, gas flow and time. The evaporation step is often combined with a solvent change-over in order to transfer analytes to an HPLC-compatible solvent. A solvent change-over opens up the possibility of achieving optimal conditions both for the SPE process and for the LC separation and

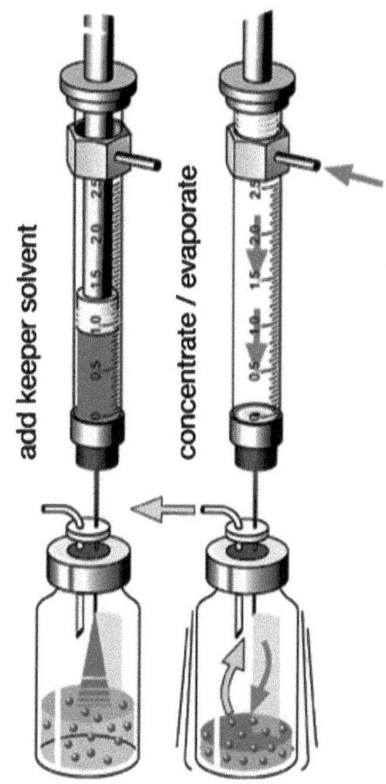

Fig. 3. Evaporation step and solvent change-over.

LC-MS ionization while also reaching lower detection limits through the concentration step.

Automating the SPE process eliminates extensive and tedious manual sample preparation steps that are known sources of errors, for example when cartridges run dry or when sample matrix restricts the flow of liquids through the cartridge. The benefits of automation are manifold: more uniformity in the handling of samples, higher recovery, improved reproducibility and robustness, higher sample throughput, and reduced exposure of laboratory staff to potentially hazardous solvents.

2.3. Automated SPE System Used for the Work Reported in This Chapter

The application examples listed under Subheading 3 were performed using GERSTEL SPE systems (GERSTEL GmbH & CO.KG, Mülheim an der Ruhr, Germany) based either on the GERSTEL MultiPurpose Sampler (MPS) or on the MPS PrepStation. Both are sample preparation and sample introduction systems equipped with automated SPE option based on commercially available modified standard cartridges. Conditioning, extraction and elution are performed in a sealed system. The MPS is configured with a single syringe; the system is shown in Fig. 4.

The GERSTEL PrepStation (Dual rail MPS) is configured with two different syringes, enabling more flexible handling of different volumes of liquid for different steps in the sample preparation process. As an example, a 1.0-mL or 2.5-mL syringe would be used for automated SPE and a 0.1-mL syringe for adding an

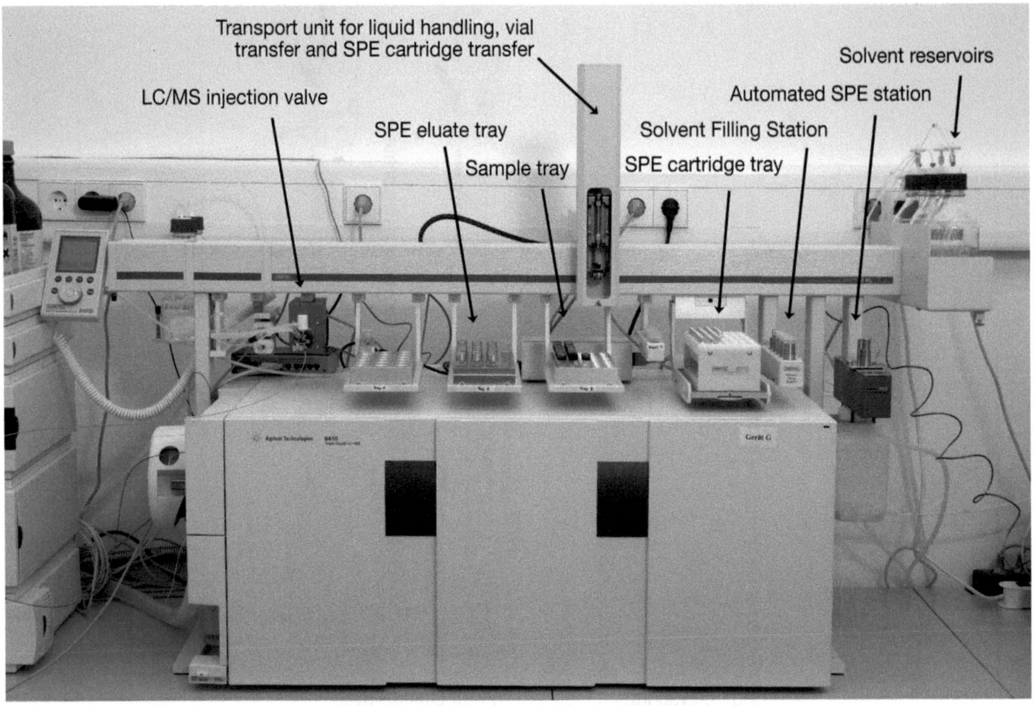

Fig. 4. GERSTEL MPS with SPE option mounted on an Agilent 6410 Triple Quad Mass Spectrometer, the system used for determination of pesticides as reported in Subheading 3.1.2.

internal standard or introducing the SPE eluate to the LC-MS/MS system. In Fig. 5, a dual rail MPS PrepStation is shown mounted on an Agilent 6460 Triple Quad MS. This is the system used for determination of PFCs reported in Subheading 3.2.1.

If fixed volume sample loop introduction to the LC-MS system is performed, and if sufficient eluate is available, it is possible to use one large syringe both for the SPE process and for sample introduction. In addition to automated SPE, the MPS can automatically perform standard sample preparation techniques that involve liquid handling and/or introduction of the SPE eluate to the LC-MS system. Method development can be automated and performed in a flexible manner and daily routine analysis tasks conveniently performed. The ability to select the optimal SPE flow within a range from 10 to 250 μL/s enables the user to set the appropriate flow for any given application. SPE is a chromatographic separation technique and the van Deemter equation therefore applies, making it important to be able to select the optimal flow for the process.

Automated SPE System details:

Capacity: 72 × 6-mL cartridges; 98 × 3-mL cartridges or 98 × 1-mL cartridges.

Sample vials: 2 mL, 4 mL or 10 mL.

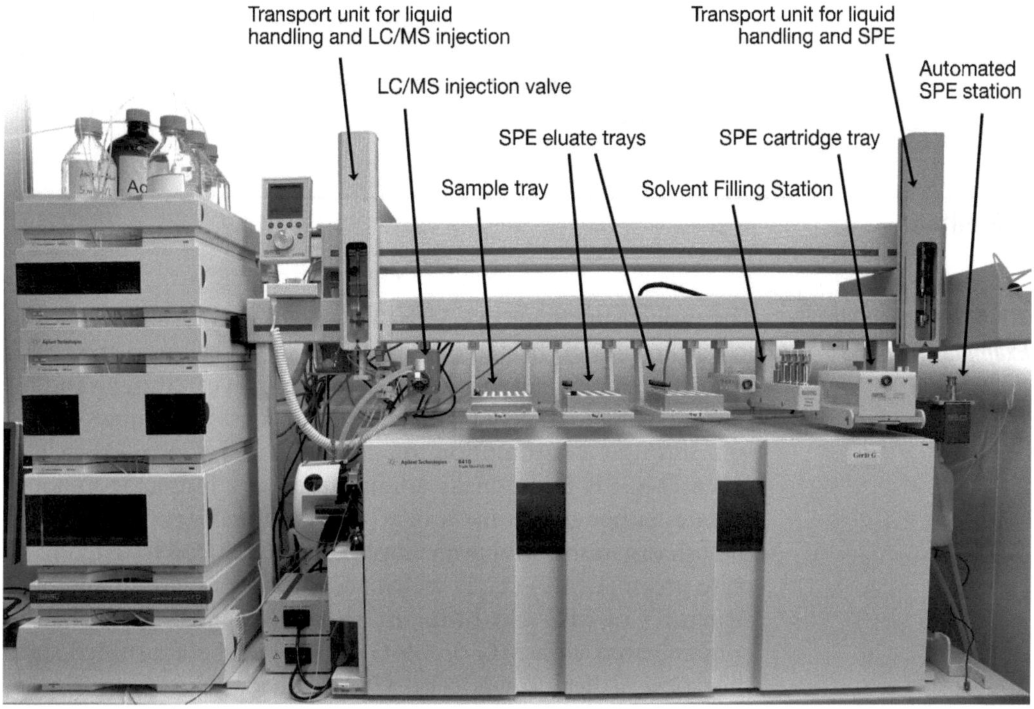

Fig. 5. Dual Rail MPS PrepStation with SPE option was used for determination of PFCs as reported in Subheading 3.2.1. The system is shown mounted on an Agilent 6460 Triple Quad Mass Spectrometer connected to an Agilent Series 1200 LC System.

Sample volume: max. 8 mL per vial (a sample can be divided between several vials).

Eluate collection vials: 2 mL, 4 mL or 10 mL.

Syringe for solvent transport: 1 mL or 2.5 mL total volume, with gas connection.

(*The syringe volume does not limit the injection volume, since multiple injections are possible.*)

Syringe for liquid additions and sample introduction: 0.01–0.25 mL.

Liquid flow/loading speed: 10–250 μL/s.

Evaporation temperature: Ambient to 120°C.

Agitation speed: 250–750 rpm.

Solvent filling station (SFS): up to two SFS units per MPS. Each SFS enables the syringe to aspirate solvent from up to four separate 1 L bottles.

The SPE systems are controlled through the GERSTEL MAESTRO software or integrated with the Agilent ChemStation or MassHunter software. One integrated sequence table operates the entire system from SPE through liquid sample prep and sample introduction to LC-MS analysis. A PrepAhead function enables the system to plan ahead delivering time-optimized sample preparation. Samples are prepared just-in-time for introduction exactly when the LC-MS system becomes ready after the previous run, helping to ensure that the LC-MS is utilized to its fullest capacity.

3. Methods

3.1. SPE Application Examples Relating to Clean-Up

3.1.1. Malachite Green in Fish

The triphenylmethane dye Malachite green (MG) is highly efficient in battling fungi, bacteria, and various single-cell parasites. MG, however, is under suspicion for being a human carcinogen and for causing damage to genetic material if it reaches the human organism through consumption of contaminated foods. Malachite green (MG) is traditionally administered as a fungicide in aquaculture, either as treatment or to prevent infections. Once inside the fish organism, MG is metabolized and reduced to leucomalachite green (LMG) (1, 2) which accumulates in fatty tissue. The structures of MG and LMG are shown in Fig. 6. Fish that are contaminated with MG or LMG should not be consumed since they pose a health risk. The USA, Canada, European Union, Japan, and Chile are among the countries that prohibit the use of malachite green in food production. The US FDA tests for malachite green at a level of 1 part per billion. In 2003, the EU Commission set a threshold value of 2 μg/kg as the upper concentration limit for the sum of MG and LMG (3).

3.1.1.1. Sample Preparation

A fish filet sample was homogenized using an immersion blender. A 5-g sample of the homogenate was then extracted with a water/acetonitrile mixture using a T 25 digital ULTRA-TURRAX® (IKA® Werke GmbH & Co. KG, Staufen, Germany). The extract was centrifuged for 10 min at 2,000 × *g* and the supernatant was collected. The extraction procedure was repeated twice, the extracts were combined and concentrated to dryness under a flow of nitrogen at 45°C before being taken up in 5 mL of a mixture of water and ethanol. Sample cleanup was performed using a GERSTEL MultiPurpose

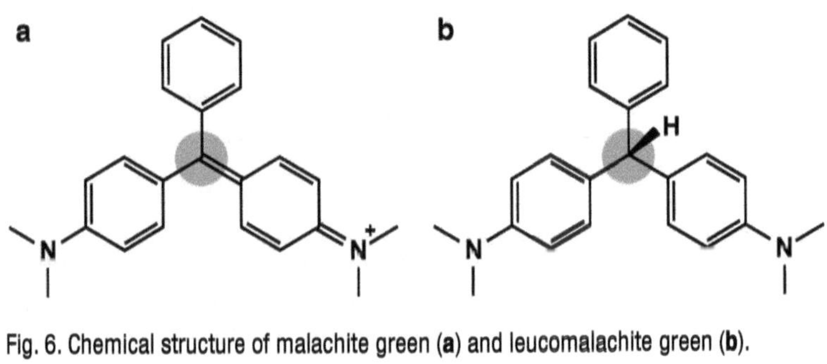

Fig. 6. Chemical structure of malachite green (**a**) and leucomalachite green (**b**).

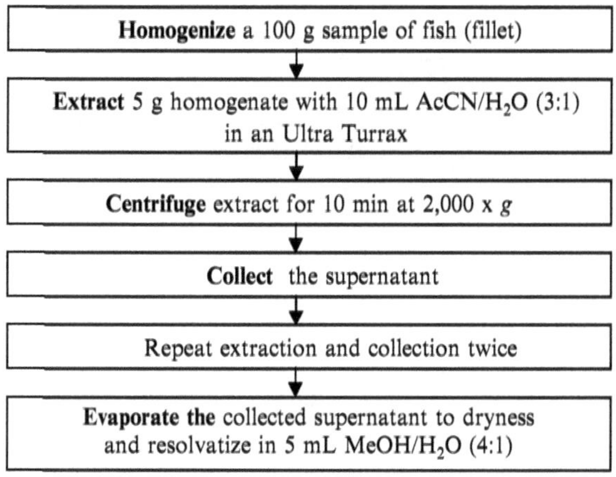

Fig. 7. Sample preparation flow chart for malachite green in fish.

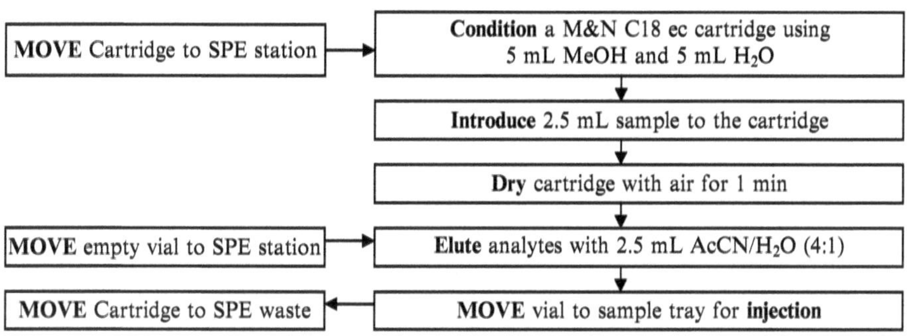

Fig. 8. Flow chart of SPE cleanup steps for malachite green in fish.

Sampler (MPS) equipped with automated SPE (GERSTEL GmbH & Co.KG, Mülheim an der Ruhr, Germany). The sample preparation procedure is delineated with a flow chart shown in Fig. 7. The SPE cleanup steps are shown in Fig. 8.

3.1.1.2. LC-MS Method

The analysis was performed on an Agilent 1100 LC-MS Ion Trap system, consisting of a binary pump, a thermostated column compartment, a GERSTEL MultiPurpose Sampler (MPS), and an Agilent XCT + Ion trap-MS (IT-MS) (Agilent Technologies, Inc., Santa Clara, CA, USA). The MPS was operated integrated with the LC-MS. For the analysis, a Zorbax SB-C18 column (50 × 2.1 mm, 1.8 μm) was used. The LC and MS operational parameters are given in Fig. 9.

3.1.1.3. Results and Discussion

Malachite green (MG) and its metabolite Leucomalachite green (LMG) are easily ionized using electrospray ionization (ESI) in positive ion mode. MG differs from LMG in that LGM forms a doubly charged ion (m/z 166) in addition to the single charged

LC method parameters Malachite Green

injection volume	5 μL	
flow rate	0.6 mL/min	
eluents	0.1 % formic acid/acetonitrile	
gradient	Time (min)	acetonitril (%)
	0	20
	12	80
	16	80
	17	100
	21	100
	23	20
HPLC column	Zorbax SB-C18 column (50 x 2.1 mm, 1.8 μm)	
Column temperature	50°C	

MS method parameters Malachite Green

Ionization mode	AP-ESI, positive
Mass transitions	malachite green: 313.1 -> 208.0 m/z leuco malachite green: 166.0 -> 158.5 m/z
nebulizer	30 psi
N_2 flow/-temperature	10 L/min/340 °C

Fig. 9. LC and MS method parameters for malachite green.

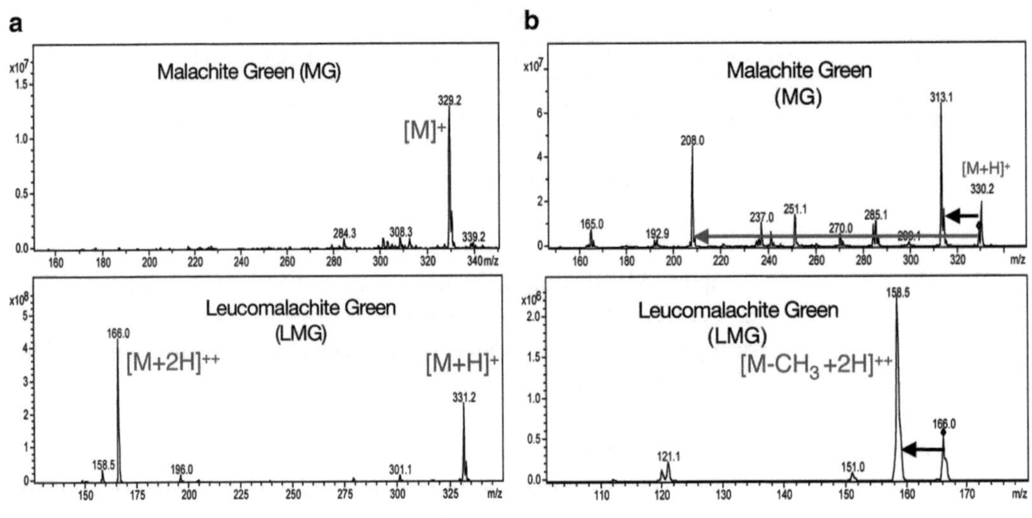

Fig. 10. MS spectra of (**a**) malachite green (MG) and Leucomalachite Green (LMG), and (**b**) MS^2 spectra of the two compounds, MG and LMG.

molecular ion $[M+H]^+$. Both their single MS and MS/MS (MS^2) spectra are shown in Fig. 10.

In MS^2 mode in the ion trap mass spectrometer, the MG precursor ion forms a product ion (*m/z* 313) and the doubly charged LMG precursor ion forms a doubly charged product ion (*m/z* 158.5) as shown in Fig. 10. The LMG transition can be used for highly sensitive determination where limits of determination of

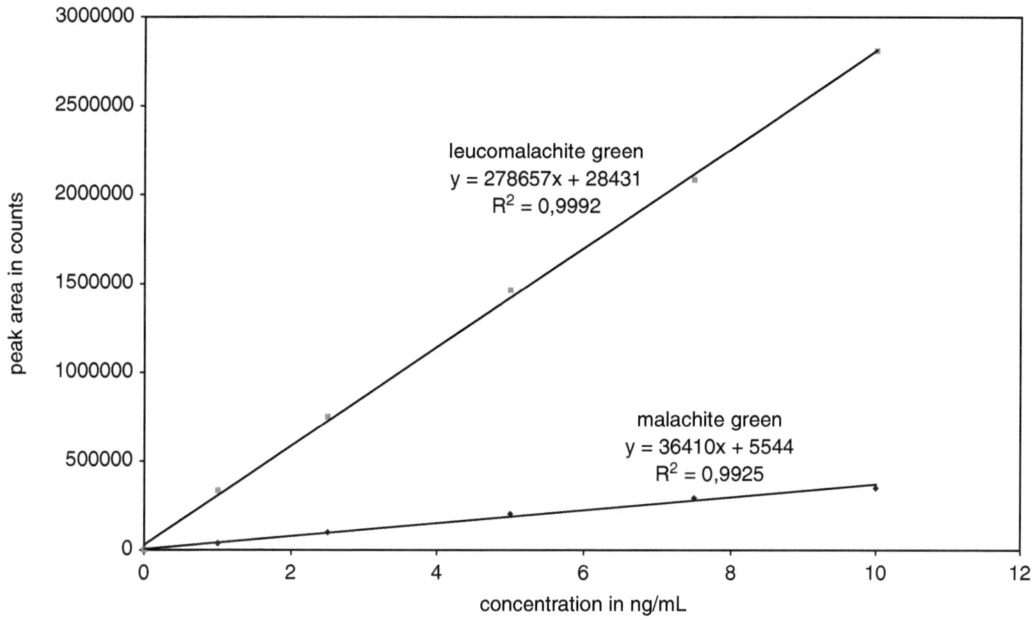

Fig. 11. Calibration curves for malachite green (*top*) and Leucomalachite Green (*bottom*) in spiked pangasius fillet.

0.5 μg/kg for MG and 0.05 μg/kg for LMG were achieved. Calibration curves for the two compounds are shown in Fig. 11.

The sample cleanup steps ensure the removal of interfering matrix residue leading to significantly better signal to noise ratios and improved detection limits for MG and LMG. RSDs range from 1.4 to 7.8% while recoveries are in the range from 90.5 to 103.2% ($n=7$) as shown in Fig. 12. Additionally, in this case, automated SPE reduces the time required for sample preparation by 50% compared with the manual procedure.

3.1.2. Pesticides in Fruit and Vegetables

Pesticides help provide an adequate and affordable supply of food to the ever-growing human population across the world, but residues in food should be kept to levels that do not affect the health of the consumer. World-wide, around 700 pesticides are in use. Fast, all encompassing analysis methods are difficult to develop since even established compound classes often cover a wide range of polarities. When fruit and vegetables are analyzed for pesticide residues, significant sample preparation, including a gel chromatography cleanup step to separate analytes from matrix, is mostly required. LC-MS is then used to determine polar to moderately apolar compounds while GC-MS is used for apolar to moderately polar compounds with some overlap between the techniques. For fruit and vegetables with low fat content, the QuEChERS extraction method (Quick, Easy, Cheap, Effective, Rugged & Safe) (4) provides a number of benefits compared with previously used methods. The sample preparation steps are much

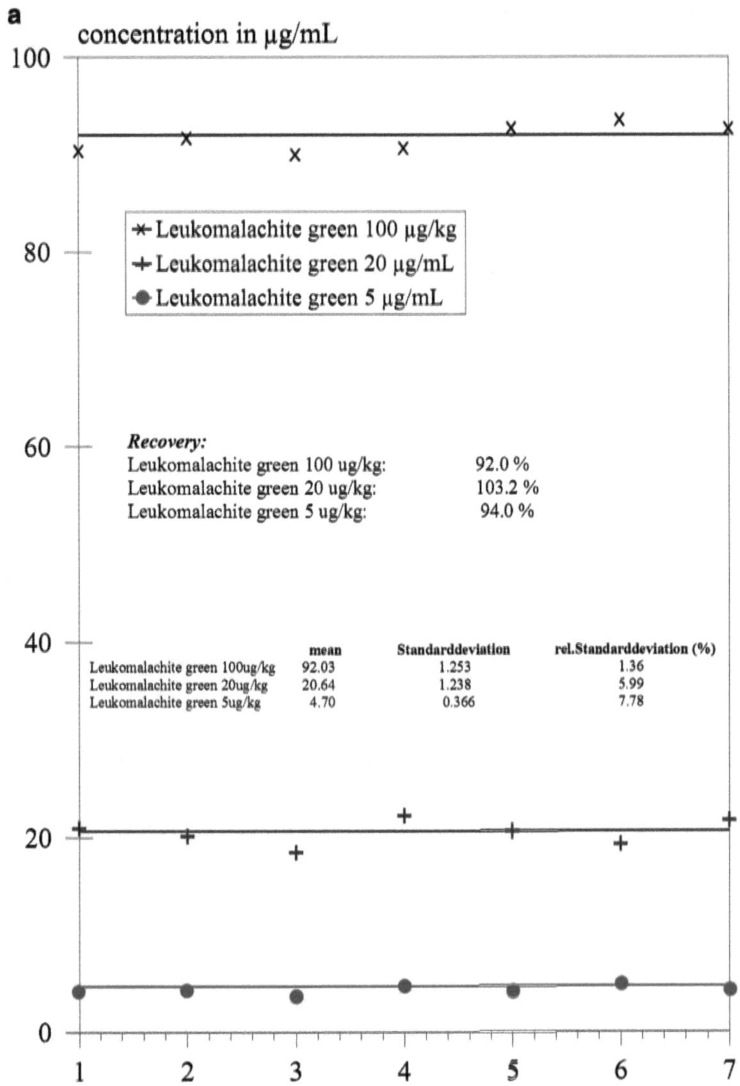

Fig. 12. Recovery of (**a**) leucomalachite green and (**b**) recovery of malachite green.

less time-consuming, a wider analyte range is covered, it is readily automated, and extracts are well suited for both GC, GC-MS, and LC-MS analysis. In addition, much smaller volumes of partly toxic organic solvents are required. The cost of materials at approximately one Euro or one US Dollar per sample is relatively low. The limits of QuEChERS and similar extraction methods are encountered whenever samples with more complex matrices need to be analyzed, such as garlic, onion, artichoke, avocado, or orange oil with much higher fat content. This can lead to problems with interferences that can especially influence quantification unless further cleanup steps such as automated SPE are performed on the extract.

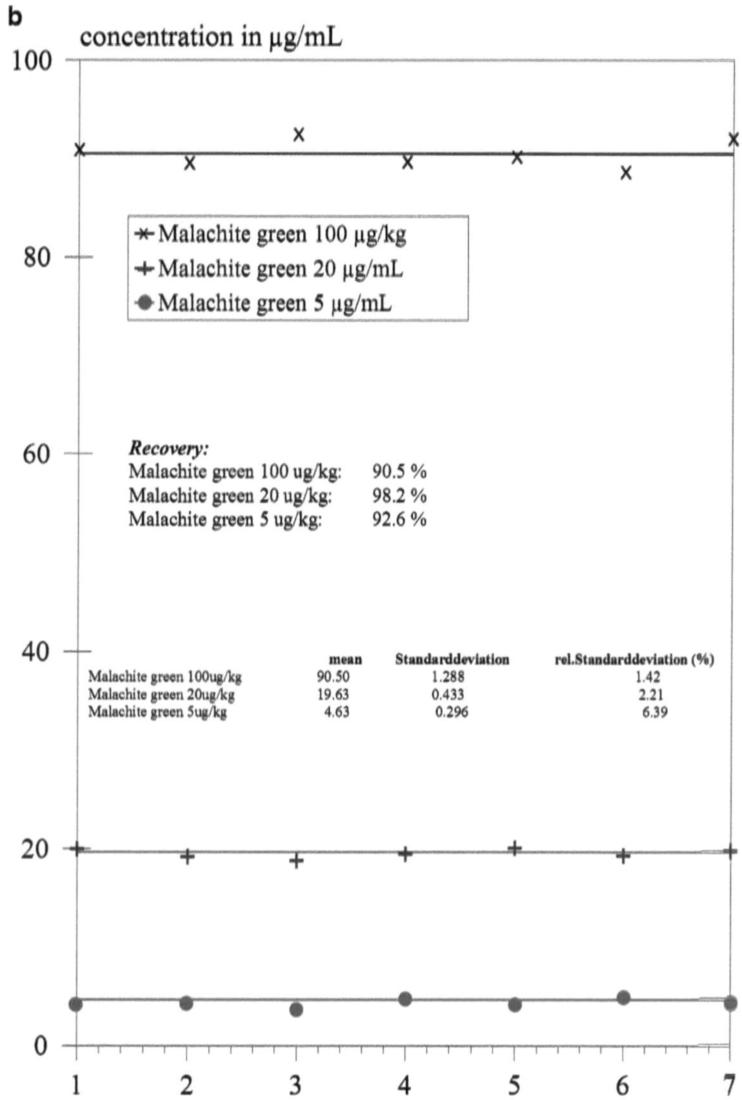

Fig. 12. (continued)

3.1.2.1. Sample Preparation

The fruit or vegetable sample was homogenized using an immersion blender. A 5-g aliquot of the homogenate was weighed into a 50-mL glass tube and extracted with 15 mL of an acetonitrile/water mixture (80:20) using an ultra turrax. This mixture was used instead of pure acetonitrile in order to improve the LC separation. The extract was mixed on a horizontal shaker for 1 h and then centrifuged for 10 min at 2,000 × *g*. The supernatant was collected in a 20-mL vial, which was placed in the sample tray of the GERSTEL MPS for SPE cleanup. The steps for cleanup are shown in Fig. 13.

SPE cleanup was performed using separate cartridges (Macherey & Nagel C18ec, 6 mL, 1 g) for each sample to eliminate

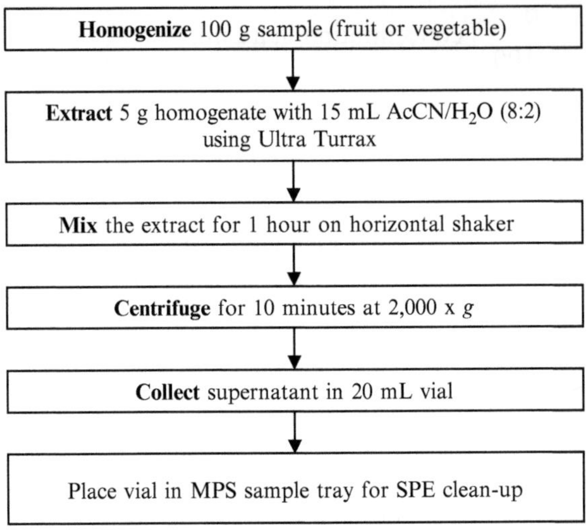

Fig. 13. Flow chart of sample preparation steps, pesticides in fruit and vegetables.

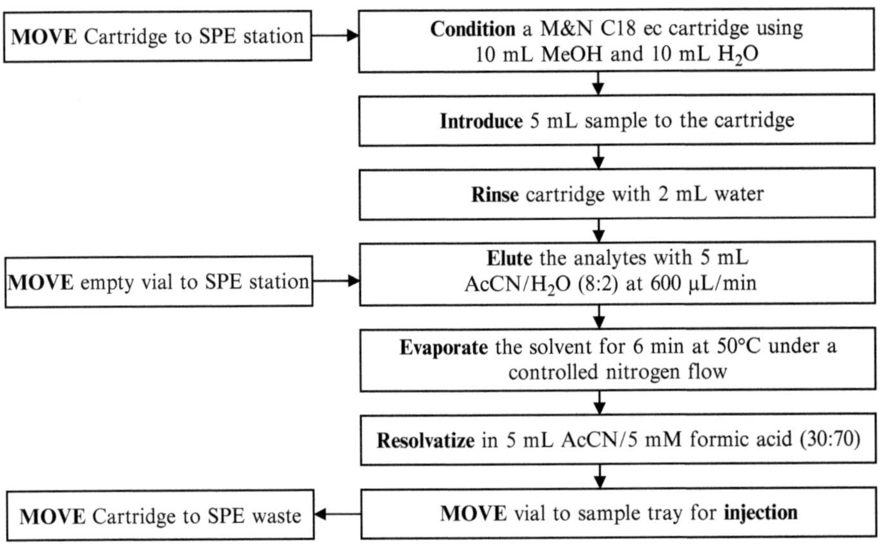

Fig. 14. Flow chart of SPE sample preparation steps, pesticides in fruit and vegetables.

cross contamination. All steps in the sample preparation procedure, including evaporation and reconstitution inside the autosampler vial were fully automated and outlined in Fig. 14

3.1.2.2. LC and MS Method

The analysis was performed on an Agilent 1200SL LC-MS QQQ system, consisting of a binary pump, a thermostated column compartment, a GERSTEL MultiPurpose Sampler (MPS) with SPE option and an Agilent 6410 Triple Quadrupole Mass Spectrometer with an electrospray ion source. The MPS was integrated into the LC-MS system and equipped with an injection valve;

LC method parameters Pesticides

injection volume	20μL	
flow rate	0.5 mL/min	
eluents	5 mmol/L formic acid/acetonitrile	
gradient	Time (min)	acetonitril (%)
	0	15
	30	95
	32	95
	35	15
HPLC column	Zorbax XDB-C18 column (100 x 2.1 mm, 1.8 μm)	
Column temperature	50°C	

MS method parameters Pesticides

Ionization mode	AP-ESI, positive
nebulizer	30 psi
N_2 flow/temperature	9 L/min/340 °C

Fig. 15. LC and MS method parameters for pesticides in fruit and vegetables.

sample introduction to the Agilent LC 1200 was performed directly by the SPE system. For the analysis, a Zorbax XDB C-18 column (100×2.1 mm, 1.8 μm) was used. The LC and MS operating conditions are given in Fig. 15.

Ionization parameters were optimized for the flow and eluant used. The triple quadrupole instrument was operated in Multiple Reaction Monitoring (MRM) mode. A total of 187 analytes were monitored in five different time segments over the LC run. For each pesticide, two transitions were monitored.

3.1.2.3. Results and Discussion

When using the QuEChERS method, additional cleanup steps, such as automated SPE, can be applied to the extracts as needed. For less complex matrices, such as lettuce or cucumber, it was shown that additional cleanup steps were not required following acetonitrile/water extraction. Complex matrices that contain fat and other challenging matrix components did require further clean-up steps. The C18 reversed phase cartridges used here produced excellent, reliable results. Automated SPE cleanup is completed in approximately 20 min during LC-MS analysis of the preceding sample, which means that SPE is performed without increasing the overall analysis time. In order to achieve good separation combined with method ruggedness, the conscious decision was made to seek only a moderate reduction of the LC-MS analysis time. The total analysis time required to determine around 187 compounds was in the order of 35 min, leaving sufficient time to prepare the following sample for introduction to the LC-MS system when it became ready after a run. Sample cleanup using SPE contributes to the ruggedness of the method while also improving reproducibility and linearity (see Note 5).

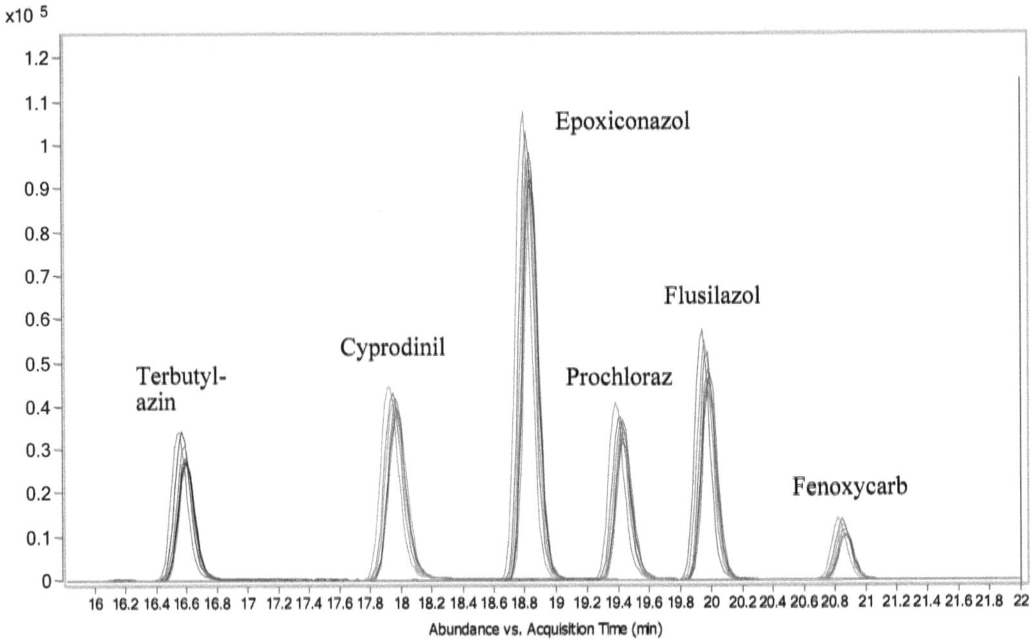

Fig. 16. Overlay medium polarity section of eight different chromatograms from eight separate sample preparations of a bell pepper.

To illustrate this, eight samples of a bell pepper were spiked with a 100 ng/mL pesticide mixture, prepared and analyzed. Overlay medium polarity sections of the eight chromatograms are shown in Fig. 16. Following SPE cleanup, retention times and peak areas of the analytes showed excellent reproducibility. Matrix matched calibration curves are shown in Fig. 17.

3.2. SPE Application Examples with Additional Analyte Concentration

3.2.1. Perfluorinated Compounds in Water and Beverages

Perfluorinated acids, most often referred to as Perfluorinated Compounds (PFCs), are produced in significant amounts and are ubiquitous in the environment. PFCs can be divided into two groups: Perfluorinated alkylsulfonates (PFAS), among which perfluorooctanesulfonate (PFOS) is the most widely known compound, and perfluorinated carboxylic acids (PFCA), whose most famous representative is perfluorooctanoic acid (PFOA). The chemical structures of PFOA and PFOS are shown in Fig. 18. Regarding long-term effects, there is no consensus even though PFCs have been reported as having cancer promoting properties and though the USEPA considers PFOA a "probable human carcinogen". In fact, mainly the effects of PFOA and PFOS have been investigated while those of other PFCs have been less extensively studied. In 2009, the US Environmental Protection Agency (EPA) issued Provisional Health Advisory (PHA) values of 0.4 μg/L for PFOA and 0.2 μg/L for PFOS. The German Commission for Drinking Water lists an upper concentration limit of 1 μg/L. The automated method presented in this work is based on the current ISO standard method (**5**).

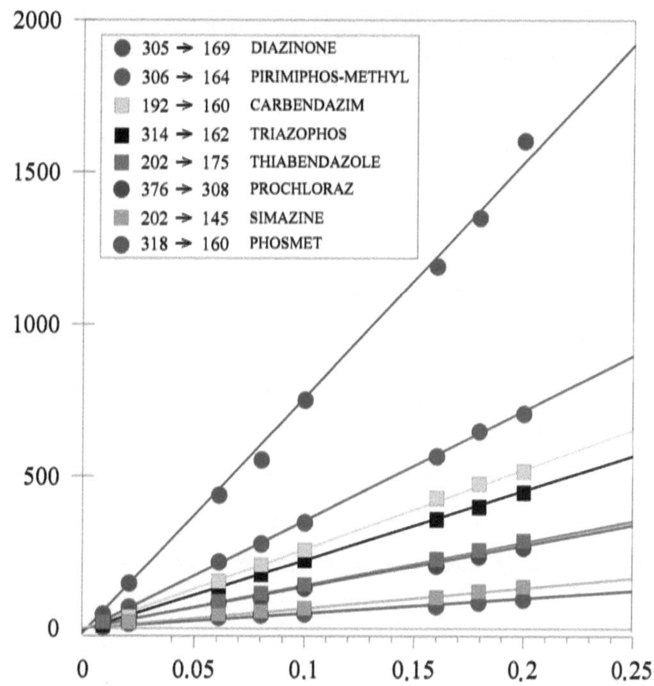

Fig. 17. Matrix matched calibration curves for eight pesticides covering nine concentration levels from 0.5 to 200 ng/ml.

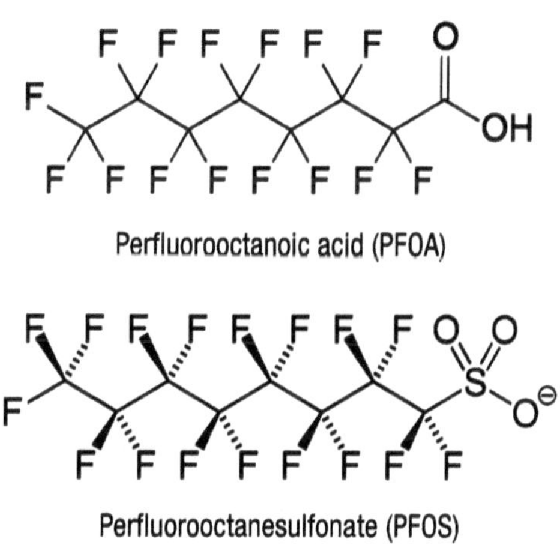

Fig. 18. Chemical structure of PFOA and PFOS.

While the ISO method focuses on PFOS and PFOA, the goal for the project reported here was to develop a simple and rugged automated SPE-HPLC-MS/MS method for the determination of a wider list of PFCs in water and beverages:

- Perfluorodecanoic acid (PFDA)
- Perfluorononanoic acid (PFNA)

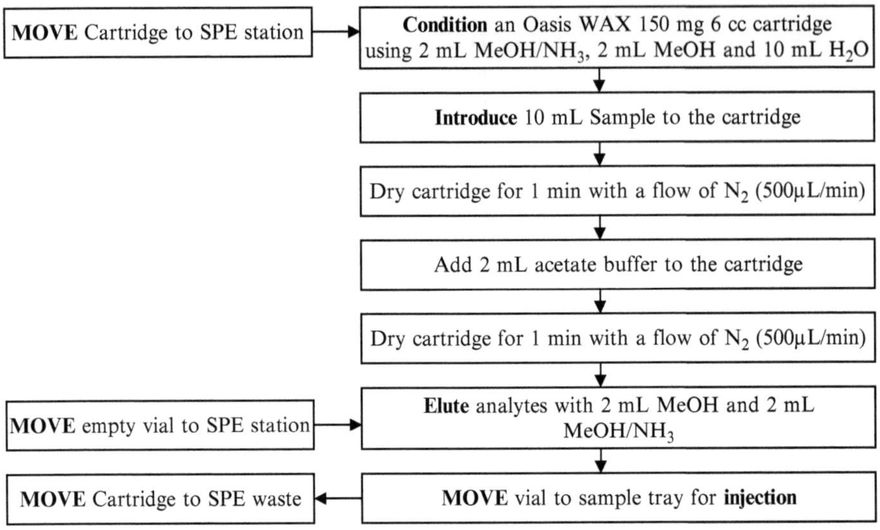

Fig. 19. Flow chart of SPE sample preparation steps, PFCs in water and beverages.

- Perfluorooctanoic acid (PFOA)
- Perfluorooctanesulfonate (PFOS)
- Perfluoroheptanoic acid (PFHpA)
- Perfluorohexanoic acid (PFHxA)
- Perfluoropentanoic acid (PFPeA)

Isotopically labeled standards are used for quantitation in the ISO 25101 standard. In this work, perfluorobutanoic acid was used since this compound was not found in the samples.

3.2.1.1. Sample Preparation

The sample (beverage or water) was filtered, transferred into a 20-mL vial and placed in the GERSTEL SPE system. The steps shown in the flow chart (Fig. 19) are completed in approximately 25 min; up to 32 samples can be processed in a single batch using a separate cartridge for each sample.

3.2.1.2. Instrumentation and Methods

The analysis was performed on an Agilent 1200SL LC-MS QQQ system, consisting of a binary pump, a thermostated column compartment, a GERSTEL Dual Rail MPS PrepStation with SPE option, and an Agilent 6410 Triple Quadrupole Mass Spectrometer with electrospray ion source. The MPS was integrated into the LC–MS system and equipped with an injection valve; sample introduction to the Agilent LC 1200 was performed directly by the SPE system. For the analysis, a Maisch Reprosil C18HD (50×2.1 mm, 3 µm) column was used. The LC-MS/MS operating parameters are given in Fig. 20.

LC method parameters PFCs

injection volume	2µL	
flow rate	0.3 mL/min	
eluents	ammonium acetate/Methanol	
gradient	Time (min)	MeOH (%)
	0	20
	10	100
	14	100
	15	20
HPLC column	Maisch Reprosil C18HD (50 x 2.1 mm, 3µm)	
Column temperature	50 °C	

MS method parameters PFCs

Ionization mode	AP-ESI, negative	
nebulizer	30 psi	
N_2 flow/temperature	10 L/min/350 °C	
Mass transitions	PFDA	513->469 m/z
	PFNA	463->419 m/z
	PFOA	413->369 m/z
	PFOS	499->99 m/z
	PFHpA	363->319 m/z
	PFHxA	313->269 m/z
	PFPeA	263->219 m/z

Fig. 20. LC and MS/MS method parameters for PFCs in water and beverages.

3.2.1.3. Results and Discussion

Aqueous samples with and without solid matrix residue that had been spiked with PFCs at different concentration levels were analyzed. For each sample, a separate cartridge was used eliminating the risk of cross contamination. The PrepAhead function was activated enabling SPE cleanup of a sample while LC-MS analysis of the preceding sample was in progress. For the PFC method, sample preparation took 25 min to complete. Calibration based on samples spiked with 5–500 ng/mL of PFCs resulted in good linearity as can be seen in Fig. 21.

An overlay of chromatograms from eight different sample preparations is shown in Fig. 22. The automated SPE-LC-MS/MS method showed excellent reproducibility performance.

Relative standard deviations were between 1 and 3.4% depending on the compound as shown in Fig. 23.

Following 2.5-fold enrichment of the SPE eluate, limits of quantitation were 0.5 ng/mL for the PFCs determined (see Note 4). The automated method produces results of significantly higher quality than the ISO 25101 method in terms of both sensitivity and reproducibility, especially when analyzing more complex samples such as beverages.

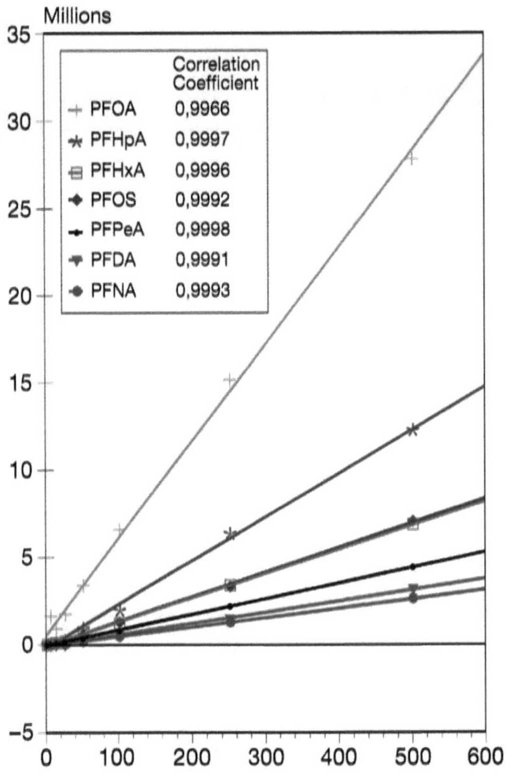

Fig. 21. PFC calibration curves covering the range from 5 to 500 ng/mL based on external standard calibration.

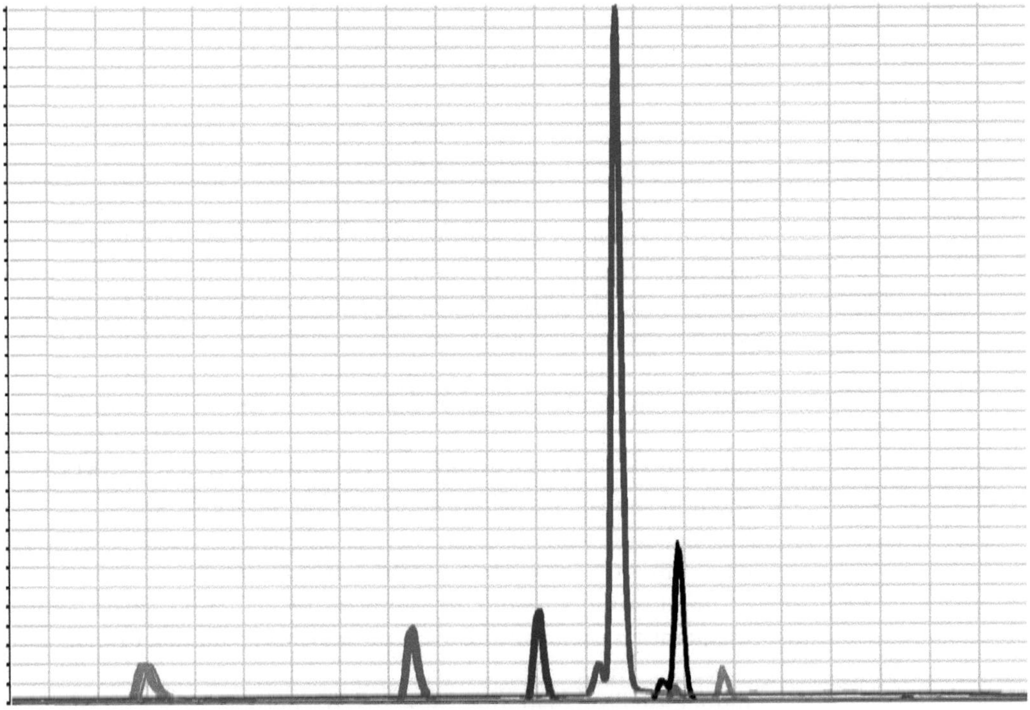

Fig. 22. Overlay MRM traces of eight different sample preparations of a water sample with matrix residue spiked with 0.5 ng/mL. Excellent reproducibility is obtained for all compounds from left to right: PFPeA, PFHxA, PFHpA, PFOA, PFNA (minor peak underneath PFOS), and PFOS, PFDA.

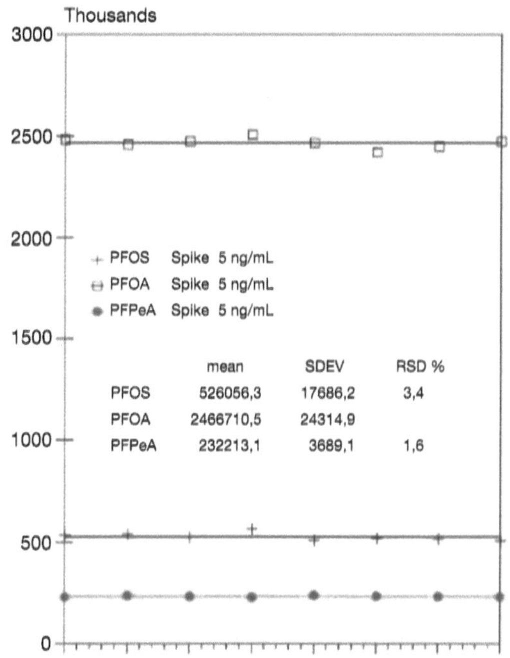

Fig. 23. Reproducibility of the method for PFOS, PFOA, and PFPeA.

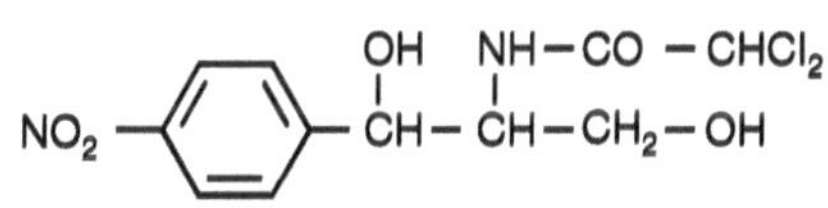

Fig. 24. Structure of chloramphenicol.

3.2.2. Chloramphenicol in Shrimp

Chloramphenicol (CAP) shown in Fig. 24 is a broad-spectrum antibiotic banned for use in food products of animal origin in the US, Europe and in many countries world-wide. CAP is a known human carcinogen suspected of causing genetic damage in human cells as well as irreversible damage to the blood-forming cells of the bone marrow. Previously, Enzyme Immunochemical Tests were widely used for the determination of CAP (6), but today this is usually performed by LC-MS. The sensitivity of the method depends greatly on the sample preparation used. A high matrix load can result in incorrect quantification of CAP, even when highly selective LC-MS/MS methods are used. SPE is the sample preparation technique of choice for such samples. In this study, the focus was on the determination of CAP in shrimp, but the method is easily adapted to other food products of animal origin.

3.2.2.1. Sample Preparation

A 100 g sample of untreated prawn tissue was homogenized using a mixer. A 5 g aliquot of the homogenate was weighed into a 25-mL glass tube and 10 mL ethylacetate as well as chloramphenicol-d5 (internal standard) were added. The sample was extracted using an Ultra Turrax and subsequently centrifuged for 10 min at 2,000 × *g*. The liquid phase was recollected and concentrated in a rotary evaporator. The residue was diluted with 5 mL methanol/water (1:10) prior to the SPE process, for which a 3 mL M&N C18ec (500 mg) SPE cartridge was used (Macherey-Nagel, Düren, Germany). All further steps are listed in Figs. 25 and 26.

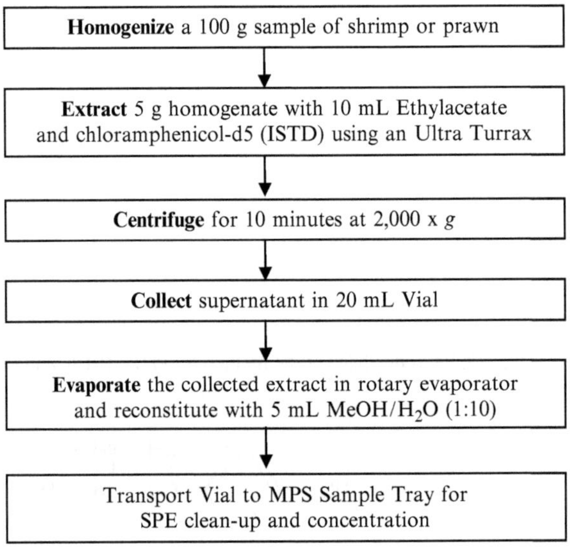

Fig. 25. Sample preparation flow chart for chloramphenicol in shrimp meat.

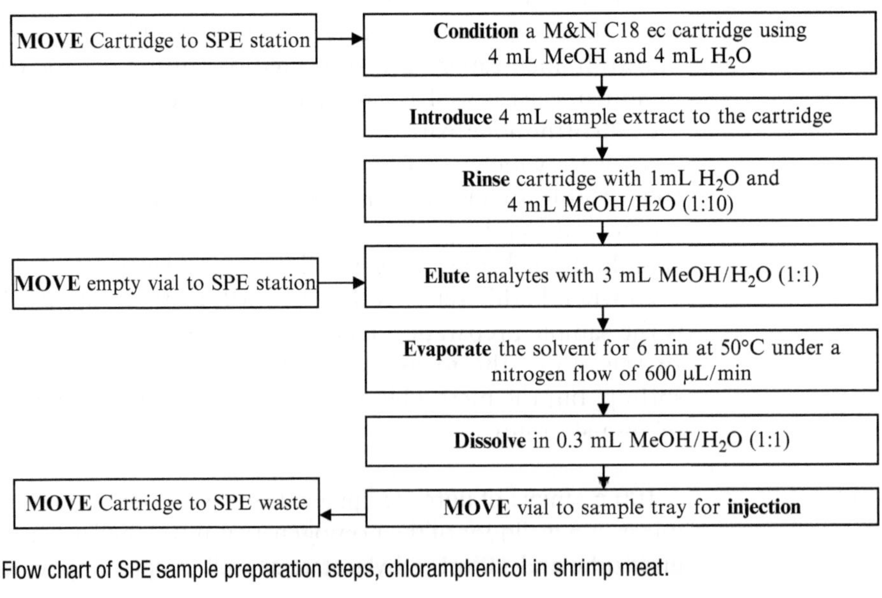

Fig. 26. Flow chart of SPE sample preparation steps, chloramphenicol in shrimp meat.

Chloramphenicol LC Method Parameters

injection volume	8 µL	
flow rate	0.28 mL/min	
eluents	Ammoniumformiate 0.005 M, pH 8.5/MeOH	
gradient	Time (min)	MeOH (%)
	0	20
	5	40
	12	90
	18	90
	20	20
HPLC column	Phenomenex maxRP (250 x 2.1 mm)	
Column temperature	30 °C	

Chloramphenicol MS Method Parameters

Ionization mode	AP-ESI, negative	
nebulizer	25 psi	
N_2 flow/temperature	8 L/min/350°C	
Mass transitions	CAP	EIC 257 + MS2 (321.0)
	CAP-d5	EIC 262 + MS2 (326.0)

Fig. 27. LC and MS/MS method parameters for chloramphenicol in shrimp (see Note 3).

3.2.2.2. Instrumental Method

The analysis was performed on an Agilent 1100 LC-MS system, consisting of binary pump, thermostated column compartment, vacuum degasser, GERSTEL MultiPurpose Sampler (MPS) with SPE option and Agilent Ion Trap XCT + Mass Spectrometer with electrospray ion source (see Note 2). The MPS was integrated into the LC-MS system and equipped with an injection valve; sample introduction to the Agilent LC 1100 was performed directly by the SPE system. For the analysis a Phenomenex maxRP column (250 × 2.1 mm) was used. The LC-MS operational parameters are given in Fig. 27.

3.2.2.3. Results and Discussion

Automated SPE sample preparation combined with LC-MS analysis as described in this method easily accomplished reaching the MRPL (Minimum Required Performance Level) for chloramphenicol of 0.3 µg/kg mandated by the EU. Evaporative concentration of the eluate by a factor of 10 after the SPE process enabled the system to reach a LOD of 0.01 µg/kg for CAP as shown in Fig. 28. The injected amount at this concentration was equivalent to 1 pg of CAP. Despite concentrating the eluate by a factor of 10 no significant interference from the accompanying matrix was observed in the quantification as can be seen in Fig. 29.

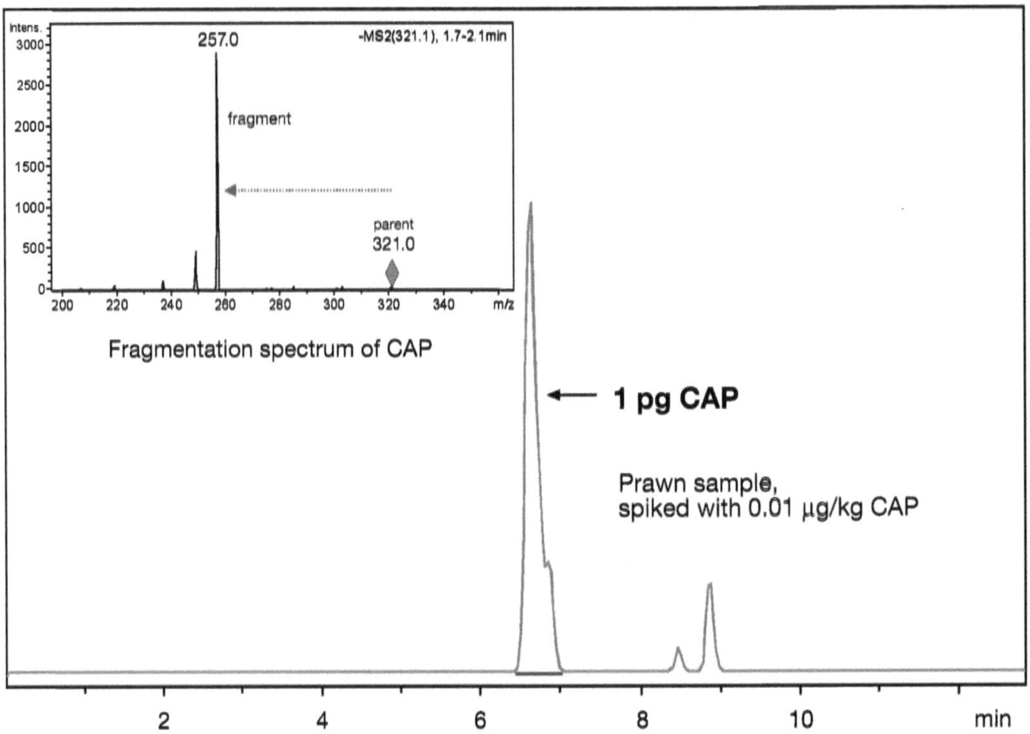

Fig. 28. Chromatogram of mass transition 321.0→257.0; detection of residual CAP at 0.01 µg/kg in prawn meat following automated SPE.

The excellent repeatability of the complete method including extraction, sample preparation, and LC-MS analysis is visualized by the signal overlay shown in Fig. 30 of chloramphenicol traces from six different prawn samples spiked with 2.0 µg/kg chloramphenicol each.

Linearity, recovery, and repeatability of the method were highly acceptable. The calibration curve is shown in Fig. 31.

For CAP a standard deviation of 2.0% for the automated and 2.2% for the manual approach using a highly experienced technician was achieved. The recovery was 92.1% (MPS) and 89.6% (manual) respectively as can be seen in Fig. 32. Manual SPE using standard cartridges can provide good results under tightly controlled conditions. Automated SPE provided slightly better results than manual SPE performed by an experienced and highly diligent technician. For CAP determinations using automated SPE the RSD was 2.0% compared with 2.2% for manual SPE. Automated SPE resulted in 92% recovery compared with 90% for manual SPE. Automated SPE provided a small improvement over the best achievable manual SPE results and a significant improvement in productivity.

The chromatographic run including syringe rinsing and equilibration time took a maximum of 25 min. During this time

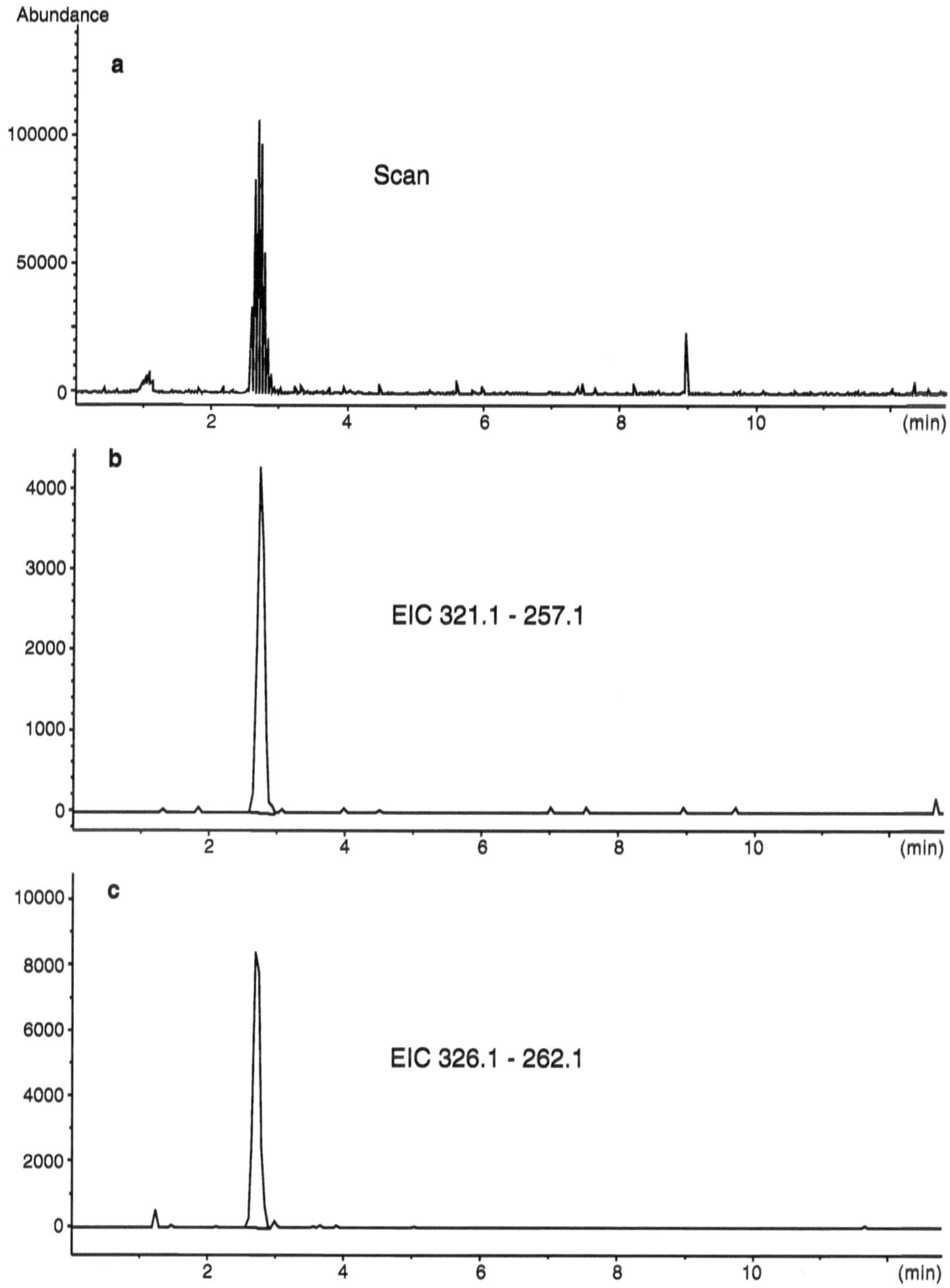

Fig. 29. LC-MS/MS chromatograms of CAP and CAP-d5 using the Agilent MSD Ion Trap XCT+. (**a**) Scan, (**b**) extracted ion chromatogram for the *m/z* 321.1 → 257.1 transition (CAP) and (**c**) the transition of *m/z* 326.1 → 262.1 (CAP-d5).

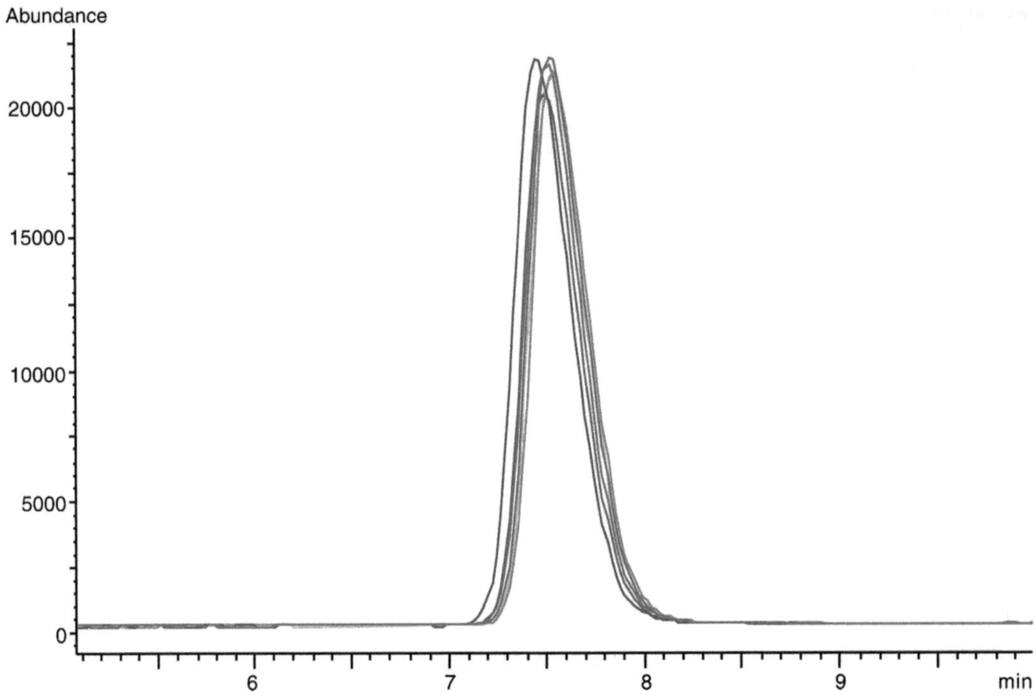

Fig. 30. Overlay of Chloramphenicol traces from six different shrimp samples spiked with 2.0 μg/kg Chloramphenicol each.

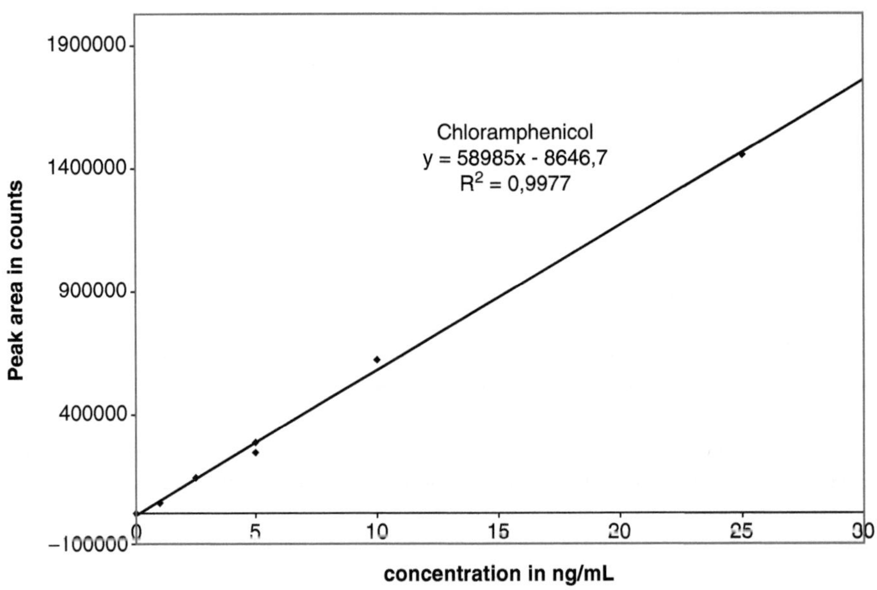

Fig. 31. Calibration curve for chloramphenicol.

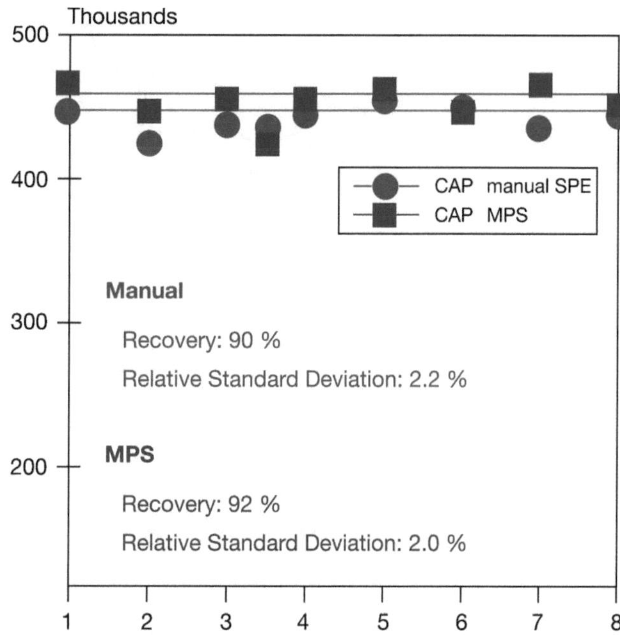

Fig. 32. Recovery and reproducibility for the determination of CAP in shrimp with manual and automated sample preparation.

automated SPE (duration about 15 min) of the following sample was performed in order to maximize throughput. Using this method, up to 50 analyses could be performed per day.

3.3. Application Example for Automated SPE Combined with Analyte Derivatization for Improved Selectivity and Lower Detection Limits

3.3.1. Aflatoxins in Food (Cereal, Nuts, and Spices)

Mycotoxin including aflatoxin contamination of food and feed is a global problem. The UN Food and Agricultural Organization (FAO) estimate that up to 25% of the world's food production is contaminated with mycotoxins. To date, more than 300 mycotoxins, formed by approximately 250 mold types, have been identified, but only a few are considered relevant for food safety. For peanuts, indehiscent fruits (mainly nuts), dried fruits, and grain intended for direct consumption or for use in food products, maximum allowable concentrations of 2 μg/kg aflatoxin B1 or 4 μg/kg total of B1, B2, G1, and G2 apply (7). Regulations limit the acceptable quantity in foods for infants and toddlers to 0.05 μg/kg aflatoxin B1. The chemical structures of aflatoxins B1, B2, G1, and G2 are shown in Fig. 33.

In order to determine the concentration of mycotoxins in foods, in pharmaceutical products, or in raw materials used in their production, most laboratories will rely on SPE (8) combined with LC-MS analysis and in many cases analyte derivatization (9). This approach ensures that detection limits will be lower than the maximum concentrations allowed by law. The LC-MS method described here was developed for the determination of B1, B2, G1, and G2

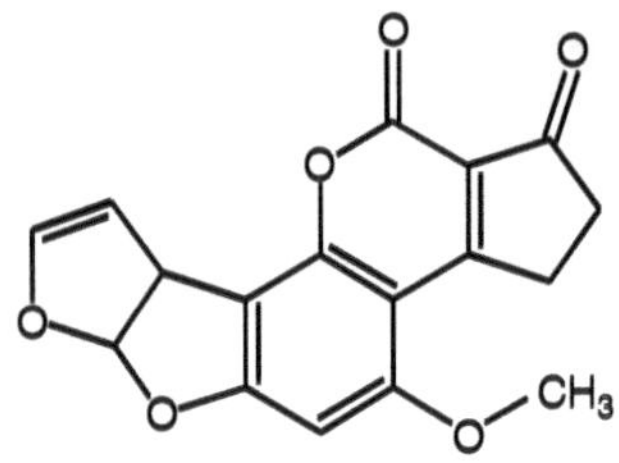

Aflatoxin B1

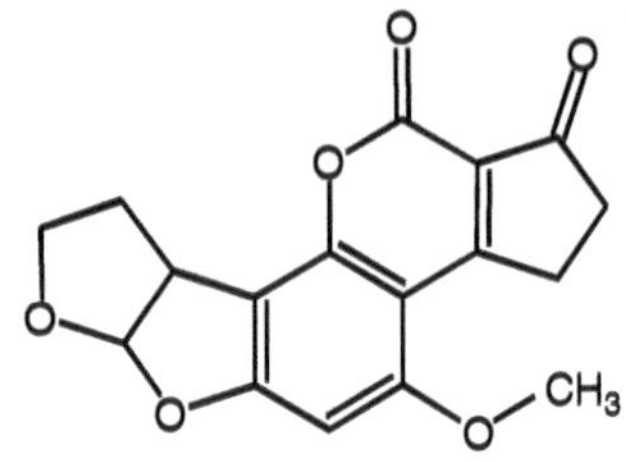

Aflatoxin B2

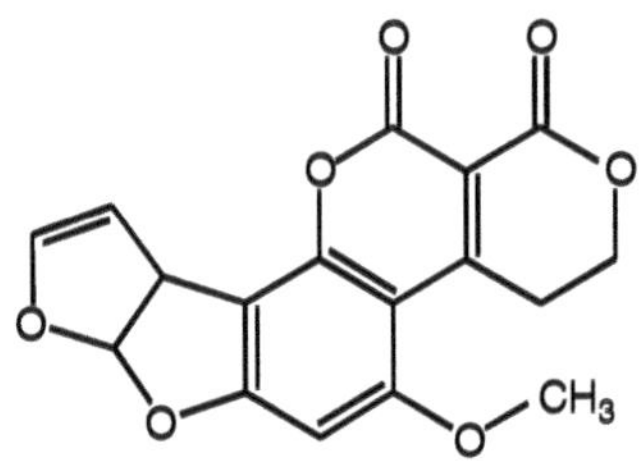

Aflatoxin G1

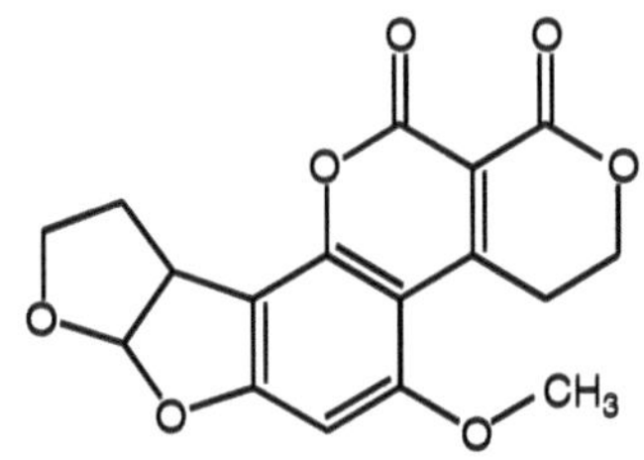

Aflatoxin G2

Fig. 33. Chemical structures of aflatoxins B1, B2, G1, and G2.

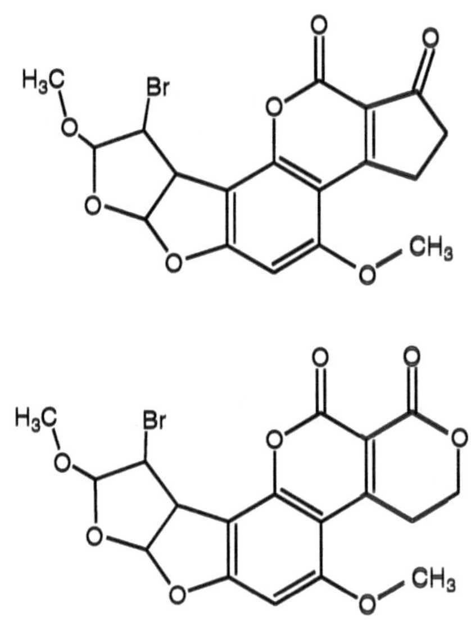

Fig. 34. Chemical structures of mono-brominated aflatoxins B1 and G1.

aflatoxins in foods such as pistachios, bell pepper seasoning and various fruits. Following cleanup on an SPE immunoaffinity column, the two aflatoxin compounds with an isolated, non-conjugated, double bond, B1 and G1, are brominated by stirring the extract with a 3% solution of bromine in chloroform. The resulting mono-brominated aflatoxins B1 and G1 are shown in Fig. 34.

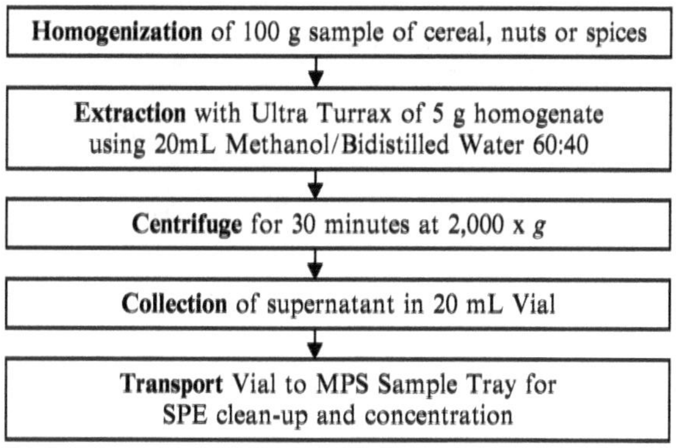

Fig. 35. Sample preparation flow chart, aflatoxins in cereals, nuts, and spices.

While longestablished manual SPE procedures may leave little room for further optimization, automation of the process can provide laboratories with more reliable results with significant time savings, while also introducing the possibility of performing automated derivatization. If analyte derivatization can be performed such that each sample is prepared directly prior to LC-MS analysis, analyte decomposition can be reduced to a minimum.

3.3.1.1. Instrumentation and Method

The sample was homogenized using an immersion mixer. A 5-g aliquot of the homogenate was weighed into a 50-mL glass tube. The extraction solution of 20 mL methanol: doubly-distilled water, 60:40, was added and the mixture extracted for 3 min using an Ultra Turrax. The extract was centrifuged for 30 min at 2,000 × *g* and the supernatant transferred to a 20-mL vial for automated sample preparation. Figure 35 shows the steps in the sample preparation. The automated SPE procedure is outline in Fig. 36.

The analysis was performed on an Agilent Technologies Series 1100 LC-MSD system, consisting of a binary pump, thermostated column compartment, vacuum degasser, diode array detector and a single quadrupole mass selective detector. The mass spectrometer was operated in selected ion monitoring (SIM) mode. The operating parameters for the LC-MS system are given in Fig. 37.

3.3.1.2. Results and Discussion

Under the chosen experimental conditions, the mass spectra indicate that bromination results only in formation of the 1-methoxy-2-bromo-substituted compounds shown in Fig. 34. These compounds yield significantly better MS responses with a characteristic bromine pattern in the mass spectra as shown in Fig. 38, providing improved differentiation from background

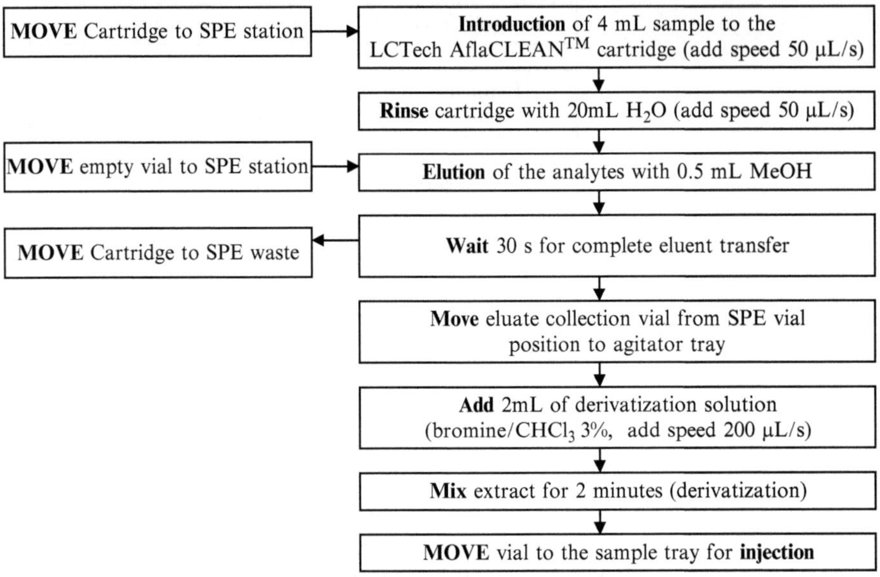

Fig. 36. Flow chart of SPE sample preparation steps, aflatoxins in cereals, nuts, and spices.

LC method parameters aflatoxins

injection volume	10 µL	
flow rate	0.3 mL/min	
eluents	formic acid 0.1 %/acetonitril (AcCN)	
gradient	Time (min)	acetonitril (%)
	0	53
	5	53
	15	100
	20	100
	21	53
HPLC column	Phenomenex Synergi Max-RP (250*2.1 mm, 4 µm)	
Column temperature	55 °C	

MS method parameters aflatoxins

Ionization mode	AP-ESI, positive	
nebulizer	45 psi	
N_2 flow/temperature	12 L/min/340 °C	
Mass transitions	B1	m/z 423
	B2	m/z 315
	G1	m/z 439
	G2	m/z 331

Fig. 37. LC and MS method parameters for aflatoxin determination.

signals and thus a better signal to noise ratio. Dibromo-substituted aflatoxins were not detected.

As can be seen in Fig. 39, the brominated aflatoxins are easily separated from each other and are significantly better retained in reversed phase chromatography than non-brominated species. Baseline separation is therefore easily achieved for all four aflatoxins

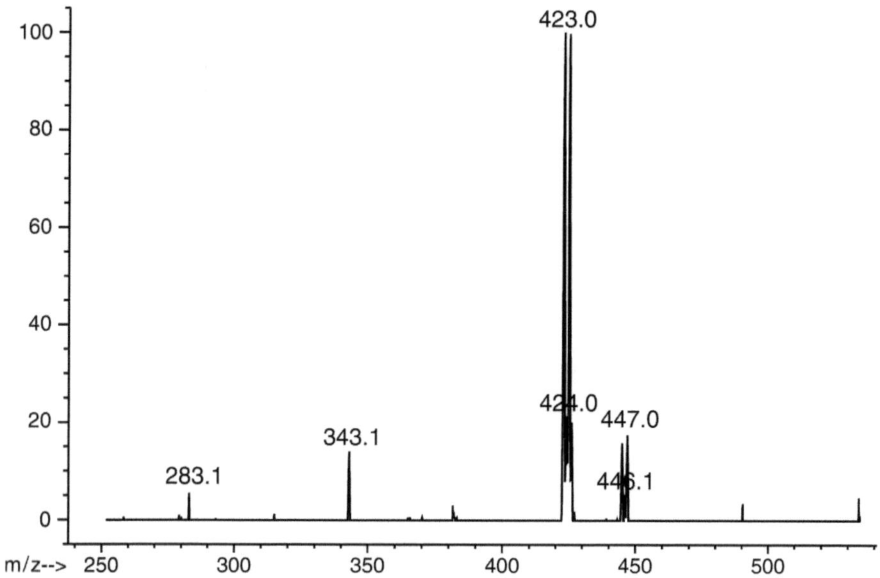

Fig. 38. Mass spectrum of mono-brominated aflatoxin B1.

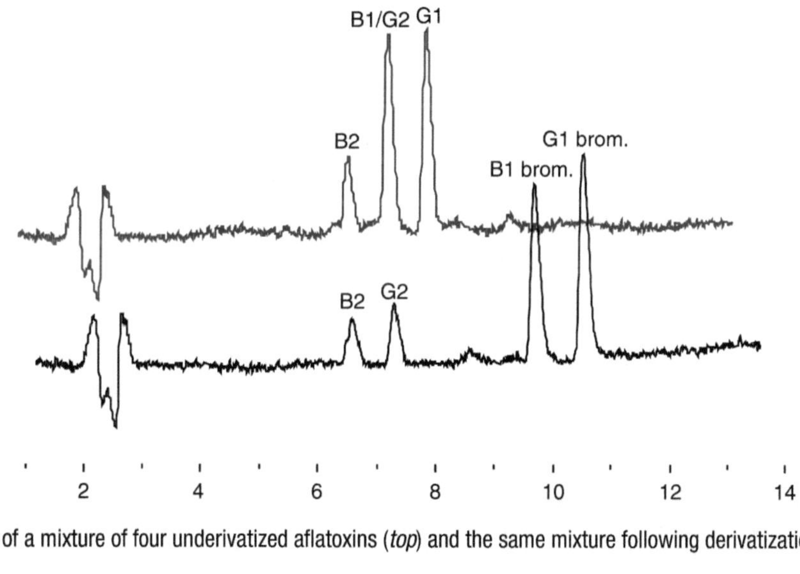

Fig. 39. TICs of a mixture of four underivatized aflatoxins (*top*) and the same mixture following derivatization (*bottom*).

with minimal interference from residual matrix. These combined advantages enable the system to reach detection limits below 0.01 µg/kg for the aflatoxins.

Reproducibility of the total analysis including sample preparation, SPE, derivatization, and LC-MS determination is illustrated in Fig. 40.

Calibration curves for the four determined aflatoxins show highly acceptable linearity as can be seen in Figs. 41 and 42.

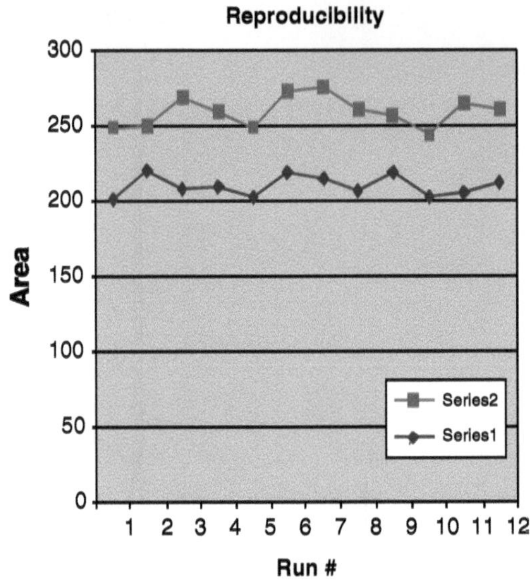

Fig. 40. Reproducibility for brominated aflatoxins over 12 runs.

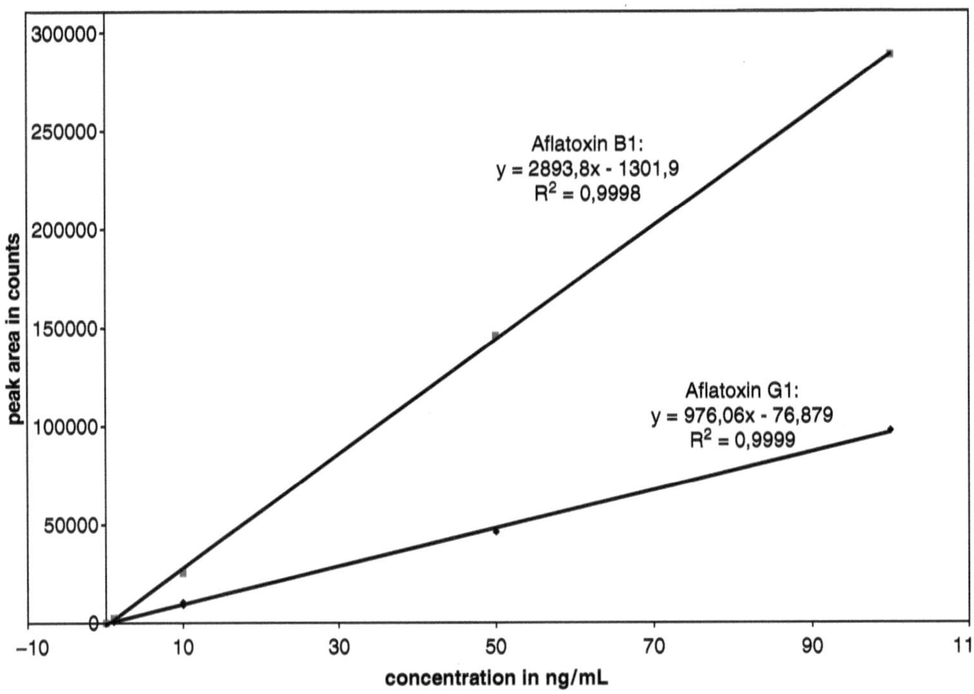

Fig. 41. Calibration curves for aflatoxins B1 and G1.

Manual sample preparation and SPE of eight samples for aflatoxin determination requires on the order of 4 h. The automated system requires only 80–95 min for preparing the same number of samples including SPE and analyte derivatization, resulting in significant time savings and increased throughput.

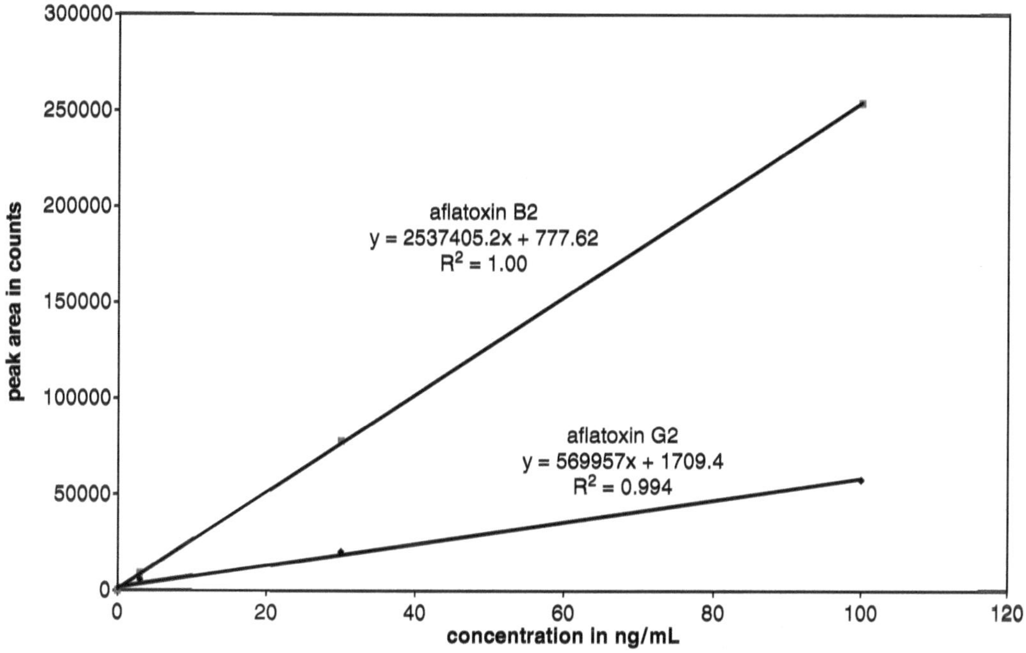

Fig. 42. Calibration curves for aflatoxins B2 and G2.

4. Notes

1. As an alternative to standard size packed bed cartridges, it is possible to perform dispersive SPE using loosely contained adsorbent. Dispersive SPE cartridges contain much less solid phase material and therefore have much less capacity, making the technique less suitable for extraction of the large sample volumes used, for example, in water analysis. Recovery in dispersive SPE is by definition determined by the equilibrium between the analyte concentration in the sample phase and the analyte concentration in the adsorbent. This means that dispersive SPE will by definition give less recovery than a robust standard SPE method in which analytes are separated from matrix based on real chromatography with one-dimensional flow of sample and eluant, similar to GPC cleanup. SPE based on a packed bed of adsorbent material has been tested, validated, and used for routine food analysis methods in laboratories around the world for more than three decades and is by many considered the standard cleanup method.
2. Ion traps can perform MS/MS using the full fragmentation spectrum in contrast to triple quadrupole instruments. This capability was used to select the best transition in the extracted ion mode. In this case, the transitions m/z 321→257 for CAP and m/z 326→262 for CAP-d5 were selected.

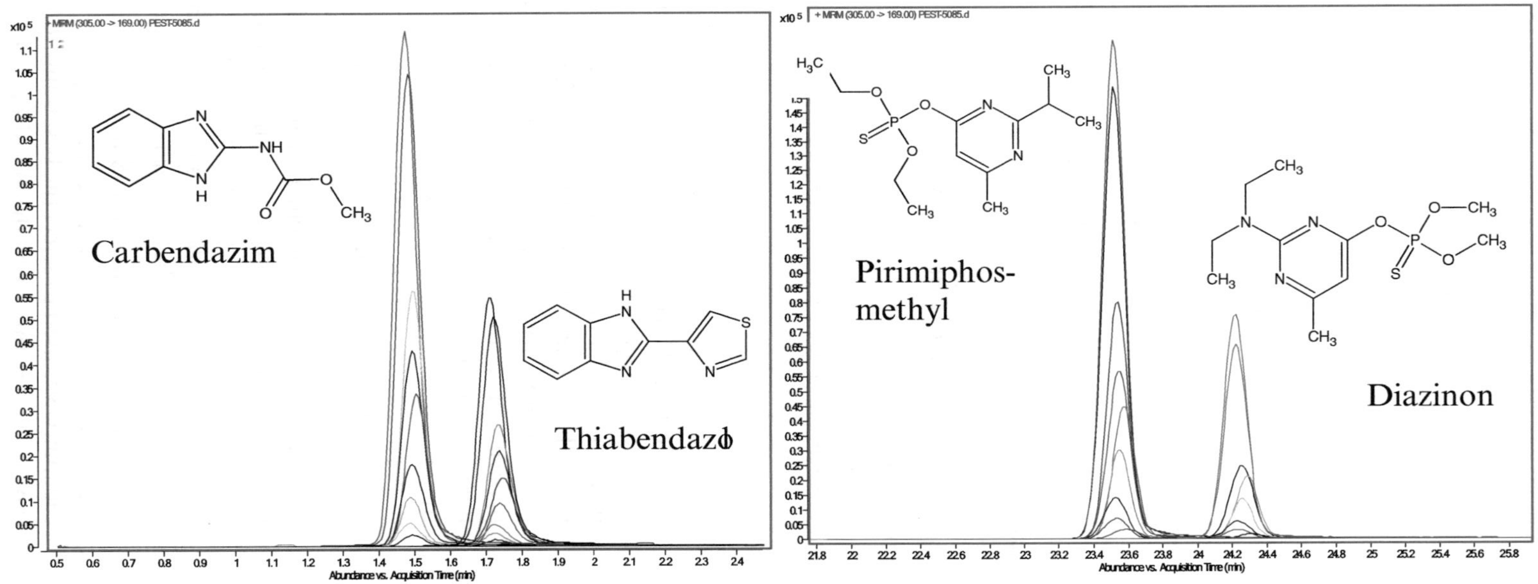

Fig. 43. Overlay of chromatograms of polar and apolar pesticides in an orange oil extract from 5 to 200 μg/mL. *Pictured left*: apolar compounds carbendazim and thiabendazol. *Pictured right*: polar compounds pirimiphos-methyl and diazinon.

3. Instead of ammonium formate (pH 8.5)/MeOH, ammonia (pH 8.5)/MeOH can be used.
4. Automated enrichment/concentration of the SPE eluate up to a factor 100 is possible, enabling a further significant reduction in the limit of quantitation (LOQ).
5. Orange oil samples were cleaned up using a slightly modified proprietary SPE method. The efficiency of SPE cleanup was illustrated by the fact that the intense yellow color of the sample was transferred to the cartridge while the resulting extract was a clear and colorless liquid. Recovery for the various compounds in this difficult matrix ranged from 70 to 90% while recoveries from fruit and vegetable samples were mainly in the range from 80 to 100%. Linearity was excellent, both for polar compounds like carbendazim and thiabendazole as well as for apolar pesticides like diazinon and pirimiphos-methyl. It is worth noting that the Zorbax SB-C18 Rapid Resolution columns provided excellent peak symmetry over the entire concentration range from 5 to 200 μg/mL as can be seen Fig. 43.
6. Acrylamide specific cartridges (M&N CHROMABOND® ABC18, 3 mL/500 mg) are available from Macherey-Nagel, Düren, Germany.

References

1. M. D. Hernando, M. Mezcua, J. M. Suarez-Barcena, and A. R. Fernandez-Alba, Liquid Chromatography with time-of-flight mass spectrometry for simultaneous determination of chemotherapeutant residues in salmon. Analytica Chimica Acta 2006, 562, (2), 176–184.
2. Edelhäuser, M., Klein, E., (1986). Determination of malachite green residues in edible fish. Deutsche Lebensmittel-Rundschau, 82 (12), 386–389.
3. 2004/25/EC: Commission Decision of 22 December 2003 amending Decision 2002/657/EC as regards the setting of minimum required performance limits (MRPLs) for certain residues in food of animal origin (Text with EEA relevance) (notified under document number C [2003] 4961). 2003.
4. M. Anastassiades, S. Lehotay, D. Stajnbaher and F. Schenck: Fast and easy multi-residue method employing acetonitrile extraction/partitioning and “dispersive solid-phase extraction” for the determination of pesticides residues in produce. J AOAC Int 86 (2) (2003): 412–31
5. ISO/DIS 25101:2009 Water quality – Determination of perfluorooctanesulfonate (PFOS) and perfluorooctanoate (PFOA) – Method for unfiltered samples using solid phase extraction and liquid chromatography/mass spectrometry.
6. Schneider, E.; Märtlbauer, E.; Dietrich, R.; Usleber, E.; Terplan, G., [Two Rapid Enzyme Immunochemical Tests for the Detection of Chloramphenicol in Raw-Milk]. Archiv für Lebensmittelhygiene 1994, 45, (2), 43–45.
7. Commission Regulation (EC) No 401/2006 of 23 February 2006 amended by Commission Regulation (EU) No 165/2010.
8. CEN/TC WI EN 14123 rev, Foodstuffs – determination of aflatoxins B1 and the sum of aflatoxin B1, B2,G1 and G2 in hazelnuts, peanuts, pistachios, figs, and paprika powder – high performance liquid chromatographic method with post column derivatization and immunoaffinity column clean-up
9. Senyuva H.Z., and Gilbert, J., (2005) Immunoaffinity clean-up with liquid chromatography using post-column bromination for determination of aflatoxins in hazelnut paste: Interlaboratory study, Journal of the AOAC International 88: No 2, 526–535

Chapter 6

Multiresidue Pesticide Analysis by Capillary Gas Chromatography-Mass Spectrometry

Jon W. Wong, Kai Zhang, Douglas G. Hayward, and Chin Kai-Meng

Abstract

A multiresidue pesticide method using a modified QuEChERS (*Qu*ick, *E*asy, *Ch*eap, *E*ffective, *R*ugged and *S*afe) procedure and capillary gas chromatography-mass spectrometry (GC-MS) is described for the determination of 166 organochlorine, organophosphorus, and pyrethroid pesticides, metabolites, and isomers in spinach. The pesticides from spinach were extracted using acetonitrile saturated with magnesium sulfate and sodium chloride, followed by solid-phase dispersive cleanup using primary-secondary amine and graphitized carbon black sorbents and toluene. Analysis is performed using different GC-MS techniques emphasizing the benefits of non-targeted acquisition and targeted screening procedures. Non-targeted data acquisition of pesticides in the spinach was demonstrated using GC coupled to a single quadrupole mass spectrometery (GC-MS) in full scan mode or multidimensional GC-time-of-flight mass spectrometery (GC × GC-TOF/MS), along with deconvolution software and libraries. Targeted screening was achieved using GC-single quadrupole mass spectrometry in selective ion monitoring (GC-MS/SIM) mode or -tandem mass spectrometry (GC-MS/MS) in multiple reaction monitoring mode. The development of these techniques demonstrates the powerful use of GC-MS for the screening, identification, and quantitation of pesticide residues in foods.

Key words: Gas chromatography-mass spectrometry, QuEChERS, Targeted analysis, Non-targeted data acquisition, Multiresidue methods

1. Introduction

Capillary gas chromatography-mass spectrometry (GC-MS) is an analytical technique, which is widely used for pesticide analysis determination in food. Many laboratories utilize GC-MS procedures because the instruments are rugged and effective to screen a variety of thermally stable and volatile or semivolatile pesticides in diverse food matrices (1, 2). The most commonly used MS detector has been the quadrupole analyzer, which can be operated

Jerry Zweigenbaum (ed.), *Mass Spectrometry in Food Safety: Methods and Protocols*,
Methods in Molecular Biology, vol. 747, DOI 10.1007/978-1-61779-136-9_6, © Springer Science+Business Media, LLC 2011

in the full scan or selective ion monitoring (SIM) mode. In the full scan mode, mass fragments over a wide range (m/z 50–500) are monitored, whereas in the SIM mode, selected ion fragments are selected to be detected. Full scan provides the entire mass spectrum of all analytes present in the sample, is the more qualitative of the two modes, and can be used to identify compounds present in the matrix. Full scan quadrupole and time-of-flight (TOF) mass spectrometers are especially effective for non-targeted screening of analytes in a matrix when deconvolution algorithms are used and compared to a database library containing mass spectra (3, 4). The computer algorithm, Automated Mass Spectral Deconvolution and Identification System (AMDIS), developed by the National Institute of Science and Technology (NIST), extracts spectra for individual components and identifies target compounds by purifying the mass spectra from the real sample and then matching the spectra against a reference library.

Another way to obtain resolved and high-quality spectra is to improve the gas chromatographic separation. Two dimensional gas chromatography coupled to a TOF-MS (GC × GC/TOF-MS) provides additional separation because coeluting peaks resulting from a one-dimensional separation can be separated by a second column possessing a different stationary phase. The resulting GC × GC chromatograms form narrow peaks and require a mass spectrometer such as TOF to be fast enough to scan enough data points for each peak (4).

Quadrupole instruments can also be useful for targeted analysis of pesticides in food matrices. Gas chromatography-mass spectrometry/selective ion monitoring (GC-MS/SIM) is used for quantitation because lower detection limits can be achieved due to the lower numbers of scans that reduce the number of mass fragments from targeted compounds. Typically, three or four ions characteristic and unique to the analyte of interest are selected for identification, one ion being the target used for quantitation and the remaining three ions used for qualification with respect to the target ion. Ratios of the abundances of the qualifier ions to the abundances of the target ion are used for identification or confirmation of the analyte. GC-MS/SIM is generally the most widely used procedure to screen and quantitate for pesticides because of its effectiveness and cost efficiency. The drawback of SIM is that components in a complex matrix can often contribute to the target and qualifier abundances and change the ratios used for identification and create false positives or negatives. The specificity of ions pertinent to the analyte can be further achieved using gas chromatography-tandem or triple quadrupole mass spectrometry. If three quadrupoles were in tandem, the analyte (target or precursor ion) of interest would be selected and permitted to pass through the first quadrupole into the second quadrupole, whereas all other analytes would not be allowed to pass. If the second quadrupole (i.e., collision chamber) were present to allow for the analyte of interest to fragment in the presence of an energetic inert gas such as Ar or N_2 (collision-induced dissociation), the resulting product ions would be permitted to enter the third quadrupole and only selected product ions would be allowed to pass and enter the detector. A tandem instrument in multiple reaction monitoring (MRM) mode is more specific and sensitive than SIM because the fragmentation reaction forms product ions characteristic of only the target compounds and the lower signal-to-noise ratios of the specific ion (target-to-product) transitions (5).

SIM and tandem mass spectrometry (MS/MS) are techniques that can be used to identify compounds from the matrix background from characteristic ions associated with these target compounds. Although target-oriented procedures such as GC-MS/SIM and GC-MS/MS are extremely sensitive and quantitative, they do not provide complete information of the screening because pesticides not targeted in either method are not screened or identified during the processes. Therefore, using both targeted and non-targeted GC-MS analysis allow for a comprehensive approach to screen pesticides in complex food matrices.

Pesticide tolerances, regulations, and food safety standards established by countries or economic unions are numerous and so surveillance and screening procedures need to be comprehensive. This chapter shall demonstrate the use of different MS techniques on spinach extracts prepared from a modified QuEChERS (*Qu*ick, *E*asy, *Ch*eap, *E*ffective, *R*ugged and *S*afe) procedure (6–9). QuEChERS has been a popular procedure because it optimizes conditions of a multiresidue method that uses a minimum number of steps, reduces solvent use and materials, and requires a small amount of glassware at a minimal cost (10). The technique is an official method of the AOAC International and the Committee of European Normalization (11, 12). Recently, the original QuEChERS procedure was modified by incorporating the use of graphitized carbon black and the addition of toluene as part of an improved cleanup step to remove pigments, sterols, and other extractives from the matrix (6–9). The method has been used effectively for multiresidue pesticide analysis using capillary gas chromatography or liquid chromatography-mass spectrometry techniques. The modified QuEChERS method, which differs from the original procedure is demonstrated on spinach samples and the resulting extracts are analyzed by GC-full scan MS, GC × GC-TOF-MS, GC-MS/SIM and GC-MS/MS techniques to demonstrate both the nontargeted and targeted analysis of pesticides in a fresh produce product. In this chapter, we demonstrated the use of four types of GC-MS on a spinach extract prepared by using a modified QuEChERS procedure.

2. Materials

2.1. Reagents

1. Solvents. Acetonitrile (Part No. 9017-33, JT Baker, HPLC grade, Mallinckrodt Baler, Phillipsburg, NJ), HPLC-grade water (Part No. W5-4, Fisher Scientific, Fair Lawn, NJ), toluene (Part No. T291-4, Fisher Scientific, Fair Lawn, NJ).
2. Magnesium sulfate (Part No. M65-3, Fisher Scientific, Fair Lawn, NJ).
3. Solid-phase dispersive tubes (900 mg anhydrous $MgSO_4$, 300 mg primary-secondary amine (PSA) sorbent, 150 mg graphitized carbon black).
4. Disposable centrifuge tubes, 50 mL, polypropylene.
5. Disposable centrifuge tubes, 15 mL, glass.
6. Nitrogen (99.999% purity).
7. Helium (99.999% purity).
8. Argon (99.999% purity).
9. Pesticide standards. A majority of the pesticide standards were obtained from the United States Environmental Protection Agency Pesticide Standard Repository, Ft. Meade, MD. Other sources were purchased from Aldrich-Sigma/Fluka/Supelco (Milwaukee, WI) and ChemService (West Chester, PA).
10. Internal standard. Tris-(1,3-dichloroisopropyl) phosphate, was purchased from TCI America (Portland, OR).
11. Quality control standards. Naphthalene-d_8, acenaphthylene-d_{10}, phenanthrene-d_{10} and chrysene-d_{12}.
12. Capillary gas chromatography columns. HP-5MS (30 m × 0.25 mm ID × 0.25 μm thickness, Agilent Technologies, Wilmington, DE, Part No. 19091S-433), HP-5MS UI. (ultra inert, 15 m × 0.25 mm × 0.25 μm thickness), VF-5MS (30 m × 0.25 mm ID × 0.25 μm thickness, Varian Inc., Walnut Creek, CA, Part No. CP8944), BPX-50 (30 m × 0.15 mm ID × 0.15 μm thickness, SGE, Darmstadt, Germany, Part No. 054741).
13. Guard columns. Siltek deactivated guard column (Restek Corp., Bellefonte, PA, 5 m × 0.25 mm ID, Part No. 10026-600 (6 pk)).

2.2. Apparatus

1. Analytical balances. Mettler AE240 balance (Mettler Instrument Corp., Hightstown, NJ), Ohaus Adventurer ARC120 balance (Ohaus Corp., Pine Brook, NJ).
2. Centrifuge (Series CR4i, ThermoElectron Corp, Milford, MA).

3. Nitrogen evaporator (N-EVAP III nitrogen evaporator, OA-SYS heating system, Organomation Associates Inc, Berlin, MA).
4. Agilent 6890 gas chromatograph equipped with a split/splitless inlet, Agilent 7673A autosampler and 5973 Agilent single quadrupole mass spectrometer (Wilmington, DE, USA).
5. Agilent 7890A gas chromatograph equipped with a multimode inlet equipped Agilent 7693A autoinjector and 5975C Agilent single quadrupole mass spectrometer (Wilmington, DE, USA).
6. Varian CP-3800 Series gas chromatograph equipped with split/splitless 1077 inlet, a CTC COMBI PAL autosampler and Varian 1200L triple quadrupole mass spectrometer (Varian Inc. Palo Alto, CA, USA).
7. Agilent 7890 gas chromatograph (Wilmington, DE, USA) equipped with a Gerstel CIS-4 PTV inlet (Linthiculm, MD, USA), Gerstel MPS 2 Multipurpose sampler (Linthiculm, MD, USA) and Pegasus 4D GC × GC-TOFMS (LECO Corp., St. Joseph, MI, USA).

3. Methods

A schematic of the QuEChERS procedure is shown in Fig. 1, listing sample preparation, extraction, and clean-up steps.

3.1. Sample Preparation

1. Dry ice is crushed using a hammer and placed at the bottom of the Robotcoupe and blended until the dry ice attains a powdery consistency.
2. The spinach is placed in the Robotcoupe blender and blended with the powdered dry ice until the spinach and dry ice form a powdery or sand-like consistency. Batches of the dry ice-spinach are pooled together and mixed thoroughly. The dry ice is allowed to evaporate and the homogenized spinach is collected in plastic freezer bags and stored at −40°C until further use.
3. Spinach (15 ± 0.1 g) samples were weighed into disposable polypropylene centrifuge tubes.

3.2. Extraction

1. Acetonitrile (15 mL) is added to the centrifuge tubes containing 15 g spinach. The internal standard (0.5 mL), Tris(1,3-dichloroisopropyl) phosphoric acid ester (3.375 μg/mL) is also added to the spinach–acetonitrile mixture. For fortification

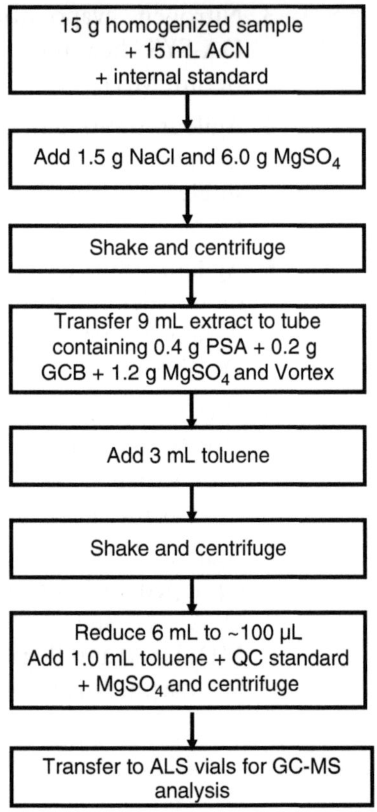

Fig. 1. Flowchart of the multiresidue analysis of pesticides in spinach using a modified QuEChERS procedure.

studies, a spiking solution containing the pesticides is also added at this step.

2. Anhydrous magnesium sulfate (6.0 g) and sodium chloride (1.5 g) are added to the centrifuge tube containing the spinach–acetonitrile mixture. The centrifuge tube is immediately shaken to easily disperse the magnesium sulfate with the spinach.
3. The centrifuge tube is centrifuged at 4,200 × *g* for 5 min. After centrifugation, three layers in the centrifuge tube are observed: the top layer consisting of the acetonitrile phase, the middle layer containing the extracted spinach matter, and the bottom layer is the water layer concentrated with the salts.

3.3. Cleanup

1. From the centrifuged extract, 12 mL of the acetonitrile are transferred to a 15 mL centrifuge tube containing 900 mg anhydrous magnesium sulfate, 300 mg primary-secondary aminopropyl (PSA) sorbent, and 150 mg graphitized carbon

black (GCB) and the acetonitrile extracts are vortexed vigorously until the original dark green color is removed.

2. To the centrifuge tubes containing the acetonitrile extracts, magnesium sulfate, PSA, and GCB, 3.0 mL of toluene is added. The centrifuge tubes are capped and vigorously shaken for 1 min.
3. The centrifuge tubes are centrifuged at 2,500 × *g* for 5 min.
4. The extracts (6.0 mL) are transferred to a clean and disposable glass centrifuge and the extracts are placed in an evaporator using a gentle nitrogen stream and a 40°C water bath until ~200 μL. Do not allow the sample to go to complete dryness.
5. To the sample, 1.0 mL of toluene, 50 μL of a 20 μg/mL quality control standard containing deuterated polyaromatic hydrocarbons (acenaphthylene-d_{10}, phenanthrene-d_{10}, and chrysene-d_{12}), and ~25 mg of anhydrous magnesium sulfate are added.
6. The extracts are centrifuged at 2,000 rpm for 5 min and transferred to ALS vials for GC-MS analysis.

3.4. Determination

1. Operating conditions for GC-MS/SIM

Gas chromatograph: Model 6890, Agilent

Sample injector: split/splitless injector, splitless mode, 22.1 psi, constant flow mode

Liner: 4 mm Siltek single gooseneck liner

Guard column: Siltek deactivated guard column, 5 m × 0.25 mm ID

Column: HP-5MS, 30 m × 0.25 mm ID × 0.25 μm film thickness

Injection temperature: 250°C

Carrier flow rate: Helium, 1.5 mL/min

Autoinjector: 7683A Agilent autosampler

Injection volume: 1 μL

Temperature program: 105°C (6 min hold) → 130°C (10°C/min) → 230°C (4°C/min) → 290°C (10°C/min, 10.5 min hold), run time = 50 min

Retention time locking program: Constant flow, 1.5 mL/min, chlorpyrifos-ethyl as the locking RTL standard at 24.54 min

Detector: 5973 mass selective detector

MS source temperature: 230°C

Table 1
Summary of pesticides found and not found in a spinach extract fortified to a final concentration 50 ng/kg and analyzed by GC×GC TOF/MS (*top results*) and GC-MS in full scan mode (*bottom results*). Four groups of spinach composites were prepared using the sample preparation procedure shown in Fig. 1, fortified with approximately 40 pesticides to a concentration of 50 μg/kg, and analyzed by GC-MS in full scan mode and GC×GC-TOF/MS. Deconvolution software was used to identify the presence of the pesticide in the mixture

Group #	1	2	3	4	
Pesticides not found by GC×GC-TOF/MS					
	Captan (degraded)	Captafol Cyfluthrin EPN Fenvalerate Folpet Iprodione Temephos Thiometon	Acrinathrin Cyanazine Cypermethrin Endrin aldehyde Methoxychlor-o,p'- *trans*-Permethrin	Azamethiphos	
Pesticides found	40 out of 41	33 out of 41	31 out of 37	38 out of 39	142 pesticides found
Pesticides not found	1	8	6	3	16 pesticides not found
Similarity range	753–958	688–939	574–913	607–931	158 total pesticides
Mean Quan ion S/N	2,400	590	390	360	
Pesticides not found by GC-MS (full scan, quadrupole MS)					
	Captan Deltamethrin Dioxathion Disulfoton Lindane (γ-BHC)	Azinphos-methyl Captafol Cyfluthrin Folpet Methidathion Tebupirimphos Temephos	Cyanazine Cypermethrin Endrin aldehyde Fensulfothion	Atrazine Azamethiphos DEF (Tribufos) Demeton-S Demeton S-methyl Edifenphos Fenamiphos Malathion	
Pesticides found	35 out of 40	33 out of 40	29 out of 33	31 out of 37	128 pesticides found
Pesticides not found	5	7	4	8	24 pesticides not found
Match factor≥30					154 total pesticides

MS quadrupole temperature: 150°C

MS transfer line temperature: 280°C

SIM Program: Refer to Table 1 for target and qualifier ions and dwell times

Software: MSD ChemStation G1701DA D.03.00.611

2. Operating conditions for GC-MS (full scan)

Gas chromatograph: Model 7890, Agilent

Sample injector: Multimode inlet injector at 17.73 psi (retention time locked to chlorpyrifos-methyl), constant pressure mode (He carrier gas)

Column: HP-5MS UI column, 15 m × 0.25 mm ID × 0.25 μm film thickness

Injection temperature program: 50°C (0.01 min), 720°C/min to 300°C (hold)

Liner: Helix double taper, deactivated

MS source temperature: 230°C

MS quadrupole temperature: 150°C

MS transfer line temperature: 280°C

Autosampler: 7683A autosampler

Injection volume: 2 μL cold splitless (fast injection)

Temperature program: 100°C (1.6 min hold) → 150°C (50°C/min) → 200°C (6°C/min) → 280°C (16 °C/min, 5 min hold), run time = 20.933 min

Retention time locking program: Constant pressure, using chlorpyrifos-methyl as the retention lock standard locked to 8.297 min

Septum purge: 3 mL/min

Purged Union: 4 psi (PCM)

Gas saver: 20 mL/min after 4 min

Cryo on: cryo use temperature 150°C; time out at 15 min

Split vent: 50 mL/min at 0.75 min

Backflush: 5 min duration during post-run; Oven 280°C, Purged union at 70 psi, multimode inlet at 2 psi, restrictor: 0.7 m × 0.15 mm deactivated fused silica tubing (from purged union to MSD)

Detector: 5975 C Mass selective detector

Solvent delay: 2.5 min

EMV mode: Gain Factor = 2

Mass range: 45–550 *m/z*

Threshold: 0

Source Temperature: 300°C

Quadrupole Temperature: 150°C

Software:

MSD ChemStation E.02.00 SP1

Deconvolution Reporting Software (DRS): 4.0.1

AMDIS: version 2.66 July 23, 2008

NIST MS Search: Version 2.0 g, build July 23, 2008

NIST08 Spectral library

3. Operating conditions: GC-MS/MS

   Gas chromatograph: Model CP-380, Varian

   Sample injector: 1077 split/splitless injector, splitless mode, 13.2 psi

   Liner: Restek Siltek Deactivated split liner with glass frit

   Guard column: Siltek deactivated guard column, 5 m × 0.25 mm ID

   Column: VF-5, 30 m × 0.25 mm ID × 0.25 μm film thickness

   Injection temperature: 280°C

   Carrier flow rate: Helium, 1.2 mL/min

   Autoinjector: CTC COMBI PAL autosampler

   Injection volume: 1 μL

   Temperature program: 105°C (6 min hold) → 130°C (10°C/min) → 230°C (4°C/min) → 290°C (10°C/min, 5.5 min hold), run time = 45 min

   Detector: Varian 1200L Triple Quadrupole Mass Spectrometer

   Electron multiplier voltage: 1,400 V

   Collision gas: Argon, 1.8 mTorr

   Source Temperature: 240 C

   Transfer line temperature: 300°C

   MS/MS program: Two ion transitions for each pesticide are listed in Table 2

   Software:

   Varian Workstation software, version 6.9

4. Operating conditions for GC × GC-TOF-MS

   Gas chromatograph: Model 7890, Agilent

   Sample injector: Gerstel CIS-4 PTV inlet

   Liner: Gerstel inserts, Part #6499-U

Table 2
Pesticide name, molecular formula and weight; GC-MS/SIM and GC-MS/MS retention times, target and qualifier ions, and percentages of qualifier-to-target ratios (%) (used for GC-MS/SIM analysis only); and transitions, collision energies (eV), MRM grouping, and ion ratios (used for GC-MS/MS analysis only) used in the study

			GC-MS/SIM						GC-MS/MS						
Pesticide	Molecular formula	Molecular weight	Retention time (min)	Target ion (T)	Q_1 (%Q_1/T)	Q_2 (%Q_2/T)	Q_3 (%Q_3/T)	Q_4 (%Q_4/T)	Retention time (min)	Quantitation transition	CE 1 (eV)	Confirmation transition	CE 2 (eV)	MRM group	Ion ratio
Acenaphthene-D_{10} (I.S.)	$C_{12}D_{10}$	164.29	12.75	164	162 (91.1)	160 (41.1)	163 (14.5)		13.50	164→162	15	164→134	25	4	33
Acrinathrin	$C_{26}H_{21}F_6NO_5$	541.45	36.64	181	208 (72.2)	209 (48.8)	289 (28.2)		37.31	181→152	15	290→93	10	29	1.9
Akton	$C_{12}H_{14}Cl_3O_3PS$	375.64	27.54	283	339 (91.6)	285 (70.6)	341 (58.8)		28.78	339→184	30	339→283	10	19	2.7
Alachlor	$C_{14}H_{20}ClNO_2$	269.77	22.56	160	188 (70.0)	146 (33.9)	237 (22.5)		23.74	188→130	35	160→77	45	13	3.5
Aldrin	$C_{12}H_8Cl_6$	364.91	23.82	263	265 (66.3)	293 (42.6)	298 (27.4)		24.92	298→193	35	263→191	40	14	1.7
Allethrin	$C_{19}H_{26}O_3$	302.41	26.70	136	123 (408.5)	168 (19.6)	107 (108.1)		28.04	123→81	5	136→108	5	18	7.5
Atrazine	$C_8H_{14}ClN_5$	215.67	18.98	200	215 (54.4)	202 (30.3)	173 (21.5)		20.27	200→94	15	200→71	15	8	2.1
Azamethiphos	$C_9H_{10}ClN_2O_5PS$	324.68	31.67	215	183 (50.5)	155 (48.7)	324 (29.7)		33.45	324→171	25	215→171	10	24	2.4
Azinphos-ethyl	$C_{12}H_{16}N_3O_3PS_2$	345.38	36.51	132	160 (68.7)	77 (66.0)	104 (19.4)	105 (42.5)	35.70	160→132	5	160→77	15	29	0.9
Azinphos-methyl	$C_{10}H_{12}N_3O_3PS_2$	317.33	35.49	160	132 (86.4)	77 (71.5)	104 (22.5)	161 (11.0)	37.77	160→132	5	160→77	15	28	0.4
α-BHC	$C_6H_6Cl_6$	290.83	17.61	181	183 (94.1)	219 (97.3)	217 (75.5)	221 (40.9)	18.89	181→146	20	219→183	10	7	83
β-BHC	$C_6H_6Cl_6$	290.83	18.89	219	183 (99.7)	181 (96.1)	217 (47.3)		20.31	181→146	20	219→183	10	8	1.2
δ-BHC	$C_6H_6Cl_6$	290.83	20.33	219	183 (93.8)	181 (97.6)	217 (80.6)		22.09	181→146	20	219→183	10	11	3.1
Benfluralin	$C_{13}H_{16}F_3N_3O_4$	335.26	17.35	292	264 (18.1)	276 (13.8)	293 (12.7)	335 (4.9)							
Bifenthrin	$C_{23}H_{22}ClF_3O_2$	422.87	34.55	181	165 (21.6)	166 (25.6)	182 (14.5)		35.73	181→165	20	181→166	15	27	2.2

(continued)

Table 2 (continued)

Pesticide	Molecular formula	Molecular weight	GC-MS/SIM Retention time (min)	Target ion (T)	Q_1 (%Q_1/T)	Q_2 (%Q_2/T)	Q_3 (%Q_3/T)	Q_4 (%Q_4/T)	GC-MS/MS Retention time (min)	Quantitation transition	CE 1 (eV)	Confirmation transition	CE 2 (eV)	MRM group	Ion ratio
Bromophos	$C_{10}H_{12}BrCl_2O_3PS$	394.05	25.27	331	329 (84.3)	333 (30.3)	125 (31.2)	316 (5.7)	26.60	331→316	15	331→286	30	16	1.5
Bromophos-ethyl	$C_8H_8BrCl_2O_3PS$	366.00	31.12	359	303 (79.6)	242 (39.2)	331 (40.1)	357 (84.2)	28.58	357→301	20	357→222	30	19	1.4
Bromopropylate	$C_{17}H_{16}Br_2O_3$	428.12	34.34	339	341 (198.7)	183 (70.4)	185 (68.3)	343 (96.0)	35.75	341→185	15	341→183	15	27	1.9
Captafol	$C_{10}H_9Cl_4NO_2S$	349.06	32.89	79	77 (21.9)	80 (28.1)	183 (8.1)		34.74	151→79	20	79→77	15	26	12
Captan	$C_9H_8Cl_3NO_2S$	300.59	26.13	79	80 (25.3)	149 (16.9)	114 (12.11)		28.01	79→77	15	107→79	15	18	9.1
Carbophenothion	$C_{11}H_{16}ClO_2PS_3$	326.74	31.75	342	157 (195.4)	199 (51.4)	121 (87.7)		33.49	342→157	5	342→143	15	25	1.1
cis-Chlordane	$C_{10}H_6Cl_8$	409.78	27.47	373	375 (94.1)	377 (50.5)	371 (41.8)		28.54	373→266	35	373→337	10	19	17
trans-Chlordane	$C_{10}H_6Cl_8$	409.78	26.80	373	375 (94.7)	377 (52.9)	371 (42.3)		29.13	373→266	15	373→232	40	20	4.0
α-Chlordene	$C_{10}H_6Cl_6$	338.87	22.64	303	230 (66.1)	338 (27.9)	301 (57.8)		24.07	303→232	30	338→232	25	13	1.9
β-Chlordene	$C_{10}H_6Cl_6$	338.87	24.10	338	230 (112.0)	303 (157.4)	301 (96.7)		25.48	230→160	20	338→230	20	14	2.7
γ-Chlordene	$C_{10}H_6Cl_6$	338.87	23.97	338	230 (171.4)	303 (125.4)	301 (86.8)		25.48	230→160	20	338→230	20	14	1.9
β-Chlorfenvinphos	$C_{12}H_{14}Cl_3O_4P$	359.57	26.51	323	267 (143.7)	269 (101.2)	295 (35.5)	325 (65.9)	27.82	267→159	20	267→123	35	17	6.2
Chlorobenzilate	$C_{16}H_{14}Cl_2O_3$	325.19	30.15	251	253 (69.4)	139 (71.2)	111 (26.9)		31.91	251→139	20	251→111	45	23	1.9
Chloroneb	$C_8H_8Cl_2O_2$	207.06	13.45	206	208 (68.2)	141 (48.1)	191 (182.4)		14.08	191→113	15	206→191	10	4	1.5
Chlorothalonil	$C_8Cl_4N_2$	265.91	20.48	266	264 (76.4)	268 (46.7)	270 (10.3)		21.47	264→133	50	264→168	20	9	1.5
Chlorpyrifos	$C_9H_{11}Cl_3NO_3PS$	350.59	24.59	314	286 (40.7)	258 (58.1)	260 (40.2)	316 (72.6)	25.63	314→258	15	197→169	15	15	2.0
Chlorpyrifos-methyl	$C_7H_7Cl_3NO_3PS$	322.54	22.28	286	288 (70.4)	125 (43.9)	197 (5.9)	290 (17.4)	23.35	286→93	20	286→241	15	12	8.8
Chlorthiophos	$C_{11}H_{15}Cl_2O_3PS_2$	361.25	31.09	269	325 (65.7)	360 (44.0)	297 (39.8)		32.47	325→269	10	325→271	20	23	2.5
Chrysene-D_{12} (I.S.)	$C_{18}D_{12}$	240.39	33.86	240	241 (20.0)	236 (23.0)	239 (12.2)		35.75	240→212	15	240→208	50	27	2.7

Coumaphos	$C_{14}H_{16}ClO_6P$	346.70	37.54	362	226 (43.9)	210 (36.3)	334 (16.8)	364 (44.8)	38.57	362→226	25	362→99	10	30	1.4
Cyanazine	$C_9H_{13}ClN_6$	240.68	24.70	225	212 (165.5)	240 (56.8)	198 (69.7)		26.03	198→91	10	138→69	15	15	6.5
Cyanophos	$C_9H_{10}NO_3PS$	243.22	19.59	243	109 (95.1)	125 (59.5)	180 (8.4)		20.83	243→127	20	243→79	25	9	5.6
Cyfluthrin 1	$C_{22}H_{18}Cl_2FNO_3$	434.29	38.01	163	206 (61.9)	226 (57.0)	199 (50.1)		39.06	163→91	30	163→127	15	30	3.0
Cyfluthrin 2	$C_{22}H_{18}Cl_2FNO_3$	434.29	38.14	163	206 (62.1)	226 (37.6)	199 (42.5)		39.19	163→91	30	163→127	15	30	3.0
Cyfluthrin 3	$C_{22}H_{18}Cl_2FNO_3$	434.29	38.23	163	206 (68.0)	226 (44.1)	199 (39.4)		39.22	163→91	30	163→127	15	30	3.0
Cyfluthrin 4	$C_{22}H_{18}Cl_2FNO_3$	434.29	38.28	163	206 (68.7)	226 (40.3)	199 (49.1)		39.25	163→91	30	163→127	15	30	3.0
λ-Cyhalothrin	$C_{23}H_{19}ClF_3NO_3$	449.86	36.26	181	197 (77.2)	208 (56.1)	199 (26.5)	209 (46.6)	35.95	197→119	10	197→91	10	27	4.6
Cypermethrin 1	$C_{22}H_{19}Cl_2NO_3$	416.30	38.40	163	181 (89.0)	165 (66.8)	209 (30.4)		39.45	163→91	30	163→127	15	30	3.0
Cypermethrin 2	$C_{22}H_{19}Cl_2NO_3$	416.30	38.53	163	181 (77.6)	165 (61.3)	209 (42.8)		39.58	163→91	30	163→127	15	30	3.0
Cypermethrin 3	$C_{22}H_{19}Cl_2NO_3$	416.30	38.61	163	181 (73.1)	165 (55.2)	209 (21.9)		39.60	163→91	30	163→127	15	30	2.3
Cypermethrin 4	$C_{22}H_{19}Cl_2NO_3$	416.30	38.66	163	181 (81.1)	165 (71.2)	209 (32.5)		39.65	163→91	30	163→127	15	30	2.4
Dacthal (DCPA)	$C_{10}H_6Cl_4O_4$	331.97	24.65	301	299 (83.0)	303 (57.7)	332 (25.8)		25.81	332→167	50	332→223	40	15	1.8
o,p′-DDD	$C_{14}H_{10}Cl_4$	320.04	28.97	235	237 (65.1)	165 (37.3)	199 (13.5)	212 (8.8)	32.33	235→165	35	235→115	50	23	7.1
p,p′-DDD	$C_{14}H_{10}Cl_4$	320.04	30.46	235	237 (64.0)	165 (37.6)	199 (10.8)	212 (8.7)	32.33	235→165	35	235→115	50	23	6.7
o,p′-DDE	$C_{14}H_8Cl_4$	320.04	27.37	246	248 (68.4)	318 (42.7)	316 (33.5)		28.76	246→176	30	318→176	40	19	50
p,p′-DDE	$C_{14}H_8Cl_4$	318.03	28.86	246	248 (62.7)	318 (93.4)	316 (73.1)		30.35	246→177	30	318→177	40	22	1.0
o,p′-DDT	$C_{14}H_9Cl_5$	354.49	30.59	235	237 (64.3)	165 (35.3)	199 (12.9)	212 (7.3)	32.33	235→165	35	235→115	50	23	6.7
p,p′-DDT	$C_{14}H_9Cl_5$	354.49	32.09	235	237 (64.6)	165 (32.3)	199 (8.7)	212 (9.0)	33.94	235→165	35	235→115	50	25	5.0
DEF (Tribufos)	$C_{12}H_{27}OPS_3$	314.52	28.81	202	169 (175.0)	226 (48.5)	258 (51.3)	314 (21.8)	30.48	202→147	5	258→112	25	22	3.7
Deltamethrin	$C_{22}H_{19}Br_2NO_3$	505.21	40.54	181	253 (100.3)	255 (48.2)	251 (50.5)		41.88	172→93	10	251→93	20	30	1.1
Demeton-S	$C_8H_{19}O_3PS_2$	258.34	18.20	88	170 (16.2)	143 (12.3)	258 (3.2)								
Demeton-S-methyl	$C_6H_{15}O_3PS_2$	230.29	15.39	88	109 (26.3)	142 (19.9)	230 (3.0)		16.92	88→60	5	142→79	10	5	3.2

(continued)

Table 2 (continued)

Pesticide	Molecular formula	Molecular weight	GC-MS/SIM Retention time (min)	Target ion (T)	Q_1 ($\%Q_1/T$)	Q_2 ($\%Q_2/T$)	Q_3 ($\%Q_3/T$)	Q_4 ($\%Q_4/T$)	GC-MS/MS Retention time (min)	Quantitation transition	CE 1 (eV)	Confirmation transition	CE 2 (eV)	MRM group	Ion ratio
Dialifor	$C_{14}H_{17}ClNO_4PS_2$	393.85	36.70	208	357 (12.0)	173 (35.6)	186 (8.9)	210 (35.3)	37.82	208→89	15	208→125	15	29	1.4
Diallate 1	$C_{10}H_{17}Cl_2NOS$	270.22	17.36	86	128 (26.8)	234 (68.7)	236 (25.9)		18.71	234→150	20	234→192	10	7	3.6
Diallate 2	$C_{10}H_{17}Cl_2NOS$	270.22	17.72	86	128 (28.2)	234 (74.5)	236 (27.4)		19.13	234→150	20	234→192	10	7	3.6
Diamidafos (Nellite)	$C_8H_{13}N_2O_2P$	200.18	21.30	107	94 (40.7)	200 (9.6)	170 (5.2)		19.64	200→107	20			7	NA
Diazinon	$C_{12}H_{21}N_2O_3PS$	304.35	20.34	304	276 (36.3)	227 (45.8)	248 (37.2)		21.24	304→179	10	179→121	25	9	4.4
Dibutyl chlorendate (I.S.)	$C_{17}H_{20}Cl_6O_4$	501.06	35.60	388	371 (68.3)	317 (45.3)	237 (31.2)		36.72	388→207	45	388→182	35	28	8.5
Dicapthon	$C_8H_9ClNO_5PS$	297.65	24.74	262	125 (38.3)	216 (10.0)	79 (13.2)	232 (19.1)	26.13	262→216	10	262→123	40	15	3.0
Dichlobenil	$C_7H_3Cl_2N$	172.01	10.18	171	173 (67.5)	100 (20.7)	136 (17.8)		10.07	171→100	20	171→126	10	2	7.1
Dichlofenthion	$C_{10}H_{13}Cl_2O_3PS$	315.16	21.89	279	223 (93.2)	162 (39.1)	251 (39.0)	281 (46.0)	23.08	279→223	30	279→205	30	12	1.1
Dichlofluanid	$C_9H_{11}Cl_2FN_2O_2S_2$	333.22	24.01	123	167 (48.4)	224 (46.1)	226 (33.1)		25.16	123→77	20	224→123	20	14	0.3
Dicloran	$C_6H_4Cl_2N_2O_2$	207.01	18.24	206	178 (70.2)	176 (108.0)	208 (77.8)		19.56	206→176	15	206→148	15	7	3.1
3,4′-Dichloroaniline	$C_6H_5Cl_2N$	162.01	11.77	161	163 (61.5)	165 (10.3)	99 (15.2)		12.21	163→90	20	161→125	10	3	2.1
4,4′-Dichlorobenzophenone	$C_{13}H_8Cl_2O$	251.11	24.45	139	250 (30.2)	111 (35.7)	141 (37.2)	252 (21.2)	26.33	139→111	20	139→75	45	16	5.2
Dichlorvos	$C_4H_7ClO_4P$	185.52	8.66	185	109 (281.0)	220 (20.0)	79 (45.6)	145 (28.6)	7.92	185→93	15	185→109	20	1	1.9
Dieldrin	$C_{12}H_8Cl_6O$	380.91	28.46	263	277 (75.6)	380 (37.6)	79 (266.9)	345 (28.5)	30.53	202→113	20	202→95	20	22	1.6
Dimethachlor	$C_{13}H_{18}ClNO_2$	255.74	21.79	134	197 (41.0)	210 (10.8)	199 (16.9)		23.09	197→120	20	197→148	20	12	1.7
Dioxabenzofos	$C_8H_9O_3PS$	216.20	20.25	216	183 (41.1)	201 (23.8)	153 (25.8)	171 (4.0)	18.43	216→201	10	216→137	25	6	10

Dioxathion	$C_{12}H_{26}O_6P_2S_4$	456.55	37.70	270	153 (115.6)	125 (149.5)	97 (205.9)	271 (148.9)	20.56	125→97	5	270→169	10	8	4.1
Disulfoton	$C_8H_{19}O_2PS_3$	274.41	20.36	88	186 (14.7)	142 (18.6)	274 (16.3)		21.63	88→60	5	88→73	5	10	2.0
Ditalimfos	$C_{12}H_{14}NO_4PS$	299.28	27.87	299	130 (248.6)	148 (114.6)	243 (85.8)	209 (82.8)	29.50	243→130	25	148→130	15	20	2.4
Edifenphos	$C_{14}H_{15}O_2PS_2$	310.38	31.89	310	173 (104.4)	109 (154.4)	201 (44.5)	218 (22.2)	33.63	310→109	20	109→65	5	25	3.3
α-Endosulfan	$C_9H_6Cl_6O_3S$	406.93	27.27	241	195 (74.5)	237 (90.0)	339 (46.3)		29.13	241→206	20	243→136	30	20	2.1
β-Endosulfan	$C_9H_6Cl_6O_3S$	406.93	29.82	195	237 (78.0)	241 (91.5)	339 (51.4)		32.08	241→206	20	243→136	30	23	2.2
Endosulfan ether	$C_9H_6Cl_6O$	342.86	20.95	241	277 (88.9)	307 (83.2)	342 (45.2)		22.61	307→69	5	272→143	30	11	6.9
Endosulfan sulfate	$C_9H_6Cl_6O_4S$	422.93	31.76	272	387 (65.3)	274 (82.7)	229 (64.7)	422 (21.3)	33.72	272→143	40	385→219	30	25	2.0
Endrin	$C_{12}H_8Cl_6O$	380.91	29.38	263	345 (42.8)	317 (80.0)	281 (64.5)	245 (58.0)	30.36	281→211	25	263→191	15	22	2.2
Endrin aldehyde	$C_{12}H_8Cl_6O$	380.91	30.75	345	279 (32.2)	347 (77.4)	250 (68.8)		32.43	248→177	30	345→253	10	23	13
Endrin ketone	$C_{12}H_8Cl_6O$	380.91	33.82	317	345 (31.8)	315 (73.8)	281 (33.6)		35.43	317→175	45	317→219	35	27	1.1
EPN	$C_{14}H_{14}NO_4PS$	323.31	34.36	157	169 (56.0)	185 (31.9)	141 (32.1)		35.75	169→77	20	157→110	20	27	3.0
Ethalfluralin	$C_{13}H_{14}F_3N_3O_4$	333.25	16.94	316	276 (123.7)	292 (45.5)	333 (31.3)		17.70	276→105	15	316→201	30	5	2.7
Ethion	$C_9H_{22}O_4P_2S_4$	384.48	30.94	231	384 (14.5)	153 (54.6)	125 (36.7)	233 (14.4)	20.78	231→129	25	231→185	10	9	4.8
Ethoprop	$C_8H_{19}O_2PS_2$	242.34	16.25	158	139 (48.1)	200 (32.4)	242 (22.6)		17.28	158→97	15	158→81	20	5	4.4
Etridiazole	$C_5H_5Cl_3N_2OS$	247.53	12.26	211	183 (78.6)	213 (68.2)	185 (54.8)		12.61	211→108	40	211→183	20	3	21
Famphur	$C_{10}H_{16}NO_5PS_2$	325.34	31.77	218	125 (21.0)	93 (17.6)	109 (8.0)	217 (20.2)	33.41	218→127	15	218→91	5	24	3.4
Fenamiphos	$C_{13}H_{22}NO_3PS$	303.36	28.30	303	154 (75.6)	288 (24.5)	260 (22.2)		29.78	303→154	15	303→195	10	21	2.1
Fenarimol	$C_{17}H_{12}Cl_2N_2O$	331.19	36.19	139	219 (74.8)	251 (61.8)	330 (37.1)		37.58	139→75	20	219→107	10	29	1.5
Fenchlorphos (Ronnel)	$C_8H_8Cl_3O_3PS$	321.55	22.93	285	287 (70.5)	125 (29.2)	167 (4.6)	289 (13.2)	24.16	285→270	15	285→240	25	13	1.2
Fenitrothion	$C_9H_{12}NO_5PS$	277.24	23.59	277	260 (75.1)	109 (75.2)	125 (98.7)	247 (43.2)	24.91	277→109	15	277→127	25	14	3.1
Fensulfothion	$C_{11}H_{17}O_4PS_2$	308.37	30.44	292	308 (6.7)	293 (38.2)	188 (9.8)	156 (48.9)	32.16	293→97	40	308→109	20	23	3.9
Fenthion	$C_{10}H_{15}O_3PS_2$	278.33	24.48	278	125 (26.0)	109 (17.6)	169 (17.6)	279 (11.2)	25.86	277→109	15	277→127	20	15	5.2

(continued)

Table 2 (continued)

Pesticide	Molecular formula	Molecular weight	GC-MS/SIM Retention time (min)	Target ion (T)	Q_1 (%Q_1/T)	Q_2 (%Q_2/T)	Q_3 (%Q_3/T)	Q_4 (%Q_4/T)	GC-MS/MS Retention time (min)	Quantitation transition	CE 1 (eV)	Confirmation transition	CE 2 (eV)	MRM group	Ion ratio
Fenvalerate 1	$C_{25}H_{22}ClNO_3$	419.91	39.50	419	167 (255.4)	181 (160.5)	225 (118.6)		40.67	167→125	10	419→167	25	30	24
Fenvalerate 2	$C_{25}H_{22}ClNO_3$	419.91	39.77	419	167 (315.1)	181 (182.7)	225 (132.7)		40.98	167→125	10	419→167	25	30	23
Fluchloralin	$C_{12}H_{13}ClF_3N_3O_4$	355.68	20.46	306	145 (14.9)	264 (46.1)	326 (84.3)		21.38	306→160	20	306→264	5	9	3.3
Flucythrinate 1	$C_{26}H_{23}F_2NO_4$	451.47	38.70	199	225 (16.6)	209 (15.5)	181 (36.1)	451 (31.7)	39.66	199→107	35	157→107	15	30	1.4
Flucythrinate 2	$C_{26}H_{23}F_2NO_4$	451.47	38.94	199	225 (15.9)	209 (9.8)	181 (34.2)	451 (27.1)	39.81	199→107	35	157→107	15	30	1.4
Fluridone	$C_{19}H_{14}F_3NO$	329.32	39.20	328	329 (46.3)	189 (1.6)	310 (2.4)	330 (8.2)	40.40	328→189	35	328→233	40	30	3.3
Fluvalinate τ- 1	$C_{26}H_{22}ClF_3N_2O_3$	502.92	39.79	250	252 (34.1)	181 (26.5)	251 (15.9)		40.82	502→250	15	181→152	30	30	1.3
Fluvalinate τ- 2	$C_{26}H_{22}ClF_3N_2O_3$	502.92	39.88	250	252 (30.4)	181 (21.8)	251 (12.4)		40.82	502→250	15	181→152	30	30	2.0
Folpet	$C_9H_4Cl_3NO_2S$	296.56	26.47	260	104 (131.6)	76 (107.3)	130 (50.5)		28.39	260→130	20	262→234	10	18	5.1
Fonofos	$C_{10}H_{15}OPS_2$	246.33	19.65	246	109 (190.43)	137 (91.1)	110 (49.1)	174 (13.0)	20.94	246→109	25	137→109	5	9	1.0
Heptachlor	$C_{10}H_5Cl_7$	373.32	23.05	272	100 (68.2)	274 (75.9)	237 (36.7)		23.89	100→65	20	272→237	15	13	1.3
Heptachlor epoxide A	$C_{10}H_5Cl_7O$	389.32	25.71	353	355 (80.4)	351 (49.0)	357 (37.9)		27.43	353→263	20	353→265	15	17	1.6
Heptachlor epoxide B									27.63	149→85	35	217→182	20	17	2.0
Hexachlorobenzene	C_6Cl_6	284.78	17.91	284	286 (82.2)	282 (52.8)	288 (34.1)		18.96	284→144	50	284→179	30	7	2.4
Iprobenfos (IBP)	$C_{13}H_{21}O_3PS$	288.34	21.11	204	123 (25.0)	288 (14.1)	246 (19.4)		29.83	377→93	25	377→157	30	21	1.0
Iprodione	$C_{13}H_{13}Cl_2N_3O_3$	330.16	34.04	314	316 (71.4)	187 (53.0)	189 (35.2)		35.57	314→245	10	314→271	10	27	1.9
Isazophos	$C_9H_{17}ClN_3O_3PS$	313.74	20.87	161	119 (75.1)	257 (55.4)	285 (27.7)	313 (14.9)	22.37	204→91	15	204→121	35	11	1.6

Isofenphos	$C_{15}H_{24}NO_4PS$	345.40	26.52	255	213 (202.7)	185 (89.5)	245 (32.0)	345 (7.4)	21.78	161→119	10	162→120	10	10	1.5
Iodofenphos	$C_8H_8Cl_2IO_3PS$	413.00	28.27	377	379 (37.2)	125 (16.7)	109 (7.5)	250 (5.3)	27.65	213→121	15	213→185	5	17	5.0
Leptophos	$C_{13}H_{10}BrCl_2O_2PS$	412.07	35.57	377	171 (90.7)	375 (75.2)	379 (26.7)		36.74	377→157	40	171→77	15	28	37
Lindane (γ-BHC)	$C_6H_6Cl_6$	290.83	19.12	181	219 (83.0)	183 (105.8)	217 (63.9)		20.51	181→146	20	219→183	10	10	1.3
Malathion	$C_{10}H_{19}O_6PS_2$	330.36	24.22	173	125 (82.8)	158 (42.9)	211 (7.2)	127 (90.1)	25.42	127→99	5	173→99	15	14	2.1
Methidathion	$C_6H_{11}N_2O_4PS_3$	302.34	27.11	145	85 (62.5)	125 (16.0)	93 (17.9)		28.70	145→85	5	145→58	15	19	1.2
o,p′-Methoxychlor	$C_{16}H_{15}Cl_3O_2$	345.65	32.96	227	121 (90.7)	228 (17.3)	152 (10.3)		35.96	227→115	50	227→141	30	27	1.4
p,p′-Methoxychlor	$C_{16}H_{15}Cl_3O_2$	345.65	34.57	227	228 (15.7)	212 (3.9)	152 (5.1)		34.24	227→121	15	227→91	30	25	5.0
Metolachlor	$C_{15}H_{22}ClNO_2$	283.79	24.44	238	162 (117.2)	240 (42.3)	146 (25.2)		25.50	238→162	10	238→133	20	14	1.6
Mevinphos	$C_7H_{13}O_6P$	224.15	11.86	127	192 (25.6)	109 (21.2)	164 (6.0)		12.06	192→127	10	192→109	20	3	3.1
Mirex	$C_{10}Cl_{12}$	545.54	35.53	272	237 (41.4)	332 (11.5)	274 (75.7)		37.19	272→237	15	272→143	35	29	3.1
Naphthalene-D_8 (I.S.)	$C_{10}D_8$	136.24	7.68	136	108 (7.7)	134 (9.0)	137 (10.4)		6.78	136→108	15	136→82	25	1	2.9
cis-Nonachlor	$C_{10}H_5Cl_9$	444.24	30.49	409	407 (86.6)	411 (61.1)	405 (35.9)		29.27	409→300	30	409→230	50	23	1.7
trans-Nonachlor	$C_{10}H_5Cl_9$	444.24	27.73	409	407 (90.5)	411 (64.5)	405 (35.0)		32.08	409→302	25	409→230	50	20	2.3
Oxadiazon	$C_{15}H_{18}Cl_2N_2O_3$	345.23	29.30	258	175 (182.8)	260 (64.0)	302 (75.9)		30.47	175→112	15	175→77	45	22	63
Parathion	$C_{10}H_{14}NO_5PS$	291.26	24.61	291	263 (13.2)	235 (12.6)	186 (22.0)	218 (9.2)	26.03	291→109	10	139→109	5	15	6.5
Parathion methyl	$C_8H_{10}NO_5PS$	263.21	22.27	263	233 (24.5)	247 (11.4)	264 (9.9)	200 (6.3)	23.69	263→109	5	263→127	10	13	1.8
Pentachloroaniline	$C_6H_2Cl_5N$	265.35	21.20	265	267 (59.7)	263 (61.7)	269 (19.2)		22.59	265→192	25	265→107	50	11	2.9
Pentachlorobenzene	C_6HCl_5	250.34	13.66	250	248 (64.5)	252 (69.8)	254 (18.1)		19.28	280→237	20	280→265	10	7	3.0
Pentachlorobenzonitrile	C_7Cl_5N	275.34	19.43	275	273 (65.6)	277 (63.0)	279 (21.8)		14.30	250→143	45	250→145	35	4	1.9
Pentachlorophenyl methyl ester	$C_7H_3Cl_5O$	280.36	18.26	280	265 (97.1)	282 (74.4)	267 (56.6)		20.41	275→205	30	275→142	30	8	1.9

(continued)

Table 2
(continued)

Pesticide	Molecular formula	Molecular weight	GC-MS/SIM Retention time (min)	Target ion (T)	Q_1 (%Q_1/T)	Q_2 (%Q_2/T)	Q_3 (%Q_3/T)	Q_4 (%Q_4/T)	GC-MS/MS Retention time (min)	Quantitation transition	CE 1 (eV)	Confirmation transition	CE 2 (eV)	MRM group	Ion ratio
Pentachlorothio-anisole	$C_7H_3Cl_5S$	296.43	23.47	296	246 (45.5)	263 (31.2)	298 (59.7)		24.92	296→263	10	246→103	30	14	6.1
cis-Permethrin	$C_{21}H_{20}Cl_2O_3$	391.29	37.23	183	165 (17.3)	163 (17.5)	184 (17.3)		38.35	183→128	30	163→127	5	30	2.1
trans-Permethrin	$C_{21}H_{20}Cl_2O_3$	391.29	37.39	183	165 (19.8)	163 (23.5)	184 (16.1)		38.53	183→128	30	163→127	5	30	2.1
Phenanthrene-D_{10} (I.S.)	$C_{14}D_{10}$	188.31	19.32	188	189 (14.5)	187 (22.7)	184 (13.5)		21.04	188→161	15	188→176	25	10	1.4
Phenothrin	$C_{23}H_{26}O_3$	350.46	35.40, 35.60	183	123 (160.5)	184 (21.5)	124 (16.3)	350 (9.2)	36.45	123→77	30	123→81	10	28	10
Phorate	$C_7H_{17}O_2PS_3$	260.38	17.64	260	121 (168.7)	97 (94.9)	231 (63.7)	75 (376.2)	18.73	260→75	10	121→93	5	7	7.4
Phosalone	$C_{12}H_{15}ClNO_4PS_2$	367.81	35.52	182	184 (35.5)	367 (28.2)	154 (25.7)		36.79	121→65	10	182→75	30	28	0.8
Phosmet	$C_{11}H_{12}NO_4PS_2$	317.33	34.17	160	161 (12.0)	133 (5.4)	317 (4.7)		35.70	160→77	15	160→133	10	27	1.6
Phenthoate	$C_{12}H_{17}O_4PS_2$	320.36	26.59	274	246 (30.8)	275 (26.8)	320 (5.1)								
Pirimiphos ethyl	$C_{11}H_{20}N_3O_3PS$	333.39	25.76	318	333 (104.6)	304 (68.1)	290 (23.3)		26.76	333→168	20	333→97	45	16	4.0
Pirimiphos methyl	$C_{11}H_{20}N_3O_3PS$	305.33	23.80	290	305 (76.0)	276 (84.1)	233 (30.8)		24.83	290→125	20	290→233	10	14	2.6
Procymidone	$C_{13}H_{11}Cl_2NO_2$	284.14	26.73	283	285 (65.0)	287 (13.0)	255 (10.2)	212 (9.4)	28.22	283→96	10	283→255	5	18	50
Profenofos	$C_{11}H_{15}BrClO_3PS$	373.63	28.61	339	208 (158.7)	295 (40.1)	337 (97.8)	374 (59.9)	30.24	337→188	35	337→269	15	22	0.8
Propachlor	$C_{14}H_{14}ClNO$	247.72	15.67	176	120 (243.8)	169 (39.2)	211 (25.6)		16.56	176→77	20	176→120	10	5	1.5
Propazine	$C_9H_{16}ClN_5$	229.70	19.15	214	172 (53.3)	229 (60.8)	216 (32.2)	187 (28.1)	20.46	214→172	10	214→79	10	8	15
Propetamphos	$C_{10}H_{20}NO_4PS$	281.31	19.81	138	194 (50.7)	222 (24.9)	236 (37.0)		20.92	138→110	10	236→138	20	9	1.7

Propyzamide	$C_{12}H_{11}Cl_2NO$	256.13	19.72	173	145 (31.4)	175 (61.0)	255 (26.2)		21.07	173→74	35	173→145	10	9	0.5
Prothiophos	$C_{11}H_{15}Cl_2O_2PS_2$	345.25	28.46	309	267 (92.0)	311 (47.9)	269 (41.9)		30.41	309→239	15	309→221	35	21	71
Pyraclofos	$C_{14}H_{18}ClN_2O_3PS$	360.80	36.75	360	194 (55.2)	362 (32.6)	138 (57.2)		37.97	360→97	30	360→138	35	29	1.9
Pyrazophos	$C_{14}H_{20}N_3O_5PS$	373.37	36.60	221	232 (35.0)	373 (24.8)	328 (4.9)	268 (10.5)	37.55	232→204	10	232→124	15	29	15
Pyridaphenthion	$C_{14}H_{17}N_2O_4PS$	340.34	34.23	340	188 (45.1)	199 (52.1)	77 (39.0)	204 (28.9)	35.48	340→91	30	340→124	35	27	28
Quinalphos	$C_{12}H_{15}N_2O_3PS$	298.32	26.58	298	146 (367.5)	157 (241.9)	270 (42.5)		27.98	146→118	15	156→103	20	18	8.0
Quintozene	$C_6Cl_5NO_2$	295.33	19.34	295	265 (68.9)	249 (60.3)	237 (139.5)		20.23	237→141	20	237→167	20	8	4.3
Resmethrin	$C_{22}H_{26}O_3$	338.45	33.64	123	143 (32.5)	171 (60.4)	128 (43.7)		34.97	171→128	10	123→81	5	26	1.4
Simazine	$C_7H_{12}ClN_5$	201.64	18.68	201	173 (43.9)	186 (63.3)	203 (33.8)		20.06	186→91	5	186→104	10	8	1.2
Sulfotep-ethyl	$C_8H_{20}O_5P_2S_2$	322.32	17.56	322	97 (34.7)	202 (40.7)	238 (24.0)	266 (28.0)	18.28	322→146	20	322→202	10	6	2.3
Sulprofos	$C_{12}H_{19}O_2PS_3$	322.45	31.40	322	156 (55.3)	140 (71.6)	125 (24.3)	280 (9.6)	33.06	322→97	30	322→156	15	24	1.5
Tebupirimfos	$C_{13}H_{23}N_2O_3PS$	318.37	21.21	318	261 (94.7)	234 (77.5)	276 (48.1)		22.31	318→152	10	318→123	40	11	1.2
Tecnazene (TCNB)	$C_6HCl_4NO_2$	260.89	15.43	261	203 (127.1)	215 (103.3)	217 (57.2)	231 (68.5)	16.10	261→203	20	261→143	35	5	6.5
Tefluthrin	$C_{17}H_{14}ClF_7O_2$	418.74	20.86	177	197 (27.2)	199 (8.5)	178 (10.5)		21.99	177→127	15	177→87	25	10	3.4
Temephos	$C_{16}H_{20}O_6P_2S_3$	466.48	43.62	466	467 (20.1)	171 (6.3)	203 (11.4)								
Terbufos	$C_9H_{21}O_2PS_3$	288.44	19.60	231	153 (22.8)	186 (16.1)	203 (9.0)	288 (9.6)							
Terbuthylazine	$C_9H_{16}ClN_5$	229.70	19.56	214	229 (29.7)	173 (43.3)	216 (39.6)		20.92	229→172	20	229→138	20	9	2.5
2,3,5,6-Tetrachloroaniline	$C_6H_3Cl_4N$	230.91	15.92	231	158 (19.0)	169 (6.7)	229 (83.5)	233 (51.5)	16.98	231→158	20			5	NA
Tetrachlorvinphos	$C_{10}H_9Cl_4O_4P$	365.96	27.68	331	329 (103.5)	333 (36.7)	109 (71.6)	240 (11.0)	29.04	331→127	10	331→126	25	19	77
Tetramethrin	$C_{19}H_{25}NO_4$	331.41	34.72	164	123 (30.2)	165 (10.8)	135 (3.3)	107 (7.0)	35.88	164→77	20	164→107	10	27	1.5
Thiometon	$C_6H_{15}O_2PS_3$	246.36	18.04	88	246 (7.9)	158 (7.8)	125 (41.4)								
Tolclofos methyl	$C_9H_{11}Cl_2O_3PS$	301.13	22.45	265	267 (37.3)	125 (18.1)	250 (11.0)		23.69	265→250	15	265→109	35	13	1.6

(continued)

Table 2 (continued)

Pesticide	Molecular formula	Molecular weight	GC-MS/SIM Retention time (min)	Target ion (T)	Q_1 ($\%Q_1/T$)	Q_2 ($\%Q_2/T$)	Q_3 ($\%Q_3/T$)	Q_4 ($\%Q_4/T$)	GC-MS/MS Retention time (min)	Quantitation transition	CE 1 (eV)	Confirmation transition	CE 2 (eV)	MRM group	Ion ratio
Tolyfluanid	$C_{10}H_{13}Cl_2FN_2O_2S_2$	347.25	26.12	137	238 (37.7)	240 (26.9)	181 (29.7)		27.59	240→137	20	237→137	10	17	2.7
Triallate	$C_{10}H_{16}Cl_3NOS$	304.67	20.62	268	270 (69.3)	145 (29.6)	86 (170.5)		21.98	268→184	25	270→186	20	10	1.1
Triazophos	$C_{12}H_{16}N_3O_3PS$	313.32	31.55	161	285 (24.5)	313 (12.5)	257 (34.2)		33.16	257→162	10	161→106	15	24	1.4
Trifluralin	$C_{13}H_{16}F_3N_3O_4$	335.26	17.39	306	264 (61.5)	290 (12.2)	335 (9.3)		18.13	306→264	5	306→148	20	6	1.9
Triphenylphosphate	$C_{18}H_{15}O_4P$	326.28	33.12	326	325 (84.4)	215 (18.7)	170 (16.0)		34.71	326→226	30	326→141	40	26	2.8
Tris-(1,3-Dichlorisopropyl) phosphate (IS)	$C_9H_{15}Cl_6O_4P$	430.91	31.96	381	383 (65.0)	321 (35.9)	303 (26.9)		33.31	379→159	20	379→123	20	24	2.1
Vinclozolin	$C_{12}H_9Cl_2NO_3$	286.11	22.28	285	212 (103.4)	187 (80.6)	198 (95.1)		23.59	285→212	10	285→172	25	13	1.4

Guard column: Siltek deactivated guard column, 5 m × 0.25 mm ID

Column 1: HP-5MS, 30 m × 0.25 mm ID × 0.25 μm film thickness

Column 2: BPX-50, 1.5 m × 0.25 mm ID × 0.15 μm film thickness

Autoinjector: Gerstel MPS 2 (Multipurpose Sampler 2)

Injection temperature: 250°C

Carrier: Helium, constant pressure, 29 psig

Inlet septum purge flow: 3 mL/min

Inlet purge time: 120 s

Inlet purge flow: 98 mL/min

Inlet solvent vent time: 9 s

Inlet solvent vent flow: 200 mL/min

Inlet solvent vent pressure: 1.4 psi

Injection volume: 1 μL

Temperature program (Oven 1): 100°C (2 min hold) → 130°C (10°C/min) → 230°C (4°C/min) → 270°C (20°C/min, 13 min hold),

Secondary temperature program: 150°C (2 min hold) → 180°C (10°C/min) → 280°C (4°C/min) → 320°C (20°C/min, 13 min hold)

Modulator temperature offset: 65°C relative to the GC oven temperature

Modulation timing (second dimension separation time): 3.0 s divided into a hot pulse time of 0.8 s and cool time between stages of 0.70 s.

Detector: LECO Pegasus 4D TOFMS

Transfer line temperature from secondary oven to mass spectrometer: 260°C

Ion source temperature: 200°C

Solvent delay: 430 s

Mass range: 40–700 *m/z*

Acquisition rate: 200 spectra/s

Detector voltage: 1,700 V

Electron voltage: −70 eV

Software: LECO ChromaTOF 4D 3.0 and NIST Mass Spectral Library

3.5. Results

1. The results from the two full scan GC-MS (time-of-flight and single quadrupole) instruments are shown in Figs. 2 and 3. Figure 2 is a chromatogram generated from GC×GC/TOF-MS showing the separation of 38 pesticide analytes in a spinach extract fortified concentration at 50 μg/kg. A typical GC-MS chromatogram of a spinach extract generated from a GC-single quadrupole MS instrument operating in full scan mode is shown in Fig. 3. The full scan quadrupole result reveals that the major components from the spinach contribute to the major peaks shown in the chromatogram. Both the GC×GC/TOF-MS and GC-MS full scan results are examples of non-targeted data acquisition screening procedures because the chromatograms generated will be subjected to the deconvolution algorithms and NIST mass spectral libraries of chemicals provided by the manufacturer. The results of screening and identifying pesticides in a fortified spinach extract at 50 μg/kg screened by GC×GC-TOF/MS and GC-MS in full scan mode are tabulated in Table 1.

2. Chromatographic results of targeted screening procedures utilizing GC-MS/SIM and GC-MS/MS are revealed in Figs. 5 and 6, respectively. In both of these procedures, ions are selected based on the ions characteristic and unique to the pesticide analytes as listed in Table 2. A mixture of 166 pesticides based on four ions (one quantitative and three qualitative ions) for GC-MS/SIM and two target-to-product transitions (one for quantitative, the other from qualitative) for GC-MS/MS are screened and shown in Figs. 5c and 6c, respectively. Ions characteristic and unique to the pesticides (m/z = 183, 184, 163, and 165) for GC-MS/SIM and two target-to-product transitions (163→127, 183→128) for GC-MS/MS can be selected or specified for the presence of permethrin shown in Figs. 5d and 6d, respectively. The chromatographic retention times and resolution are important in the separation of the pesticide from the sample matrix and other pesticides and this is demonstrated by the separation of the *cis*- and *trans*-isomers of permethrin, which cannot be resolved by mass spectrometry.

3. An incurred spinach sample is analyzed to demonstrate each of the four GC-MS techniques. Figures 3b and 4 demonstrate the identification of permethrin generated by GC-TOF/MS and GC-MS using full scan mass spectra by comparison to the library spectra. Figures 5 and 6 demonstrate the use of a selected number of four ions or two precursor-to-product ion transitions to identify permethrin using GC-MS/SIM and GC-MS/MS techniques. Three GC-MS techniques, GC-TOF/MS, GC-MS/SIM, and GC-MS/MS have been

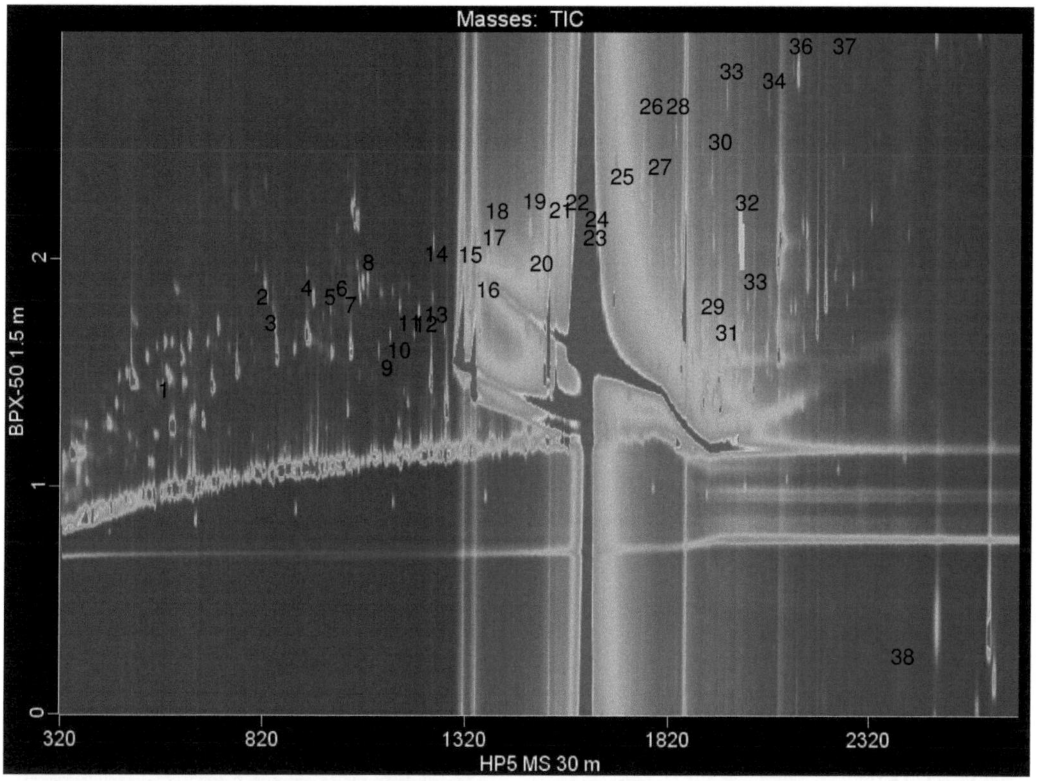

Fig. 2. GC × GC-TOF MS chromatogram of a spinach extract containing 50 μg/kg of the following pesticides: 1: Mevinphos, 2: Demeton S-methyl, 3: Ethoprop, 4: α-BHC, 5: Demeton-S; 6: Atrazine, 7: Pentachlorobenzonitrile, 8: Fonofos, 9: Diazinon, 10: Triallate, 11: Iprobenfos, 12: Dichlofenthion, 13: Tolclofos-methyl, 14: Heptachlor, 15: Aldrin, 16: Malathion, 17: Chlorpyrifos, 18: Dicapthon, 19: Tolylfluanid, 20: Isofenphos, 21: Allethrin, 22: o,p′-DDE, 23: Tetrachlorvinphos, 24: Fenamiphos, 25: DEF (Tribufos), 26: Endrin, 27: p,p'-DDD, 28: Ethion, 29: Edifenphos, 30: Resmethrin, 31: Pyridaphenthion, 32: Bifenthrin, 33: Phosalone, 34: λ-Cyhalothrin, 35: Dialifor, 36: Coumaphos, 37: Flucythrinate, 38: Fluvalinate-tau.

a

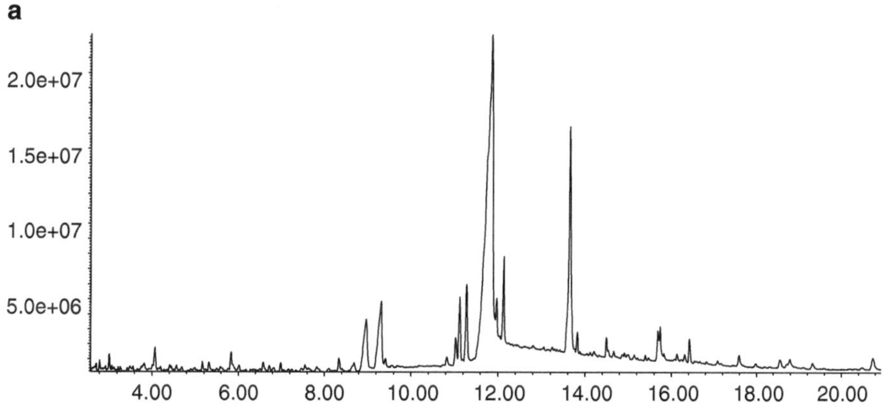

Fig. 3. (**a**) Full scan GC-MS chromatogram of an incurred spinach extract. (**b**) Extracted full scan mass spectra of *cis*- (*left*) and *trans* (*right*)-permethrin found in the spinach extract by AMDIS software. Mass spectra shown are the spectra from the full scan GC-MS analysis (*top*), spectra processed from the AMDIS software (*middle*) and spectra of *cis*- and *trans*-permethrin isomers from the NIST library (*bottom*).

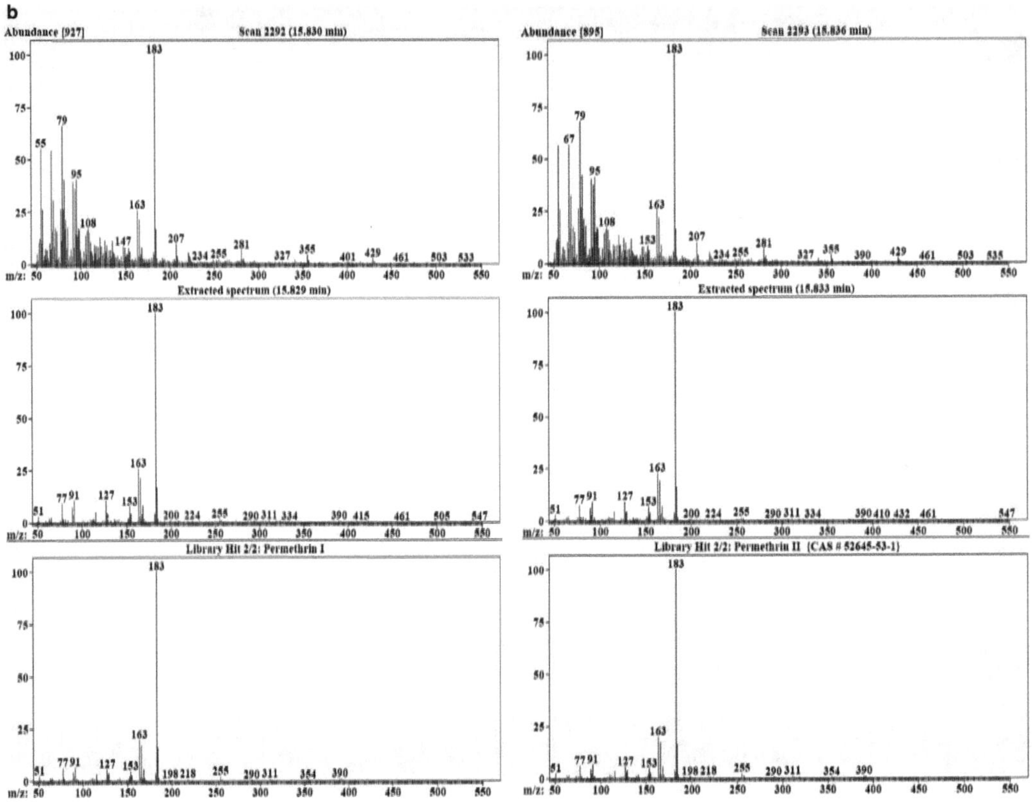

Fig. 3. (continued)

demonstrated by other researchers (4, 5, 7, 9, 11, 14–23) to also provide quantitative results. Recovery data were generated from fortified spinach extracts and determined using matrix-matched calibration standards and analyzed by GC-MS/MS and GC-MS/SIM. These results are used to assess the accuracy and precision of the method and are tabulated in Table 3.

4. Notes

1. For the cleanup in Subheading 3.4, step 1, the official QuEChERS procedure established by the Association of Official Analytical Chemists (AOAC) (11) and the Comité Européen de Normalisation (European Committee for Standardization, CEN) (12), graphitized carbon black is not used as the primary procedure in the dispersive clean-up step. However, other acetonitrile based extraction procedures for multiresidue pesticide analysis have utilized both tandem

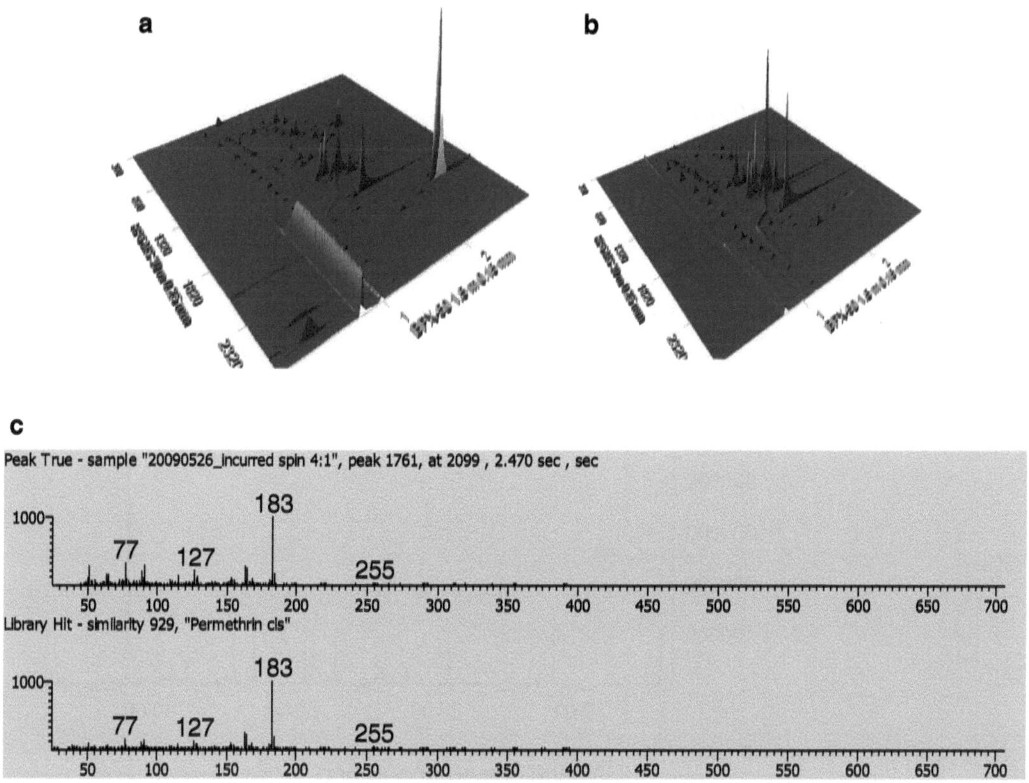

Fig. 4. GC × GC-TOF/MS chromatogram of an incurred spinach extract. (**a**) Extracted ions *m/z* = 183, 184, 163, 165 of *cis*- and *trans*-permethrin, (**b**) total ion chromatogram of the spinach extract. (**c**) spectra from incurred spinach extract (*top spectra*) and from the library (*bottom spectra*) of *cis*-permethrin.

graphitized carbon black (GCB) and primary-secondary amine (PSA) solid-phase extraction cartridges for effective cleanup (1, 2) The CEN (12), Walorczyk (13) and Li et al. (14) have utilized GCB/PSA without the toluene by using smaller amounts of GCB. GCB/PSA dispersive clean-up with toluene addition has been demonstrated by Mol et al. (6), Wong et al. (7, 9), and Schenck et al. (8, 19).

2. The official QuEChERS procedures usually involve the addition of acetic acid or citrate-based buffer for extraction (11, 12). The reason for the acid addition is that certain basic-based pesticides have better stability or extractability into the acetonitrile solvent. However, the addition of acid in the sample may also protonate the PSA or aminopropyl-based sorbents, thereby decreasing the effectiveness of the PSA sorbent to adsorb certain matrix co-extractives, such as organic acids. Many of the compounds that benefit from the buffered

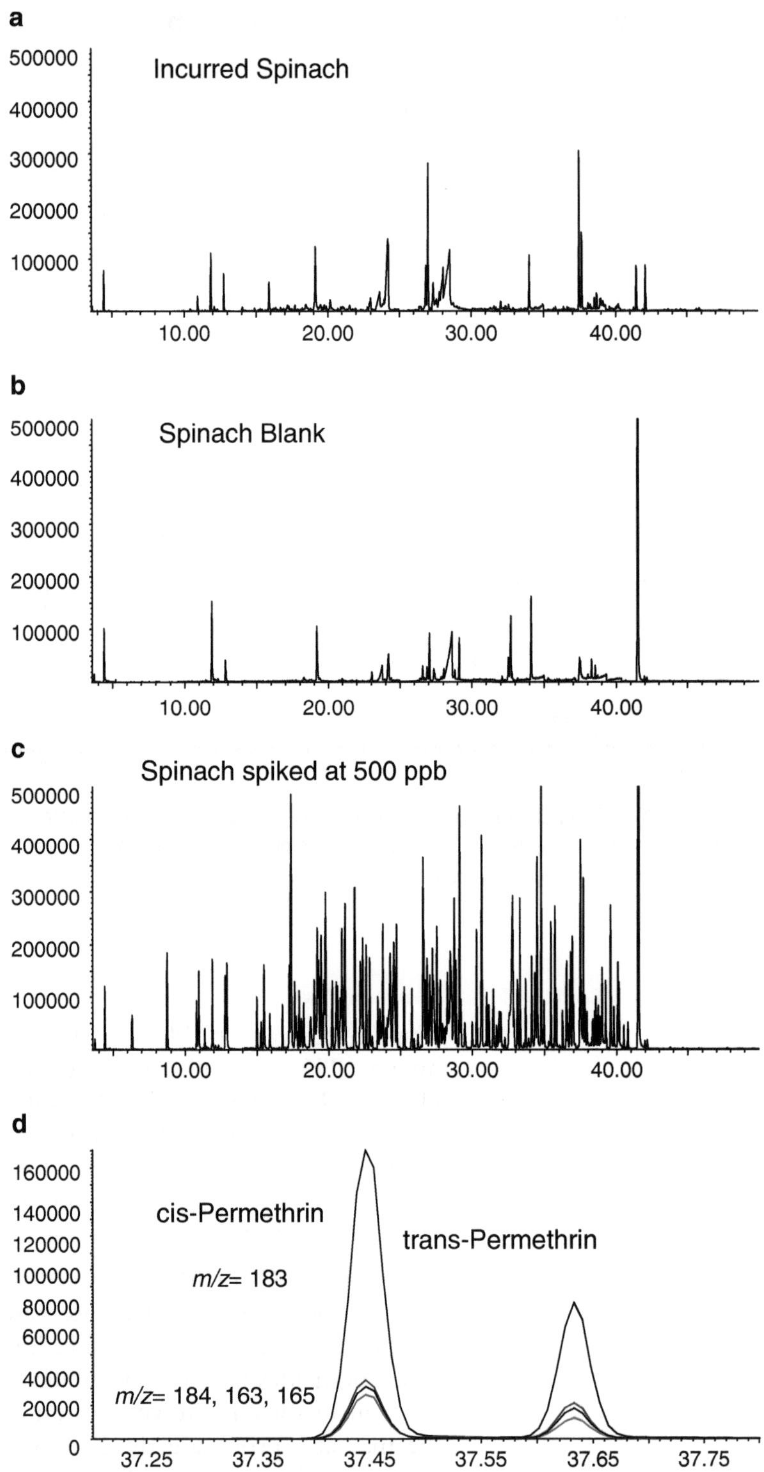

Fig. 5. GC-MS/SIM chromatograms of a spinach extract. (**a**) an incurred spinach extract, (**b**) organic spinach blank, (**c**) organic spinach extract fortified with pesticides at a concentration of 500 μg/kg and (**d**) extracted ions at m/z = 183, 184, 163, and 165 used to identify *cis*- and *trans*-forms of permethrin present in the incurred spinach extract in (**a**).

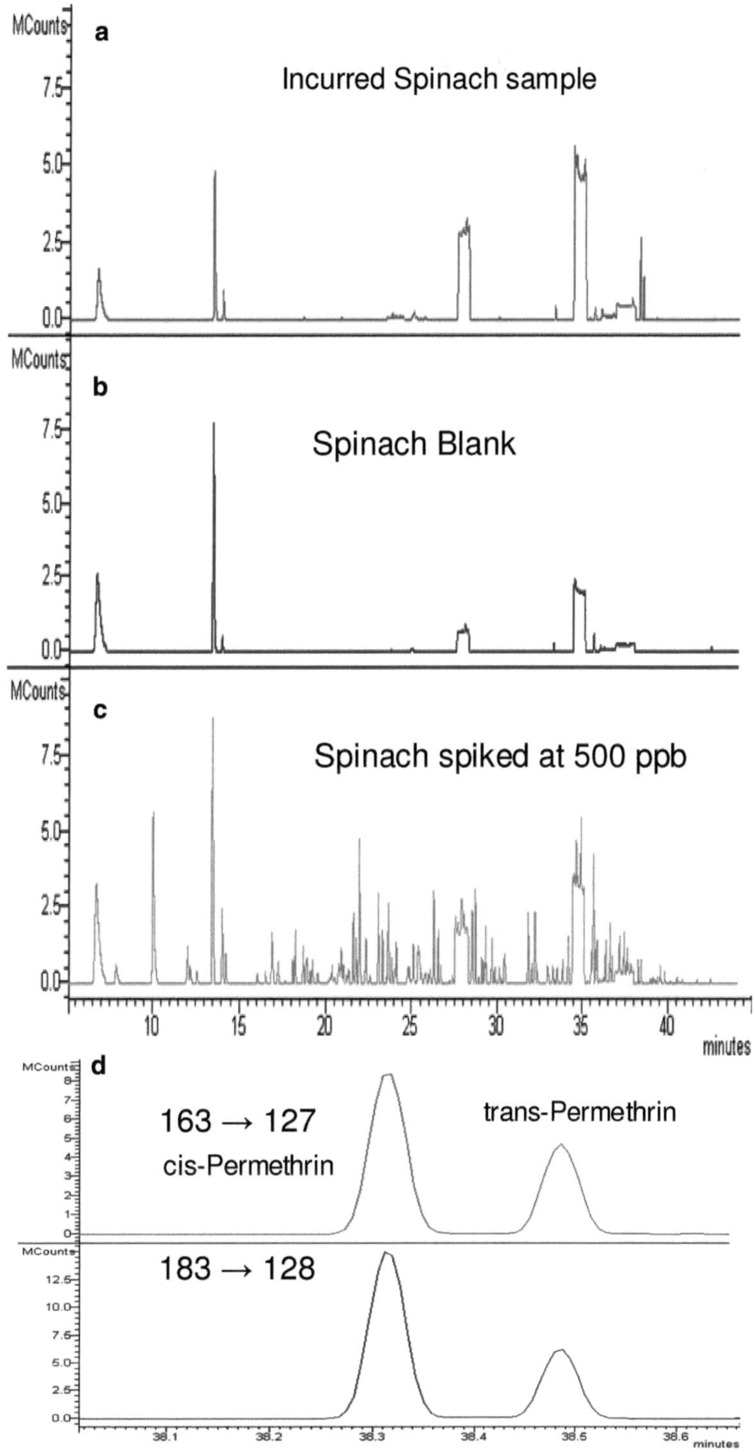

Fig. 6. GC-MS/MS chromatograms of a spinach extract. (**a**) An incurred spinach extract, (**b**) organic spinach blank, (**c**) organic spinach extract fortified with pesticides at a concentration of 500 μg/kg and (**d**) extracted ion transitions 163→127 and 183→128 used to identify *cis*- and *trans*-forms of permethrin present in the incurred spinach extract in (**a**).

Table 3
Recoveries (% average ± standard deviation) of pesticides extracted from spinach fortified at 10, 25, 100 and 500 ng/g (ppb) levels ($n = 4$ at each level) using modified QuEChERS procedure and analyzed by GC-MS/SIM and GC-MS/MS using matrix (spinach)-matched standards. *ND* not detected and *NA* not analyzed

	10 ng/g				25 ng/g				100 ng/g				500 ng/g			
	GC-MS/MS		GC-MS/SIM		GC-MS/MS		GC-MS/SIM		GC-MS/MS		GC-MS/SIM		GC-MS/MS		GC-MS/SIM	
Pesticide	AVE	RSD	AVE	RSD	AVE	RSD	AVE	RSD	AVE	RSD	AVE	RSD	AVE	RSD	AVE	RSD
Acrinathrin	ND		98	7	91	14	100	1	98	4	90	4	99	1	88	2
Akton	97	5	95	5	95	4	97	1	91	3	98	2	91	3	96	1
Alachlor	102	1	104	6	98	3	99	1	98	3	98	4	98	4	97	2
Aldrin	64	10	98	6	60	7	96	4	69	3	93	2	70	5	93	2
Allethrin	97	12	ND		103	5	ND		104	4	97	3	98	3	97	2
Atrazine	89	9	96	2	102	4	96	3	110	3	99	6	111	7	96	1
Azamethiphos	95	3	ND		98	1	ND		97	2	98	3	97	2	102	13
Azinphos-ethyl	96	3	92	12	93	4	99	1	92	3	99	2	93	3	98	1
Azinphos-methyl	102	6	ND		97	5	ND		96	4	93	9	95	2	92	7
BHC, alpha-	91	3	96	4	93	4	90	4	91	4	90	6	92	4	91	2
BHC, beta-	94	7	93	5	37	7	91	8	90	7	77	3	98	5	81	3
BHC, delta-	107	5	100	4	43	5	98	3	92	3	93	1	99	2	92	3
BHC, gamma- (Lindane)	92	3	ND		99	5	ND		102	5	93	1	106	4	90	3
Bifenthrin	95	5	89	3	90	2	97	2	87	2	97	2	87	3	95	1

Bromophos	94	3	93	3	85	3	93	2	86	3	92	2	80	3	91	1
Bromophos ethyl	87	5	91	6	93	1	89	1	82	3	89	2	85	3	87	2
Bromopropylate	102	8	94	1	96	3	99	0	90	2	99	1	91	2	99	2
Captafol	ND		ND		ND		ND		64	5	ND		35	14	ND	
Captan	ND		ND		79	49	ND		ND		ND		74	9	ND	
Carbophenothion	90	6	ND		121	2	102	2	91	3	103	1	95	3	100	0
Chlorbenzilate	103	6	101	4	91	8	98	2	93	3	100	1	93	2	100	1
Chlordane, *cis*-	98	4	96	2	96	2	94	1	88	3	94	2	88	2	94	1
Chlordane, *trans*-	98	4	74	2	96	2	75	3	88	3	75	1	88	2	73	2
Chlordene, alpha-	100	4	101	4	103	3	101	4	99	3	101	3	99	3	101	2
Chlordene, beta-	84	3	98	3	89	7	100	2	88	4	100	2	89	3	99	2
Chlordene, gamma-	95	5	ND		94	0	87	5	94	3	75	8	95	3	66	11
Chlorfenvinphos, beta-	119	2	122	8	99	3	119	3	114	3	120	4	116	2	119	1
Chloroneb	95	3	ND		98	2	97	3	93	4	92	6	97	4	90	2
Chlorothalonil	27	19	ND		17	7	ND		29	12	23	13	45	6	42	9
Chlorpyrifos	95	7	92	6	98	2	96	2	93	3	96	4	93	3	95	2
Chlorpyrifos-methyl	104	3	96	4	99	2	97	2	95	2	94	5	95	4	94	3
Chlorthiophos	95	6	94	10	95	3	95	2	92	3	99	1	91	3	96	1
Coumaphos	86	26	107	7	85	5	96	2	90	10	94	3	92	5	96	2

(continued)

Table 3
(continued)

Pesticide	10 ng/g				25 ng/g				100 ng/g				500 ng/g			
	GC-MS/MS		GC-MS/SIM		GC-MS/MS		GC-MS/SIM		GC-MS/MS		GC-MS/SIM		GC-MS/MS		GC-MS/SIM	
	AVE	RSD	AVE	RSD	AVE	RSD	AVE	RSD	AVE	RSD	AVE	RSD	AVE	RSD	AVE	RSD
Cyanazine	90	10	ND		89	6	ND		92	5	96	3	95	3	97	3
Cyanophos	98	9	ND		100	3	96	1	96	4	96	4	97	3	94	2
Cyfluthrin I	104	8	ND		117	4	98	5	93	7	93	3	95	2	92	3
Cyfluthrin II	104	8	ND		117	4	97	6	93	7	96	3	95	2	94	2
Cyfluthrin III	104	8	ND		117	4	96	4	93	7	93	5	95	2	95	2
Cyfluthrin IV	104	8	ND		117	4	96	5	93	7	98	2	95	2	92	2
Cyhalothri, lambda-	97	6	106	7	117	4	95	3	96	2	97	2	99	1	93	1
Cypermethrin I	99	8	ND		117	4	ND		93	7	97	3	95	2	101	2
Cypermethrin II	99	8	ND		117	4	ND		93	7	94	5	95	2	95	2
Cypermethrin III	98	7	ND		97	3	ND		94	4	98	2	97	1	99	4
Cypermetherin IV	99	8	ND		87	8	ND		93	7	94	4	95	2	93	2
Dacthal (DCPA)	97	8	98	3	98	2	101	2	95	1	101	4	96	4	99	2
DDD, o,p′-	98	6	102	1	93	2	98	1	89	2	99	2	88	2	98	2
DDD, p,p′-	98	6	101	4	93	2	94	3	89	2	88	1	88	2	87	1
DDE, o,p′-	97	3	101	4	94	1	98	0	90	3	97	2	90	3	95	2
DDE, p,p′-	123	4	116	4	122	2	112	2	119	3	112	1	121	3	120	1
DDT, o,p′-	98	6	101	4	93	2	95	3	89	2	88	1	88	2	88	1

DDT, p,p′-	87	10	ND		82	2	87	5	82	2	92	1	86	3	94	4
DEF (Tribufos)	96	3	ND		96	4	102	10	93	2	96	2	92	3	92	1
Deltamethrin	100	9	ND		90	6	90	8	91	7	92	7	95	6	94	2
Demeton-S-methyl	95	4	90	6	NA		93	5	92	5	92	6	94	4	92	2
Demeton-S	NA		ND		96	3	99	4	NA		93	9	NA		95	2
Dialifor	96	4	102	8	94	3	99	5	94	3	97	1	97	2	95	1
Diallate 1	99	2	ND		98	4	ND		97	3	93	7	98	4	97	3
Diallate 2	98	5	ND		100	4	ND		96	4	97	3	97	4	100	3
Diazinon	99	4	101	5	101	2	100	2	95	4	101	4	97	3	99	2
Dicapthon	86	3	95	3	95	1	88	4	85	3	88	2	86	3	88	2
Dichlofluanid	9	1	ND		88	2	13	26	23	22	22	18	30	10	30	7
Dichlofenthion	97	6	96	5	9	8	98	2	90	3	96	5	91	3	96	2
Dichloroaniline, 3,4-	67	15	80	5	57	29	77	4	74	10	70	10	69	5	71	5
Dichlorobenzophenone, 4,4′-	98	3	ND		99	2	92	3	96	3	96	1	96	3	97	2
Dichlorvos	83	4	93	9	90	5	94	3	87	8	87	8	95	4	93	3
Diclobenil	47	9	90	5	94	2	89	1	85	8	84	8	90	5	89	2
Dicloran	89	6	95	6	91	3	87	2	89	3	88	3	93	4	90	3
Dieldrin	96	5	ND		91	1	ND		94	4	ND		92	3	98	2
Dimethachlor	102	4	94	4	101	3	96	2	99	2	97	5	99	3	97	2

(continued)

Table 3
(continued)

Pesticide	10 ng/g				25 ng/g				100 ng/g				500 ng/g			
	GC-MS/MS		GC-MS/SIM		GC-MS/MS		GC-MS/SIM		GC-MS/MS		GC-MS/SIM		GC-MS/MS		GC-MS/SIM	
	AVE	RSD	AVE	RSD	AVE	RSD	AVE	RSD	AVE	RSD	AVE	RSD	AVE	RSD	AVE	RSD
Dioxathion	92	8	ND		98	4	103	5	99	2	100	2	101	4	100	1
Disulfoton	90	3	90	8	92	2	89	4	89	5	91	8	91	4	88	3
Ditalimfos	56	18	ND		56	6	57	9	59	8	60	10	62	6	61	4
Edifenphos	93	4	ND		96	3	ND		92	2	91	4	92	3	91	4
Endosulfan ether	94	2	ND		86	7	92	2	89	2	94	2	92	3	95	2
Endosulfan I	81	5	ND		94	3	ND		95	1	95	3	95	3	94	6
Endosulfan II	89	7	ND		85	6	ND		100	3	96	5	99	2	97	4
Endosulfan sulfate	88	10	100	7	87	4	96	3	90	1	87	5	95	4	87	2
Endrin	99	7	ND		115	10	ND		116	3	84	4	121	2	84	2
Endrin aldehyde	70	8	ND		50	3	ND		38	19	42	22	46	12	43	8
Endrin ketone	79	17	ND		83	4	71	17	91	4	64	16	90	3	60	8
EPN	95	3	98	9	89	5	97	3	92	4	100	2	94	4	99	1
Ethalfluralin	85	7	94	4	88	4	99	4	92	3	96	6	98	5	97	1
Ethion	100	4	104	5	100	2	100	2	95	3	102	1	97	3	100	1
Ethoprop	101	5	107	9	100	4	107	3	98	3	100	3	99	4	100	2
Etridiazole	80	21	ND		93	3	85	7	86	6	63	7	93	5	69	7
Famphur	92	7	ND		93	3	ND		93	2	100	2	94	4	99	2

Fenamiphos	84	5	93	3	34	5	90	2	92	2	90	5	96	3	88	2
Fenarimol	114	28	83	5	75	3	107	4	95	4	99	5	97	3	99	3
Fenchlorphos (Ronnel)	100	3	96	4	93	5	97	3	91	2	96	5	91	3	95	2
Fenitrothion	98	4	99	12	103	2	100	3	98	3	98	3	97	4	98	2
Fensulfothion	109	1	100	4	107	2	99	4	111	2	108	2	112	3	105	1
Fenthion	107	5	99	5	105	4	100	3	96	4	98	4	97	3	98	2
Fenvalerate 1	98	11	ND		88	5	96	3	90	7	95	4	91	4	92	2
Fenvalerate 2	107	18	ND		88	5	99	4	90	7	97	3	91	4	94	2
Fluchloralin	82	8	95	4	87	5	96	2	88	3	95	3	96	4	94	2
Flucythrinate 1	112	5	ND		105	5	98	2	111	6	97	3	115	3	100	1
Flucythrinate 2	90	8	ND		82	5	98	2	87	7	97	3	87	4	93	2
Fluridone	349	64	158	3	151	10	148	4	154	11	148	1	160	15	144	3
Fluvalinate 1	86	8	97	2	79	4	95	3	90	9	94	3	94	4	93	1
Fluvalinate 2	98	16	98	2	79	4	91	2	90	9	97	4	94	4	93	1
Folpet	22	0	ND		31	6	ND		48	10	53	34	50	9	38	12
Fonofos	100	5	ND		99	3	96	2	95	3	95	5	97	3	94	2
Heptachlor	112	4	101	10	112	3	89	7	105	3	82	3	107	3	84	3
Heptachlor epoxide	93	4	NA		97	1	NA		92	3	NA		93	3	NA	

(continued)

Table 3 (continued)

	10 ng/g				25 ng/g				100 ng/g				500 ng/g			
	GC-MS/MS		GC-MS/SIM		GC-MS/MS		GC-MS/SIM		GC-MS/MS		GC-MS/SIM		GC-MS/MS		GC-MS/SIM	
Pesticide	AVE	RSD	AVE	RSD	AVE	RSD	AVE	RSD	AVE	RSD	AVE	RSD	AVE	RSD	AVE	RSD
Hexachlorobenzene	70	1	78	1	69	3	75	5	69	5	72	3	69	5	72	2
Iprobenfos	94	7	89	4	92	2	91	3	91	3	90	5	92	3	90	2
Iprodione	91	18	87	7	85	4	88	1	89	5	89	3	92	2	87	2
Isazophos	96	2	ND		98	2	101	2	95	3	97	6	95	3	94	2
Isofenphos	104	5	ND		103	3	107	7	99	4	105	4	100	3	101	1
Iodofenphos	104	3	105	2	99	3	101	1	93	2	99	1	95	3	99	1
Leptophos	84	3	90	4	84	4	85	2	85	3	88	2	85	3	87	2
Malathion	102	4	ND		103	2	106	10	101	3	100	1	101	3	101	1
Methidathion	101	5	ND		101	3	100	8	96	2	71	58	97	3	101	1
Methoxychlor,o,p′-	88	8	ND		84	3	85	10	83	2	63	15	88	4	60	8
Methoxychlor,p,p′-	95	6	ND		92	4	80	17	88	1	56	24	89	5	57	13
Metolachlor	103	3	99	4	104	2	101	2	95	3	103	2	96	3	102	1
Mevinphos	98	5	98	4	101	3	98	4	97	5	96	3	97	4	94	1
Mirex	87	6	84	4	91	2	74	4	80	2	63	9	75	2	62	2
Nonachlor trans	93	6	97	7	80	2	90	3	88	2	89	1	89	2	89	2
Nonachlor, *cis-*	93	6	95	3	92	2	95	2	89	3	93	1	87	2	91	2

Oxadiazon	105	7	104	4	100	1	99	2	100	3	101	1	99	3	101	1
Parathion	94	5	ND		96	1	98	1	95	2	99	2	98	3	99	1
Parathion methyl	100	3	92	3	98	2	93	2	94	3	92	4	96	4	93	3
Pentachloroaniline	77	6	75	6	90	2	74	3	72	3	74	3	72	3	75	2
Pentachlorobenzene	84	3	89	4	81	3	85	1	77	4	83	4	82	4	84	3
Pentachloroben-zonitrile	75	5	78	6	74	2	77	4	75	4	77	2	75	4	76	1
Pentachlorophenyl methyl ester	90	1	95	5	75	5	91	2	86	3	90	2	87	4	88	3
Pentachloro-thioanisole	67	5	75	7	70	3	74	5	72	3	73	2	69	4	72	2
Permethrin, *cis-*	100	6	ND		93	3	94	2	92	3	96	1	95	1	94	8
Permethrin, *trans-*	98	3	ND		91	6	98	1	92	4	97	2	94	1	95	1
Phenothrin	22	19	101	9	72	9	98	3	93	2	98	2	92	3	96	2
Phenthoate	NA		106	3	NA		100	4	NA		100	3	NA		100	1
Phorate	97	4	ND		98	2	96	3	94	3	96	5	97	4	96	1
Phosalone	78	3	99	5	90	3	97	5	97	3	98	4	97	2	97	1
Phosmet	96	3	ND		93	4	95	3	92	3	94	5	93	3	95	3
Pirimiphos ethyl	96	5	101	4	98	1	101	2	92	3	100	3	93	3	98	1
Pirimiphos-methyl	100	6	102	3	102	4	100	2	96	4	99	4	97	3	99	1
Procymidone	104	3	98	4	99	3	98	1	97	3	100	2	97	3	99	1

(continued)

Table 3 (continued)

Pesticide	10 ng/g				25 ng/g				100 ng/g				500 ng/g			
	GC-MS/MS		GC-MS/SIM		GC-MS/MS		GC-MS/SIM		GC-MS/MS		GC-MS/SIM		GC-MS/MS		GC-MS/SIM	
	AVE	RSD	AVE	RSD	AVE	RSD	AVE	RSD	AVE	RSD	AVE	RSD	AVE	RSD	AVE	RSD
Profenofos	94	7	ND		92	2	92	10	91	1	87	4	91	3	91	2
Propachlor	95	5	98	5	97	5	96	2	96	3	97	4	99	4	97	1
Propazine	78	5	105	5	95	5	99	1	102	2	99	4	104	4	96	2
Propetamphos	97	4	91	8	101	2	97	3	98	4	98	3	100	4	98	3
Propyzamide	100	3	96	2	99	2	96	1	96	3	96	4	98	4	95	1
Prothiophos	93	6	ND		95	3	97	2	89	3	97	2	89	2	96	1
Pyraclofos	106	9	102	3	99	6	98	6	100	5	102	4	102	3	100	3
Pyrazophos	102	4	98	1	93	4	100	2	93	2	98	1	94	1	96	2
Pyridaphenthion	84	9	109	1	93	6	99	1	97	3	99	1	98	2	99	1
Quinalphos	98	4	ND		98	2	101	4	96	2	101	4	96	3	98	2
Quintozene	80	7	89	6	83	5	93	4	86	2	91	3	89	3	89	2
Resmethrin	80	6	78	9	98	2	81	4	74	3	81	7	67	2	75	2
Simazine	76	13	105	8	75	10	103	5	77	19	98	3	87	7	95	1
Sulfotep-ethyl	96	5	96	5	97	3	98	2	92	3	96	5	94	4	96	2
Sulprofos	97	6	97	3	95	2	98	1	92	4	98	1	93	3	97	2

Tebupirimfos	94	5	105	2	102	4	103	2	95	3	101	5	96	4	99	3
Tecnazene	89	5	99	1	91	3	95	3	90	2	95	4	93	4	93	3
Tefluthrin	94	3	95	5	NA		95	2	90	3	94	4	91	3	94	2
Temephos	NA		ND		NA		ND		NA		NA		NA		ND	
Terbufos	NA		98	4	95	2	98	3	NA		98	4	NA		96	2
Terbuthylazine	88	1	ND		92	5	98	2	97	2	99	4	99	2	96	2
Tetrachloraniline, 2,3,5,6-	92	4	94	4	92	1	92	2	87	3	91	4	89	4	91	2
Tetrachlorvinphos	89	6	97	5	89	3	95	2	91	2	94	2	96	3	94	3
Tetramethrin	98	6	102	5	98	3	97	2	97	3	98	1	97	2	97	1
Tolclofos methyl	98	3	96	4	95	1	97	1	87	2	94	4	89	3	94	2
Tolylfluanid	12	0	ND		17	6	ND		34	17	27	18	41	7	35	9
Triallate	102	5	81	2	96	2	76	2	94	3	84	5	94	3	106	2
Triazophos	105	1	ND		100	4	96	13	102	2	105	4	100	3	103	1
Trifluralin	80	3	99	2	82	7	96	2	89	1	95	3	93	3	96	2
Vinclozolin	97	5	102	3	99	4	96	4	96	3	97	3	96	2	97	2

solution, such as captafol and captan can be difficult to analyze by GC-MS because they fragment poorly using electron impact ionization. Therefore, the non-buffered QuEChERS procedure was selected to utilize the full benefit of the PSA effectiveness for a better clean-up. In this procedure, the original extraction procedure from Anastassiades et al. (10) was used, utilizing anhydrous magnesium sulfate and sodium chloride, rather than sodium acetate or sodium citrate.

3. Spinach extracts were prepared using the modified QuEChERS procedure outlined in Fig. 1. The four extracts were concentrated by a factor of 4.5 and fortified with approximately 40 pesticides at a concentration 50 ppb (equivalent to 10 ppb at a 1 g/mL basis). The extracts were analyzed evaluated by GC-MS/SIM and GC×GC TOF-MS to screen for the pesticides in the extracts. An example of the GC×GC-TOF/MS chromatogram of Group 4 pesticides is shown in Fig. 2. The extract was separated using a non-polar column (HP-5MS) in the first separation and a polar column (BTX-50) in the second. 38 of 39 pesticides were evaluated using the ChromaTOF software and NIST library. The remaining results are provided in Table 1 and reveal 142 of 158 (89.9%) and 124 of 154 (80.5%) of the pesticides were found by GC-full scan MS and GC×GC TOF-MS, respectively. The pesticides that could not be found in the spinach extracts by either method were captan, captafol, cyfluthrin, folpet, temephos, azamethiphos, and endrin aldehyde.
4. The experiment described in Table 1 illustrates GC-full scan MS and GC×GC-TOF/MS as examples of non-targeted acquisition screening. Non-targeted screening implies that the analyst does not have any prior knowledge of what is present in the sample but by using the appropriate software tools and a database or library of compounds, a tool for the analysts is provided to screen for potential unknown pesticides or contaminants in the sample. GC-full scan techniques are not very sensitive but potentially can be used for quantitative purposes. The potential problem is whether the peak corresponding to the analyte can be properly identified in full scan spectra or with the AMDIS and other software algorithms that are used for peak purity. GC×GC-TOF/MS and other GC-high mass resolution TOF instruments have been successfully shown to quantitate pesticide levels in plant food matrices (15, 23) but they have limited linear dynamic ranges (15–18).
5. Targeted GC-MS methods such as GC-MS/SIM (1, 2, 7) and GC-MS/MS (5, 17, 19–22) are very effective for pesticide screening. These targeted methods have the advantage of screening, quantifying, and identifying the target analytes in a single injection. However, the disadvantage of targeted

procedures is that analytes that are not suspected of being present in the food sample would go through the detection unnoticed and unidentified. Therefore, any new pesticide or contaminant present not recognized by the screen would not be detected in the food sample. Table 2 lists targeted parameters for 166 pesticides to be screened by GC-MS/SIM and GC-MS/MS. These parameters include retention time, target and qualifier ions and percentages of qualifier-to-target ratio used for GC-MS/SIM as well as retention times, transitions from precursor-to-product ions, and ion ratios between the quantitative and confirmatory transitions used for GC-MS/MS.

6. Recovery studies were determined by fortifying spinach samples with spike solutions to final concentrations of 10, 25, 100, and 500 μg/kg (ppb). The samples were then processed according to the procedures listed in Subheading 3.1 and 166 organohalogen, organophosphorus and pyrethroid pesticides, metabolites and isomers targeted and analyzed by GC-MS/SIM and GC-MS/MS. Recovery results at the four fortification levels are listed in Table 3. Of those examined, 65 analytes could not be analyzed or detected by GC-MS/SIM at the 10 ppb level compared to three analytes by GC-MS/MS. A larger number of pesticides could be detected by GC-MS/SIM at the 25 ppb fortification level and higher. This difference between the two techniques is due to the selectivity of GC-MS/MS over GC-MS/SIM. Compounds that could not be detected by GC-MS/MS are most likely due to the lack of stable fragments, such as captan and captafol. For GC-MS/SIM, the major problem is the lack of identification due to interferences from the spinach matrix. At the 100 and 500 μg/kg levels, there seem to be no sig nificant difference between the two MS techniques. Mean recoveries were 93 ± 26, 91 ± 18, 91 ± 14, 92 ± 14 and 97 ± 10, 94 ± 11, 92 ± 15, 92 ± 14 at the 10, 25, 100 and 500 μg/kg fortification levels for GC-MS/SIM and GC-MS/MS, respectively. From the mean recoveries and relative standard deviations and the results of the individual pesticides listed in Table 3, most of the pesticides were in the preferred 70–120% recovery range with relative standard deviations <20%. The recovery results reveal that the method was effective for analyzing pesticides at the four fortification levels.

7. The planar and aromatic compounds such as hexachlorobenzene and pentachlorobenzene had recoveries ranging from 69 to 78% and 77 to 89%, respectively, at the four fortification levels using GC-MS/MS or GC-MS/SIM. This indicates that the use of toluene was effective to desorb these compounds from the graphitized carbon. As expected, the pesticides that

had poor recoveries (<50% at all levels) were captan, captafol, chlorothalonil, dichlofluanid, endrin aldehyde, folpet, temephos, and tolylfluanid. The one pesticide that had high recoveries (>120%) at all fortification levels was fluridone, which cannot be fully explained. These compounds are either unstable during the extraction process, interact or adsorb with the clean-up sorbents, fragment poorly under electron impact ionization, or chromatograph poorly under the conditions provided.

8. Suspected spinach samples containing incurred residues were analyzed using the procedures listed in Subheading 3.1 and the extracts of a spinach sample were subjected to non-targeted acquisition analysis using GC-full scan MS and GC×GC-TOF/MS. Chromatograms and findings for each of these methods are shown in Figs. 3 and 4. Figure 3a shows the total ion chromatogram from a GC equipped with single quadrupole mass spectrometer and operated in full scan mode with the deconvolution software and NIST library. Figure 3b illustrates the mass spectra generated from the AMDIS/NIST results to match that of the *trans*- and *cis*-forms of the pyrethroid insecticide, permethrin. Figure 4 illustrates the use of GC×GC TOF-MS analysis to identify permethrin in the same spinach sample using extracted ions (m/z = 183, 184, 163 and 165) (Fig. 4a) and full scan chromatogram (Fig. 4b) as well as the matching mass spectrum of one of the permethrin peaks (Fig. 4c).

9. Results of targeted approaches such as GC-MS/SIM and GC-MS/MS are illustrated in Figs. 5 and 6, respectively for the same spinach extract screened by non-targeted approaches. GC-MS/SIM and GC-MS/MS programs using the parameters in Table 2 are used to screen and quantitate for 166 pesticides that could be present in the spinach sample. In the GC-MS/SIM result in Fig. 5, *cis*- and *trans*-permethrin in the first spinach extract were determined and identified by their retention times and m/z values of each pesticide's target and qualifying ions, and the ratios of the qualifier ions to the target ion. In Fig. 5, the reconstructed GC-MS/SIM chromatograms of the suspected spinach (5a), a pesticide-free spinach blank (5b), the blank spiked with 166 pesticides at 500 μg/kg (5c) and extracted ions m/z = 183, 184, 163, and 165 of the *cis*- and *trans*-permethrin are shown (5d). The same spinach extract was subjected to GC-MS/MS analysis as shown in Fig. 6 for the suspected sample (6a), blank (6b), fortified spinach extract (6c), and extracted ion transitions, 163→127 and 183→128 and ion ratio used to identify both permethrin isomers (6d). Both procedures were effective to identify permethrin in the incurred spinach samples.

10. GC-MS/SIM and GC-MS/MS were used to quantitate the presence of *cis*- and *trans*-permethrin in the spinach. The levels of *cis*- and *trans*-permethrin in spinach were 214 ± 7 and 96 ± 4 μg/kg and 212 ± 4 and 104 ± 4 μg/kg as determined by GC-MS/SIM and GC-MS/MS, respectively.

11. The use of a modified QuEChERS method to spinach and non-targeted acquisition and targeted procedures using different GC types were shown to be effective. The combined use of these techniques in this chapter could lead to a comprehensive approach to pesticide screening of a large number of pesticides in a wide variety of fresh produce commodities.

References

1. Fillion, J.; Sauve, F.; Selwyn, J. Multiresidue method for the determination of residues of 251 pesticides in fruits and vegetables by gas chromatography/mass spectrometry and liquid chromatography with fluorescence detection:chromatographic pesticide residue analysis. *J. AOAC Int.* **2000**, **83**, 698–713.
2. Pang, G. F.; Fan, C. L.; Liu, Y. M.; Cao, Y. Z.; Zhang, J. J.; Li, X. M.; Li, Z. Y.; Wu, Y. P.; Guo, T. T. Determination of residues of 446 pesticides in fruits and vegetables by three-cartridge solid-phase extraction–gas chromatography–mass spectrometry and liquid chromatography–tandem mass spectrometry *J. AOAC Int.* **2006**, **89**, 740–771.
3. Gómez. M. J., Gómez-Ramos, M. M., Agüera, A., Mezcua, M., Herrera, S., Fernández-Alba, A. R. (2009) A new gas chromatography/mass spectrometry method for the simultaneous analysis of target and non-target organic contaminants in waters. *J. Chromatogr. A.* **1216**, 4071–4082.
4. Zrostlíková, J., Hajšlová, J., Čajka, T. (2003). Evaluation of two-dimensional gas chromatography-time-of-flight mass spectrometry for the determination of multiple pesticide residues in fruit. *J. Chromatogr. A.* **1019**, 173–186.
5. Bolaños, P. P., Moreno, J. L. F., Shtereva, D. D., Frenich, A. G., Vidal, J. L. M. (2007). Development and validation of a multiresidue method for the analysis of 151 pesticide residues in strawberry by gas chromatography coupled to a triple quadrupole mass analyzer. *Rapid Commun. Mass Spectrom.* **21**, 2282–2294.
6. Mol, H. G. J.; Rooseboom, A.; van Dam, R.; Roding, M.; Arondeus, K.; Sanarto, S. Modification and re-evaluation of the ethyl acetate-based multi-residue method for pesticides in produce. *Anal. Bioanal. Chem.* **2007, 389**, 1715–1754.
7. Wong, J. W.; Hennessy, M. K.; Hayward, D. G.; Krynitsky, A. J.; Cassias, I.; Schenck, F. J. Analysis of organophosphorus pesticides in dried ground ginseng root by capillary gas chromatography-mass spectrometry and -flame photometric detection. *J. Agric. Food Chem.* **2007**, **55**(4), 1117–1128.
8. Schenck, F., Wong, J., Lu, C., Li, J., Holcomb, J. R., Mitchell, L. M. (2009) Multiresidue Analysis of 102 Organophosphorus Pesticides in Produce at Parts-per-Billion Levels Using a Modified QuEChERS Method and Gas Chromatography with Pulsed Flame Photometric Detection. *J. AOAC Int.* **92**, 561–573.
9. Wong, J. W., Zhang, K., Tech, K., Hayward, D. G., Makovi, C. M., Krynitsky, A. J., Schenck, F. J., Banerjee, K., Dasgupta, S., Brown, D. (2010) Multiresidue pesticide analysis in fresh produce by capillary gas chromatography-mass spectrometry/selective ion monitoring (GC-MS/SIM) and –tandem mass spectrometry. *J. Agric. Food Chem.* **58**, 5868–5883.
10. Anastassiades, M., Lehotay, S. J., Stajnbaher, D., Schenck, F. J. 2003. Fast and easy multiresidue method employing acetonitrile extraction/partitioning and "dispersive solid-phase extraction" for the determination of pesticide residues in produce. *J. AOAC Int.* **86**, 412–431.
11. AOAC Official Method 2007.01. Pesticide residues in foods by acetonitrile extraction and partitioning with magnesium sulfate. Gas chromatography/mass spectrometry and liquid chromatography/tandem mass spectrometry. 2007 AOAC International.
12. prEN 15662-version 2007-10-24. Foods of plant origin – Determination of pesticide residues using GC-MS and/or LC-MS/MS following acetonitrile extraction/partitioning

and cleanup by dispersive SPE – QuEChERS-methods.

13. Walorczyk, S. Application of gas chromatography/tandem quadrupole mass spectrometry to the multi-residue analysis of pesticides in green leafy vegetables. *Rapid Commun. Mass Spectrom.* **2008**, **22**(23), 3791–3801.
14. Li, L., Li, W., Qin, D., Jiang, S., Liu, F. Application of graphitized carbon black to the QuEChERS method for pesticide multiresidue analysis in spinach. J. AOAC Int. 2009, **92**(2), 538–547.
15. van der Lee, M. K., van der Weg, G., Traag, W. A., Mol, H. G. J. Qualitative screening and quantitative determination of pesticides and contaminants in animal feed using animal feed using comprehensive two-dimensional gas chromatography with time-of-flight mass spectrometry. *J. Chromatogr. A.* **2008,** 1186, 325–339.
16. Patil, S. H., Banerjee, K., Dasqupta, S., Oulkar, D. P., Patil, S. B., Jadhay, M. R., Savant, R. H., Adsule, P. G., Deshmukh, M. B. (2009) Multiresidue analysis of 83 pesticices and 12 dioxin-like polychlorninated biphenyls in wine by gas chromatography-time-of-flight mass spectrometry. *J. Chromatogr. A.* **1216**, 2307–2319.
17. Leandro, C. C., Hancock, P., Fussell, R. J., Keely, B. J. 2007. Quantification and screening of pesticide residues in food by gas chromatography-exact mass time-of-flight mass spectrometry. *J. Chromatogr. A.* **1166**, 152–162.
18. Hayward, D. G., Wong, J. W. 2009. Organohalogen and organophosphorus pesticide method for ginseng root-a comparison of gas chromatography-single quadrupole mass spectrometry with high resolution time-of-flight mass spectrometry. *Anal. Chem.* **81**(14), 5716–5723.
19. Walorczyk, S. (2007) Development of a multiresidue screening method for the determination of pesticides in cereals and dry animal feed using gas chromatography-triple quadrupole tandem mass spectrometry. *J. Chromatogr. A.* **1165**, 200–212.
20. Leandro, C. C., Fussell, R. J., Keely, B. J. (2005) Determination of priority pesticides in baby foods by gas chromatography tandem quadrupole mass spectrometry. *J. Chromatogr. A.* **1085**, 207–212.
21. Moreno, J. L. F., Frenich, A. G., Bolaños, P. P., Vidal, J. L. M. (2008). Multiresidue method for the analysis of more than 140 pesticide residues in fruits and vegetables by gas chromatography coupled to triple quadrupole mass spectrometry. *J. Mass Spectrom.* **43**, 1235–1254.
22. Payá, P., Anastassiades, M., Mack, D., Sigalova, I., Tasdelen, B., Oliva, J., Barba, A. 2007. Analysis of pesticide residues using the Quick Easy Cheap Effective Rugged and Safe (QuEChERS) pesticide multiresidue method in combination with gas and liquid chromatography and tandem mass spectrometric detection. *Anal. Bioanal. Chem.* **389**, 1697–1714.
23. Banerjee, K., Patil, S. H., Dasqupta, S., Oulkar, D. P., Patil, S. B., Savant, R., Adsule, P.G. (2008). Optimization of separation and detection conditions for the multiresidue analysis of pesticides in grapes by comprehensive two-dimensional gas chromatography-time-of-flight mass spectrometry. *J. Chromatogr. A.* **1190**, 350–357.

Chapter 7

Targeted Pesticide Residue Analysis Using Triple Quad LC-MS/MS

Lutz Alder

Abstract

The determination of pesticide residues by HPLC-MS/MS requires decisions on a multitude of analytical parameters. This includes the selection of eluents, columns and ion sources, but also the optimization of the tandem mass spectrometer for the selected target analytes. Another aspect is the use of the restricted acquisition time between two chromatographic data points. An appropriate selection of all these parameters as well as the measures to avoid interference by cross talks and wrong quantitative results by matrix effects is discussed in this chapter.

Key words: Pesticides, Multi-residue methods, Electrospray, ESI, Atmospheric pressure chemical ionization, APCI, Mass spectrometry, MS/MS, SRM, Dwell time, Cycle time, Scheduled MRM, Matrix effects

1. Introduction

Based on the compilation of the British Crop Protection Council, approximately 880 active substances are formulated in pesticide products currently (1). These substances belong to more than 100 substance classes. Azoles, benzoylureas, carbamates, organophosphorous compounds, pyrethroids, sulfonylureas or triazines are the most important groups. The chemical and physical properties of pesticides may differ considerably.

Probably, no other use of chemicals is regulated more extensively than that of pesticides. Maximum residue levels (or tolerances) have been established for pesticides in foodstuffs and drinking water in most countries to avoid any adverse impact on public health, and to insist on good agricultural practice.

Depending on the purpose, determination of pesticide residues may be targeted analysis or non-target analysis. An example

Jerry Zweigenbaum (ed.), *Mass Spectrometry in Food Safety: Methods and Protocols*,
Methods in Molecular Biology, vol. 747, DOI 10.1007/978-1-61779-136-9_7, © Springer Science+Business Media, LLC 2011

of targeted analysis is the inspection of maximum residue limits (MRLs) in food. All analytes, which have to be quantified, are fixed by the residue definitions given in the MRL regulation. Such targeted analysis does not require a complete analysis of all compounds extracted from the sample, which is possible with, e.g., the combination of liquid chromatography and time-of-flight mass spectrometers (LC-TOF-MS) or mass spectrometers based on orbitrap technology (LC-Orbitrap).

In targeted analysis using LC-TOF-MS or LC-Orbitrap the reduction of background signals produced from sample matrix and HPLC solvent clusters is obtained by the use of very narrow mass ranges for the reconstruction of ion chromatograms (e.g., ±5 milli amu around the mass of the target analyte). However, this type of background reduction needs a very high mass resolution to avoid false positives from coeluting matrix components of similar mass.

In tandem mass spectrometry, the chemical background can be reduced with low resolution quadrupoles, if an acquisition in selected-reaction-monitoring mode (SRM) is applied. Even if a co-extracted matrix component and a pesticide have the identical molecular mass, usually both (nominal) isobaric ions can be separated in SRM experiments, because their fragmentation in the collision cell most often results in different product ions. Therefore, tandem mass spectrometers offer excellent sensitivity and selectivity. For this reason, triple quadrupole mass analyzers have been the most often applied MS detectors in targeted analysis (2).

However, multi-residue methods based on tandem mass spectrometry require a huge number of acquisition parameters. Some aspects of the selection of best parameters will be explained in this chapter.

2. Materials

In pesticide residue analysis using triple quad MS/MS usually no unique materials are used. However, compared to GC-MS the applicability of this technique on more polar and more labile pesticides requires a reasonable care on the stability of standard solutions. A short discussion of matrix-matched standards is needed due to the regular occurrence of significant matrix effects in electrospray ionization.

2.1. External Standards (Native Compounds)

2.1.1. Pesticide Stock Solutions

Individual stock solutions of analytical standards are prepared at concentrations that are sufficiently high to allow the preparation of complex pesticide mixtures. The solvent used should not negatively influence the stability of the pesticides employed.

2.1.2. Pesticide Mixtures

Because of the broad applicability of the targeted pesticide residue analysis using triple quad MS/MS and due to the partly divergent pH-stability of pesticides, analyte mixtures of different composition may be needed. These are prepared by mixing together defined volumes of the required analyte stock solutions and appropriately diluting them with a suitable solvent (e.g., methanol, acetonitrile). The analyte concentrations in the mixtures should be sufficient to allow the preparation of matrix-matched standards with moderate dilution of the blank sample extracts (e.g., less than 20%).

2.1.3. Working Standard Solutions (in Solvent)

Solvent-based standards are prepared by mixing a certain volume of an appropriate solvent (e.g., methanol, acetonitrile or its mixtures with water) with known amounts of pesticide mixtures. The preparation of multiple standards of different pesticide concentration is useful to cover a broad concentration range.

2.1.4. Matrix-Matched Standards

Matrix-matched standards are prepared in the same way as the solvent-based standards. But instead of pure solvents final extracts of blank samples are used. To minimize differences caused by matrix induced effects during ionization, it is best to choose the same commodity (e.g., apple for apple samples, carrots for carrot samples, etc.).

2.2. Internal Standards

2.2.1. Surrogate Standards

A surrogate standard is added to the test portion before or during extraction and may be used to correct for deviations in the volume of extraction solvent, to correct for losses of solvent by uncontrolled evaporation or in steps, which require dilution to a fixed volume.

Typical surrogate standards used in pesticide residue analysis are triphenyl phosphate (for ESI+), tris-(1,3-dichlorisopropyl)-phosphate (for ESI+ and ESI−) and bis-nitrophenyl urea (common name: nicarbazin; for ESI−).

2.2.2. Stable Isotope-Labeled Standards

Stable isotope-labeled standards are mainly used to compensate for matrix effects; but also losses of an analyte during extraction and cleanup may be corrected by stable isotope-labeled standards. In both cases a relatively high concentrated solution of these standards is needed. A small volume of this solution is either added to the final extract or to the test portion before extraction. In the first case matrix effects are corrected only, whereas addition before extraction will result in complete correction. In both cases, a fixed concentration of the stable isotope-labeled standard(s) in the final extracts is aspired. The same fixed concentration of stable isotope-labeled standards is needed in all working standard solutions which are used for calibration.

2.3. HPLC Solvents and Buffers

The quality of HPLC solvents may influence analytical results in several ways. Most solvent quality aspects well known from HPLC-UV are valid for HPLC-MS, too. A significant difference

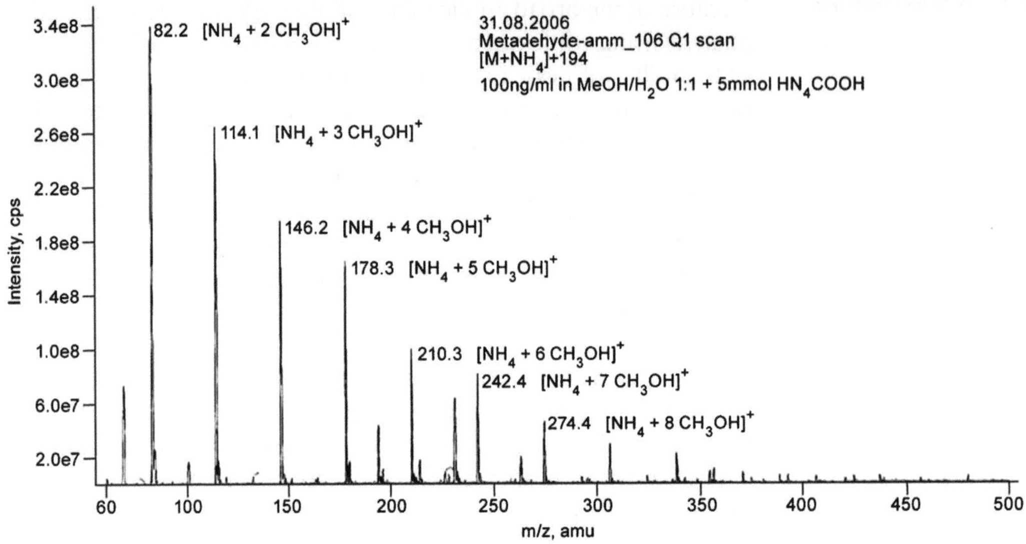

Fig. 1. Formation of solvent clusters in a binary methanol/water system containing ammonium formate.

to HPLC with traditional detectors (e.g., UV, DAD or fluorescence) is the need of volatile buffers. If this is not considered, residues of the non-volatile buffer will be accumulated in the mass spectrometer, which may inhibit the proper function of the instrument. In addition the lowest levels possible of sodium are required to reduce the formation of $[M+Na]^+$ quasimolecular ions, which most often do not form characteristic fragments (less important with buffers containing NH_4^+ ions).

Instead of low UV absorbance of solvents in HPLC-UV the general absence of ionisable contaminants is more important to obtain low background signals. If contaminants are present, they are more easily detected in LC-MS full scan mode than by registration of SRM traces. But signals from solvent clusters should not be mistaken for contaminants in LC-MS full scan spectra (Fig. 1).

Since any unessential manipulation (e.g., dilution, other bottles, filtration) may introduce additional contaminants, ready-to-use HPLC solvents should be preferred.

3. Methods

3.1. Which Pesticides Should Be Covered by an Individual LC-MS/MS Method

As noted in the introduction, approximately 880 active substances are currently used in pesticide formulations. The majority of these substances can be detected and quantified by liquid chromatography combined with mass spectrometry using electrospray ionization (3). Nevertheless, LC-MS/MS is not able to analyze all these analytes. For instance, all dithiocarbamates, biological agents (bacteria, fungi, viruses, etc.) and inorganic compounds are usually out of the scope of LC-MS/MS. In addition, several pheromones are less important because their application does not result in residues.

In addition to the remaining pesticides, many metabolites, degradation products and "old" (persistent) pesticides have to be analyzed by residue analytical chemists. Consequently, more than 800 unwanted substances from pesticide use are candidates for the target screening by triple quad MS/MS. The final decision on the selection of individual pesticides/metabolites covered by an individual LC-MS/MS method should be based on non-analytical and analytical criteria.

Non-analytical criteria are:

1. The selected pesticides should be either produced actually or should be persistent (e.g., atrazine, isoproturon or metolachlor),
2. The selected pesticide should be regulated in that country where crops are marketed (otherwise data cannot be compared to tolerances)
3. Those pesticides/metabolites should be preferred, which are more often found in food monitoring programs
4. The occurrence of specific pesticides in individual crops should be observed (data are available from an EU database (4))
5. The scope of a method should include all pesticides which are used in the place/country where crops are grown

Analytical criteria for the selection of pesticides are:

6. The selection should consider different ionization efficiencies in ESI+/- and APCI+/-
7. The HPLC eluents and their pH must be appropriate for all analytes

3.2. Selection of Basic LC Conditions

3.2.1. Selection of the Appropriate Column Type

From all stationary phases available the reversed-phase (RP) column type is the most common in pesticide residue analysis. Other alternatives (e.g., normal phase HPLC, hydrophilic interaction, size exclusion or ion chromatography) are very seldom used. To cover more polar analytes which do not show effective retention on RP phases, RP-HPLC gradients should start with a very high percentage of water. This makes polar endcapped RP columns more favorable. Polar endcapping results in stable straight chain alkyl groups which do not collapse in water and offers more retention for the very polar compounds.

Beside this water stability, other column properties are less important. However, short columns (e.g., 5 cm) and small particles (1.7–2.4 μm) offer faster chromatography. Special separation efficiency offered by strong (or weak) cation or anion exchange columns, or modification with cyano or phenyl groups is seldom required. The selectivity of tandem mass spectrometers operated in SRM mode is so high that chromatographic separation from a coeluting analyte or matrix component is most often not required. Even analytes present at concentrations of 0.01 mg/kg in sample result in "empty" chromatograms.

3.2.2. Solvent Composition

Generally, the selection of HPLC solvents strongly depends on the column type used for separation. For pesticide residue analysis most often RP phases are applied. To elute polar and non polar analytes in one run, gradient elution is essential. An appropriate way is to work with binary solvent systems. In this case solvent systems based on acetonitrile/water or methanol/water are very common. Mixtures with acetonitrile offers lower viscosity. Consequently, lower column pressure and sharper peaks are obtained. Mixtures with methanol are sometimes preferred for toxicological reasons.

Another criterion is the response of analytes. Some comparisons are made for a limited number of pesticides. Such comparison depends on the type of ion source, on flow rate, on type and concentration of buffer, et cetera. Therefore, general rules do not exist. But there are some results which support the use of methanol/water, if electrospray ionization is applied (5).

3.2.3. Influence of Flow Rate on Sensitivity

Atmospheric pressure chemical ionization (APCI) and electrospray ionization (ESI) are influenced by the HPLC flow rate in a different way. In APCI a higher flow rate combined with a larger injection volume results in an increase of signal intensity. Due to the evaporation of solvent before ionization APCI is mass flow dependent and the mass flow rises with injection volume. In contrast, the signal intensity obtained with electrospray depends on the concentration of analytes. Typically, this concentration is not significantly enhanced, if a larger volume is injected at a higher flow rate.

Consequently, APCI performs often better with standard HPLC columns (ID 4.2 mm, particle size 5 μm), while ESI should be preferred, if narrow columns with small particles (2 mm, 2 μm) are used.

3.2.4. Importance of Buffers on ESI Response

In electrospray ionization the total ion current can be influenced mainly by two parameters. The first is the field strength in the source (i.e., the electrospray voltage), a higher voltage results in a higher ion current. Beside the electrospray voltage the enhancement of eluent's conductivity by addition of buffer offers the second chance to enhance the ion current and therewith sensitivity. Consequently, acidic buffers may also enhance the response of acidic analytes in ESI−, even if analytes are less deprotonated in solution in that case. Typical buffers used in pesticide residue analysis are ammonium formate, ammonium acetate, formic acid and acetic acid in concentrations between 2 and 20 mmol/L.

To obtain better retention on the HPLC column, acidic pesticides are more often analyzed with formic acid or acetic acid.

3.3. Solvent Composition of Final Extracts

Most often reversed-phase HPLC gradients start with a high concentration of water and a low concentration of the organic solvent. In that case, all polar analytes suffer from peak distortion, if the final sample extract is prepared in pure organic solvent (methanol, acetonitrile, etc.), as shown in Fig. 2.

On the other hand, mixing of final sample extracts with water often results in turbid solutions, if the water content exceeds 30%. This turbidity is caused by the formation of a suspension of fatty/oily matrix components in the water containing solvent. Since oil particles "extract" analytes with high octanol/water partition coefficients (log $K_{OW} > 3$) and tend to form an upper layer in the sample vial, low recoveries of non-polar analytes are observed.

To avoid both disadvantages, some options exist. The first option is the injection of very small volumes (<1.5% of flow in 1 min). This strategy requires high analyte concentrations in the

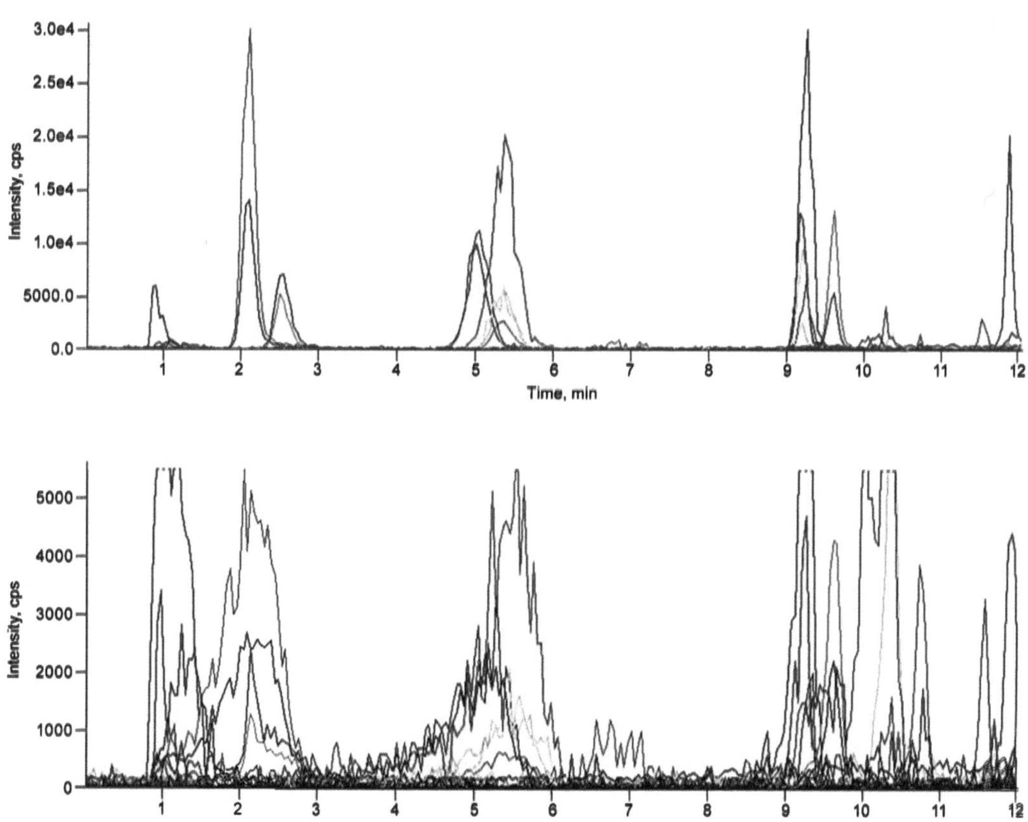

Fig. 2. Injection of 4 μl spiked sample extract obtained in pure methanol after dilution with 16 μl water (*upper trace*) and without dilution (*lower trace*). Polar endcapped RP column 50 mm × 2.1 mm × 5 μm; start gradient: 80% water and 20% methanol, flow rate: 200 μl/min.

final extract. A second option is the use of longer HPLC columns. In that case mixing with eluent A is obtained in the first part of HPLC column. The associated disadvantage is some longer run times. A last, elegant option is the mixing of extract and water in the HPLC injector. This can be achieved by the alternate uptake of small amounts of sample extract and pure solvent into the injector needle (sandwich technique). In that case, any suspension is formed not earlier than in the injector and all parts of suspension reach the column. Oily particles are later re-dissolved as soon as the organic fraction in the gradient is sufficient high.

3.4. Selection of the Appropriate Ion Source

Most manufacturers offer at least three types of ion sources. These are electrospray ionization (ESI), atmospheric pressure chemical ionization (APCI) and atmospheric pressure photo ionization (APPI). In a practical test it could be shown that all three types are able to ionize most pesticides with comparable intensity (Fig. 3). Nevertheless, in practice electrospray ionization is the technique most widely used for pesticide residue analysis. ESI requires no additional dopant (an extra pump is required for APPI) and offers better sensitivity at low flow rates compared to APCI.

However, if the dynamic range of analyte detection is an important criterion for the selection of the ion source, APCI may be the better choice, because the ESI source type sometimes suffers from saturation phenomena at high analyte concentrations.

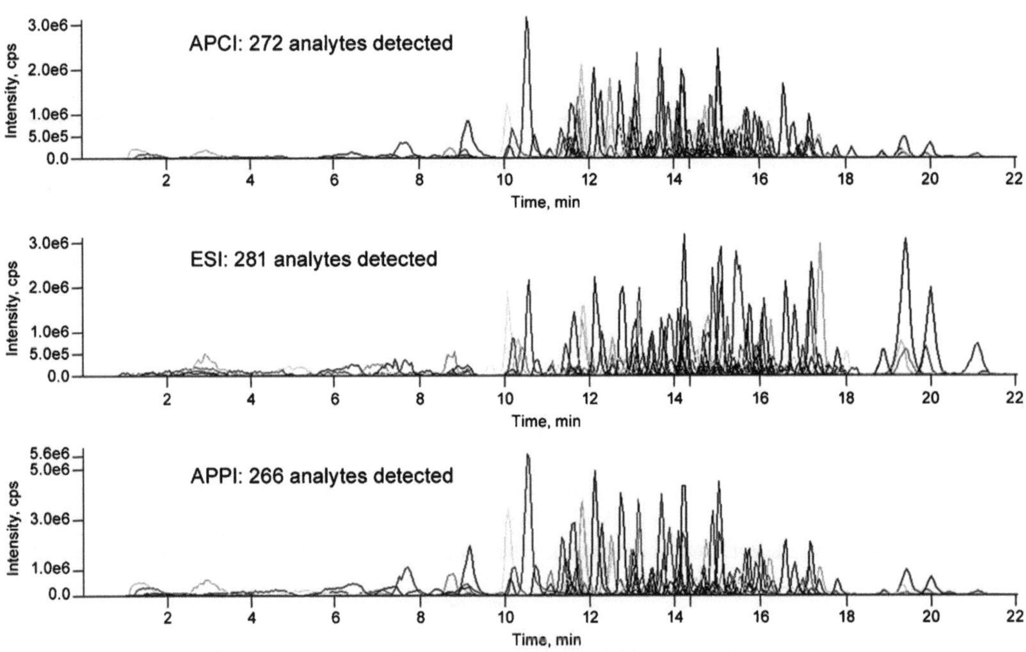

Fig. 3. Repeated injection of a pesticide mixture containing 281 pesticides/metabolites into a LC-MS/MS system (Agilent 1100 LC/Applied Biosystems API 2000 MS) after exchange of ion source.

To avoid contamination of the mass spectrometer, several spray geometries were developed (off axis spray, orthogonal spray, Z-spray). Today, the robustness of ion sources is no matter of discussion. Compared to ion source cleaning of GC-MS instruments all these modern spray geometries offers less contamination of the mass spectrometer and ion source cleaning is very easy.

3.5. Optimization of General Ion Source Conditions

Before tandem mass spectrometers may be used in SRM mode, hundreds of parameters have to be optimized to obtain best sensitivity. Most of these parameters depend on the targeted analytes and may be changed in milliseconds. But a smaller portion of ion source parameters (e.g., electrospray/needle voltage, nebulizing gas flow/pressure, drying gas flow/pressure, source/drying gas temperature) cannot be switched with each new compound. The optimum of gas flows or pressures and the temperature in the ion source depends mainly on HPLC flow rate and solvent composition. In most cases gradient elution is used and the final compromise should be dictated by the less sensitive compounds. If these elute in the beginning of the chromatogram, gas flows and temperature should be optimized for an eluent with high water content. A practical way to optimize all source dependent parameters is to prepare a mixture of analytes with low response factors. This mixture is used in a series of flow injection experiments. In each new experiment one parameter will be changed (Fig. 4).

3.6. Tuning of Analyte Specific Parameters

Today, analyte specific parameters are optimized in automatic procedures provided by the instrument's software. Nevertheless, users sometimes forget the influence of correct conditions for the

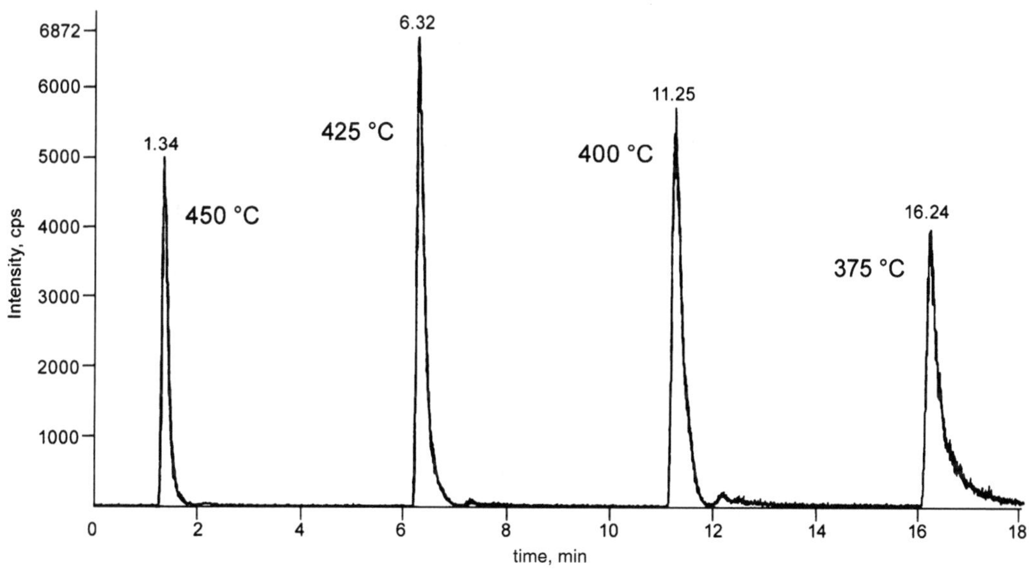

Fig. 4. Optimization of APCI temperature by flow injection using temperature steps of 25°C.

permanent infusion of analyte solutions, which is needed for this optimization. A too low declustering potential may be obtained if analytes are infused in pure organic solvent or with lower flow rate. If buffer is not added in infusion experiments, instead of $[M+H]^+$ or $[M+NH_4]^+$ ions quasimolecular ions with sodium $[M+Na]^+$ may be formed, which do not offer intense fragment ions and which disappear in the presence of buffer. When using negative electrospray the presence of formic acid or the formate ion is a prerequisite for the formation of quasimolecular ions of the $[M+COOH]^-$ type.

Due to small but not generally insignificant deviations between measured and calculated (theoretical) masses of ions, only the calculated masses of precursor and product ions (if structure is known of the latter) should be used for SRM acquisition methods. Otherwise, the center of mass peaks may be missed, resulting in higher variation of quantitative results.

Usually, the collision induced fragmentation observed in tuning experiments does not differ generally between tandem mass spectrometers from different providers, even if different types and pressures of collision gas is used. Therefore, if comparable collision energies are used, published SRM transitions obtained with other instruments (3, 6) may be used for a first test without optimization. Later, only those SRM transitions that suffer from insufficient sensitivity have to be optimized.

3.7. Determination of the Lowest Acceptable Dwell Time

In the selected reaction monitoring mode, the signal produced from the reaction of an intact analyte ion (precursor ion) to a collision induced fragment (product ion) has to be measured for a certain amount of time. The parameter, which describes the duration of product ion counting, is called "*dwell time.*"

However, product ion counting starts with some delay. Before detection at the multiplier, the ions of each transition have to pass through the ion path (length approx. 1 m) of the mass spectrometer. For example, the $[M+H]^+$ ion of diazinon (mass: 305 amu) reaches a velocity of >9,000 km/h or $2,5\times10^3$ m/s, if a typical acceleration voltage of 10 V is applied. That means the travel of analyte ions through the mass spectrometer needs approximately 0.4 ms. Consequently, a switching to the next transition after an ion counting phase ("dwell time") of smaller than 0.4 ms would generally result in the loss of any signal.

In addition to the speed of parent ions, the time needed to clean out the collision cell has to be considered. This second time is usually called "*pause time*" or "*interscan delay*" and it is required to wipe out ions of the former transition and to stabilize the voltages for the new transition. This second time mainly depends on the additional acceleration of slower ions formed by collision activated dissociation within the collision cell. Without any acceleration in the collision cell, often more than 10 ms are needed to remove all ions of a transition from the collision cell. Shorter times are obtained only, if the collision cell uses either semiconducting rods providing linear acceleration or if it contains a lens stack (travelling wave cell®) and non-parallel quadrupole rods (LINAC®), respectively. Some examples of such newer collision cells are shown in Figs. 5 and 6.

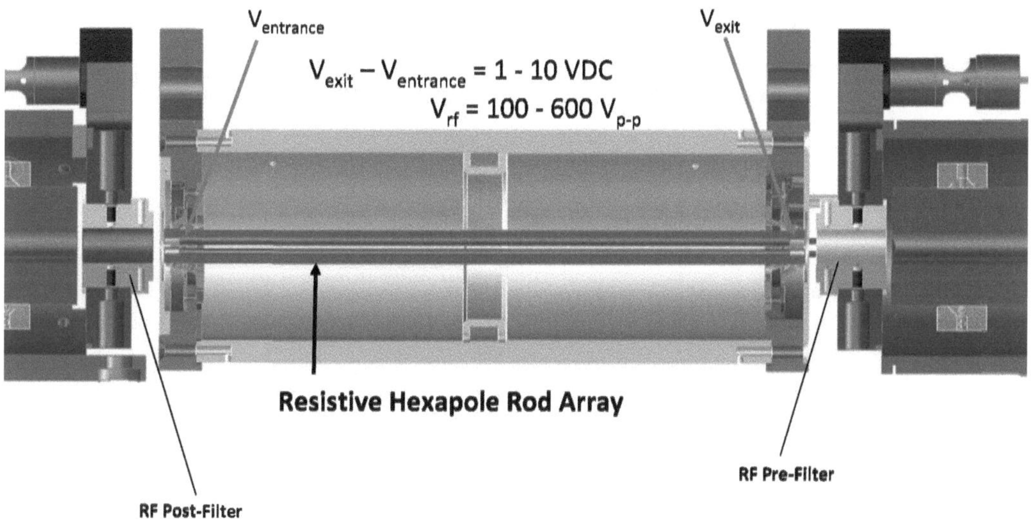

Fig. 5. Agilent's hexapole collision cell with semi conductive rods (electrodes).

Fig. 6. Travelling wave (T-wave) cell® containing 122 ring electrodes used in instruments of Waters. ©2009 Waters Corporation. Used with permission. The traveling wave device described here is similar to that described by Kirchner in US Patent 5,206,506; 1993.

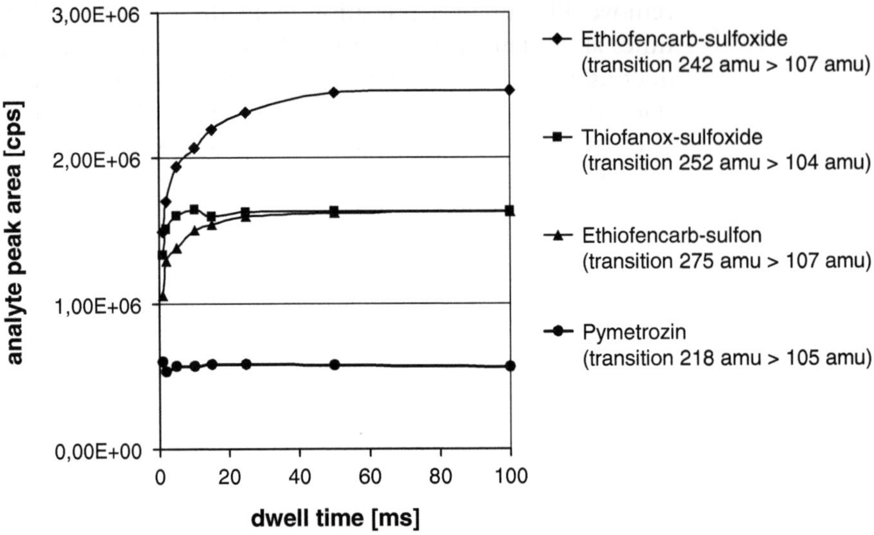

Fig. 7. Influence of dwell time on analyte peak area of four analytes.

The lowest acceptable dwell time may be determined experimentally in a simple experiment. This experiment requires a series of acquisition methods. The starting method should contain a limited number of SRMs (e.g., 5) with equal dwell time for each transition (e.g., 100 ms each). In this example (five SRMs, 100 ms each), the resulting sum of dwell times is 5×100 ms = 500 ms. Based on such an acquisition method, further methods are needed with reduced dwell times for the first four SRMs (e.g., 50, 25, 15, 10, 5, 2, or 1 ms, each). With reduction of dwell times, the fifth SRM shall have corresponding longer dwell times (e.g., 300, 400, 440, 460, 480, 492, or 496 ms) to obtain a constant total dwell time and consequently the same number of data points for each analyte peak.

Since peak intensities are measured in counts per second dwell time (cps), the intensities of all analyte signals in the above measurement should remain constant as long as the traveling time across the collision cell is negligible compared to the dwell time. In contrast, if switching between SRMs is too fast, the peak intensity declines. The result of such experiment is shown in Fig. 7, which demonstrates intensity losses less than 50% even at 1 or 2 ms dwell time. It should be noted that in this example a collision cell with internal acceleration is used.

3.8. Cross Talk

In some cases, the expected product ions of an analyte may be also formed from other precursor ions. If the mass spectrometer changes to a new SRM transition without waiting for an empty collision cell, the product ions from the former transition may feign the presence of the analyte. This phenomenon is called "*cross talk*."

Consequently, in addition to a short dwell time, also the clean out and stabilization time ("pause time" or "interscan delay") should be set at the lowest appropriate value. This is especially important if many SRM transitions have to be acquired in a multi residue method. Appropriate means that this time is sufficient to avoid the occurrence of signals produced by ions of the former transition remaining in the collision cell.

The complete cleanout of the collision cell can be tested with an acquisition method which contains the SRMs of two analytes. These analytes shall have precursor ions of different mass and shall form product ions of identical mass. Such requirement is fulfilled e.g., by an acquisition method established for the two component mixture of carbendazim ($[M+H]^+=192$ amu) and linuron ($[M+H]^+=249$ amu). Both analytes provide an intense product ion at m/z 160.

In a LC-MS/MS or flow injection experiment with one of both analytes (e.g., carbendazim), only the corresponding transition (192 amu → 160 amu) should provide a visible peak. The chromatogram of the transition of the second analyte (in this example: 249 amu → 160 amu) shall not show any deviation of the baseline. If a peak of the second transition is detected also, not all fragment ions of carbendazim have left the collision cell before the second transition (for linuron) was recorded. If short dwell times (between 5 and 20 ms) are used, the intensity of cross talk signals is apparently larger, because a fixed number of remaining ions from the former transition are divided by a shorter dwell time.

To avoid cross talks, a sufficient large pause time (or interscan delay) has to be used. In addition, the order of SRM transitions in an acquisition method can be rearranged to avoid the consecutive measurement of two analytes, which form fragments of identical nominal mass.

3.9. Optimization of the Number of Simultaneously Monitored Transitions

If the selected reaction monitoring mode is used for multi-residue analysis, usually many analyte transitions (reactions) have to be recorded either during the entire run time or within a given time window. If for example 50 transitions have to be recorded in a given time window, the mass spectrometer consecutively has to measure the intensity for each separate transition. The series of 50 measurements in this example is called a "cycle." The time needed for one cycle or one chromatographic data point is the "*cycle time*."

In multi residue acquisition methods a fast switching between different SRM transitions within a limited cycle time is necessary. The upper limit of the cycle time depends on the HPLC peak width and should not be higher than 20% of the peak width at half maximum (FWHM) to obtain at least 10 data points from each HPLC peak. Consequently, for columns with 5 μm particles the cycle time should not be lower than 2,500 ms. If UHPLC columns (>2.0 μm particles) are used, the required cycle time may be as low as 200 ms.

Even the fastest tandem mass spectrometers require a pause time of ≥1 ms. The appropriate dwell time depends on the ionization efficiency and should also not be smaller than 1 ms. Often, in total more than 5 ms are needed for each transition. Consequently, not more than 20 analytes can be recorded simultaneously within one time window under optimized UHPLC conditions, if two SRM transitions are acquired for each analyte.

Even with columns containing 5 μm particles, the cycle time of 2,500 ms may limit the number of simultaneously recorded transitions. Therefore, as mentioned before, the pause time (interscan delay) should be set at the lowest appropriate value. The allocation of dwell time should strictly consider the ionization efficiency of each analyte. A good allocation is found, if most analytes show similar signal-to-noise ratios (S/N) after injection of equal amounts. However, such analytes which exhibit a bad signal-to-noise even with 100 ms dwell time should not be recorded using considerably longer dwell times. The reason is the link between noise and the square root of the reciprocal dwell time. That means, a doubled dwell time usually improves the S/N ratio by a factor of 1.41, only, or four times longer dwell times are needed to improve the S/N ratio by a factor of two.

Please note that peak intensity, if counted in counts per seconds, is an inappropriate parameter to check the adequate allocation of dwell times. However, it can be used to sort analytes into groups that require similar dwell times.

The proper use of the restricted dwell time can be improved, if each analyte receives its own "time window." If the retention time (RT) of an analyte is known, modern software is able to record each transition in a separate time window around the peak maximum (e.g., RT ± 20 s). Available "scheduled MRM" or "dynamic MRM" software allows the recording of up to 1,000 analytes in separate time periods during one run. In contrast to the well known time windows, the acquisition of SRM transitions apart from the expected retention time is avoided. However, this advantage of scheduled MRM has to be paid by the negative aspect of inflexible dwell times. In scheduled MRM, all simultaneously eluting analytes are recorded with an identical dwell time. In "empty" regions of the chromatogram very long dwell times are used, but in "overcrowded" regions dwell times are much shorter. Nevertheless, at least 11 ms dwell time are available for each analyte in the most loaded chromatographic regions, if more than 500 pesticides are quantified in a single chromatographic run of 12 min (Fig. 8).

3.10. Detection, Reduction, and Compensation of Matrix Effects

In quantitative LC-ESI-MS/MS sample matrix frequently alters the signal intensity of a target analyte. Such influence of coeluting matrix constituents on the intensity of the MS/MS signal is called "*matrix effect.*" Matrix effects have been found for all kinds of

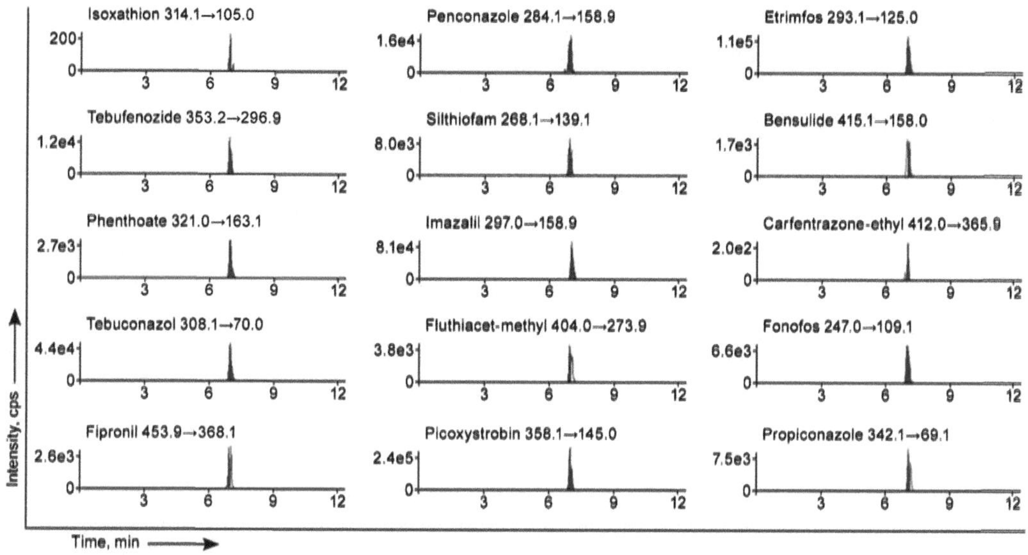

Fig. 8. Chromatography of 507 pesticides on a 5 cm × 2 mm RP column using 2.5 μm particles by scheduled MRM. Shown are 15 pesticides, which elute in a region (6.95–7.08 min) where 58 analytes had to be acquired in parallel.

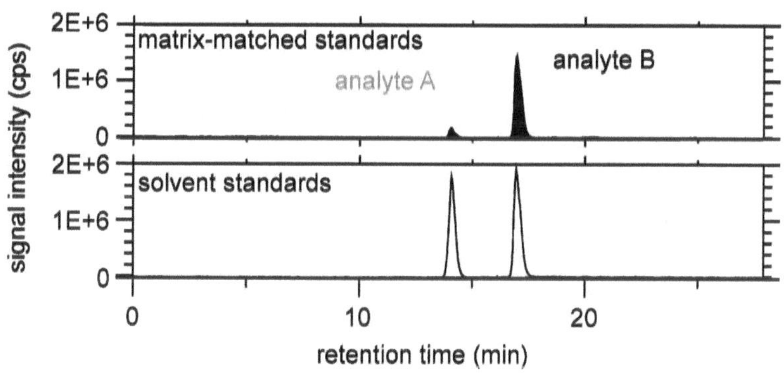

Fig. 9. Ninety percent suppression of azoxystrobin's transition 404.1 → 371.9 amu (analyte A) and weak matrix effect on furathiocarb (383.2 → 195.0 amu, analyte B) by matrix components of orange fruit.

sample materials, from plasma and urine, over plant and animal tissues, to soil or water extracts. Thus all fields of quantitative LC-MS/MS application, including biomedical analysis, life sciences, metabolomics, environmental trace analysis, as well as residue analysis of food and feed suffer from matrix effects. Correspondingly, matrix effects have recently been called the Achilles heel of quantitative LC-MS/MS (7).

Such matrix effects may be quantified by comparison of the peak area obtained for a matrix-matched standard (spiked blank extract) with those of an equally concentrated standard in solvent (Fig. 9). Alternatively, the slope of calibration graphs prepared in blank matrix extract can be compared to the slope obtained for working standards prepared in solvent. If both slops are significantly different, most often the ionization efficiency is suppressed or enhanced by matrix.

Several mechanisms have been suggested by which a matrix compound could interfere with the ionization and transfer of a co-eluting analyte in the electrospray process. For example, competition may occur between matrix and an analyte for the limited elemental charges on a droplet (8) or for the droplet surface from which the ionized analyte has to escape into the gas phase (9, 10). Matrix compounds may alter physical properties of the droplets like its surface tension or they may enclose the droplets in micelle like structures. Both effects would reduce the formation of offspring droplets. Matrix may also alter the viscosity of the droplet solution, thus affecting the transport of analytes from the droplets interior towards its surface (11).

Generally matrix effects may be reduced or avoided by an improved separation of analytes from matrix compounds before ionization. A more selective extraction, a more thoroughly cleanup or an improved chromatography may avoid matrix effects. In multiresidue methods, however, when a large number of analytes (up to several hundreds) with a broad range of physico-chemical properties is analyzed such measures are not applicable.

Therefore matrix effects in multiresidue methods have to be compensated. One approach is the use of matrix-matched standards (12). This type of compensation of matrix effects has one important drawback. For a reliable compensation it requires the availability of samples without any residues, which are identically composed to the samples which contain the residue. This requirement often causes problems in food monitoring, but is easily to fulfill for method validation experiments. Therefore, matrix-matched standards are widely applied in method validation, where fruits or vegetables without any residue may be used to prepare both spiked samples and blank extracts for the preparation of matrix-matched standards. However, the complete compensation of matrix effects caused by substances of an individual apple is not possible with any other apple.

A second way for compensation is the use of isotopic-labeled internal standards (13, 14). However, this approach is limited to pesticides, which are available as labeled compounds. The doubled number of SRM transitions for each analyte pair requires extra dwell time. In addition, in some cases it has been shown that even very small differences in retention time between native and labeled compound (0.02 min!) may avoid a complete compensation of matrix effects by the labeled standard (15, 16). For the same reason unlabeled internal standards (e.g., homolog compounds or compounds from the same compound class) usually fail to compensate for matrix effects.

Since matrix effects are dependent on the concentration of matrix components in the final extracts, a very simple approach is the dilution of final extracts or the reduction of volume injected to HPLC (dilute-and-shoot). However, this way of reduction of matrix effects requires more sensitive tandem mass spectrometers.

If identical matrices for matrix-matched standards or labeled standards are not available and if instrument sensitivity is not sufficient for dilute-and-shoot, the laborious standard addition technique has to be used at least for samples with residues above MRLs or tolerances (13). In such case, the reduced effort that LC MS/MS offers with respect to sample preparation is consumed by the increased analytical work to compensate matrix effects. To avoid this effort, the detection and compensation of matrix effects by post column infusion may be an interesting alternative (17). However, matrix effects continue to be a problem and a subject of continued research.

4. Notes

1. In Subheading 2.1.1, usually stock solutions should be stored at −18°C. Check the stability of stock solutions during storage regularly. In some cases the addition of acids or bases can be helpful to enhance stability and extend the acceptable storage period.
2. In Subheading 2.1.2, usually pesticide mixtures should be stored at ≤ −18°C. Since the stability of the pesticides in a mixture may be lower than in stock solutions, stability has to be checked regularly. In some cases the addition of acids or bases can be helpful to enhance stability and extend acceptable storage times. A blank sample extract is a sample of the same commodity (fruit, vegetable, grain, etc.) being analyzed. This blank is known to contain none of the pesticides being analyzed.
3. In Subheading 2.1.3, usually working standards are replaced with each new batch of analysis.
4. In Subheading 2.1.4, usually matrix-matched standards are replaced with each new batch of analysis
5. In Subheading 2.2.1, when using surrogate standards for correction of peak areas it is important to know that any shift in the surrogate standard signal will directly influence the calculated concentration of the analytes. Ideally, the surrogate standard signal should only shift due to volume differences and thus improve the accuracy of measurement. But there are also other factors that may affect the signal intensity of the surrogate standard thus introducing errors in the analyte quantification. A specific suppression of the surrogate standard signal, potentially occurring due to co-eluting matrix components,

will result in analyte overestimations. In LC-MS applications such matrix effects will depend on whether the commodity extract contains specific components that will co-elute with the surrogate standard and affect its ionization process.

6. In the analytical criteria for selection of compounds to be analyzed, Subheading 3.1, it should be considered whether the analytes are amenable to LC/MS ionization techniques. For example, organohalide pesticides such as eldrin, DDT, and hexachlorobenzene do not ionize by any LC/MS methodology and should not be considered for monitoring by this technique.
7. In Subheading 3.2.1 the reference to an empty chromatogram denotes one where the only peak is that of the analyte devoid of any other signal.
8. It should be noted in Subheading 3.2.4 that the use of buffers has one adverse effect. If the pH of the buffer in the eluent is equal to the pK_a of an analyte, a broad peak is observed for this analyte. In multi residue methods which cover large numbers of pesticides, such distorted peaks cannot be avoided. There is no gap in pK_a values of pesticides which can be used for a special buffer.
9. The sandwiching of sample referred to in Subheading 3.3 can be accomplished with most autosamplers that allow programable injection. The program would consist of drawing from a vial containing water, then drawing from the sample vial, and then drawing from the water vial again. It would be best to have a separate water vial for each sample to avoid contamination of one sample to the next.
10. In Subheading 3.8 it should be noted that most manufacturers of LC-MS/MS instruments use the unit "counts per second" for intensity!
11. In Subheading 3.9 it should be noted that with 5-μm particle size packings in LC columns, a 2,500 ms cycle time assumes that the peak width is 25 s, thus providing 10 data points across the peak. UHPLC is the generic term for ultra high pressure liquid chromatography and is used with columns containing sub 2 μm particles. These columns give very high pressures under normal operating conditions and provide very sharp peaks requiring fast cycle times. In the example given where 200 ms cycle time is needed, the chromatographic peak width at its base is 2 s. Also in that section the description of "scheduled MRM" where each overlapping transition being monitored has the same dwell time is descriptive of one manufacturer's application of this concept. Other manufacturers handle overlapping peaks in different ways.

12. In Subheading 3.10 the use of labeled isotope internal standards is typically used for a select number of pesticides where they are easily acquired without great cost. The ability and cost to obtain a labeled internal standard for every pesticide to be analyzed would be prohibitive. Also, the term MRL refers to the maximum residue limit set by the regulating body for which the pesticide is being monitored. Exceeding MRLs can cause rejection of the food requiring that the analytical result be as accurate as possible.

References

1. Tomlin C.D.S. (ed) 2006. *The Pesticide Manual - A World Compendium. Fourteenth edition*, British Crop Protection Council (BCPC), Hampshire GU34 2QD, UK
2. Pico Y, Blasco C, Font G. (2004) Environmental and food applications of LC-tandem mass spectrometry in pesticide-residue analysis: an overview. *Mass Spectrom. Rev.* **23**, 45–85
3. Alder L,. Greulich K, Kempe G, and Vieth B (2006) Residue Analysis of 500 High Priority Pesticides – better by GC-MS or LC-MS/MS? *Mass Spectrom. Rev.* **25**, 838–865
4. Data Pool of the EU Community Reference Laboratories for Residues of Pesticides: http://www.crl-pesticides-datapool.eu/
5. Kostiainen, R.; Kauppila, T.J. (2009) Effect of eluent on the ionization process in liquid chromatography-mass spectrometry. *J. Chromatogr. A* **1216**, 685–699
6. http://www.bfr.bund.de/cd/5832
7. Taylor, P. J. (2005) Matrix effects: the Achilles heel of quantitative high-performance liquid chromatography–electrospray–tandem mass spectrometry. *Clinical Biochemistry* **38**, 328–334
8. Bruins, A. P. (1998) Mechanistic aspects of electrospray ionization. *J. Chromatogr. A* **794**, 345–357
9. Enke, C. G. (1997) A predictive model for matrix and analyte effects in electrospray ionization of singly-charged ionic analytes. *Anal. Chem.*, **69**, 4885–4893
10. Cech, N. B.; Enke, C. G. (2001) Practical implications of some recent studies in electrospray ionization fundamentals. *Mass Spectrom. Rev.* **20**, 362–368
11. King, R.; Bonfiglio, R.; Fernandez-Metzler, C.; Miller-Stein, C.; Olah, T. (2000) Mechanistic investigation of ionization suppression in electrospray ionization. *J. Amer. Soc. Mass Spectrom.* **11**, 942–950
12. Method validation and quality control procedures for pesticide residue analysis in food and feed; Document N° SANCO/2007/3131; http://ec.europa.eu/food/plant/protection/resources/qualcontrol_en.pdf
13. Rychlik M and Asam S (2008) Stable isotope dilution assays in mycotoxin analysis. *Anal. Bioanal. Chem.* **390**, 617–628
14. Benijts T, Dams R, Lambert W, and De Leenheer A (2004) Countering matrix effects in environmental liquid chromatography-electrospray ionization tandem mass spectrometry water analysis for endocrine disrupting chemicals. *J. Chromatogr. A* **1029**, 153–159
15. Lindegardh N, Annerberg A, White NJ, and Day NPJ (2008) Development and validation of a liquid chromatographic-tandem mass spectrometric method for determination of piperaquine in plasma Stable isotope labeled internal standard does not always compensate for matrix effects. *J. Chromatogr. B* **862**, 227–236
16. Wang S, Cyronak M, and Yang E (2007) Does a stable isotopically labeled internal standard always correct analyte response?: A matrix effect study on a LC/MS/MS method for the determination of carvedilol enantiomers in human plasma. *J. Pharm. Biomed. Anal.* **43**, 701–707
17. Stahnke H, Reemtsma T, Alder L (2009) Compensation of Matrix Effects by Postcolumn Infusion of a Monitor Substance in Multiresidue Analysis with LC–MS/MS. *Anal. Chem.* **81**, 2185–2192

Chapter 8

LC/TOF-MS Analysis of Pesticides in Fruits and Vegetables: The Emerging Role of Accurate Mass in the Unambiguous Identification of Pesticides in Food

Imma Ferrer, E. Michael Thurman, and Jerry Zweigenbaum

Abstract

The detection, identification, confirmation, and quantitation of pesticides in fruits and vegetables are typically performed from a list of suspect compounds or targets. However, there is mounting concern that pesticides not targeted are finding their way into the food supply. This chapter describes the use of LC with time-of-flight mass spectrometry (LC/TOF-MS) for the detection and identification of pesticides that are not targeted. The use of accurate mass measurement and its implication for the identification of non-targeted compounds are discussed. The need for unambiguous identification and requirements therein are evaluated in detail.

Key words: Non-targeted analysis, Identification of unknowns, Accurate mass databases, Confirmation

1. Introduction

In today's market place we receive strawberries from Morocco, tomatoes and peppers from Spain, and salmon from Chile, to name only a few sources. How does our society answer simple questions about the quality and safety of our food from global sources? One way is through monitoring programs in both the U.S. and in Europe, whose role is to detect pesticides and banned substances in food. The U.S. Department of Agriculture has recently published a report on the occurrence of pesticides in food (1). Twelve thousand assays were performed during the past 3 years to monitor pesticides and regulations exist about the quantity of each pesticide that is allowed for the consumer. In Europe, the European Union operates a country-by-country monitoring program that oversees the quality of imported vegetables, fruits, and seafood commodities (2).

Jerry Zweigenbaum (ed.), *Mass Spectrometry in Food Safety: Methods and Protocols*,
Methods in Molecular Biology, vol. 747, DOI 10.1007/978-1-61779-136-9_8, © Springer Science+Business Media, LLC 2011

Mass spectrometry, and specifically gas and liquid chromatography/mass spectrometry, play an important role in the analysis of pesticides in food (3). Typically GC/MS uses single quadrupole analysis and selected ion monitoring (SIM) as the tool of choice for pesticide analysis although GC-MS/MS is becoming more prominent. LC/MS favors the use of triple quadrupole MS/MS methods using multiple reaction monitoring (3). In these applications, the monitored pesticides are selected a priori and the precursor and product ions of interest are entered into the analysis procedure. The methods are highly effective for monitoring pesticides at trace levels, with detection limits at the determined safety level of 0.05 ppm for banned substances in the EU (2).

However, what if banned substances or older pesticides are applied and no ions are selected? Then these compounds may very well be missed by current monitoring techniques. Thus, there is a need for MS techniques to screen, detect, and to identify, the so-called, "non-target" pesticides and banned substances (Fig. 1). The pyramid in Fig. 1 shows the target pesticides at the bottom. This symbolizes the simplest and most common use of mass spectrometry to identify pesticides in food. The instruments for this analysis include single quadrupole GC/MS and LC/MS, as well as the MS/MS applications of both of these instruments. The next level in the pyramid is the non-target pesticide. This is typically a more difficult analysis and requires different instrumentation, for example, the LC/TOF-MS analysis with databases using accurate mass. This method is a powerful approach and

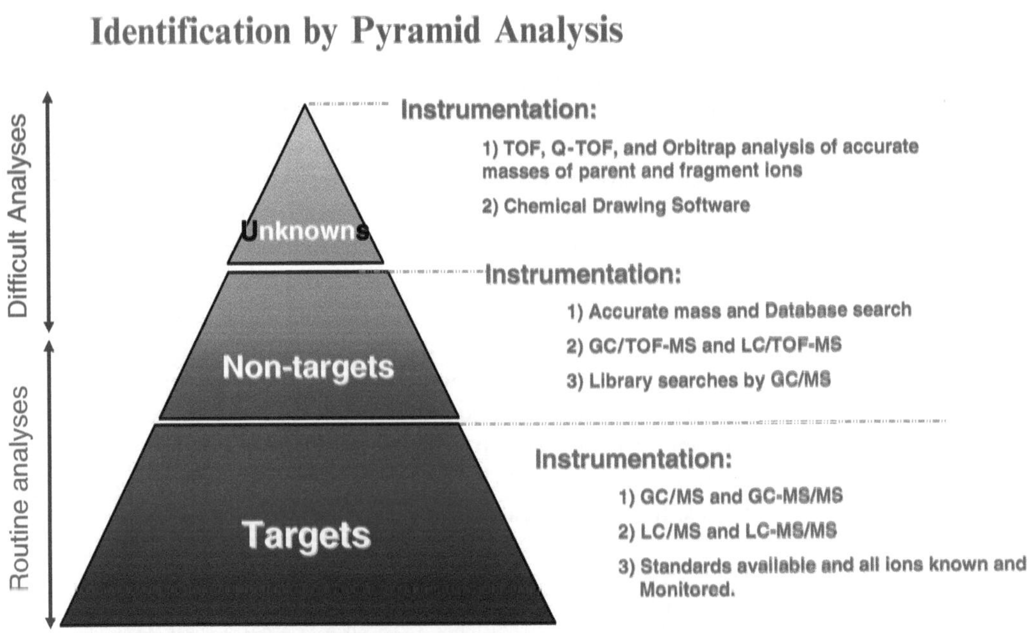

Fig. 1. Identification of targets, non-targets, and unknown compounds by mass spectrometry techniques.

works well for non-target but known pesticides, thus, a standard is available for identification. The final and peak of the pyramid is the unknown substance, which could be a banned substance no longer monitored or an unknown that is not available by database analysis. In this case, it represents the most difficult and challenging of analysis procedures, which requires accurate mass of both the protonated molecule and its fragment ions. It is a good example of the need for tandem full spectral MS techniques such as LC/Q-TOF-MS instrumentation or other types of accurate mass fragment-ion analysis. All three of these applications will be discussed in this chapter. Another aspect of this type of analysis is timeliness of results and accuracy of the analysis (that is the probability of correct results).

For example, there is a need for quality assurance about the detection of pesticides in food, because of the financial impact of banning shiploads of vegetables or fruits based on a chemical analysis. This point brings us to several basic questions of chemical analysis and the difference between confirmation and identification and the role of selectivity and specificity in the chemical analysis of pesticides.

Zeeuw (4) writing on the philosophical topic of substance identification in analytical toxicology explains that the term "confirmation" gained widespread acceptance in analytical toxicology after it appeared in the Mandatory Guidelines for Workplace Drug Testing in 1988 (5). Paraphrasing Zeeuw (4),

- *Confirmation* presumes the presence of a substance *x* in a sample, based on initial tests (screening) or prior information. The presence of *x* can then be "confirmed" by further tests, such as MS.
- *Identification* does not make a priori presumptions based on initial tests (screening) or other information but uses method(s) that are rigorous, specific, and unambiguous to identify substance *x*.

Thus, it should be realized that a positive confirmation thus obtained is *not* an unambiguous identification of *x*. Unambiguous identification of *x* requires that all other (relevant) substances can be excluded, so that *x* remains as the *only* possible candidate for identification (4). Thus, this term of unambiguous identification brings us to the importance of selectivity and specificity in chemical analysis (6), which is commonly defined as:

- *Selectivity* is the capacity of an analytical method to produce signals that depend almost exclusively on the target analyte.
- *Specificity* is the ultimate selectivity level that a method(s) is capable of, thus producing, in the best case, no interfering substances.

Valcarcel (6) notes that various strategies may be used to increase selectivity, such as chemical reactivity (derivatization schemes, including optically active reagents), separation techniques (sample preparation, solid phase extraction, and chromatography both GC and LC), and selectivity of detection (MS/MS, and high resolution MS). Thus, the topic of specificity may seem a pedantic one but, as Zeeuw (4) points out, it is grounded in the social, judicial and economical issues of correct analysis. Thus, the importance of specificity in mass spectral analysis of pesticides in food is a key topic for this chapter, emphasizing the role of high resolution and accurate mass analysis in the quest for unambiguous identification.

1.1. Mass Spectrometry for Identification

Mass spectrometry is the main analytical tool used worldwide by regulatory authorities for doping analysis and trace substance identification (5, 7–10). The main organizations include the European Council (EC) for animal drug residues, U.S. Food and Drug Administration (FDA) for animal drug residues, the Association of Official Racing Chemists (AORC) for equine doping, the International Olympic Committee (IOC) for human doping control, and the World Anti-Doping Agency (WADA) also for human doping control (7). The criteria of all of these agencies for unambiguous identification of banned substances by GC/MS and LC/MS methods are summarized by Van Eenoo and Delbeke (7). Furthermore, the American Society of Mass Spectrometry (ASMS) has published criteria from a workshop on the identification of trace substances by mass spectrometry (8). Although the criteria for identification are not identical they are quite similar.

The most clearly defined criterion for analyte identification by all of the regulatory agencies, including the ASMS workgroup, is the so-called "3-ion criterion" first proposed by Sphon (11). The 3-ion criterion includes either a full-scan spectrum or selected ion monitoring (SIM). The ion ratios are also monitored within certain tolerance limits, which vary from one regulatory agency to another, but are generally between 10 and 20% (7). There are some examples where four ions are used for identification in single quadrupole operation (7, 9, 10), such as with the European Commission (EC) for banned substances, but three ions are used for routine monitoring by the EC in single quadrupole of permitted substances.

Retention time of the monitored substance and the standard is also used as a criterion for identification. Again there are differences among the groups but generally the values for GC/MS must be between 0.5 and 2%, and for LC/MS from 1 to 5% (7). The retention time may be envisioned as a significant increase in selectivity and is a powerful identification tool. For example, it will separate isomers, which often share the same mass spectral pattern.

Thus, the use of a standard for retention-time comparison and ion ratios is also part of the identification procedure of all groups (7). Generally, all of the regulatory agencies discourage the use of library searching for identification (7), especially for LC/MS where variation exists in fragmentation and no search libraries are standard. However, recent advances using deconvolution software, library searching, and a retention time locking (normalized retention times) promises to be an important new tool in pesticide screening in food (3) when using standard GC/MS methods and the NIST02 library.

The 3-ion identification is also used in both GC-MS/MS and LC-MS/MS (7), where the precursor ion is counted and two product ions are included. The MS/MS method begins as a more selective method than the single quadrupole methods because of ion selection followed by fragmentation of a narrow ion window, typically 1–2 mass-units wide. The LC-MS/MS triple quadrupole coupled with retention time has been called by some the "gold standard" of identification of trace environmental substances (12). In fact, the trend in analytical analysis of trace substances by LC/MS is the triple quadrupole with retention time matching and the 3-ion criterion (precursor plus two transition ions) (12). The LC/MS ion trap is also widely used and may be operated in MS^3 mode to generate the 3-ion criteria or in MS^2 with two product ions.

1.2. Emerging Role of High Resolution and Accurate Mass

High resolution and accurate mass refer to four basic types of instrumentation in mass spectrometry. They are magnetic sector, time-of-flight, Orbitrap, and Fourier transform ion cyclotron mass spectrometric instruments. Although all four instruments will work readily with either GC or LC systems, the magnetic sector instruments were developed initially with GC/MS (13) and the TOF and Orbitrap systems have been developed principally with LC/MS (12, 14). The FTICR MS systems are currently considered powerful research tools (15), only, and are seldom used in the screening and monitoring area of mass spectrometry for pesticide identification (12). The emerging trend in environmental trace analysis is the use of LC/TOF-MS, LC/Q-TOF-MS, and bench top Orbitrap systems because of the ease of operation and power of accurate mass (12, 14, 16).

Typically the TOF instruments have resolving power of approximately 10,000–40,000 with routine accuracy of 1–5 ppm. Magnetic sector instruments offer resolving power from 10,000 to 40,000 with 2–5 ppm accuracy, Orbitrap instruments are from 10,000 to 100,000 resolving power with 1–2 ppm mass accuracy, and the FTICR MS systems have the highest resolving power and accuracy (>100,000 with 1–2 ppm accuracy). For comparison, a single quadrupole MS will have a resolving power of 1,000 and accuracy of 500 ppm. Therefore, the concept of resolving power and accuracy are important in high resolution mass spectrometry and refer to the instrument's ability to measure the mass of two closely related ions precisely and accurately.

- *Resolving power* in mass spectrometry is defined as $M/\Delta M$ at full peak width and one-half maximum (FWHM). Where M is mass being measured and ΔM is the peak width in mass units at a given peak height (typically at one-half maximum, FWHM). A second calculation of resolving power, less commonly used, is that the ΔM term is the measured difference in mass units of two closely related peaks expressed as a 10% valley. The difference in the two definitions is approximately a factor of 2 (17). Thus, 10,000 resolving power by the 10% valley method equals 20,000 resolving power by FWHM. (Note that the 10% valley has been classically associated with the resolving power of double focusing magnetic sector instruments, whereas the FWHM definition has been classically used for quadrupole, time-of-flight, and trapping instruments.)
- *Resolution* in mass spectrometry is defined as the inverse of resolving power or $\Delta M/M$. This term is confused with resolving power in the literature, and is often used interchangeably with resolving power. Thus resolution is a small number and defines the ability to resolve two peaks of nearly equal mass. For example, an instrument with 10,000 resolving power (FWHM) at a mass of m/z 300 could separate masses that differ by 0.03 mass units.
- *Accuracy* is the deviation of the measured mass from the true calculated mass (using the exact mass of the elements) divided by the calculated mass and expressed as ppm error.

The recent reviews of regulatory methods by Van Eenoo and Delbeke (7) and Thevis and Schazer (18) show that the use of high-resolution GC/MS, LC/MS and MS/MS methods have received little attention for unambiguous identification of trace drug analysis and monitoring. One reason for this has been the high cost of instrumentation (18) and complexity of analysis. However, this is changing and recent advances in both LC/TOF-MS and LC/Q-TOF-MS have reduced instrument costs and made the analysis less complicated (12) and more accurate (19).

The EC regulations (7, 8) include high resolution and accurate mass. High resolution is defined by the 10% valley method and a resolving power of 10,000, which was a protocol developed from the analysis of dioxin (2,3,7,8-tetrachlorodibenzo[*b,e*][1,4] dioxin) in the environment during the 1980s by HRGC/MS, where the molecular ion and chlorine isotope clusters were monitored with accurate mass (no fragment ions). Figure 2 shows an

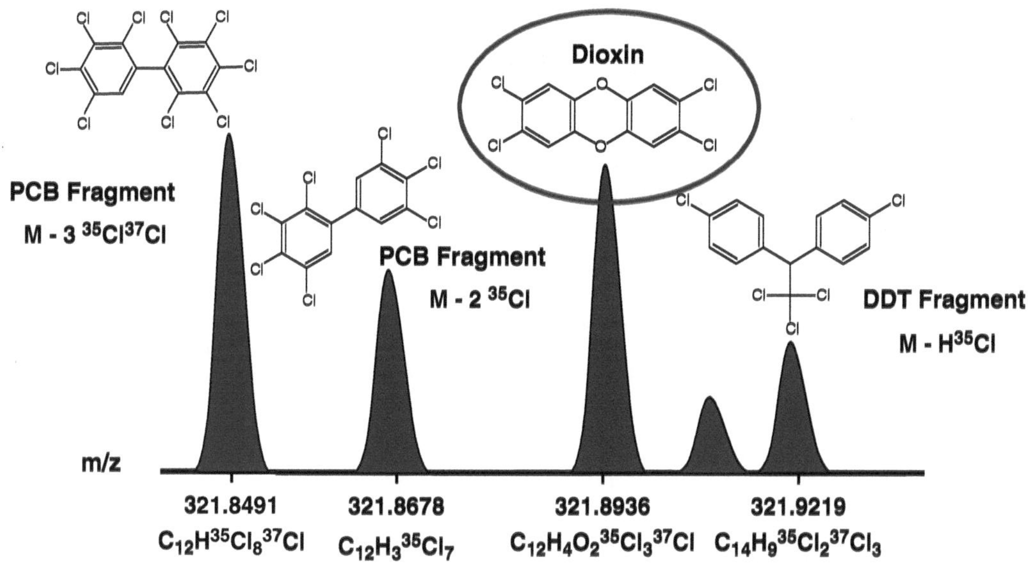

Fig. 2. GC/HRMS at resolution >10,000, 10% valley. Published with Permission of Analytical Chemistry (28).

example of how the accurate mass of dioxin may be separated from PCB fragment and DDT fragments ions, both of which can co-elute in the gas chromatograph. The ions are separated due to the high resolving power (10,000) at 10% valley definition.

Recent advances in LC/TOF-MS and Orbitrap, however, offers bench top instrumentation that are amenable to screening and identification of pesticides in food, not only of the selected analytes, but also non-target pesticides (20–23). Furthermore, rapid screening and confirmation of over 500 drugs in human urine has been reported recently by LC/TOF-MS (24). Because bench top LC/TOF-MS instrument collects data in full-spectrum mode at all times and with accurate mass routinely at less than 3 ppm (20), it is possible to use a combination of elemental composition and database searching to determine unknown or non-target pesticides in food, even initially without standards (21, 22). Furthermore, recent advances in LC/TOF-MS show that the analysis of pesticides in food may be done quantitatively on a routine basis (20). These abilities lend great strength to accurate mass technology that has seen several recent technical advances over the past few years (12, 20–23).

1.3. Identification Points and Mass Spectrometry

There is an identification-point system in effect in the European Union for the detection of banned pharmaceuticals in meat (7, 8), although such point systems have been advocated for pesticides in vegetables and water (12, 25), there are none. Table 1 shows the points assessed for identification using mass spectrometry (7, 8). The point system uses a 4-ion criterion for single quadrupole

Table 1
IPs earned by Mass Spectral Identification

MS technique	IPs earned per ion
Low resolution mass spectrometry LR-MS	1.0
LR-MS^n precursor ion	1.0
LR-MS^n transition products	1.5
High resolution mass spectrometry HR-MS	2.0
HR-MS^n precursor ion	2.0
HR-MS transition products	2.5

Based on refs. 9, 10

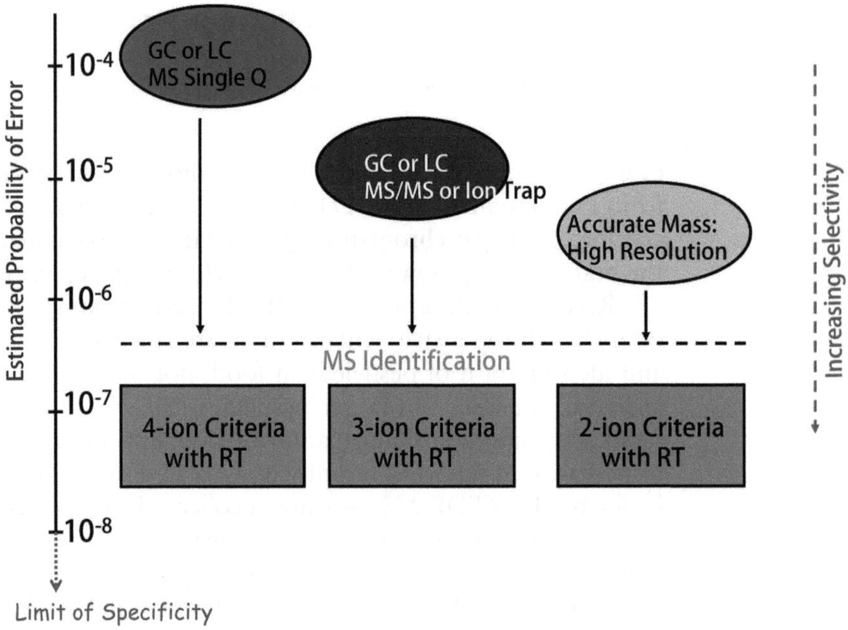

Fig. 3. Selectivity and MS instrumentation coupled with the idea of identification Points (IP), where all three methods have the same IP value. Published with permission of Analytical Chemistry (28).

analysis, since 1 point is added for each ion and 4 points are required for unambiguous identification of banned substances. Only 3 points are required for permitted substances (i.e., the 3-ion criterion).

Figure 3 shows how the EC point system could be visualized when applying each of the types of mass spectrometry. In the EC system of points, MS/MS is considered more selective than single quadrupole and only three ions are required to reach the 4-point identification, this would include triple quadrupole and ion trap identification. MS^3 is equivalent to monitoring the precursor ion and two fragment ions with the triple quadrupole.

High resolution and accurate mass identification is considered more specific yet than MS/MS and only two ions are required to reach the 4-point identification, for example, the protonated molecule and one fragment ion.

The definition of high resolution for the EC system is 10,000 resolving power at 10% valley at all masses. Thus, with this criterion, many of the bench-top LC/TOF-MS instruments of today cannot reach this standard, and therefore, the identification point system is invalid by definition of Table 1! In fact, for fragment ions at *m/z* 150, resolving power >30,000 resolution (FWHM) would be needed because of the fact that in TOF resolving power is defined at *m/z* 1,000 and there is a decline in resolving power with lower masses (19). This definition of high resolution of the EC brings up again the discussion point about resolving power and how much is needed for correct identification of pesticides. This question, addressed earlier, will probably require more discussion between regulators, scientists, and instrument manufacturers.

2. Examples of High Resolution and Accurate Mass in Food Analysis

The selectivity gained by high resolution and accurate mass moves us toward the goal of specificity and unique analysis. Figure 4 shows a mass spectrum for an unknown pesticide in a cucumber extract. The accurate mass improves the probability of a correct analysis when monitoring ions in the mass spectrum. For example, if we consider the number of possible compounds available with the mass of *m/z* 331.0435 within a 3-ppm window, there are only five formulas that match this ion. This statement is also true for the fragment ions at *m/z* 285.0016 and 257.0056. Compare this result with the nominal mass measurements that are used with single quadrupole and triple quadrupole analysis where the *m/z* 331.0 ± 0.1 ion gives a possibility of ~500 compound formulas that are consistent with this mass. Thus the accurate mass calculation at 3 ppm adds about 100 times more selectivity (two orders of magnitude) to the mass measurement than a nominal mass measurement. This calculation is approximately valid for fragment ions that are large mass ions (>*m/z* 200).

Because these probabilities are related, it is apparent that selectivity by mass spectrometry may be improved by as much as two to four orders of magnitude with accurate mass considering the 3-ion criteria! These considerations of selectivity are valid for accurate mass MS/MS as well, and in fact are greater, because of the selectivity of the Q in Q-TOF-MS.

Furthermore, high resolution removes interferences from other compounds and their fragment ions, adducts, and stable isotopic peaks (Fig. 4). Thus, it would seem that the more resolution,

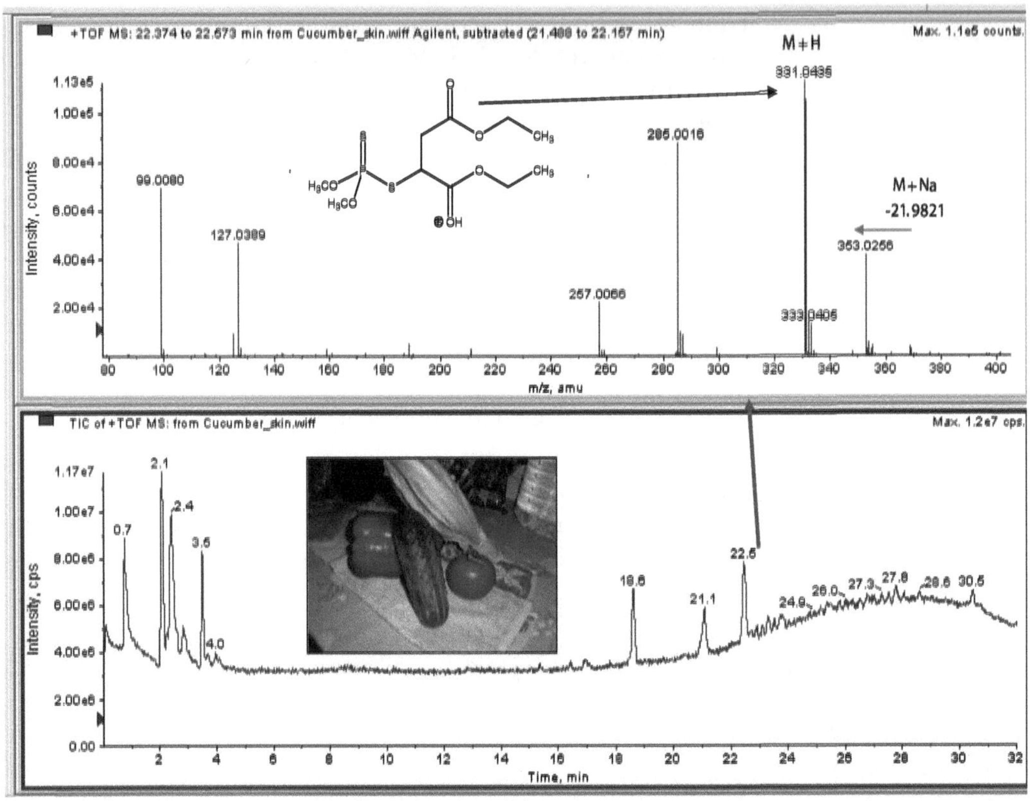

Fig. 4. LC/TOF-MS of malathion in cucumber. Published with permission of Analytical Chemistry (28).

the better, at least for interferences. However, there is a hidden problem coming from accuracy constraints. That is, a unique molecular formula separated from every possible interference is not a unique compound! Many compounds may have the same formula, even those that are not structural isomers. Thus, fragment ion information and chromatography are needed for truly unambiguous identification.

The limit of selectivity gained by accurate mass and high resolving power is the elemental formula for the trace compound being identified. Once that formula is reached then additional resolving power and mass accuracy are of no value. This limit is a function of the molecular mass of the compound. At a mass of *m/z* 300 (many pesticides are in this range) and accuracy of ~1 ppm, it is possible to have a unique formula. If isotope information is included for the *A*+1 and *A*+2 isotopes, this accuracy window can be stretched to 2–3 ppm and still have a unique formula.

The amount of resolving power needed is a function of two things: the resolving power required for 1–3 ppm accuracy (~10,000 resolution (15)) and the resolving power needed to remove mass interferences based on the complexity of the food matrix. Our previously published results (20–22) have shown that for many vegetable matrices a resolving power of 6,000–10,000

(FWHM) is sufficient for this purpose. However, future studies will continue to clarify these results.

If we now return to the problem of a unique formula, how many compounds may share this same formula, and same exact mass? This question is not easy to answer but we may estimate it from available databases. The example in Fig. 4 gives the formula of malathion, $C_{10}H_{19}O_6PS_2$. If this formula is run through the Merck Index database, it is the only match out of 10,000 choices in the database. The ChemIndex also gave only one match from ~80,000 compounds. This suggests that for organophosphate pesticides the probability of natural compounds and pharmaceuticals to interfere with identification is less than 1 chance in 10^4–10^5.

To check this probability further, we examined a pesticide database of 500 compounds known to ionize in positive ion electrospray for identical elemental formula. To our surprise five pairs of compound matches were found, including an organophosphate insecticide. This database included some degradates of the pesticides as well. Thus, this simple chemometric consideration suggests that in some cases the probability of identical elemental composition for pesticides could be as high as once chance in ~100.

Thus, the surprise of these chemometric considerations is that chromatographic resolution is quite important, even using high resolution and accurate mass. Furthermore, fragment ions are essential to unambiguous identification and add selectivity. It is estimated that each accurate mass fragment ion adds from 10 to 100 times more selectivity to the analysis (based on the same likelihood of identical empirical formulas).

The last hurdle for unambiguous identification is the identification of isomers that have identical fragment ions. For example, let us return to Fig. 4, and the identification of malathion in the cucumber sample by high resolution LC/TOF-MS. This sample also contained an identical mass spectrum in the sample at a retention time 5 min earlier. This result suggested an isomer of malathion, and in fact was identified later as the sulfur substituted isomer of malathion. Thus for increased selectivity one should include resolution and standard matching as an important consideration for “unambiguous” identification. In the ultimate case, one might consider LC/NMR coupled with LC/Q-TOF-MS as our ultimate in specificity and unambiguous mass spectrometry identification.

3. Database Analysis

In order to use accurate mass in rapid screening of many pesticides in food, a database of exact masses is needed (and even of exact mass fragment ions) in order to rapidly look for these substances via the data system of the mass spectrometer. Thus, the creation of the database is fundamental in this type of analysis.

The creation of a database involves three steps: selection of the pesticide and its fragment ion, calculation of the exact masses of the molecules, and the creation of the csv (comma separated value) file. A description of each molecule and its retention time may be included but is not necessary for database operation. The csv file is then the database that is searched. The process involves the selection of a method file that uses the LC/TOF-MS sample data-file, which has already been analyzed on the instrument, and the actual processing of the work list to generate the search and data report. Typically on a complex food sample a database of 100 compounds is searched in less than 1 min. (Note that different vendors have their own format for creation and maintenance of databases with this content. The .csv file format is a more generic form.)

The csv file may be formed by use of a spreadsheet supplied by the instrument software that calculates the exact mass of the molecule of interest. One enters the number of carbon, hydrogen, oxygen, nitrogen, chlorine, or sulfur atoms and the spreadsheet calculates the exact mass of the neutral compound. The specified adduct (H^+, NH_4^+, etc.) is then used for the accurate mass comparison of the ions found in the data file searched with an ± error of 5 ppm (set by the user). An example of a csv file searched as an accurate mass database is shown in Fig. 5. The csv file contains the formula, exact mass, retention time, and name of the compounds.

3.1. Searching Accurate Mass Databases of Pesticides

Table 2 shows the results of a library search of six fruit and vegetable samples from a local grocery store (apple, pear, tomato, potato, pepper, cucumber) and one commercial brand of olive oil for the 100 pesticides in the database. The csv file database search found from 50 to 300 accurate mass peaks in the sample chromatograms. The least complicated sample matrix was the tomato, and the apple and pepper were the most complex samples. The sensitivity of the search was set at a signal-to-noise of 10:1. The quantity of peaks found approximately doubles with decreasing the signal-to-noise ratio from 20:1 to 10:1. The value of 10:1 is chosen in order to obtain good $X+1$ and $X+2$ isotope signatures of the compound with the maximum instrument sensitivity. The accuracy window of the search is set at 5 ppm in order to be well within the mass accuracy of the LC/TOF-MS system. The only criterion to be included in this match was that the MH^+ ion was within 5-ppm of the database value.

On the pepper sample, only three formulas were identified based on the correct isotope signature, and the correct retention-time match. They were the compounds imazalil, diazinon, and buprofezin. The identification was checked not only in the print-out of the automated database match but also by manual confirmation of the data file.

The confirmation of the molecular formula varied from no detections in the potato sample, one pesticide in olive oil, three

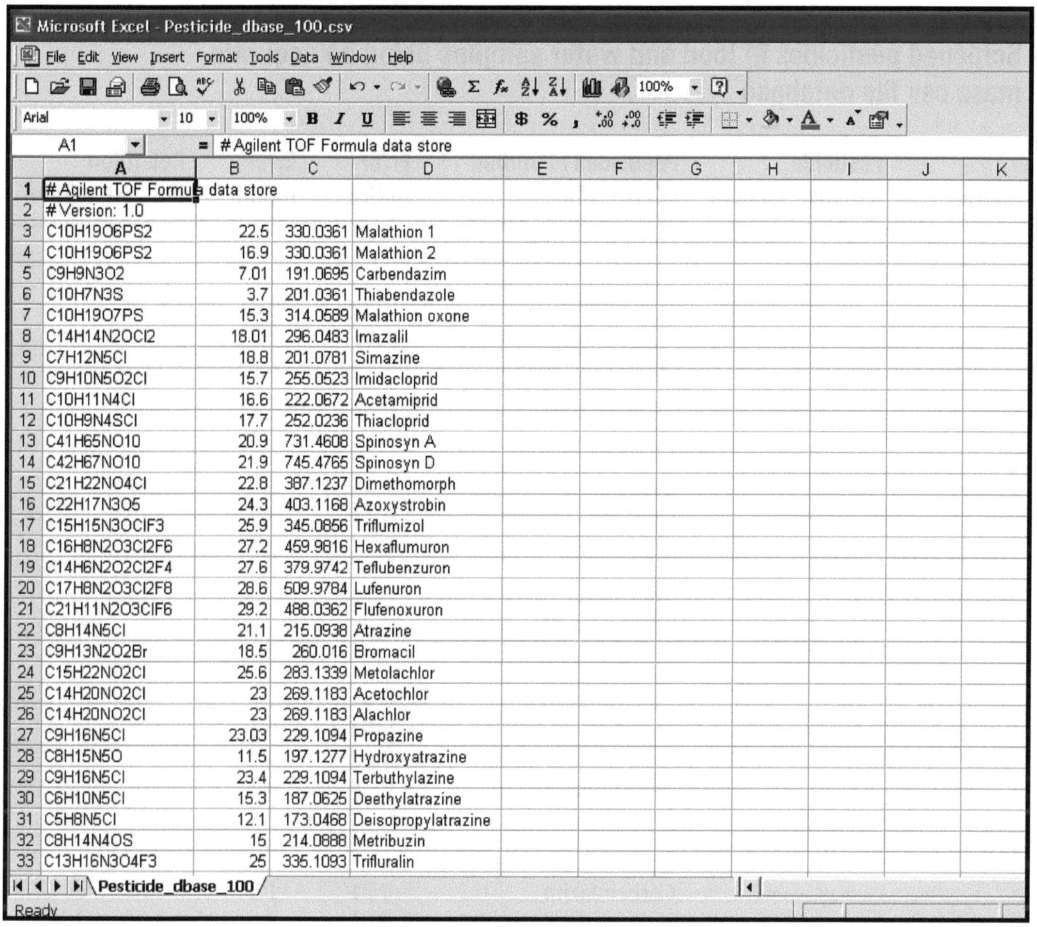

Microsoft Excel - Pesticide_dbase_100.csv

	A	B	C	D
1	#Agilent TOF Formula data store			
2	#Version: 1.0			
3	C10H19O6PS2	22.5	330.0361	Malathion 1
4	C10H19O6PS2	16.9	330.0361	Malathion 2
5	C9H9N3O2	7.01	191.0695	Carbendazim
6	C10H7N3S	3.7	201.0361	Thiabendazole
7	C10H19O7PS	15.3	314.0589	Malathion oxone
8	C14H14N2OCl2	18.01	296.0483	Imazalil
9	C7H12N5Cl	18.8	201.0781	Simazine
10	C9H10N5O2Cl	15.7	255.0523	Imidacloprid
11	C10H11N4Cl	16.6	222.0672	Acetamiprid
12	C10H9N4SCl	17.7	252.0236	Thiacloprid
13	C41H65NO10	20.9	731.4608	Spinosyn A
14	C42H67NO10	21.9	745.4765	Spinosyn D
15	C21H22NO4Cl	22.8	387.1237	Dimethomorph
16	C22H17N3O5	24.3	403.1168	Azoxystrobin
17	C15H15N3OClF3	25.9	345.0856	Triflumizol
18	C16H8N2O3Cl2F6	27.2	459.9816	Hexaflumuron
19	C14H6N2O2Cl2F4	27.6	379.9742	Teflubenzuron
20	C17H8N2O3Cl2F8	28.6	509.9784	Lufenuron
21	C21H11N2O3ClF6	29.2	488.0362	Flufenoxuron
22	C8H14N5Cl	21.1	215.0938	Atrazine
23	C9H13N2O2Br	18.5	260.016	Bromacil
24	C15H22NO2Cl	25.6	283.1339	Metolachlor
25	C14H20NO2Cl	23	269.1183	Acetochlor
26	C14H20NO2Cl	23	269.1183	Alachlor
27	C9H16N5Cl	23.03	229.1094	Propazine
28	C8H15N5O	11.5	197.1277	Hydroxyatrazine
29	C9H16N5Cl	23.4	229.1094	Terbuthylazine
30	C6H10N5Cl	15.3	187.0625	Deethylatrazine
31	C5H8N5Cl	12.1	173.0468	Deisopropylatrazine
32	C8H14N4OS	15	214.0888	Metribuzin
33	C13H16N3O4F3	25	335.1093	Trifluralin

Pesticide_dbase_100

Fig. 5. Example of csv file used for database search of pesticides in food by accurate mass searching of the protonated molecule.

pesticides in pepper and tomato, and five pesticides in the cucumber and apple. The most common compound identified by the screen in the fruit and vegetable samples was imazalil, which is a post-harvest fungicide used for transport and storage of fruits and vegetables before their sale. Other compounds included organophosphate insecticides, such as diazinon, phosmet, and malathion and the oxon of malathion, which is a pesticide degradation product. The insect growth regulator buprofezin was found in a tomato and pepper sample.

The accuracy of all confirmed samples had an absolute-value average of 0.3 mDa or ~1.2 ppm and a standard deviation of 0.25 mDa and ~1.0 ppm, respectively. The absolute-value average for retention time match was 0.07 min and standard deviation of 0.09 min. Thus, the windows chosen for the database search are chosen with enough margin of error to find 99% of the pesticides in samples based on two standard deviations of the mean for mass accuracy and retention time.

Table 2
Screened pesticides in food and water samples using an accurate mass csv file database

Sample	Pesticide matches <5 ppm	Pesticides identified LC/TOF-MS	Error (mDa)	Error (ppm)	Retention time error (min)
Apple	5	Imazalil			
		Imazalil degradate	-1	-3.9	-0.08
		Iprodione	0.22	0.7	-0.12
		Fluquinconazole	0.05	0.1	
		Difenoconazole	-0.74	-1.8	
Olive oil	1	Terbuthylazine (deisopropylatrazine)	0.06	0.2	0.09
Pepper	3	Imazalil	0.3	1	0.07
		Diazinon	0.11	0.3	0.12
		Buprofezin	1	3.3	0.1
Tomato	3	Buprofezin			
		Carbendazim			
		Thiophanate methyl			
Cucumber	5	Thiabendazole	0.2	1	0.01
		Malathion isomer 1	0.04	0.1	0.05
		Malathion isomer 2	0.22	0.7	-0.03
		Malathion oxon	0.25	0.8	0.08
		Imazalil	-0.1	0.3	0.05
Pear	4	Imazalil			
		Carbendazim	-0.21	-1.1	0.05
		Imazalil degradate	0.51	2	0.03
		Phosmet	0.31	1	0.04
Potato	11	None			

3.2. Probability of Database Match with One Fragment Ion

The concept of confirmation in accurate mass has been addressed by Stolker et al. (10) as part of the EU program for banned substances in animal products. The identification point system is used to find target compounds using accurate mass and high resolution. For accurate mass the requirement is a retention-time match and two ions at accurate mass for a total of 2.0 identification points for each ion and a total of four identification points, as well as correct ion ratios (26). Further-more the requirement is that the resolution (resolving power) be 10,000 for each of the ions (at 10% valley or 20,000 FWHM). The requirement of resolving power of 20,000 (FWHM) for the low mass-to-charge ions found in pesticides (less than 300 m/z) is currently possible by current LC/TOF-MS instrumentation based on instrument manufacturer information. However, using this as a guide for match quality, it is possible to calculate approximate probabilities of an error of the pesticide database.

Examining the data in Table 2, it was noted that there are nine sets of compounds that have the same molecular formula and thus the same exact mass. This is approximately 18% of the pesticides examined in this study. Furthermore, the number of fragment ions that have the same exact mass was six sets or 12% of the total ions. These fragment ions are often diagnostic of a class or set of compounds and are useful for identification of degradation products of pesticides in food and water samples (19, 20). Combining these probabilities for the possibility that a pesticide might have the same exact mass MH^+ ion and the same fragment ion (the ion comes from in-source CID fragmentation) gives a 2.2% chance, or approximately one compound in the database:

$$0.18 \times 0.12 = 0.022$$

There are two hidden assumptions in this calculation. First is that these probabilities are independent, which they are not because of similar structures in nearly every set. Second is that the compounds would have the same exact retention time, which of course they do not. For example, see the separation of isomers of three compounds in Table 2. This represents the maximum probability of an error (the worst case scenario). Thus, a database match with the exact mass and the exact mass fragment ion results in a match of ~0.98 based on these assumptions (1–0.022=0.978 or ~0.98).

This match may be improved by the use of a deconvolution program and re-examination of the assumption of retention time. The software is able to match the mass peak as a function of retention time with a reproducibility of ±0.02 min for the MH^+ ion and its fragment ion. The 100 pesticides are separated over a time window of 7–30 min with the majority of the compounds separating in a 10-min window. To be conservative the 10-min window may be divided into 250 increments of ±0.02 min. Thus, the ±0.02 window gives another level of error probability of approximately 0.004, or four chances in 1,000 of over-lapping ions and the final probability match is lowered only a slight amount,

$$0.98 - 0.004 = 0.98,$$

thus the probability of error of the database match is 0.98 or a match quality of 98. The accurate mass database approach is quite useful for identification purposes, even without the use of standards, because the retention-time match is based on the overlapping of not only the two ions but also that the compound would have the same chromatographic behavior using a standard 1.8 μm C-8 or C-18 column with approximately 15,000 theoretical plates. Thus, chromatographic resolution is used for added match quality without the use of an authentic standard. This makes the accurate mass database with a fragment ion available for all instruments, although it may be necessary to use it in a more manual approach if software is not available.

The calculations of probability above do not consider the error in making the mass measurement and assigning a molecular formula (i.e., the error of ±5 ppm of the search). The match probability is based on the exact mass being measured correctly against the database. Some skill in mass spectrometry is required here in order to use the $X+1$ and $X+2$ isotope information for the correct formula and to lower the probability of error from the 5-ppm of the search to less than 2-ppm of instrument capability. To help in this manual process, the database program also gives the isotope table and the isotope abundances for each match, which is quite useful for correct formula identification. Most instrument manufacturers of accurate mass instruments provide software to help with isotope identification and matching.

There are a number of compounds containing the common elements of C, H, N, O, S, P, Cl, Br, and Na, between the masses of 200 and 300 (within a 5-ppm mass accuracy) that may occur. For example, there are approximately 100–150 possible elemental formulae within the 5-ppm mass accuracy. Note that the number of possibilities increases exponentially with increasing mass. This number of possibilities may be reduced quite easily to a manageable number of two to three possibilities by examining the isotope cluster of the $X+1$ and $X+2$ ions. Both the accurate mass and the intensity of the isotope cluster may be used to simplify the number of correct formulas.

An example is shown for imazalil in Fig. 6, which was identified in the pepper extract (Table 2). Note that the accurate mass defect when going from the X peak to the $X+2$ and the $X+4$ peak was 1.997 mass units, which is characteristic of the addition of ^{37}Cl and is accounted for by the difference in mass between ^{35}Cl and ^{37}Cl or 1.997 mass units. Also note that the abundances of the $X+2$ and $X+4$ isotopic pattern match a compound containing two Cl atoms. Likewise, the $X+1$ isotope shows a positive mass defect of 1.0022 mass units (298.0586 - 297.0564) for the carbon-13 isotope and a peak area consistent with 14 carbons. Thus, the isotopic pattern is a good match for imazalil with a formula of $C_{14}H_{14}N_2O_2Cl_2$. Software is available on most accurate mass instruments to help with this process including the accurate masses of the ^{13}C and ^{37}Cl isotopes. This analysis is a manual operation at this time, and the automation of this process is the last hurdle before the accurate-mass database is totally automated.

3.3. Weakness and Strengths of Accurate-Mass Database

A major weakness of a database search is the occurrence of false negatives created by compounds of high concentration that skew the mass accuracy beyond the 5-ppm mass limit. Many of today's

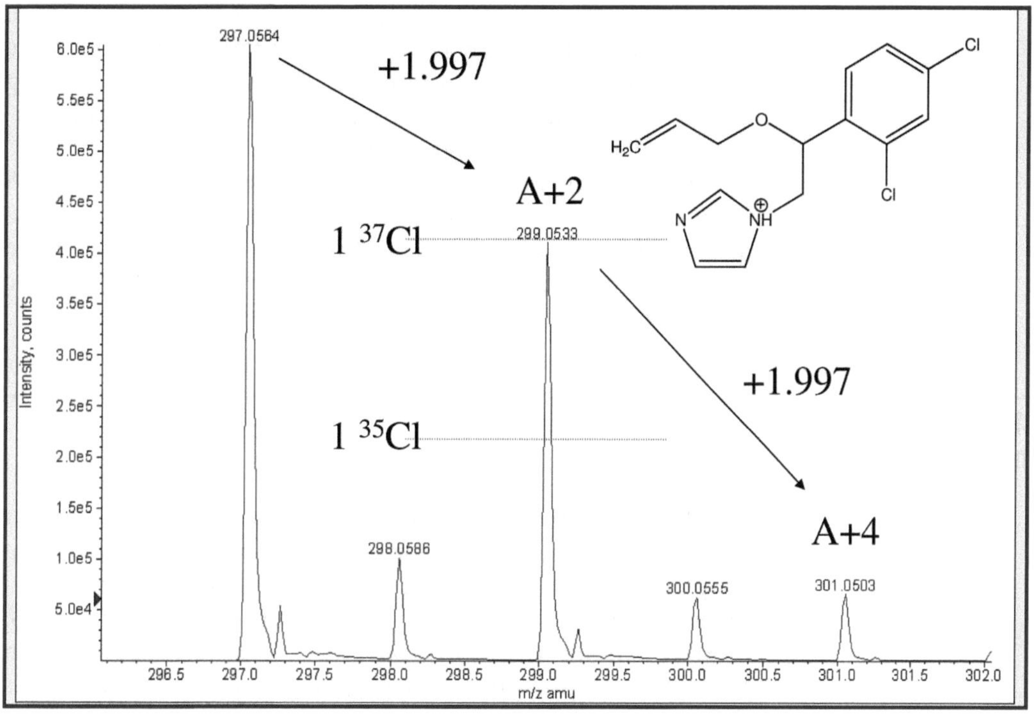

Fig. 6. Isotope cluster of the *m/z* 297 ion and imazalil structure in a pear extract.

instruments have the capability to note that this problem may occur and have the ability to repair this problem and report it as part of the database search (Agilent software, for example, has this feature).

A second cause of false negatives is a mass interference that causes loss of mass accuracy. This problem is more difficult and calls for the use of higher resolving power instruments. When the mass defect of the MH^+ is approximately +0.2 mDa there can be problems from the matrix because this value is similar to the mass defect of the food matrix, which consists of compounds containing C, H, N, and O. Most pesticides have a mass defect less than this value because of the addition of halogen atoms. There are a number of possible ways of solving this problem with the current instrumentation, by increasing the resolving power of the accurate mass instrument, by increasing chromatographic resolution, or by reducing the number of possible formulae by use of isotope-matching software.

The strength of the library search is the speed of screening hundreds of pesticides in minutes, including both their protonated molecules and fragment ions. The use of classical fragmentation libraries with comparison to fragmentation patterns probably is often not needed when using accurate mass because the match is based on presence of the ion rather than the intensity of the ion and its fragmentation pattern. The problem of instrument variation and matrix effects on fragmentation is partially eliminated with accurate mass. However, it may be necessary to monitor more than one fragment ion depending on the type of pesticide being determined. Pesticides without characteristic $X+2$ isotopic patterns need at least two fragment ions.

It is the view of the authors that a large problem in LC/MS libraries is on the verge of being solved with the use of accurate mass databases for pesticide screening in food and water using the protonated molecule, a major fragment or product ion(s), isotopic matching, and chromatographic retention-time matching.

4. Identification of Unknowns

There are multiple definitions of the unknown. Figure 1 shows the concept of the pyramid for unknown analysis. At the base of the pyramid is the Target compound. This is the most basic analysis, which may use several types of mass spectrometry techniques, most commonly is the LC/MS/MS triple quadrupole. When there are a small number of targets, let us say less than 20 this is the simplest and most sensitive instrument for their analysis. A standard is available and the ions and retention time of the standard are used in the conventional analysis of pesticides in a food sample. This is not unknown analysis. Next up the pyramid is the non-target compound. This is a compound that is not sought after initially and for which a standard may be available but has not been used for identification. The non-target is typically identified by either LC/MS ion trap or LC/TOF-MS or LC/Q-TOF-MS using accurate mass and database searching, which we have already discussed. This is an example of an initially unknown compound in the sample.

The last category is the unknown for which we have no standard but we attempt to identify the unknown using accurate mass and another set of tools available to the analyst. This is our definition of an unknown analysis. It is possible in the analysis that once a structure has been identified that a standard may be available.

The concept of unknown identification using the LC/TOF-MS and ion trap consists of four steps, which are outlined in detail below and shown in Fig. 7. They are:

1. Analyze the vegetable extract with LC/TOF-MS, which is always in full spectrum mode looking for large unknown peaks using a mild in-source CID fragmentation typically with positive ion electrospray. Determine the accurate mass and the most likely molecular formula.
2. Search Merck Index, ChemIndex, or similar database for unknowns using the generated molecular formulas and any $A+2$ isotopes, such as Cl, Br, or S, if present.

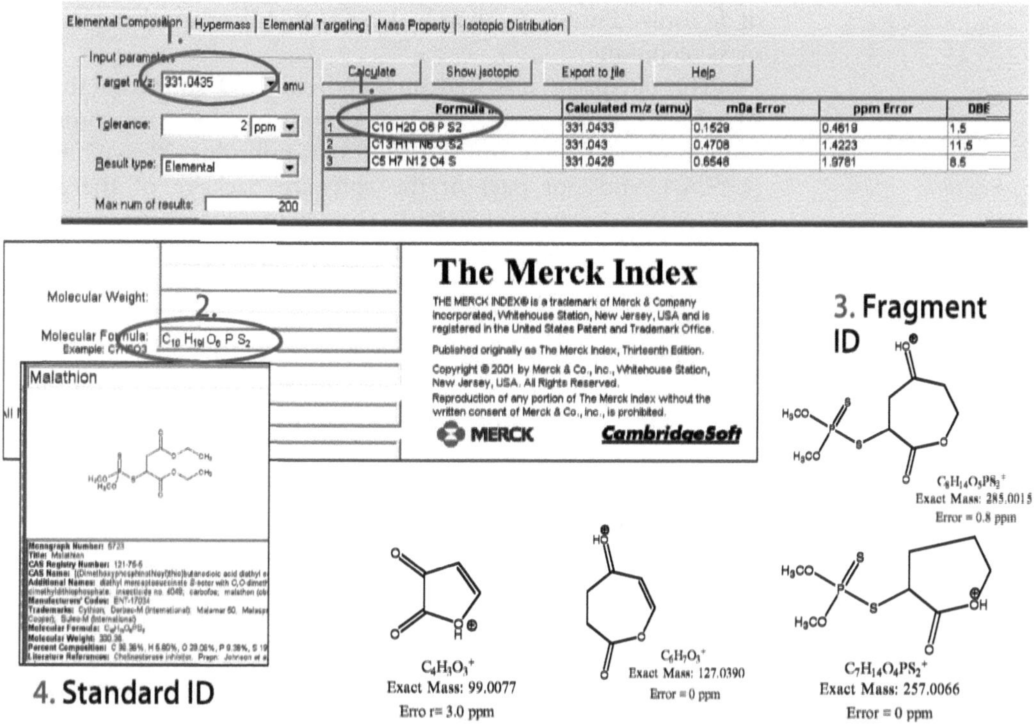

Fig. 7. The 5-step procedure for identification of unknowns.

3. Proceed to MS/MS with proposed structures and check accurate mass ions. This may be done by Q/TOF-MS, TOF-MS with in-source fragmentation, or Orbitrap MS. Use a chemical-structure drawing program to identify ion fragments and their accurate masses. Then, combine with accurate mass data of fragment ions (chemical formula of fragment ion), if available. Make tentative identification.
4. Obtain and analyze standard for final confirmation, if available.

This chapter gives a detailed example of this process using store purchased tomatoes, which contained "unknown white powders" that were subsequently identified by the above process for various "unknown, non-target pesticides" on the skin, as the next section shows.

4.1. An Example of Unknown Analysis

Selected tomatoes containing white powder from a commercial market place were extracted as follows. Carefully wash the skin of the tomato three times with methanol to remove the white powder, from 2 to 5 mL, depending on the size of the vegetable. Capture the solvent in a 150-mL Pyrex beaker. After mixing, transfer the methanol to a 5-mL syringe and filter through a Millex®-FH PTFE filter and aliquot 0.3 mL. Dilute with 0.6 mL of de-ionized water. Analyze by LC/TOF-MS or another accurate mass technique.

The databases searched included two CD ROM databases, The Merck Index and ChemIndex, both of which are commercially available from CambridgeSoft in Cambridge, Massachusetts (U.S.A.), and not part of the software package of the LC/TOF-MS instrument. Likewise, the chemical drawing software was ChemDraw also from CambridgeSoft. The chemical drawing software has the capability to do accurate mass analysis for either GC/MS (electron ionization) or LC/MS (electrospray), including adducts and protonated molecules. The database and chemical drawing software are sold as a package called ChemOffice (CambridgeSoft, Cambridge, MA, USA) and run on a windows environment (e.g., laptop).

Figure 8 shows the total ion chromatogram (TIC) for the rapid extraction of a white powder on a store-purchased tomato (skin) using LC/TOF-MS. The simplicity of the extraction of the tomato skin results in a clean chromatogram without interferences. Several recent studies have shown that the skins of vegetables contain high concentrations of pesticides (21, 22); thus, this extract is a good medium for unknown pesticide identification. Furthermore, it is a food safety relevant extract since the skins of tomatoes are eaten in salads and extracted for many food uses.

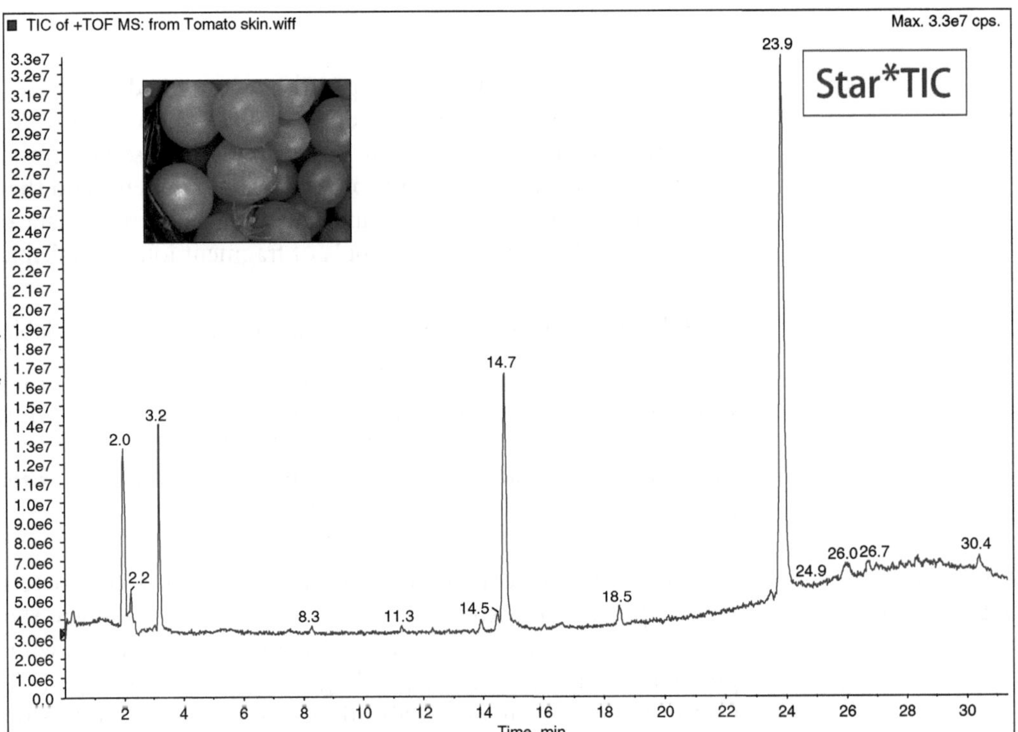

Fig. 8. LC/MS TOF analysis of a tomato extract.

The LC/TOF-MS instrument of this study (Agilent) is one of the first of its type to use an analog-to-digital converter (ADC) instead of a time-to-digital converter (TDC) for mass-spectral peak analysis. Discussions of ADC to TDC state generally that there is a wider window of sample intensity and, by inference, mass accuracy across a wider concentration range, before saturation of the detector with the ADC type. The ADC detector was used in this study and it was found that slicing the peak was of no advantage over taking the entire peak for mass accuracy. Thus, the method used here was to take the center 95% of the peak for mass-accuracy measurements up to an intensity equal to the calibration ions (intensity of 200,000 counts). If intensities exceeded the calibration ions by greater than 50% the accuracy measurement was taken off center to reduce counts or the sample was diluted and re-analyzed. We did notice a deviation from the 3-ppm accuracy limit of the manufacturer if the intensity of the unknown ion was greater than 5–10 times the intensity of the calibration ion (a point of saturation of the detector). This does indicate that saturation of even the ADC detector can occur with large peak intensities. The remedy is to dilute the sample at least ten times and reanalyze. Intensities as much as ten times less than the calibration ion may also cause deviation from the 3-ppm specifications and it may be important to concentrate the sample before analysis for the highest accuracies (<3 ppm).

The chromatogram (Fig. 8) is much simpler than a whole tomato extract because of the lack of natural product peaks, while remaining effective on pesticides on the surface of the tomato. For example, there are four major peaks in the chromatogram at retention times of 2.0, 3.2, 14.7, and 23.9 min. The peak at 2.0 min is the void volume of the LC column. For purposes this example, let us examine the peak at 23.9 min, which is the "Star" in the chromatogram otherwise known as the largest peak in the TIC.

This major peak in the chromatogram of Fig. 8 at 23.9 min resulted in the accurate mass of *m/z* 306.1642, which gave the molecular formula of $C_{16}H_{24}N_3OS$ at 2.4 ppm error. The molecular formula was then searched in the Merck Index and the insect growth regulator, buprofezin, was found. Buprofezin is used extensively on white flies according to the Merck Index and according to a recent publication on tomatoes in Spain (21). Thus, this compound was a good candidate for a positive identification by fragment-ion analysis. Note in Fig. 9 the ion at *m/z* 201.1059. LC/MSD ion trap MS/MS of the *m/z* 306 ion gave the 201 and further MS3 gave the *m/z* 116 ion (Fig. 10). It was possible to draw reasonable chemical structures for the 201 and 116 product ions that resulted from fragmentation of buprofezin. Furthermore, the accurate mass from the *m/z* 201.1059 fragment ion matched the formula from the chemical drawing software quite closely, which gave a higher certainty for identification.

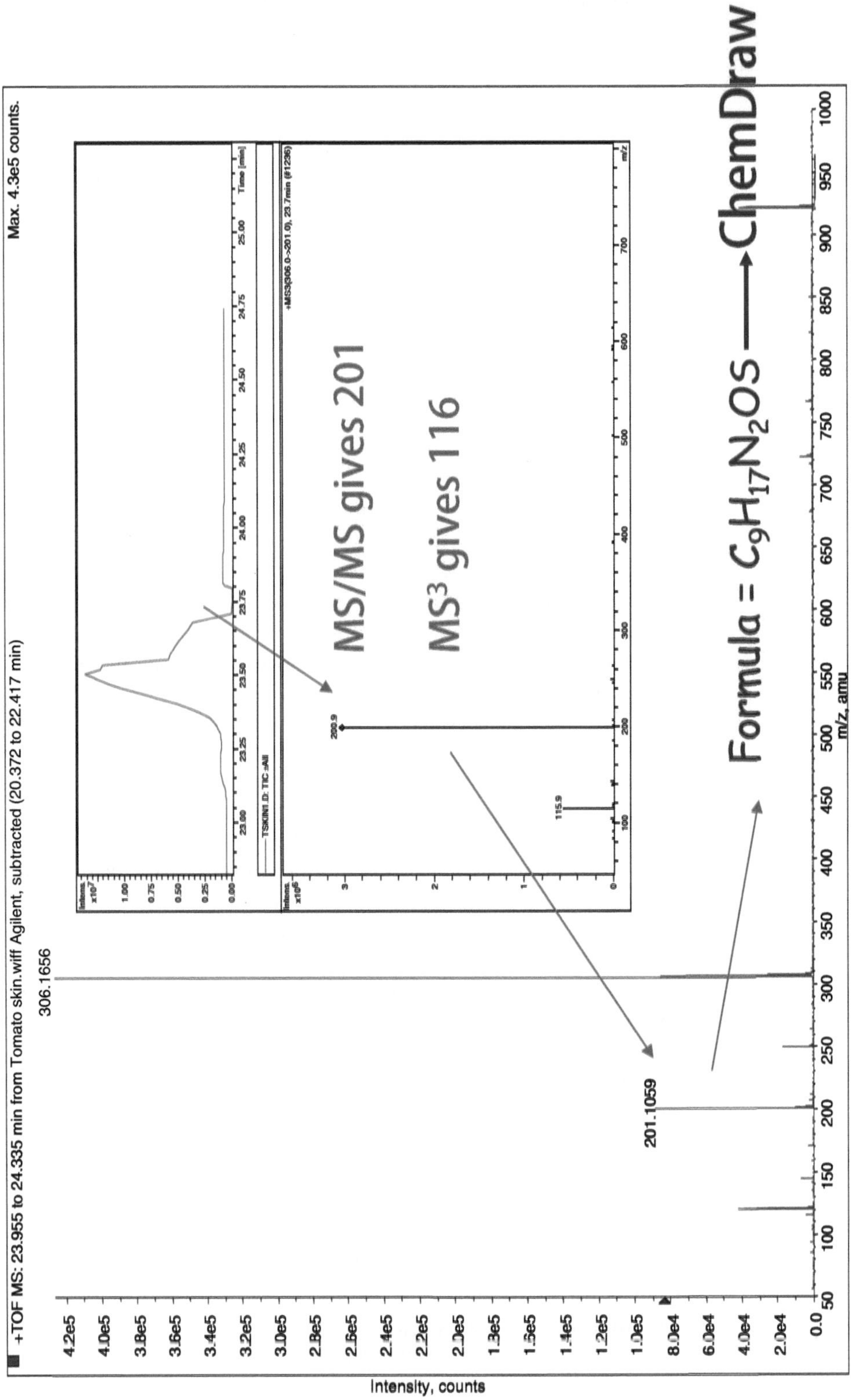

Fig. 9. LC/MS TOF spectrum and LC/MS/MS Ion Trap spectrum for m/z 306.

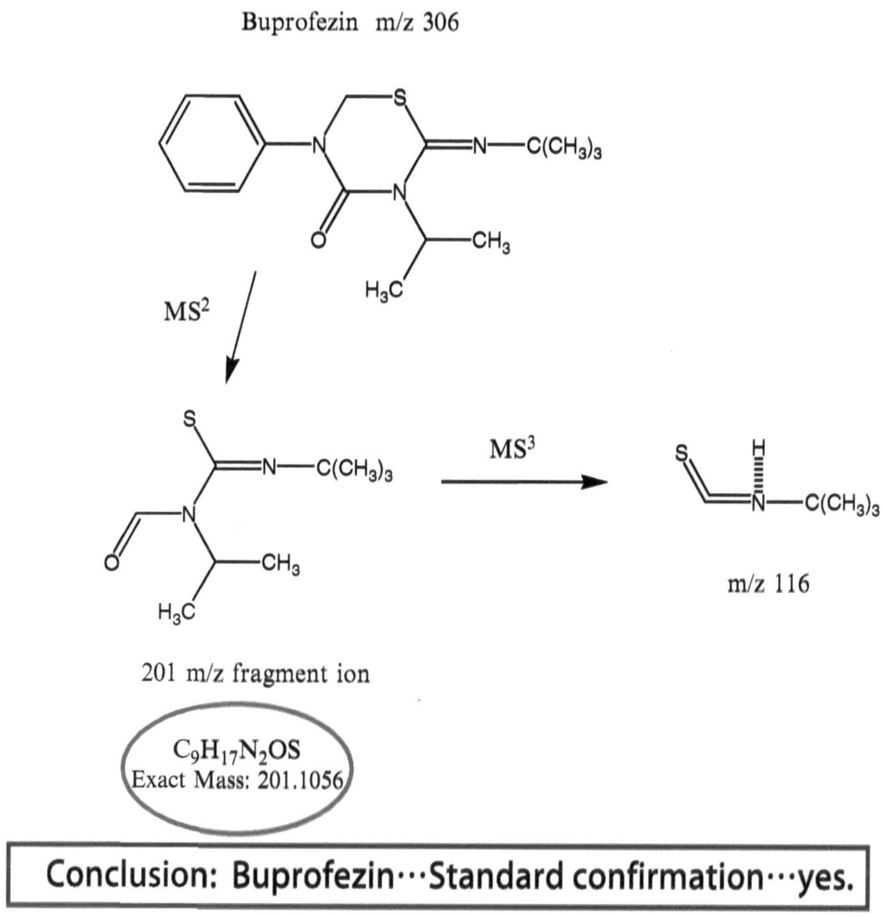

Fig. 10. Reasonable structures of fragment ions consistent with the structure of buprofezin.

After obtaining the buprofezin standard, the final data (retention time match and accurate masses) show a perfect match, which shows the ability of combining the LC/TOF-MS and LC/MS ion trap to identify unknowns (27).

Thus, it appears that this four-step procedure is a new approach and a powerful method for unknown analysis and is substantially different than the checking of library spectra that is commonly used in GC/MS identification methods or the use of selected ion monitoring or multiple reaction monitoring of LC/MS methods. The biggest liability of the approach is the lack of databases available to search molecular formulas. It is important to realize that accurate mass LC/MS techniques are ushering in a new approach to unknown identification especially when combined with LC/MS ion trap (or even better LC/Q-TOF-MS) and chemical drawing software with accurate mass capabilities.

Finally, in summary, the complementary nature of the two instruments, LC/TOF-MS and LC/MS ion trap, is shown.

Other combinations of mass spectrometers that can be combined with TOF include Q/Trap and triple quadrupole for MS/MS verification, (and of course Q/TOF by itself or Orbitrap with MS/MS capabilities).

Finally, it has not escaped the authors attention that the well-known environmental analysis quote of Lynn Roberts (28) concerning the hunt for emerging contaminants (e.g., pesticides, pharmaceuticals in water, soil, food, etc.) by mass spectrometry (28), *As any analytical chemist knows, what you see depends on what you look for.* is not *always* true (i.e., you don't always need standards and selected ions a priori to make identifications). The combined power of these two instruments (accurate mass within 3 ppm and MS/MS), chemical drawing software (including elemental calculators), and good databases make "a priori" unknown identification possible. Furthermore, it is important to take into account the mass of the electron in accurate mass calculations that are less than 2 ppm. Some of the instrument manufacturers have left this consideration out. Both the Agilent TOF and the Cambridge Software take the electron mass into account. It must be noted, however, that the identification of a "true unknown," that is a compound detected that is not in any accessible database, is a formidable task and may require the combination of accurate MS measurement combined with intense sample preparation for NMR or other powerful structure elucidation techniques. In addition, chemical knowledge of the sample and the circumstances of its procurement along with the purpose of the determination is always important.

5. Future of LC/TOF-MS in Food Analysis

The future of LC/TOF-MS in food analysis is one of growing importance for the detection of non-target pesticides (29). There are some limitations to be overcome before routine analysis is possible. Limitations include improved databases for identification of unknowns. Now we have at our disposal indexes, such as the Merck Index, Chemical Index, and on-line databases. Clearly not all pesticides that are in the Market Place are included. Thus, a major effort should be made to increase the availability of searchable elemental databases. Second is the use of accurate mass fragment ions and their identification. Typically this is not a problem when one combines CID fragmentation with chemical drawing software (21, 22). The use of Q/TOF and ion trap combined with LC/TOF-MS and Orbitrap MS may also answer this problem and thus, the combination of ion trap and TOF is powerful (21–23, 27) or the use of Q/TOF analysis of unknowns (25).

The use of high resolution liquid chromatography with sub 2 μm particles is another approach to increase the number of theoretical plates and the peak capacity of a chromatographic analysis. The use of LC/NMR too when combined with LC/Q-TOF-MS represents the most powerful selectivity (and thus specificity) that is currently available. Thus, the use of accurate mass by TOF will be an important tool in the years to come as resolving power, accuracy, and sensitivity continues to increase with instrument improvements.

References

1. Department of Agriculture Web Report 2005 for Pesticides in Food.
2. EU Food Directives, 2002, 91/414/EEC.
3. Analysis of Food by Mass Spectrometry, 2005, Elsevier, ed. Fernandez-Alba, A. 356p.
4. Zeeuw, R.A. de, 2004, Substance identification: the weak link in analytical toxicology: Journal of Chromatography A, v. 811, pp. 3–12.
5. Mandatory Guidelines for Federal Workplace Drug Testing Programs, Substance Abuse and Mental Health Service Administration, US Department of Health and Human Services, Federal Register 53 (1988), nr 69, Di-D12.
6. Valcarcel, M., Gomez-Hens, A., Rubio, S., 2001, Selectivity in analytical chemistry revisited: Trac Trends in Analytical Chemistry, v. 20, pp. 386–393.
7. Van Eenoo, P.; Delbeke, F.T., 2004, Criteria in chromatography and mass spectrometry-A comparison between regulations in the field of residue and doping analysis: Chromatographia Supplement, v. 59, pp. S39–S44.
8. Bethem, R.A.; Boyd, R.K., 1998, Mass spectrometry in trace analysis: American Society of Mass Spectrometry v. 9, pp. 643–648.
9. Andre, F.; et al., 2001, Trends in the identification of organic residues and contaminants: EC regulations under revision, Trends in Analytical Chemistry, v. 20, pp. 435–445.
10. Stolker, A.A.M.; Brinkmann, U.A.Th., 2005, Analytical strategies for residue analysis of veterinary drugs and growth-promoting agents in food-producing animals-a review, Journal of Chromatography A, v. 1067, pp. 15–54.
11. Sphon, J.A., 1978, Use of mass spectrometry for the confirmation of animal drug residues: Journal of the Association of Official Analytical Chemists, v. 61, pp. 1247–1252.
12. Thurman, E. M., Ferrer, Imma, 2003, Comparison of Quadrupole, Time-of-flight, triple quadrupole, and ion-trap mass spectrometry for the analysis of emerging contaminants, Chapter 2 in: Liquid Chromatography/Mass Spectrometry/Mass Spectrometry and Time-of-Flight Mass Spectrometry for the analysis of emerging contaminants, American Chemical Society Symposium Volume 850, pp. 14–31.
13. Budde, W.L., 2001, Analytical Mass Spectrometry: Strategies for Environmental and Related Applications, American Chemical Society, Washington, D.C., 386 p.
14. Ferrer, I., Thurman, E.M., 2009, Liquid Chromatography Time-of-Flight Mass Spectrometry: Principles, Tools, and Applications for Accurate Mass Analysis, John Wiley & Sons, Inc., 261 p.
15. Rogers, R.P.; Schaub, T.M.; Marshall, A.G., 2005, Petroleomics: MS returns to its roots, Analytical Chemistry, v. 77, pp. A21–A27.
16. Ferrer, Imma; Thurman, E.M., 2003, Analysis of Emerging Contaminants by LC-MS TOF: TrAC v. 35, pp. 304–312.
17. Sparkman, O.D., 2000, Mass Spec Desk Reference, Global View Publishing, Pittsburgh, 106 p.
18. Thevis, M.; Schanzer, W., 2005, Mass spectrometry in doping control analysis: Current Organic Chemistry v. 9, pp. 825–848.
19. Ferrer, Imma; Thurman, E.M., 2005, Measuring the mass of an electron by LC/TOF-MS: A study of "twin ions", Analytical Chemistry, v. 77, pp. 3394–3400.
20. Ferrer, Imma; Thurman, E.M.; Fernandez-Alba, A.R., 2005, Quantitation and accurate mass analysis of pesticides in vegetables by LC/TOF-MS: Analytical Chemistry, v. 77, pp. 2818–2825.
21. Thurman, E.M., Ferrer, Imma, and Amadeo Fernandez-Alba, 2005, Matching unknown empirical formulas to chemical structure using LC/MS TOF accurate mass and database searching: example of unknown pesticides on tomato skin: Journal of Chromatography A., v. 1067, pp. 127–134.

22. Thurman, E.M.; Ferrer, Imma; Garcia-Reyes, J.F.; Zweigenbaum, J.A.; Woodman, Michael; Fernandez-Alba, A.R., 2005, Discovering metabolites of post harvest fungicides in citrus with liquid chromatography/time-of-flight mass spectrometry and ion trap tandem mass spectrometry: Journal of Chromatography A, v. 1082, pp. 71–80.
23. Kellmann, M., Muenster, H., Zomer, P., Mol, H., 2009, Full scan MS in comprehensive qualitative and quantitative residue analysis in food and fee matrices: How much resolving power is required?, Journal of the American Society for Mass Spectrometry, v. 20, pp. 1464–1476.
24. Pelander, A.; Ojanpera, I.; Laks, S.; Rasanen, L.; Vuori, E., 2003, Toxicological screening with formula-based metabolite identification by liquid chromatography/time-of-flight mass spectrometry, Analytical Chemistry, v. 75, pp. 5710–5718.
25. Sanchez, J.V., 2004, Analysis of unknowns in water by LC/Q/TOF-MS: Rapid Communications in Mass Spectrometry, v. 19, pp. 143–148.
26. Blokland, M. H.; Zoontjes, P. W.; Sterk, S. S.; Stephany, R. W.; Zweigenbaum, J.; van Ginkel, L. A., Confirmatory analysis of Trenbolone using accurate mass measurement with LC/TOF-MS. Analytica Chimica Acta 2008, 618 (1), 86–93.
27. Ferrer, I., Heine, C.E., Thurman, E.M., 2004, Combination of LC/TOF-MS and LC/Ion trap MS/MS for the identification of diphenhydramine in sediment samples: Analytical Chemistry, v. 76, pp. 1437–1444.
28. Roberts, L., 2000, Personal Communication, Oral Presentation, American Chemical Society Meeting.
29. Thurman, E.M., Ferrer, I., Zweigenbaum, J.A., 2006, High-resolution and accurate mass analysis of xenobiotics in food: Analytical Chemistry, v. 78, pp. 6703–6708.

Chapter 9

Hormone Analysis in Food Products

Marco H. Blokland and Saskia S. Sterk

Abstract

The administration of growth-promoting compounds, to food producing animals is banned within the European Union. We developed several methods whereby the cleanup is based on LLE and SPE and detection based on GC–MSMS or LC–MSMS to identify and confirm the identity of different growth promoting agents in several food products. This chapter describes methods to isolate and identify these growth promoting agents.

Key words: Growth promoting agents, Gas-chromatography, Liquid-chromatography, Mass-spectrometry, Banned, Illegal

1. Introduction

The abuse of steroids for growth promotion in cattle has been a problem for at least 30 years within the European Union. A broad range of different steroids were used (1–4). The use of steroids and other growth promoting agents is banned in the EU since 1988 (96/23/EC)(5). To detect these growth promoters, specific analytical methods are necessary, which are capable to detect these compounds at residue levels, around 0.5–1.0 μg/kg. These growth promoters are analyzed in food products from within the European Union and also in imported food products from outside the European Union. We developed several methods to detect growth promoters at residue levels in various food products. Clean-up procedures and mass spectrometric detection for different growth promoters have often similar steps in the cleanup, which are therefore interchangeable between methods. In this chapter these common steps will be first described. From these common steps a few methods for the analysis of meat, shrimps, and liver will be

Jerry Zweigenbaum (ed.), *Mass Spectrometry in Food Safety: Methods and Protocols*,
Methods in Molecular Biology, vol. 747, DOI 10.1007/978-1-61779-136-9_9, © Springer Science+Business Media, LLC 2011

described in detail for several classes of compounds. Validation results of these methods will be given.

2. Materials

2.1. Equipment

Any reference to type and/or product is only to inform the user and identify the equipment and does not imply exclusion of others.

1. Vortex.
2. Centrifuge.
3. Sample concentrator for SPE.
4. Ultrasonic bath.
5. Glass test tubes, 10 mL.
6. Electric water bath with thermostat adjustable with nitrogen facility (Turbo Vap).
7. Rotating apparatus.
8. Glass injection vials with glass 50 μL inserts and caps.
9. Gas chromatograph equipped with automatic injector coupled to a mass spectrometer.
10. Ultrasonic finger with microtip.
11. Polypropylene tubes 50 mL.
12. Ultra high pressure liquid chromatograph coupled to a tandem quadrupole mass spectrometer (UHPLC-MS/MS).
13. SPE Oasis HLB 1 mL 30 mg extraction columns or equivalent.
14. SPE C_{18} column, 1 g, 6 mL.
15. SPE NH_2 column, 1 g, 6 mL.
16. Injection vials, 12 × 32 mm, glass screw neck vial, Quick Thread, LectraBond cap, preslit, PTFE/silicone.
17. IAC (immunoaffinity columns) for trenbolone (containing five columns, 20× concentrated wash-buffer, and 5× concentrated storage-buffer).

2.2. Chemicals

1. Suc d'Helix Pomatia juice (S.H.P.), Enzymatic activities, β-glucuronidase: 100,000 FU/mL, Sulfatase: 1,000,000 FU/mL.
2. Sodium hydroxide 1 M, dissolve 40 g sodium hydroxide in water and add water to a final volume of 1,000 mL.
3. Hydrochloric (HCl) solution, 5 M, add 490 g hydrochloric acid 37% to 510 g of water.

4. MSTFA++: consists of a mixture of *N*-methyl-*N*(trimethylsilyl) trifluoroacetamide (MSTFA): ammonium iodide: dithioerythreitol (1,000:2:4, v/w/w%).
5. Buffer of 2 M acetate, pH 5.2. Dissolve 25.2 g acetic acid and 129.5 g sodium acetate in 800 mL of water. Adjust the pH to 5.2 ± 0.1 and add water to a final volume of 1,000 mL.
6. Buffer of 0.1 M phosphate, pH 7.0: Prepare a solution of 14.2 g Na_2HPO_4 in 500 mL water. Prepare a solution of 13.6 g KH_2PO_4 (MW 136 g/mol) in 500 mL water. Add ca. 360 mL of the KH_2PO_4 solution to the Na_2HPO_4 solution. Fill up to 1,000 mL with water.
7. Buffer of 50 mM phosphate, pH 5. Dissolve 6 g of NaH_2PO_4 in 900 mL water; adjust to pH 7 with KOH, NaOH or HCl as necessary; adjust to 1 L with water.
8. A solution of 1 M acetic acid. Dilute 6.0 g acetic acid with water to a final volume of 100 mL.
9. Buffer,0.1 MTris,pH9.5.Dissolve 12.1 gTris(hydroxymethyl)-amino-methane in 800 mL of water. Adjust the pH to 9.5 ± 0.1 and add water to a final volume of 1,000 mL.
10. Heptane/butanol (80:20 v/v%). Add 200 mL butanol to 800 mL heptane.
11. IAC-eluting solution 70:30 v/v% ethanol:water. Add 300 mL water to a final volume of 1,000 mL.
12. IAC wash buffer: Dilute 50 mL column wash buffer concentrate with water to a final volume of 1,000 mL.
13. IAC storage buffer: Dilute 200 mL of column storage buffer concentrate with water to a final volume of 1,000 mL.
14. A solution of 5 mM ammonium acetate. Dissolve 0.39 g ammonium acetate in 1,000 mL water.
15. A solution of 5% ammonia in 70% methanol/acetonitrile (1:1): mix 70 mL methanol/acetonitrile (1:1) with 20 mL ammonia 25% and 10 mL water.
16. β-glucuronidase from *Eschericha coli*.

3. Methods

3.1. Common Extraction Steps

3.1.1. Homogenization and Destruction of Tissue

It is not known if the analytes are equally distributed in the tissue samples. Samples have to be homogenized before extraction and growth promoters have to be extracted from tissue before further processing. It is not always known if the compounds are bound e.g. proteins, or are present within cells. If so, it is necessary to destruct the tissue. Destructing, in general, is performed in two ways, enzymatic or by ultrasonic disruption.

- Homogenization
  - Tissue samples are crushed and homogenized in a blender. A homogeneous test portion is weighed into a 50-mL centrifuge tube.
- Enzymatic destruction
  - Make a fresh solution of 1 mg Subtilisin/2 mL Tris-buffer pH 9.5 and add 2 mL of this solution to each sample, vortex.
  - Incubate during 2 h at 55°C, shake/vortex every 30 min.
  - Allow the samples cool down for 10–20 min to room temperature.
- Ultrasonic destruction
  - Add 10 mL of water and vortex tissue/water mixture for 30 s.
  - Disintegrate the tissue using a Branson 250 Sonifier equipped with a microtip under the following conditions: duty cycle 50%, output control 6, and 20–25 cycles per sample.

3.1.2. Deconjugation

After administration, anabolic steroids undergo extensive metabolism, in which they are converted to more polar compounds and are excreted in urine. The main phase I metabolic pathways are oxidation, hydrolysis, and reduction. This phase I metabolism often brings a more polar group to the steroid structure offering a site for the conjugation in phase II metabolic reactions. The most common phase II conjugation reaction in animals is glucuronidation in which the steroid is coupled to glucuronic acid. Analysis of steroids is based on the free compounds. To break up the conjugates an enzymatic hydrolysis is performed. For meat, this deconjugation step is not necessary since the compounds will be in their free form. However, in liver, conjugates of steroids can be present.

- Add 10 mL of 2 mol/L acetate buffer, pH 5.2 and 50 μL of Suc d'Helix Pomatia. The pH is checked with pH paper. When necessary the pH is adjusted to 5.2 with diluted acetic acid or NaOH. Hydrolysis is performed overnight at 37°.

3.1.3. Primary Extraction and Removal of Fat

Extraction from the destructed matrix is in general performed with an organic solvent. Together with the growth hormones other compounds are also extracted such as fats. To remove these compounds a washing step with hexane is performed.

- Primary Extraction
  - To the destructed and/or hydrolyzed samples, 10 mL of the organic solvent is added. Vortex for 1 min. Centrifuge for 10 min at 2,600 × *g*.
  - The organic layer is transferred to a clean glass tube with a disposable pasteur pipette and evaporated to dryness under a stream of nitrogen in a heating block or electric water bath at 55°C.
- Defattening
  - Redissolve the dry residue in a mixture of 5 mL methanol:water and wash twice with 5 mL heptane (each time centrifuged for 3 min at 2,600 × *g*). The heptane layer is discarded.

3.1.4. Purification and Concentration

Further purification is performed by solid phase extraction (SPE), in general C_{18} material is the extraction sorbent of choice. Sometimes the SPE step is combined with a liquid–liquid extraction (LLE) step.

- SPE Oasis, basic wash
  - Condition the SPE columns with one column volume ethanol from a dispension flask.
  - Wash the SPE columns with one column volume water.
  - Transfer the extract to the SPE column.
  - Wash with 1 mL of 30% Methanol.
  - Wash with 1 mL of a mix of 400 mL methanol/600 mL 0.1 M NaOH.
  - Wash with two column volumes of water and empty the columns by means of vacuum.
  - Elute with 1 mL of 80% methanol, use vacuum.
  - Evaporate to dryness under a stream of nitrogen at 55°C.
- SPE C_{18}, neutral wash
  - Pre wash a C_{18} cartridge with 5 mL methanol followed by 5 mL of water.
  - Pass the extract over the column.
  - Wash the column with 5 mL of water.
  - Wash the column with 5 mL of methanol/water 40/60 (v/v-%).
  - Elute the column with 5 mL of methanol/water 80/20 (v/v-%).
  - Evaporate at 55°C under a gentle stream of nitrogen.
- SPE NH_2
  - Precondition the NH_2 column with 5 mL of 80/20 v/v-% acetone/methanol.

- Pass the sample through the column and collect the eluate.
- Dry the eluate at 50°C under a gentle stream of nitrogen.
- Redissolve the residue in 0.5 mL ethanol and transfer it into a Waters injection vial.

- SPE Immuno Affinity Chromatography
  - The IAC-column is conditioned by washing it twice with 5 mL of wash-buffer. The total sample mixture is applied to the IAC-column.
  - After sample application, the column is washed twice with 5 mL of wash-buffer, 5 mL of water and eluted with 5 mL of IAC-eluting solution.
  - The eluate is evaporated to dryness under a stream of nitrogen in a heating block or electric water bath at 55°C. The residue is dissolved in 500 µL of ethanol.
- Liquid–liquid extraction with pentane
  - Dissolve the extract in 3 mL of water and extract with 5 mL of *n*-pentane.
  - Centrifugate for 10 min at 2600 × g and transfer the *n*-pentane layer into a clean tube.
  - Repeat the *n*-pentane extraction and combine the *n*-pentane layers, evaporate.
  - Dissolve the dried extract in 0.4 mL ethanol.

3.2. Detection

Detection is mostly performed by LC–MS/MS, methods have to fulfill the criteria as described in Commission Decision 2002/657(6). In Table 1 relevant LC–MSMS tune parameters for the different classes of compounds measured are given.

3.3. Methods for Analysis of Growth Promotors in Food Products

In this chapter, several methods are described for the analysis of different growth promoters in food products using the previously described cleanup steps. Steps that are different from the standard steps are described in detail. Validation results of each method are given.

3.3.1. Chloramphenicol

This method describes the analysis of Chloramphenicol (CAP) in samples of shrimp and meat. Chloramphenicol is a bacteriostatic antibiotic originally derived from the bacterium *Streptomyces venezuelae*. The presence of chloramphenicol in food presents a potential risk for human health. It is a potent, broad-spectrum antibiotic drug, which is banned for use in food producing animals in the EU since 1994.

3.3.1.1. Procedure

1. Weigh 1 ± 1% gram of meat or shrimp and add 0.25 ng of internal standard CAP-d5.
2. Enzymatic destruction.

Table 1
LC–MS/MS tune parameters

	Chloramphenicol	Resorcyclic acid lactones	Trenbolone
Ionization mode	**ES negative**	**ES negative**	**ES positive**
Capillary (kV)	3	2.5	3.5
Cone (V)	60	50	50
RF lens 1	0	30	30
Aperture (V)	0	0.5	0.5
RF lens 2	0.3	0.5	0.5
Source temperature (°C)	120	130	110
Desolvation temperature (°C)	300	350	350
Cone gas flow (L/h)	109	95	109
Desolvation gas flow (L/h)	795	800	825
LM 1 resolution	15	14	12.5
HM 1 resolution	15	14	12.5
Ionenergy 1	1	0	0
Entrance	1	-1	20
Exit	0	2	10
LM 2 resolution	15	14	14.5
HM 2 resolution	15	14	14.5
Ion energy 2	1.5	2	7.1
Multiplier	650	650	650
Collision cell pressure (mbar)	3.66e-3	3.0e-3	2.70e-3

3. Primary extraction is performed with ethylacetate.
4. Defattening step with heptane, ratio of water:methanol is 100:0 v/v-%.
5. After evaporation, 150 μL of water is added.
6. SPE Oasis basic wash.
7. The residue is dissolved in 100 μL water, ultrasonic and vortexed.
8. From the extract is 10 μL on the LC–MSMS injected. The LC Column used was: Waters acquity UPLCTM BEH C_{18}, 1.7 μm, 2.1 × 100 mm with filterholder, temperature column thermostat: 65°C, autosampler temperature: 10°C, eluens A: methanol:water = 20:80 v/v-%, eluens B: methanol, flow: 0.5 mL/min. The gradient starts at 100% B and is increased to 25% B in 2.5 min followed by a rapid increase to 75% B in

Table 2
MRM transitions measured

Component	Transition 1	Transition 2	Collison energy
Chloramphenicol	321.1 > 152	321.1 > 194	11
Chloramphenicol-d5	326.1 > 157.1	–	11

Table 3
Performance characteristics of the method

Compound	CCα (pg/mL)	CCβ (pg/mL)	Measurement uncertainty (%)
Chloramphenicol	9.7	16.5	47

0.1 min and 100% B in 0.01 min, the gradient remains till 4.5 min at 100% B. In Table 2, an overview is given of the transitions measured.

In Table 3 an overview of the performance characteristics is given.

3.3.2. Resorcylic Acid Lactones

This procedure describes a quantitative method of analysis of α-zearalanol (Zeralanol), β-zearalanol (Taleranol), α-zearalenol, β-zearalenol, zearalanone and zearalenone in liver.

3.3.2.1. Procedure

1. Weigh 2 ± 1% gram of liver and add 10 ng of internal standard mix (Zearalenone-D6, α/β-Zearalenol-D4, and α/β-Zearalanol-D4).
2. For ultrasonic destruction, use 5 mL of acetate buffer at pH 5.2 instead of 10 mL water.
3. Hydrolysis.
4. Primary extraction with TBME.
5. Reconstitution in water:methanol 70:30 v/v-%.
6. The extract is passed over a SPE C_{18} (neutral wash).
7. Reconstitution in 80/20 v/v-% acetone/methanol and passed over SPE NH_2 column.
8. From the extract, 20 μL is injected on the LC–MSMS. The LC Column used was: Waters acquity UPLCTM BEH C_{18}, 1.7 μm, 2.1 × 100 mm with filterholder, temperature column thermostat: 40°C, autosampler temperature: 5°C, eluens A: methanol/water, 5 mM NH_4Ac, 10:90 v/v-%, eluens B: methanol/water, 5 mM NH_4Ac, 10:90 v/v-%, flow: 0.35 mL/min. The gradient starts at 30% B and is increased to 55% B in 1 min followed by a increase to 65% B in 6.5 min and 100% B in 0.01 min, the gradient remains till 9 min at 100% B. In Table 4, an overview is given of the transitions measured.

Table 4
MRM transitions measured

Component	Transition 1	Transition 2	Collision energy
Zearalenone	317.15 > 131.15	317.15 > 175.15	25
α/β-Zearalenol	319.15 > 160.15	319.15 > 275.15	25
Zearalanone	319.15 > 275.15	319.15 >205.15	25
α/β-Zearalanol	321.15 > 277.15	321.20 > 259.20	25
Zearalenone-D6	323.15 > 134.15	–	25
α/β-Zearalenol-D4	323.15 > 160.15	–	25
α/β-Zearalanol-D4	325.15 > 281.15	–	25

Table 5
Performance characteristics of the method

Compound	CCα (pg/mL)	CCβ (pg/mL)	Measurement uncertainty (%)
Zearalenone	0.11	0.18	5
α/β-Zearalenol	0.27	0.47	52
Zearalanone	0.19	0.33	17
α/β-Zearalanol	0.35	0.59	19
Zearalenone-D6	0.13	0.23	34
α/β-Zearalenol-D4	0.18	0.30	11

In Table 5 an overview of the validation characteristics is given.

3.3.3. Trenbolone

This procedure describes the analysis of 17α- and 17β-trenbolone (17-hydroxy-19-norandrosta-4, 9, 11-trien-3-one) in samples of meat, liver, and fish.

3.3.3.1. Cleanup Procedure

1. Weight 5 ± 1% gram of meat, fish, and liver and add 5 ng internal standard, 17β-Trenbolone-D3.
2. Meat is enzymatically destructed and the liver is hydrolysed.
3. To the samples, 2 mL of 37% HCl is added.
4. Primary extraction is performed with TBME.
5. Defattening with heptanes from a mixture of water:methanol 30:70 v/v%.
6. The methanol–water layer is passed over an IAC column. After evaporation, 25 μL methanol:water (40/60 v/v%) is added. The vial is vortexed for 30 s. The mixture is transferred into an insert.

Table 6
MRM transitions measured

Component	Transition 1	Transition 2	Col. energy
17α-Trenbolone	271.3>253.3	271.3>199.11	20
17β-Trenbolone	271.3>253.3	271.3>199.11	20
17β-Trenbolone-D3	274.4>256.4	–	20

Table 7
Performance characteristics of the method

Compound	CCα (pg/mL)	CCβ (pg/mL)	Measurement uncertainty (%)
17α-Trenbolone	0.04	0.07	24
17β-Trenbolone	0.03	0.05	8

7. From the extract 10 μL is injected on the LC–MSMS. The LC Column used was: Agilent Zorbax Eclipse XDB C_{18} 2.1×100 mm 3.5 μm, temperature column thermostat: 40°C, autosampler temperature: 5°C, eluens A: methanol/water, 5 mM NH_4Ac, 10:90 v/v-%, eluens B: methanol/water, 5 mM NH_4Ac, 10:90 v/v-%, flow: 0.3 mL/min. The gradient starts at 40% B and is increased to 80% B in 10 min followed by an increase to 100% B in 0.1 min, the gradient remains till 12.5 min at 100% B. In Table 6, an overview is given of the transitions measured.

In Table 7, an overview of the validation characteristics is given.

3.3.4. Multimethod

Here, we present a method which is capable to detect and confirm the identity of a large number of different growth promoters in meat at the suggested control level of 0.5 μg/kg.

3.3.4.1. Cleanup Procedure

1. A 1 g portion of meat was transferred to a 50-mL tube. Internal standard mixture (2.5 ng) were added.
2. Ultrasonic destruction.
3. LLE with TBME.
4. The residue was dissolved into 4 mL of methanol:water (80:20 v/v-%) and defatted.

5. The water/methanol phase was evaporated until the volume was less than 0.5 mL, after which 4 mL of water was added.
6. Extraction by SPE C_{18} column.
7. To the evaporated residue, 3 mL of water was added after which the sample was extracted twice with 5 mL of pentane. The pentane layer was transferred into a 15-mL glass tube and evaporated.
8. The extract was dissolved in ethanol and transferred into a derivatization vial. The ethanol was evaporated until dryness. The dried extract was reconstituted in 25 μL of derivatization reagent and incubated for 1 h at 60°C. After 1 h the derivatization reagent was evaporated. The dried residue was reconstituted in 50 μL of iso-octane and transferred into an injection-vial.
9. Gas chromatography coupled to mass-spectrometer (GC–MS/MS) was carried out on a Varian 1,200 L. A GC capillary column, 30 meter VF-17MS (Varian) i.d. 0.25 mm, 0.25-μm film thickness with a constant flow of 1.0 mL helium/min was used. Injection was performed in splitless mode at 250°C. Injection volume was 2 μL. The oven temperature was kept constant at 110°C for 1 min and was increased, 20°C/min, to 340°C and was kept constant at this temperature for 4 min. The MRM transitions and their collision energies for each compound are given in Table 8.

In Table 9 an overview is given of the validation results.

Table 8
Parameters GC–MS/MS, between brackets the collision energy is given

Analyte	MRM1 (screening)	MRM2 (confirmation)
Cis-Diethylstilbestrol-d6	418 > 220 (−25 V)	–
Cis-Diethylstilbestrol	412 > 217 (−20 V)	412 > 383 (−15 V)
Hexestrol-d4	209 > 180 (−5 V)	–
Hexestrol	207 > 179 (−10 V)	207 > 191 (−10 V)
Trans-Diethylstilbestrol-d6	418 > 220 (−25 V)	–
Trans-Diethylstilbestrol	412 > 217 (−20 V)	412 > 383 (−15 V)
Dienestrol-d2	412 > 397 (−20 V)	–
Dienestrol	410 > 381 (−5 V)	410 > 395 (−5 V)
Benzestrol	207 > 179 (−10 V)	207 > 191 (−10 V)
α-Nortestosterone	418 > 313 (−12 V)	418 > 328 (−14 V)
α-Boldenone	430 > 206 (−18 V)	430 > 325 (−12 V)

(continued)

Table 8 (continued)

Analyte	MRM1 (screening)	MRM2 (confirmation)
α-Testosterone	432 > 209 (−10 V)	432 > 327 (−8 V)
β-Nortestosterone-d3	421 > 316 (−20 V)	–
β-Nortestosterone	418 > 313 (−12 V)	418 > 328 (−10 V)
β-Boldenone-d3	433 > 206 (−15 V)	–
β-Testosterone-d2	434 > 211 (−11 V)	–
β-Boldenone	430 > 206 (−18 V)	430 > 325 (−12 V)
β-Testosterone	432 > 209 (−10 V)	432 > 327 (−5 V)
α-Estradiol	416 > 285 (−16 V)	416 > 326 (−18 V)
β-Estradiol-d3	419 > 285 (−28 V)	–
β-Estradiol	416 > 285 (−16 V)	416 > 326 (−18 V)
Norethandrolone	432 > 287 (−25 V)	432 > 342 (−15 V)
Methyltestosterone-d3	449 > 301 (−30 V)	–
Methyltestosterone	446 > 301 (−30 V)	446 > 356 (−5 V)
Methylboldenone	444 > 206 (−15 V)	444 > 339 (−10 V)
Ethinylestradiol-d4	429 > 233 (−15 V)	–
Ethinylestradiol	425 > 231 (−14 V)	425 > 205 (−14 V)
Norethandrolone	446 > 356 (−10 V)	446 > 287 (−20 V)
Megestrol	453 > 273 (−17 V)	Not present
Megestrol-d3	456 > 276 (−15 V)	–
Medroxyprogesterone-d3	563 > 331 (−25 V)	–
Medroxyprogesterone	560 > 328 (−15 V)	560 > 315 (−15 V)
Progesterone-d5	463 > 448 (−15 V)	–
Progesterone	458 > 443 (−5 V)	458 > 157 (−20 V)
Melengestrol	570 > 480 (−10 V)	570 > 465 (−15 V)
Ml-d3	573 > 483 (−10 V)	–
Norclostebol	452 > 417 (−5 V)	452 > 321 (−5 V)
Ct-d3	469 > 338 (−5 V)	–
Chlorotestosterone	466 > 335 (−15 V)	466 > 431 (−10 V)
37-Chloromadinone	580 > 231 (−5 V)	580 > 490 (−10 V)
Chloromadinone	578 > 231 (−15 V)	Not present
Norclostebol acetate	422 > 216 (−10 V)	422 > 387 (−5 V)
Chlorotestosterone acetate-d3	439 > 404 (−5 V)	–
Chlorotestosterone acetate	436 > 401 (−15 V)	436 > 230 (−20 V)

Table 9
Overview of the validation results

Analyte	CCα (μg/kg)	CCβ (μg/kg)	U (%)
Hexestrol	0.07	0.12	52
Dienestrol	0.14	0.23	87
Diethylstilbestrol	0.06	0.10	48
Benzestrol	0.07	0.12	50
α-Nortestosterone	0.07	0.11	73
α-Estradiol	0.05	0.09	63
α-Testosterone	0.11	0.18	50
β-Nortestosterone	0.13	0.22	55
β-Estradiol	0.05	0.09	45
Norethandrolone	0.07	0.12	116
β-Testosterone	0.07	0.11	47
Ethinylestradiol	0.12	0.21	52
Methyltestosterone	0.05	0.08	35
Norethandrolone	0.10	0.17	83
Norclostebol	0.22	0.37	123
Progesterone	0.12	0.21	81
Chlorotestosterone	0.16	0.27	93
Norclostebol acetate	0.13	0.22	145
Chlorotestosterone acetate	0.09	0.16	56
Medroxyprogesterone	0.22	0.38	110
Chloromadinone	0.36	0.61	181
α-Boldenone	0.16	0.27	181
β-Boldenone	0.12	0.21	50
Methylboldenone	0.18	0.31	128
Melengestrol	0.57	0.97	470
Megestrol	1.43	2.43	1,375

4. Notes

1. In section 3.1.2 it must be noted that deconjugation of steroids is not effective for steroid-esters, so in case of detection of steroid-esters in meat, other methods have to be applied.
2. In the common extraction seciton 3.3 be aware that you are working at residue level, so cross over contamination can easily occur; use always disposables and clean pipettes.
3. In section 3.3.3.1 for the analysis of trenbelone, IAC columns can be used more often than what the manufacture claims, which can save costs.
4. For the mulit-residue method in section 3.3.4 DES has two isomers *trans*-DES and *cis*-DES. These isomers are in dynamic balance, and this balance depends on light, temperature, etc. For a standard DES in ethanol this balance is approximately *trans*-DES: *cis*-DES, 95:5. In this method *trans*-DES-D6 is used as internal standard; during cleanup, *trans*-DES-D6 isomerizes to *cis*-DES-D6. This isomerization will also occur to *trans*-DES. In this procedure the sum of *trans*+*cis*-DES peak area's and *trans*+*cis*-DES-d6 peak area's are used for quantification.

References

1. A. Lommen, R. Schilt, J. Weseman, A.H. Roos, J.W. van Velde, M.W. Nielen, J. Pharm. Biomed. Anal. 28 (2002) 87–96.
2. K. Vanoosthuyze, E. Daeseleire, A. Van Overbeke, C. Van Peteghem, Ermens, Analyst 119 (1994) 2655–2658.
3. K. De Wasch, H. De Brabander, D. Courtheyn, C. Van Peteghem,123 (1998) 2415–2422.
4. J. Sabbe, T.V. Beken (Eds.), BUFALAW-2001. European Commission, Falcone Programme, Maklu, Antwerpen-Apeldoorn, BE-NL. ISBN 9062156215803X, 2002, p. 226.
5. Council Directive 96/22/EC of 29 April 1996 concerning the prohibition on the use in stockfarming of certain substances having hormonal or thyrostatic action and of beta-agonists and repealing Directives 81/602/ EEC, 88/146/EEC and 88/299/EC. Off. J. Europ. Commun. 1996, L125, 3
6. Commission Decision No. 2002/657/EC of 12 August 2002 implementing Council Directive 96/23 concerning the performance of analytical methods and the interpretation of results. Off. J. Europ. Comm. (2002): L221, 8

Chapter 10

Analysis of Multiple Mycotoxins in Food

Jana Hajslova, Milena Zachariasova, and Tomas Cajka

Abstract

Mycotoxins are secondary metabolites of microscopic filamentous fungi. With regard to the widespread distribution of fungi in the environment, mycotoxins are considered to be one of the most important natural contaminants in foods and feeds. To protect consumers' health and reduce economic losses, surveillance and control of mycotoxins in food and feed has become a major objective for producers, regulatory authorities, and researchers worldwide. In this context, availability of reliable analytical methods applicable for this purpose is essential. Since the variety of chemical structures of mycotoxins makes impossible to use one single technique for their analysis, a vast number of analytical methods has been developed and validated. Both a large variability of food matrices and growing demands for a fast, cost-saving and accurate determination of multiple mycotoxins by a single method outline new challenges for analytical research. This strong effort is facilitated by technical developments in mass spectrometry allowing decreasing the influence of matrix effects in spite of omitting sample clean-up step. The current state-of-the-art together with future trends is presented in this chapter. Attention is focused mainly on instrumental method; advances in biosensors and other screening bionanalytical approaches enabling analysis of multiple mycotoxins are not discussed in detail.

Key words: Mycotoxins, Liquid chromatography, Mass spectrometry, Food

1. Introduction

Mycotoxins are natural toxic secondary metabolites produced by microscopic filamentous fungi, which grow on various agricultural commodities in the field, and/or during post-harvest period (transport, processing, and storage). The toxinogenic fungi belong mainly to genera *Aspergillus*, *Fusarium*, *Penicillium*, and *Alternaria* (1–4). Currently, more than 500 different mycotoxins are known; however, sufficient knowledge has been collected only for a limited number of them. With regard to the health hazard posed by mycotoxins to the end consumers (and farm animals), many countries have set up regulations for their control in food chain. In Table 1,

Jerry Zweigenbaum (ed.), *Mass Spectrometry in Food Safety: Methods and Protocols*, Methods in Molecular Biology, vol. 747, DOI 10.1007/978-1-61779-136-9_10,

Table 1

Overview of most common mycotoxins together with their producers, typical food commodities, major health adverse effects, and current maximal legislative limits (reproduced and updated from (19) with permission from Springer)

Mycotoxins	Main producers/origin	Food commodity	Maximum level (EC 1881/2006 amended by EC 1126/2007)
Fumonisins			
Fumonisins A1, A2, A3, B1, B2, B3, C1, C2, C3, P1, P2, P3	*Fusarium verticillioides*, *F. proliferatum*, *F. anthophilum*, *F. moniliforme*, *F. dlamini*, *F. napiforme*, *F. nygamai*, *Alternaria alternata*	Maize, maize based products, sorghum, sorghum, asparagus, rice	Sum of fumonisins B1 and B2: 200–4,000 μg/kg (infant foods, processed maize-based foods, unprocessed maize)
Hydrolyzed and partially hydrolyzed fumonisins	Product of food processing		
Trichothecenes			
Type A trichothecenes: T-2 toxin, HT-2 toxin, diacetoxyscirpenol, neosolaniol, verrucarol	*Fusarium sporotrichioides*, *F. poae*, *F. culmorum*, *F. equiseti*, *F. graminearum*, *F. moniliforme*, *Cephalosporium sp.*, *Myrothecium sp.*, *Trichodermasp.*, *Trichothecium sp.*, *Phomopsis sp.*, *Stachybotrys sp.*, *Verticimonosporium sp.*	Cereals, cereal based products	In discussion for T-2 and HT-2 toxin
Type B trichothecenes–nivalenol, deoxynivalenol, 3-acetylDON, 15-acetylDON, fusarenon-X	*Fusarium graminearum*, *F. culmorum*, *F. sporotrichioides*, *F. cerealis*, *F. lunulosporum*	Cereals, cereal based products	Deoxynivalenol: 200–1,750 μg/kg (infant food, processed cereal-based foods, unprocessed cereals)
Deoxynivalenol-3-glucoside	Metabolite of deoxynivalenol		
Zearalenones			
Zearalenone	*F. graminearum*,*F. culmorum*, *F. crookwellense*, *F. equiseti*, *F. sporotrichioides*	Barley, oats, wheat rice, sorghum, sesame, soy beans, cereal based products	20–400 μg/kg (maize-based infant food, processed cereal-based and maize-based foods, unprocessed maize, refined maize oil)
α- and β-zearalenol, α- and β-zearalanol	Metabolites of zearalenone		

(continued)

Table 1 (continued)

Mycotoxins	Main producers/origin	Food commodity	Maximum level (EC 1881/2006 amended by EC 1126/2007)
Ochratoxins			
Ochratoxins A, B, C	*Aspergillus ochraceus*, *A. niger*, *A. melleus*, *A. alutaceus*, *A. alliaceus*, *A. albertensis*, *A. citricus*, *Neopetromyces muricatus*, *Penicillium viridicatum*, *P. verrucosum*, *P. cyclopium*, *P. carbonarius*	Cereals, dried fruit, raisins, wine, coffee, oats, spices, rye	Ochratoxin A: 0.5–10 μg/kg (infant foods, processed cereal-based foods, unprocessed cereals, dried vine fruits and instant coffee)
Ochratoxin α	Metabolite of ochratoxin A		
Aflatoxins			
Aflatoxins B1, G1, B2, G2	*Aspergillus flavus*, *A. nomius*, *A. parasiticus*, *A. arachidicola*, *Emericella astellata*, *E. venezuelensis*, *E. olivicola*	Maize, wheat, rice, spices, almonds, oilseeds, dried fruits, cheese	Sum of aflatoxins B1, B2, G1 and G2: 4–15 μg/kg, aflatoxin B1: 0.1–8 μg/kg; (nuts, ground nuts, dried fruits, cereals, maize)
Aflatoxins M1 and M2	Metabolites of aflatoxin B1 and B2	Milk, eggs, meat	Aflatoxin M1: 0.025–0.05 μg/kg (infant and dietary foods, milk)
Ergot alkaloids			
Ergocornine/inine, ergocristine/inine, ergocryptine/inine, ergosine/inine, ergotamine/inine	*Claviceps purpurea*, *C. africanana*, *C. fusiformis*, *C. fusiformis*, *C. paspali*, *Neotyphodium coenophialum*	Wheat, rye, hay, barley, millet, oats, sorghum, triticale	
Alternaria toxins			
Altenuene, alternariol, alternariolmonomethyl ether, altertoxin I, altertoxin II, altertoxin III, tenuazonic acid	*A. alternata*, *A. dauci*, *A. cucumerina*, *A. solani*, *A. tenuissima*, *A. citri*	Wheat, rice, rye, olives, sorghum, tobacco, apples, peppers, sunflower seeds, oilseed rape, pecan nuts, tomatoes, mandarins	

(continued)

Table 1 (continued)

Mycotoxins	Main producers/origin	Food commodity	Maximum level (EC 1881/2006 amended by EC 1126/2007)
Enniatins			
Enniatin A, enniatin A1, enniatin B, enniatin B1	*Fusarium Avenaceum*, *F. orthoceras*, *some Alternaria*, *Halosarpheia*, *Verticillium ssp.*	Wheat, corn, barley, bread mill, oat flour, rice	
Patulin	*Aspergillus clavatus*, *A. longivesica*, *A. terreus*, *P. expansum*, *Penicillium griseofulvum*, *Byssochlamys sp.*	Apples, apple juice, cherries, cereal grains, grapes, pears, bilberries	10–50 μg/kg (infant foods, apple juice, solid apple, spirit drinks derived from apples or containing apple juice, fruit juices)
Beauvericin	*F. bulbicola*, *F. denticulatum*, *F. lactis*, *F. phyllophilum*, *F. pseudocircinatum*, *F. succisae*	Wheat, corn, barley, bread mill, oat flour, rice	
Fusaroproliferin	*Fusarium proliferatum*, *F. concentricum*, *F. antophilum*, *F. begoniae*, *F. succisae*, *F. bulbicola*, *F. circinatum*, *F. udum*, *F. subglutinans*	Wheat, corn, barley, bread mill, oat flour, rice	

there is presented an overview of the major mycotoxins, which are currently under focus (5–7). Those, for which maximum limits based on exposure and toxicity data have been established by the European Union, are indicated by asterisk (8, 9). Aflatoxins, patulin, deoxynivalenol, fumonisins, and ochratoxin A are also included by the Food and Drug Administration Compliance program guidance manual (10).

While relatively extensive information is available on occurrence of regulated mycotoxins, the requirements for more comprehensive information on food crops contamination by toxins such as ergot alkaloids, beauvericin, or enniatins have been raised only recently. In addition to free mycotoxins, also occurrence of mycotoxin conjugates in cereals represents an emerging issue in food safety. Nowadays, most attention has been paid to deoxynivalenol-3-glucoside and zearalenone-4-glucoside originating in food plants as a result of detoxification process (11–13). Supposing such compounds are, at least partly bioavailable, then, dietary exposure might be underestimated.

As Fig. 1 documents, mycotoxins introduced in Table 1 represent largely differing structure classes, and consequently, their

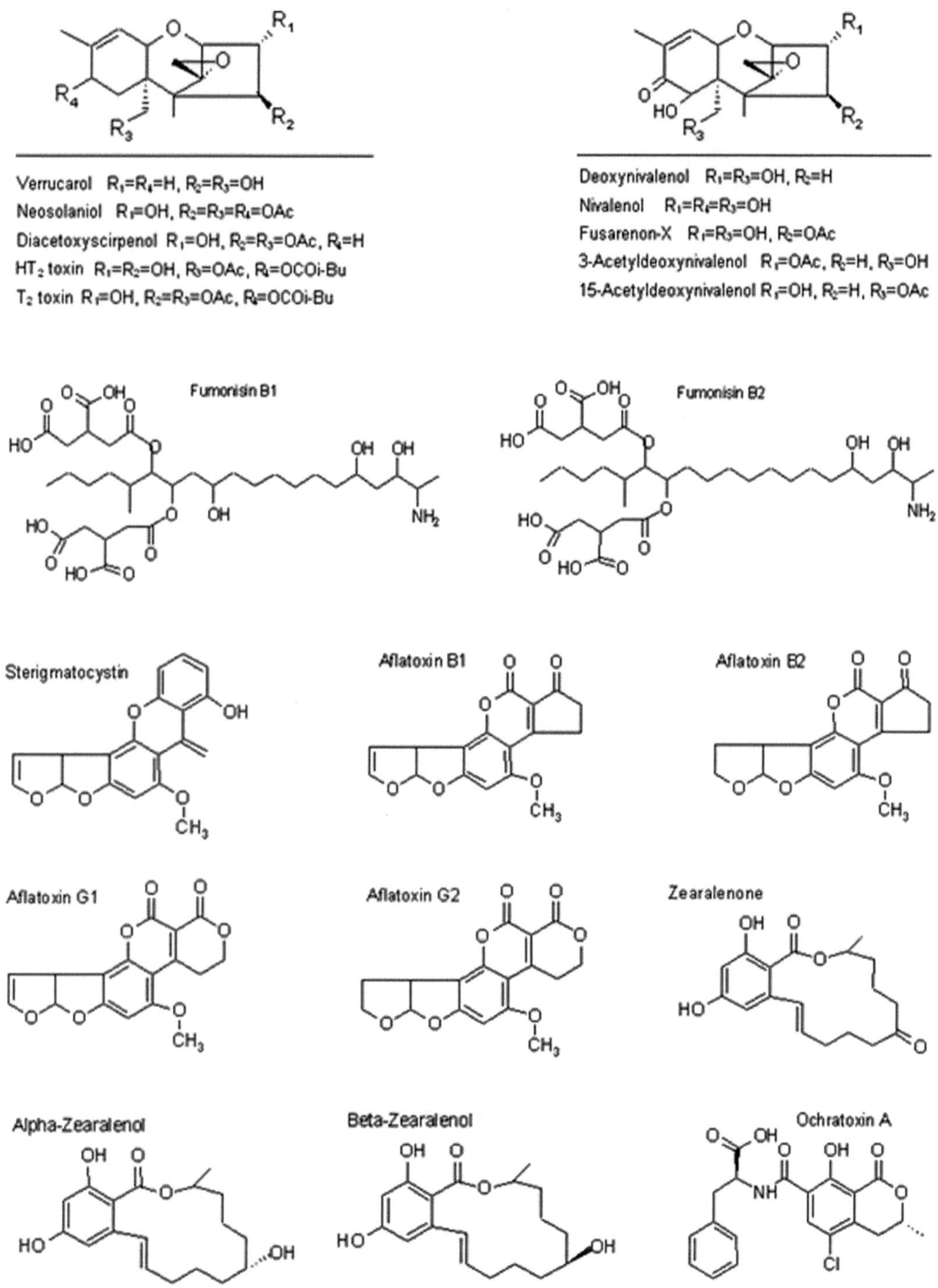

Fig. 1. Structures of selected mycotoxins.

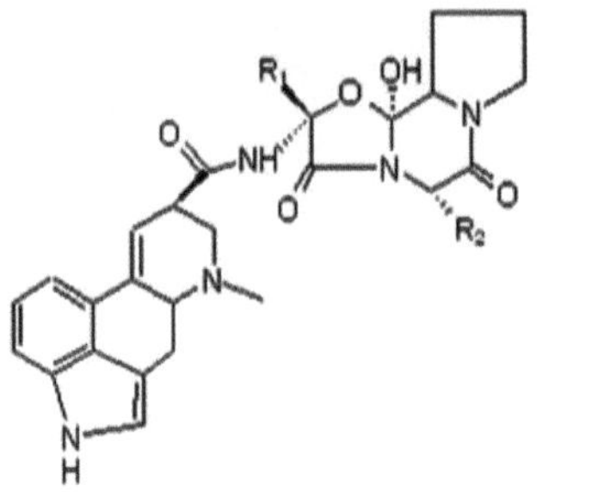

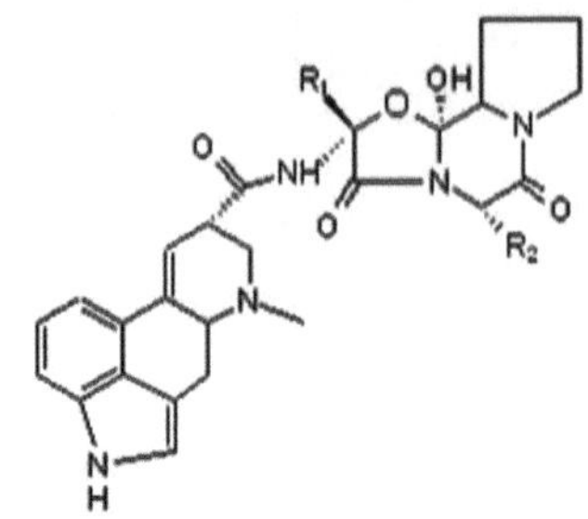

Ergosine	R_1=CH_3, R_2=$CH_2CH(CH_3)_2$
Ergotamin	R_1=CH_3, R_2=$CH_2C_6H_5$
Ergocornin	R_1=$CH(CH_3)_2$, R_2=$CH(CH_3)_2$
Ergocristin	R_1=$CH(CH_3)_2$, R_2=$CH_2C_6H_5$

Ergosinine	R_1=CH_3, R_2=$CH_2CH(CH_3)_2$
Ergotaminine	R_1=CH_3, R_2=$CH_2C_6H_5$
Ergocorninine	R_1=$CH(CH_3)_2$, R_2=$CH(CH_3)_2$
Ergocristinine	R_1=$CH(CH_3)_2$, R_2=$CH_2C_6H_5$

Altenuene

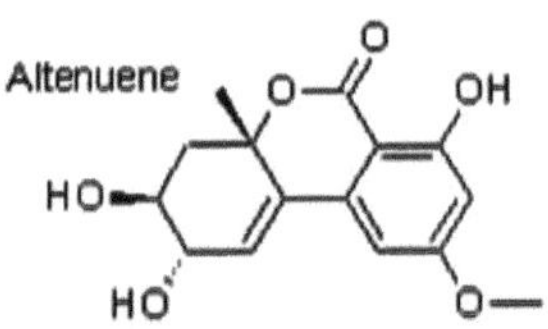

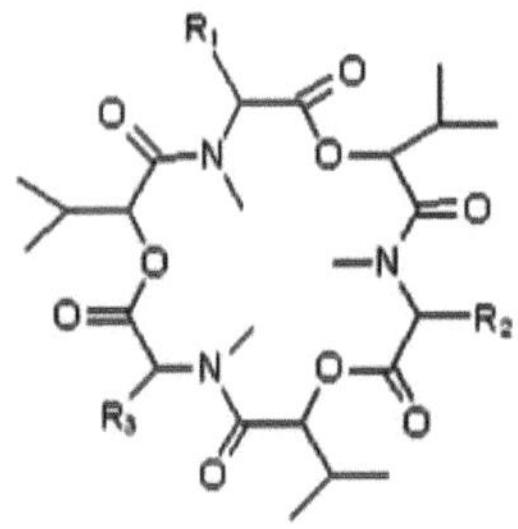

Alternariol

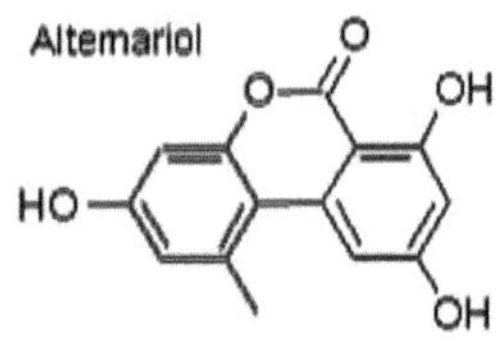

Alternariolmethylether

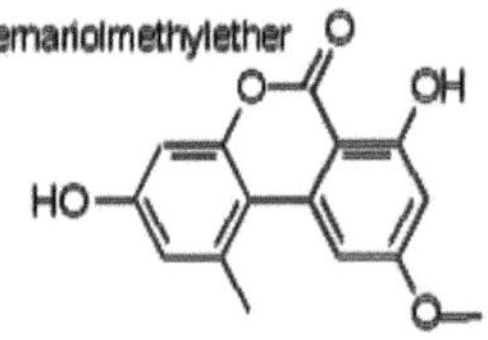

Enniatin A	R_1=R_2=R_3=$CH_2C_6H_5$
Enniatin A_1	R_1=R_2=R_3=$CH(CH_3)CH_2CH_3$
Enniatin B	R_1=$CH(CH_3)CH_2CH_3$, R_2=R_3=$CH(CH_3)_2$
Enniatin B_1	R_1=R_2=R_3=$CH(CH_3)_2$

Fig. 1. (continued)

physicochemical properties vary in a wide range. Most of mycotoxins are relatively stable compounds thus surviving under various conditions employed in food processing including thermal treatment. On this account, they can be, at least in some extent, transferred from contaminated raw material into final product (14, 15). Some of them, like fumonisins, might be transformed into bound forms (to starch or to proteins) after thermal processing (16), or can be partially and/or fully hydrolyzed when alkaline treatment is performed (17). Another interesting finding of the recent years is a significant increase of deoxynivalenol-3-glucoside in fermented cereal-based products, such as malt and beer (15).

To support extensive preventive efforts made by establishing simultaneous restriction of the allowed amounts of certain mycotoxins in foods (and feedstuffs), but also to enable collecting information on other, until now not regulated mycotoxins, reliable and accurate analytical methods, which allow their unambiguous identification and confirmation, as well as an accurate quantification at very low concentration levels in various matrices, have to be available. In following paragraphs, strategies to control multiple mycotoxins within a single analytical run will be discussed. In addition to meeting desired performance criteria (18), also laboratory throughput and workload are taken into consideration.

2. Methods for Mycotoxins Analysis

Analysis of mycotoxins in food is generally a multistep process comprised of (1) sampling, (2) extraction of analytes from the matrix (usually with mixtures of water and polar organic solvents) possibly followed by an extract purification, and (3) final detection and quantitative determination. Due to a large diversity of extraction, clean-up procedures, and respective detection steps available within analysis of mycotoxins and their conjugated forms, a comprehensive discussion of all existing methods would exceed the scope of this chapter. For this reason, we will focus just on the most common trends and recent advances in mycotoxins analysis.

2.1. Sampling

Distribution of mycotoxins in most of agricultural commodities is very heterogeneous; in most cases, the microscopic filamentary fungi and their secondary metabolites occur in so-called "hot spots." Thus, sampling is the largest source of variability associated with the mycotoxins analysis procedure, and the most crucial step in obtaining reliable results (19). In the past, a lot of papers related to sampling of aflatoxins were published (20–22). Recently, the sampling strategies have been set-up also for other mycotoxins, e.g. ochratoxin A, patulin, and *Fusarium* toxins. The European Commission issued the Commission Regulation (EC) 401/2006 laying out the sampling methods and the performance criteria for the methods of analysis to be used for the official control of mycotoxins in foodstuff (18). This Regulation provides sampling plans for groups of food commodities taking into account the heterogeneous distribution of mycotoxins. Different sampling plans were also established in other countries, e.g. in the USA for aflatoxins in peanuts (23).

Generally, it is possible to recommend that the most effective way to reduce the overall variability of results is to increase the size of the laboratory sample, ensure the proper milling, and homogenization (19, 24, 25).

2.2. Extraction and Crude Extract Purification

Mycotoxins are usually extracted from ground solid matrices by shaking with aqueous acetonitrile (*liquid–solid extraction*). Aqueous methanol or ethyl acetate has also been used to a lesser extent.

The most extensively used extraction mixture for a simultaneous co-isolation of a wide range of mycotoxins, is an acetonitrile:water (84:16, *v/v*) mixture. Rarely, mycotoxins can be isolated from samples by employing ***accelerated solvent extraction*** (ASE), in which the extraction efficiency is increased by enhanced pressure and temperature (26, 27). However, in spite of its automation, this technique may become rather laborious and time-consuming, since obstruction of extraction cells due swelling of starches in cereals often occurs. Moreover, performing a thorough clean-up of ASE extract is typically needed due to more co-extracted impurities compared to traditional shaking.

The choice of the extraction medium is closely related to the selected clean-up procedure. In mycotoxins analysis, purification of extracts is important, especially in case of their determination at trace levels. Commonly, procedures for mycotoxins clean-up are based either on ***solid-phase extraction*** (SPE) or use ***immunoaffinity columns*** (IACs). Among commercially available SPE columns, MycoSep cartridges are the most frequently used (28–30). Currently, multifunctional columns containing, e.g. charcoal, celite, and alumina are available for trichothecenes, zearalenone, aflatoxins, ochratoxins, moniliformin, fumonisins, and ergot alkaloids analysis. Employing SPE based on polymeric reversed-phase columns (*N*-vinylpyrrolidone/divinylbenzene columns Oasis HLB) is also possible obtaining good recoveries for both type-A and type-B trichothecenes (31).

Regarding the IAC-based clean-up, its main advantage includes, in addition to the purification effect, also the possibility of analytes pre-concentration what results in decreasing of detection limits. Another advantage is its applicability for complex matrices and reduced usage of organic solvents. The highly appreciated feature of this type of purification approach is its specificity, which is, however, limiting for simultaneous determination of different groups of analytes (32). Depending on the type of antibody, some cross-reactivity may be encountered potentially leading to results overestimation. This phenomenon can be successfully exploited in analysis of masked mycotoxins since, in addition to the target compound, also structurally related metabolites can be bound. For instance, thanks to cross-reactivity of DON dedicated DONprep columns also deoxynivalenol-3-glucoside, the major "masked" ***Fusarium*** toxin, can be isolated together with free DON (33). Additionally to DON, the IACs are commercially available for T-2 and HT-2 toxins, fumonisins, zearalenone, aflatoxins, and ochratoxin A. It is worth mentioning that combined multimycotoxin immunoaffinity columns capable to purify a broader range of mycotoxins, in particular HT-2 and T-2 toxins, deoxynivalenol, and zearalenone, (34), are currently available in the market.

Since all mycotoxins vary considerably in their polarities, in the case of multi-mycotoxin analysis, an optimal extraction and purification step for each group of analytes is not possible to perform, and, unavoidably, some compromises have to be made. An example of very fast generic extraction/purification is the QuEChERS approach (Quick, Easy, Cheap, Effective, Rugged, and Safe), currently widely used in pesticide residue analysis. The key principle is partitioning of an acetonitrile:water mixture induced by addition of inorganic salts. While the analytes are transferred into an organic phase, more polar matrix impurities are left in an aqueous layer. As in the case of pesticides, the residual impurities in acetonitrile (some sugars and fatty acids) can be removed by dispersive SPE realized by addition of primary secondary amine (PSA) sorbent. However, due to the acidic nature of some mycotoxins (e.g. fumonisins) and the risk of their binding on the sorbent, this approach is not recommended (35).

2.3. Examination of Sample Extracts

Depending on the purpose of analysis, either simple semiquantitative (immunochemical) screening assays, or accurate instrumental methods, namely when compliance with legislation is to be checked, are used. Figure 2 shows trends in the mycotoxins analysis area during the last 50 years. A growing employment of bioanalytical methods from the beginning of the 1990s such as Enzyme Linked Immunosorbent Assay (ELISA) as well as the biosensors in the subsequent decade was noticed. Concerning the instrumental analysis, liquid chromatography coupled with mass spectrometry (LC–MS) revolutionizes the mycotoxins analysis area, enabling quantitative and confirmatory analysis of multiple mycotoxins, independent of their chemical structure or biological activity.

Due to the inherent complexity of food matrices and the impossibility to get samples free of co-extracts, most common

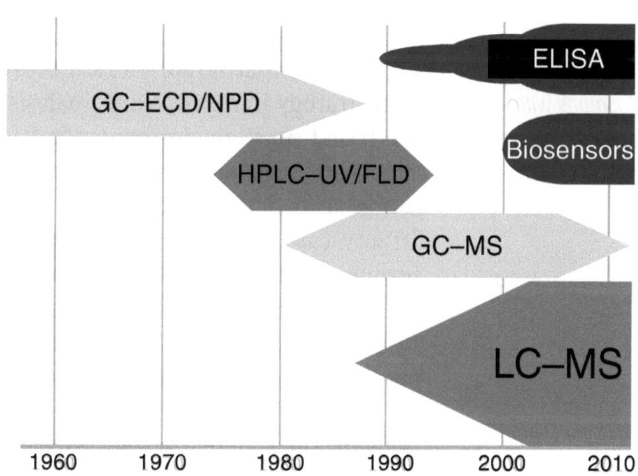

Fig. 2. Trends in the analysis of mycotoxins from the time perspective (years 1960–2010).

instrumental analytical procedures involve some sample separation step prior to identification/quantification of the analytes to reduce interferences within these processes. However, recent availability of highly selective detection tools opened the door to applications in which the separation step is eliminated. In the following paragraphs, a brief introduction of conceivable analytical approaches and their development during the time is summarized.

2.3.1. Gas Chromatography

In the past, methods based on a gas chromatographic (GC) approach were routinely used for determination of trichothecenes, zearalenone, ochratoxin A, patulin, and citrinin (36–40). However, GC-based methods suffer from some significant drawbacks; the major one is the need to carry out derivatization of analytes prior to sample analysis. Most of the mycotoxins are small nonvolatile and polar molecules, which require breaking of hydrogen bridges to become amenable to GC–MS analysis. For this purpose, silylation and acylation agents are preferably employed. Moreover, for detection of mycotoxins with the electron capture detector (GC–ECD), brominating or fluoroacylating agents have to be used to take advantage of detector specificity (28, 30, 41, 42). In addition to labor and time demands of these procedures, problems such as double peaks of analytes caused by the incomplete derivatization can appear (43). Other analytical problems encountered with procedures employing GC included non-linearity of calibration curves, over-estimation of results due to matrix effects (when using pure standards for calibration), poor repeatability, and memory effects from previous sample injections (30, 44). Except of the study of Jelen and Wasowicz reporting the use of comprehensive two-dimensional GC with time-of-flight mass spectrometry (GC×GC–TOFMS) for the trichothecene analysis in wheat (45), no other advances in the GC area have been recently published. LC–MS is becoming the most effective tool for the mycotoxins analysis.

2.3.2. Liquid Chromatography with Conventional Detectors

Liquid chromatography (LC) represents the dominating separation strategy in mycotoxins analysis. Current "classic" procedures are based on high performance LC coupled to the conventional detectors such as fluorescence detector (FLD), UV detector, diode-array detector (DAD), or photodiode array detector (PDA). In any case, sample pretreatment for minimizing matrix interferences, thus unbiased results, is a task of major importance (it should be noted that, contrary to mass spectrometric detection mentioned below, correction of results by using isotopically labeled internal standards is not feasible for optical detection). An overview of the latest methods for mycotoxins analysis using conventional detectors is presented in Table 2. Fluorescence detection is often employed for the analysis of ochratoxin A, aflatoxins, and zearalenone. However, in the absence of natural

Table 2

Overview of latest analytical methods for mycotoxins determination employing liquid chromatography coupled with conventional detectors

References	(46)				(47)	(48)	(49)	(50)	(51)	(52)		(53)	(54)	(55)	(56)	(57)	(58)	(59)
Matrices	Cereals, cereal products				Rice	Medical plants	Red paprika, black pepper	Wheat flour	Milk	Paprika		Honey, natural sweetners, vinegars, apple juice	Rye, rye products	Dried apple rings	Apple juice	Oats, cereal foods	Green coffee, roasted coffee	Blue cheese
Analyte	FB1, FB2	OTA	ZEA	DON	Aflatoxins	OTA	OTA	DON	Aflatoxin M1	Aflatoxins	OTA	PAT	Ergot alkaloids	PAT	PAT	HT2, T2	OTA	OTA
LODs (μg/kg)	41 (FB1); 31 (FB2)	0.014	5.5	30–55	0.1–0.16	0.3	3	0.03[a]	0.01	0.23–0.45[b]	0.8[b]	0.09 (μg/L)	up to 3.3	n.p.	0.23 (μg/L)	8 (HT2); 8 (T2)	0.032	0.02
Extraction solvent	MeCN: MeOH: water	MeCN: water	MeCN: water	MeCN: water	MeOH: water	MeOH: water	MeCN: water	MeCN: water/ water/ MeOH: water	None	MeOH: water	MeOH: $NaHCO_3$	Water: MeCN: perchloric acid	Ethyl acetate: MeOH: aqueous ammonia	Water	Ethyl acetate: Na_2CO_3 solution	MeOH: water	MeOH:water containing sodium hydrogen carbonate and PEG	$CHCl_3$, NaCl, H_3PO_4
Purification	IAC	IAC	IAC	IAC	IAC	IAC	Mycosep	MycoSep/ IAC/Oasis HLB	IAC	IAC	IAC	no	SPF with basic alumina	Silica gel SPE	no	IAC	IAC	IAC
Detection method	FLD[c]	FLD	FLD	DAD	FLD	FLD	FLD	PDA	FLD	FLD	FLD	DAD	FLD	DAD	DAD	FLD[b]	FLD	FLD

Abbreviations of analytes: *FB1* fumonisin B1, *FB2* fumonisin B2, *OTA* ochratoxin A, *ZEA* zearalenone, *DON* deoxynivalenol, *PAT* patulin, *HT2* HT-2 toxin, *T2* T-2 toxin

n.p. – information not provided

[a]LOD presented in μg/L

[b]LOQ (limit of quantification)

[c]Pre-column derivatization employed

fluorescence for trichothecenes and patulin, UV/DAD detection is available.

2.3.3. Liquid Chromatography–Mass Spectrometry

Considering the need for efficient food safety control, the speed of analysis and the applicability to as wide as possible range of mycotoxin/matrix combinations are nowadays obviously the driving forces in multi-mycotoxin analysis development. Robustness, selectivity, sensitivity, as well as flexibility regarding the method scope, are the key criteria for selection of optimal detection methods enabling identification/quantification following sample separation. In this context, mass spectrometry is currently the only powerful detection tool providing satisfactory solutions for accurate analysis including confirmation of target, and, in some cases, nontarget analytes.

High and ultra-high performance liquid chromatography (HPLC/U-HPLC) coupled with various mass spectrometric platforms are outstandingly qualified for multi-toxin analyses. Triple-quadrupole (QqQ) tandem mass spectrometry (MS/MS) is currently considered as a "gold standard," although the benefits of other mass analyzers mentioned below have been recognized by many laboratories concerned with control of natural toxicants in the food chain.

Besides its high sensitivity, MS/MS also provides a high degree of certainty in analytes identification (especially in the case of poor chromatographic resolution). Under common conditions, obtaining a sufficient number of identification points, in accordance with the EU guidelines for obtaining unambiguous data (60), is easily possible. Confirmation of target analytes is usually achieved by recording at least two mass transitions in selected reaction monitoring (SRM) mode.

One of the first quantitative LC–MS/MS methods for multi-mycotoxin analysis allowing simultaneous determination of mycophenolic acid, griseofulvin, roquefortine C, chaetoglobosin B, verruculogen, and penitrem A in food and feed matrices was published by Rundberget and Wilkins in 2002. The extraction step performed by an acetonitrile:water:formic acid mixture (900:99:1, *v/v/v*) was followed by defatting with hexane. The atmospheric pressure chemical ionization (APCI) was used for quantification by an ion trap MS instrument (61). Another method for determination of type A trichothecenes (T-2 and HT-2 toxin, acetyl T-2 toxin, diacetoxyscirpenol, monoacetoxyscirpenol (15-acetoxyscirpenol), and neosolaniol) in oats after MycoSep purification was published in 2002. Analytes were separated on a reversed-phase narrow-bore column and detected in positive APCI-MS/MS (62). Other tandem MS method for the determination of four trichothecenes type B in maize was published in 2003 by Lagana et al. Nivalenol, deoxynivalenol, fusarenon-X, and 3-acetyl deoxynivalenol were determined under negative electrospray ionization and multiple reaction monitoring mode (MRM) of a triple-quadrupole mass spectrometer (63). One year later, Royer et al. reported a method for the determination of deoxynivalenol, fumonisin B1, and zearalenone in maize. ASE was used for sample extraction. For quantification, isotopically labeled internal standards were employed for obtaining accurate results. Detection of target analytes was carried out by APCI-ion trap-MS/MS (26). LC–ESI-MS/MS method for detection and quantification of beauvericin, enniatins, and moniliformin in grain-based foods was published (64). Similar triple quadrupole LC–MS/MS methods for the quantification of trichothecenes and zearalenone in cereals were presented in 2005. After extraction with an acetonitrile:water (84:16, *v/v*) mixture and MycoSep clean-up, analytes were detected by using of ESI (65) and APCI (66) interfaces. The MycoSep purification was enabled also by Tanaka et al. who, additionally to trichothecenes and zearalenone, included aflatoxins B1, B2, G1, and G2 into their method. The APCI-TOFMS ionization/detection was found to be suitable for the screening of multiple mycotoxins in cereals and cereal-based products (67). In 2005, Cavaliere et al. presented the method for the determination of trichothecenes, fumonisins, zearalenone, and α-zearalenol in corn, and the ESI-MS/MS technique was employed for detection (68). Furthermore, the LC–ESI-MS/MS method for the determination of mycotoxins and their metabolites in milk was introduced by Sorensen et al. in 2005. Aflatoxin M1, deoxynivalenol, deepoxynivalenol, 3- and 15-acetyldeoxynivalenol, HT-2 and T-2 toxins, T-2 triol, diacetoxyscirpenol, monoacetoxyscirpenol, fumonisins B1 and B2, ochratoxin A, zearalenone, and its α- and β- metabolites (zearalenols and zearalanols) were extracted with an acetonitrile:hexane mixture (16:10, *v/v*), and purified by employing *N*-vinylpyrrolidone/divinylbenzene co-polymer columns (69). Kokkonen et al. published an MS/MS method for the determination of aflatoxins, ochratoxin A, mycophenolic acid, penicillic acid, and roquefortine C in blue cheese by triple quadrupole with ESI ionization. For fats removal, hexane was added to an acetonitrile:formic acid extract (70).

Continuous advances in technical parameters of modern LC–MS instrumentation offered new possibilities to increase sample throughput and expand methods scope. Both introduction of U-HPLC and improving MS detection sensitivity (modification of ion sources and mass analyzers performance) enabled, approximately in mid of first decade of this century, application of the "dilute-and-shoot" approach.

In 2006, Spanjer et al. presented an ESI triple-quadrupole MS/MS method for the simultaneous determination of aflatoxins B1, B2, G1, and G2, ochratoxin A, DON, 3-acetyl-DON, 15-acetyl-DON, fumonisins B1 and B2, diacetoxyscirpenol, ZON, T2-toxin, HT2-toxin, roquefortine, and sterigmatocystin in various foodstuffs like peanuts, cornflakes, wheat, and figs. An acetonitrile:water extract was diluted in a ratio of 1:3 and analyzed directly, without any clean-up (71). Further ESI-MS/MS method omitting the clean-up step was published by Sulyok et al. in 2006. Altogether 39 mycotoxins (in addition to common mycotoxins represented by A- and B-trichothecenes, zearalenone, patulin, fumonisins, aflatoxins, ochratoxins, and their metabolites along with the ergot alkaloids, *Alternaria* toxins, enniatins, and moniliformin) potentially occurring in cereals were determined in diluted an acetonitrile:water:acetic acid extract (72). The LC–ESI(+)-MS/MS chromatogram of diluted wheat extract was spiked with a multi-mycotoxin mixture is presented in Fig. 3

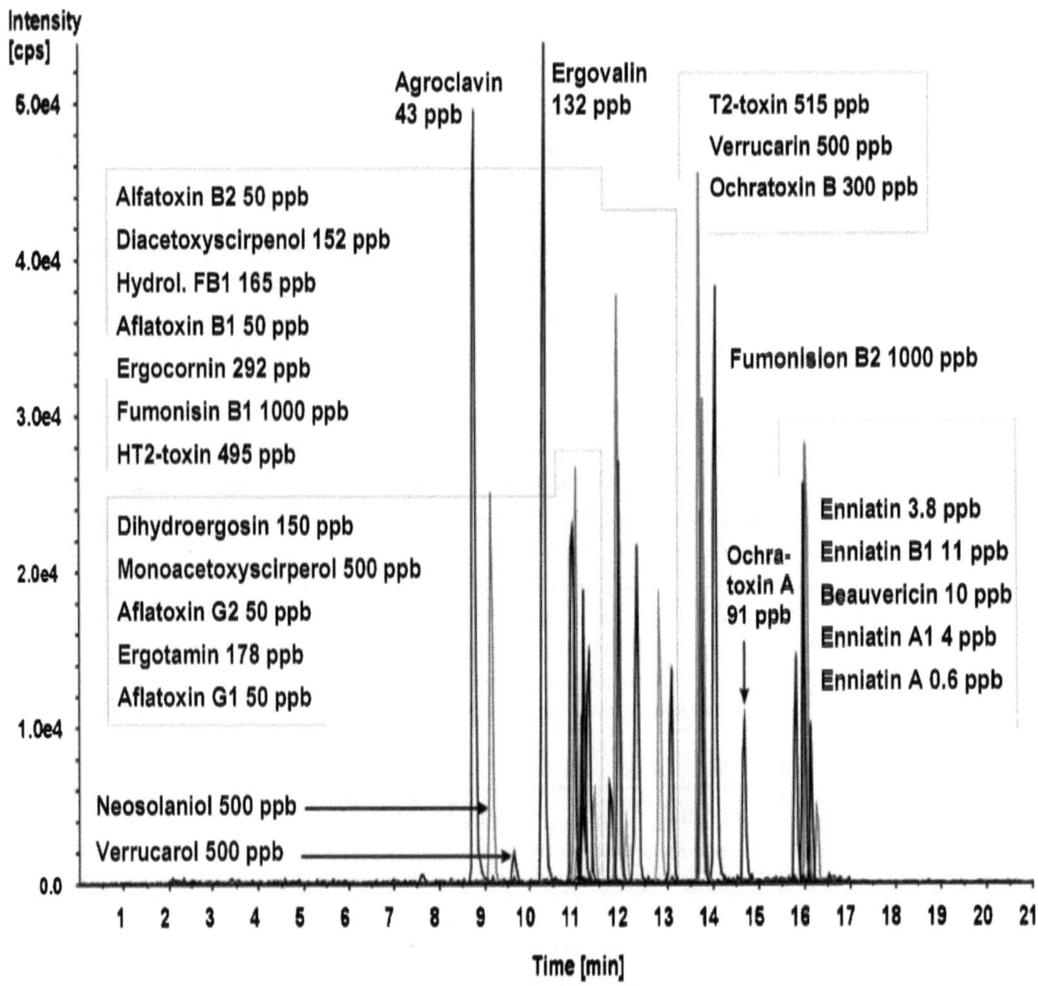

Fig. 3. The LC–ESI(+)-MS/MS total ion current chromatogram (sum of all MRM transitions) of a mixture of mycotoxins. The diluted wheat extract was spiked with a multi-mycotoxin standard and injected directly (reproduced from (73) with permission from Springer).

(73). In the follow-up study, the method was extended up to 87 analytes and fully validated (74).

It is worth to note that the generic sample preparation allows the simultaneous determination of mycotoxins with other food contaminants, including pesticides, plant toxins, and veterinary drugs (75). The detailed characterization of the most interesting multi-mycotoxin methods published in the recent 5 years is summarized in Table 3.

The key limitation of MS/MS-based methods is that due to monitoring only specific mass transitions, neither post acquisition data reprocessing nor screening of unidentified unknowns is possible. In this context, the growing interest in employing high-resolution mass analyzers is not surprising; they represent, indeed, a challenging option in the field of LC–MS mycotoxins analysis: full scan experiments make possible not only non target screening, but also retrospective data mining (35, 81, 82). The mass resolving power of currently marketed instruments with high performance time-of-flight (TOF) analyzers is around 15,000 FWHM (full width in half maximum). In some very recent instruments, the FWHM value might be up to 30,000, and mass accuracies are typically <5 ppm. A mass resolving power as high as 100,000 FWHM and higher is achievable by orbitrap MS technology. Figure 4 illustrates the benefits of high mass resolving power for the discrimination of isobaric interferences. The importance of an optimal choice of the extraction window width is demonstrated here. Employing a wide mass extraction window may result in worsened selectivity, however, too narrow mass extraction window, especially in case of employing lower mass resolving power, may imply the occurrence of false-negative results.

As mentioned above, in LC–MS analysis, the ionization efficiency can be affected by the presence of co-extracted matrix compounds co-eluting with the analyte of interest leading to the signal suppression or enhancement. For compensation of those matrix effects, matrix-matched standards are often used. Other possible strategy to ensure good trueness of generated data is employing of isotopically labeled internal standards. Nowadays, ^{13}C-labeled standards of deoxynivalenol, 3-acetyldeoxynivalenol, zearalenone, HT-2 and T-2 toxin, fumonisin B1 and B2, ochratoxin A, and aflatoxin B1 are commercially available (19, 83).

2.3.4. Ambient Ionization Mass Spectrometry

Another challenging development in the field of MS is associated with introduction of novel ambient desorption ionization techniques, represented mainly by (commercialized) desorption electrospray ionization (DESI) and direct analysis in real time (DART) (84). As in other ambient MS techniques, (chromatographic) sample separation is omitted.

Table 3
Overview of recent LC–MS based methods for multi-mycotoxin analysis with none of minimal sample clean-up

Reference	(74)	(75)	(76)	(77)		(27)	(78)	(79)	(35)	(80)	(81)
Matrices, which is method validated for	Wheat, maize, bread-crumbs	Maize, meet, eggs, milk, honey, horse feed	Peanut slurry, pistachio slurry, wheat slurry, maize slurry, dry-miller cornflakes, raisin slurry, fig slurry	Wheat, maize	Wheat, maize	Wheat, barley, oats	Maize, walnuts, biscuits, breakfast cereals	Maca, soy isoflavones, garlic, black radish, St John's wort, ginko biloba	Wheat, barley, maize	Beer	Beer
Example of LOD (μg/kg) for particular analyte/matrix combination[a]	Bread-crumbs	Horse feed	Maize slurry	Wheat	Wheat	Wheat	Maize	Maca	Wheat	Beer[b]	Beer[b]
DON	20[c]	>250	50	10	2,000	35	1.1	6	25	0.14	3
HT2	20	50	25	8	100	100	1	1	12.5	0.06	4
T2	2	20	25	1	20	15	0.1	3	5	0.07	2
ZEA	0.4	250	10	4	100	20	1.5	6	5	0.1	1
FB1	8	50	100	35	80	20	0.1	1	10	0.07	n.d.
FB2	7	20	100	30	80	15	0.2	0.3	5	0.09	n.d.
OTA	1	50	1	4	12	10	0.3	1	n.d.	0.02	60
AFB1	0.8	10	0.5	0.5	20	10	0.02	6	n.d.	0.04	2
AFB2	0.7	10	1.0	30	20	10	0.1	6	n.d.	0.05	0.5
AFG1	0.5	10	1.0	1	20	10	0.2	6	n.d.	0.03	3
AFG2	1	20	0.5	10	20	20	0.2	6	n.d.	0.08	2
PAT	100	n.d.	n.d.	800	2,000	15	n.d.	n.d.	n.d.	n.d.	n.d.

Extraction solvent (purification)	MeCN: water: acetic acid	MeCN: water: formic acid	MeCN:water	MeCN: water: acetic acid	MeCN: water: acetic acid	MeCN: water	MeCN: water	Ethyl acetate: formic acid	MeCN: water: formic acid (NaCl, $MgSO_4$)	Sonication (C18 clean-up)	Sonication, (MeCN precipitation)
Total number of targeted mycotoxins	87	23	33	30	32	31	12	23	11	12	32
Run time/no of chromatographic runs necessary for determination	21 min/2	22 min/2	35 min/1	33 min/2	33 min/1	35 min/2	8.5 min/1	25 min/1	18 min/2	6.5 min/1	18 min/1
Matrix equivalent per 1 mL of injected sample/matrix equivalent in the injected volume	0.125/ 0.000625 g	0.125/ 0.000625 g	0.0625/ 0.00125 g	0.25/ 0.00125 g	0.25/ 0.0025 g	5/0.05 g	0.5/ 0.0025 g	5/0.1 g	0.2/0.001 g	5/0.025 mL	1/0.005 mL
Number of steps within the sample preparation[d]/timing estimate of the preparation of 1 sample	2/98 min	1/190 min	2/130 min	1/93 min	1/93 min	3/40 min	1/15 min	5/155 min	3/18 min	2/45 min	3/15 min
Type of MS detection	ESI-MS/MS (Qtrap)	ESI-MS/MS (triple quadrupole)	ESI-MS/MS (triple quadrupole)	ESI-MS/MS (triple quadrupole)	ESI-MS (orbitrap)	ESI-MS/MS (triple quadrupole)	ESI-MS/MS (tandem quadrupole)	ESI-MS/MS (triple quadrupole)	ESI-MS (time-of-flight)	MS/MS (tandem quadrupole)	APCI-MS (orbitrap)
MS detector	Qtrap 4000 (Applied Biosystems)	Quatro Premiere XE (Waters)	Quatro Ultima (Waters Micromass)	TSQ Quantum Ultra AM (Thermo-Finnigan)	LTQ Orbitrap (Thermo Scientific)	Micromass Quatro Micro (Waters)	Acquity TQD (Waters)	Micromass Quatro Micro (Waters)	LCT Premier XE (Waters)	Acquity TQD (Waters)	Exactive (Thermo Fisher Scientific)

Abbreviations of analytes: *DON* deoxynivalenol, *HT2* HT-2 toxin, *T2* T-2 toxin, *ZEA* zearalenone, *FB1* fumonisin B1, *FB2* fumonisin B2, *OTA* ochratoxin A, *AFB1* aflatoxin B1, *AFB2* aflatoxin B2, *AFG1* aflatoxin G1, *AFG2* aflatoxin G2, *PAT* patulin

n.d. not determined

[a]Examples of LODs (limits of detection) for selected mycotoxins (mostly regulated, maximum limits established by (EC) No 1881/2006 implemented by (EC) No 1126/2007) in selected matrices

[b]LOD in beer in μg/L

[c]LOD in the solvent standard due to the lack of blank

[d]Operations considered as the sample preparation step: sonication, extraction, dilution, evaporation, liquid–liquid extraction, solid phase extraction, and partitioning

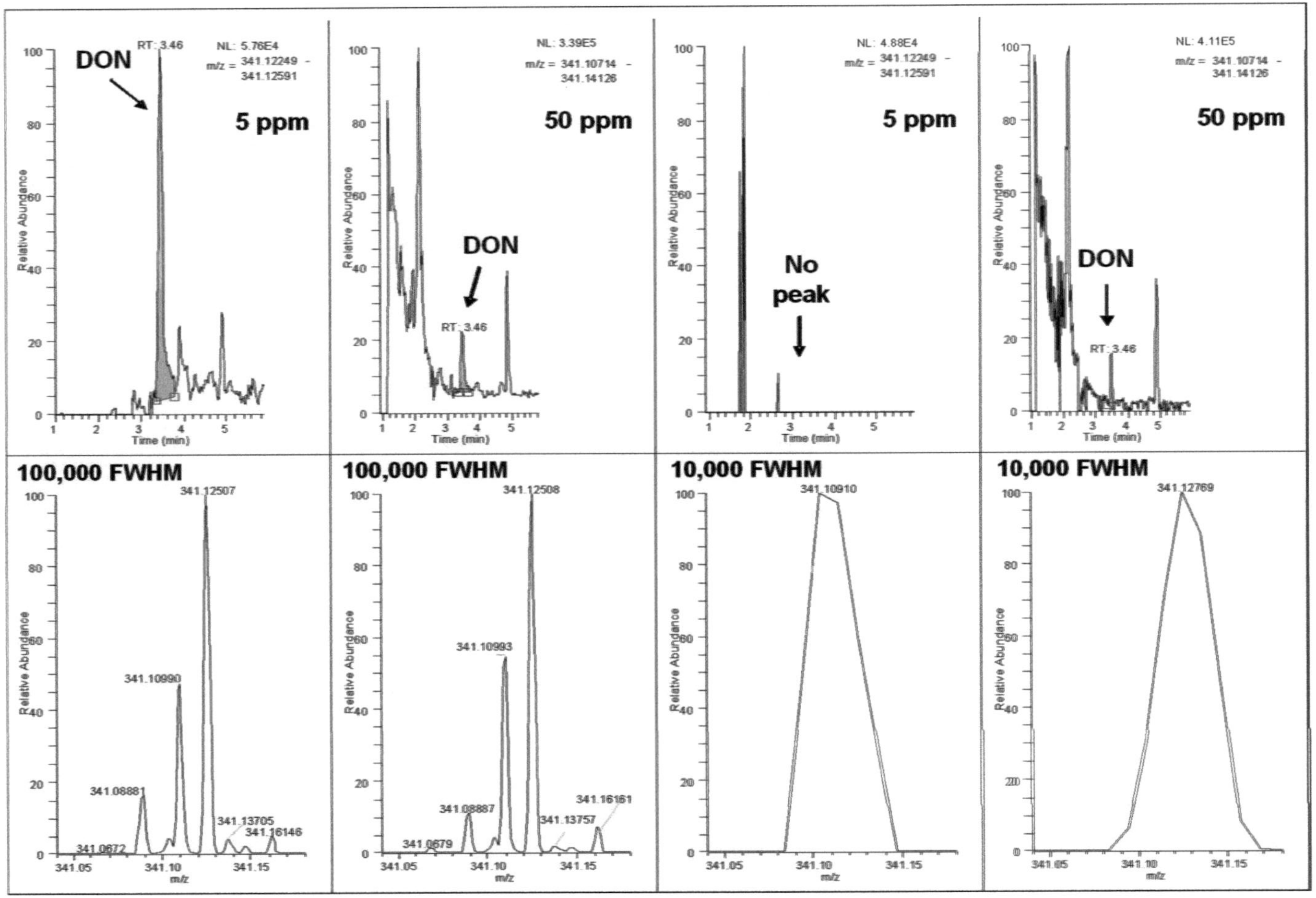

Fig. 4. Extracted ion chromatograms and mass spectra of deoxynivalenol (m/z 341.1242) in beer (10 μg/L) obtained at two different mass resolving power settings of orbitrap MS (10,000 and 100,000 FWHM) and two different mass extraction windows (±5 and ±50 ppm).

In the DESI source, ionization takes place by directing an electrically charged mist to the sample surface. Created ions travel through the atmospheric pressure interface into the mass spectrometer. The only, until now, reported study concerned with application of DESI (ionization resembles ESI) for mycotoxins analysis was focused on fumonisins in maize (85).

DART technology employs excited-state helium atoms to produce reactive species for APCI-like ionization of analytes that occurs in the vapor phase following their thermo-desorption from the sample (86). The first study concerned with analysis of multiple mycotoxins has been reported only recently. Vaclavik et al. demonstrated the potential of DART coupled to ultra high resolving power orbitrap MS to quantify selected trichothecenes, *Alternaria* toxins, zearalenone, and sterigmatocystin in a QuEChERS-based extract prepared from cereals (87). Figure 5 shows the DART–orbitrap MS spectra obtained for particular mycotoxins in wheat extract spiked at a level of 500 μg/kg. The lowest calibration levels (LCLs) ranged from 50 to 150 μg/kg. The method was shown to be applicable for high-throughput control of maximum limits of ZEA and DON established in EU regulation [(EC) 1126/2007] for unprocessed wheat/maize. Improved reproducibility of the measurement was achived by employing of matrix-matched calibration together with isotope dilution-based quantification.

2.3.5. Bio-Analytical Tools

Immunochemical techniques, represented by ELISA, are a widely established technology employed mainly for rapid and sensitive screening of mycotoxins in unprocessed commodities/raw materials. The most common microtitre-plate format has found a place in routine laboratories. It is easy to use, typically, no clean-up or analyte enrichment steps are required. In most cases, the endpoints are colorimetric or fluorimetric, hence only very simple devices are needed to run the assay. Also other formats (some of them portable) of bio-analytical tools have become available during this time; nevertheless, many of the biosensors, immunosensors, and test strips/dipsticks are essentially modifications of the two basic forms of ELISA where either the antigen or the (anti-toxin) antibody is immobilized. In recent years, membrane-based immunoassay methods, such as flow-through immunoassays and lateral flow devices (LFDs) have been introduced into the market. This innovative approach is of growing interest since it allows rapid on-site (pre)-screening. More detailed discussions of advantages and limitations is available in recent reviews (85, 88, 89).

Substantial developments reflecting the demand for multiple mycotoxins measurement have also occurred. Biosensor arrays employing parallel simultaneous assays, physically separated from one another, seem to be a very challenging option (90). The most pertinent for routine applications appear to be those based upon fluorescence or surface plasmon resonance (SPR). The later technique

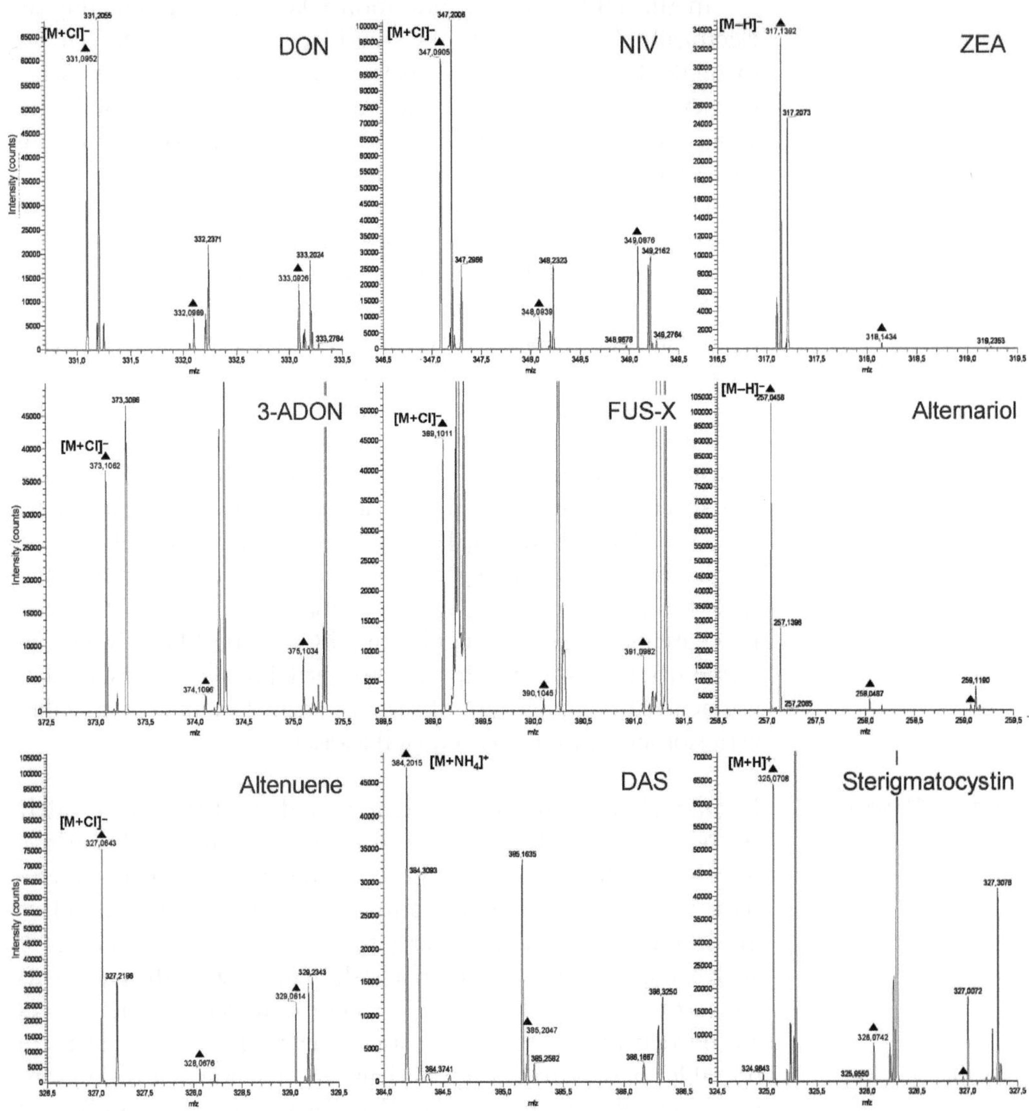

Fig. 5. Mass spectra of examined mycotoxins obtained by DART–orbitrap MS analyses of wheat extract (spike 500 μg/kg) at a mass resolving power setting of 50,000 FWHM. Ions yielded by target analytes (filled triangle) (Analytes abbreviations: DON deoxynivalenol, NIV nivalenol, ZEA zearalenone, 3-ADON 3-acetyl deoxynivalenol, FUS-X fusarenon-X, DAS diacetoxyscirpenol) (Reproduced from (87) with permission from Elsevier).

is based on measuring the impact of mass concentration changes on angle, or intensity, of internally reflected light at metal film liquid interface in respective flow chip where binding/dissociation event between analytes (mycotoxin) and (bio)recognition element (antibodies, molecular imprinted polymers, MIPs) takes place.

Only recently, challenges offered by RNA fingerprinting assay or Luminex xMAP technology (which comprise existing technologies – flow cytometry, microspheres, lasers, digital signal processing, and traditional chemistry) in the analysis of multiple mycotoxins have been addressed in EU funded projects Biocop (91) and Conffidence (92).

3. Notes

Following recommendations (Notes 1–5) should always be considered within multi-mycotoxin analysis:

1. *Blank matrix.* In mycotoxins analysis, for compensation of matrix induced ion suppression/enhancement, matrix-matched standards should be used whenever possible. Although the use of absolutely blank matrix is an ideal solution, unfortunately, obtaining it in practice is hardly achievable (most of wheat-based matrices contain at least traces of DON, similarly for maize-based matrices, presence of fumonisins traces is typical). Higher background mycotoxin concentration tends to increase the analytical bias of the results. Hence, samples with no or very low mycotoxins levels should be kept in the laboratory for analytical purposes (matrix-matched calibration for matrix effects correction, as well as spiking experiments for the recovery assessment).
2. *Internal standards.* As a general rule, internal standard employed for mycotoxins analysis must not be present in the sample, and should combine physiochemical properties chromatographically similar to those of target mycotoxins. Use of internal standards as surrogates (known amount of internal standard added at the beginning of the sample preparation procedure) is recommended for compensation of the analytes losses throughout the analytical procedure. During recent years, the number of internal standards available in mycotoxins analysis has rapidly increased, especially in case of ^{13}C-labeled mycotoxins, which are also often employed for matrix effects correction.
3. *Clean-up.* When immunoaffinity columns are used for purification of sample extract and/or pre-concentration of analytes, exceeding of the column capacity (this information, usually in nanograms of analyte, should be given by column producer) has to be avoided. Breakthrough of analytes may occur when antibodies binding sites are saturated.

4. *LC determinative steps.* For checking the signal stability during the sequence, running of analytical standards at the beginning and the end of each (longer) sequence is recommended. Analyses have to be performed within the linear range. In case of highly contaminated samples possibly exceeding the calibration range, they have to be diluted before the analysis, and the diluting factor has to be included in the results calculation.
5. *Instrument's maintenance.* When a significant decrease in signal of analytes is observed, instrument's maintenance including cleaning of the ion source and ion optic is required. As far as decreasing of the quality of chromatographic data is registered (poor peak shape), replacing of a pre-column or the LC column is recommended. The LC–MS analyses, especially the U-HPLC (ultra-high performance LC with sorbent particles less than 2 μm) should always include filtration of the final extract by a syringe filter (0.22 or 0.45 μm for U-HPLC or HPLC, respectively). This simple procedure significantly prolongs the lifetime of a particular LC column.

Acknowledgments

This chapter was financially supported by the Ministry of Education, Youth and Sports of the Czech Republic (projects MSM 6046137305 and NPV II 2B08049) and the Ministry of Agriculture of the Czech Republic (projects QI 111B044 and QI 111B154).

References

1. Moss, M. O. (1992) Secondary metabolism and food intoxication – molds. *J. Appl. Bacteriology* **73**, S80–S88.
2. Sweeney, M. J., Dobson, A. D. W. (1999) Molecular biology of mycotoxin biosynthesis. *FEMS Microbiol. Lett.* **175**, 149–163.
3. Placinta, C. M., D'Mello, J. P. F., Macdonald, A. M. C. (1999) A review of worldwide contamination of cereal grains and animal feed with *Fusarium* mycotoxins. *Anim. Feed Sci. Technol.* **78**, 21–37.
4. Medina, A., Valle-Algarra, F. M., Mateo, R., Gimeno-Adelantado, J. V., Mateo, F., Jimenez, M. (2006) Survey of the mycobiota of Spanish malting barley and evaluation of the mycotoxin producing potential of species of *Alternaria*, *Aspergillus* and *Fusarium*. *Int. J. Food Microbiol.* **108**, 196–203.
5. Bennett, J. W., Klich, M. (2003) Mycotoxins. *Clin. Microbiol. Rev.* **16**, 497–516.
6. Uhlig, S. Torp, M., Heier, B. T. (2006) Beauvericin and enniatins A, A1, B and B1 in Norwegian grain: a survey. *Food Chem.* **94**, 193–201.
7. Meca, G., Zinedine, A., Blesa, J., Font, G., Manes, J. (2010) Further data on the presence of *Fusarium* emerging mycotoxins enniatins, fusaproliferin and beauvericin in cereals available on the Spanish markets. *Food Chem. Toxicol.* **45**, 1412–1416.
8. EC (2006) Commission Regulation (EC) No 1881/2006 of 19 December 2006 setting maximum levels for certain contaminants in foodstuffs. *Off. J. Eur. Union*, **L 364**, 5–24.
9. EC (2007) Commission Regulation (EC) No 1126/2007 of 28 September 2007 amending Regulation (EC) No 1881/2006 setting maximum levels for certain contaminants in foodstuffs as regards *Fusarium* toxins in maize and maize products. *Off. J. Eur. Union*, **L 255**, 14–17.

10. FDA Compliance program guidance manual 7307.001 (2008) Mycotoxins in domestic and imported foods.
11. Berthiller, F., Dall'Asta, C., Schuhmacher, R., Lemmens, M., Adam, G., Krska, R. (2005) Masked mycotoxins: Determination of a deoxynivalenol glucoside in artificially and naturally contaminated wheat by liquid chromatography-tandem mass spectrometry. *J. Agric. Food Chem.* **53**, 3421–3425.
12. Berthiller, F., Sulyok, M., Krska, R. Schuhmacher, R. (2007) Chromatographic methods for the simultaneous determination of mycotoxins and their conjugates in cereals. *Int. J. Food Microbiol.* **119**, 33–37.
13. Vendl, O., Berthiller, F., Crews, C., Krska, R. (2009) Simultaneous determination of deoxynivalenol, zearalenone, and their major masked metabolites in cereal-based food by LC–MS–MS. *Anal. Bioanal. Chem.* **395**, 1347–1354.
14. Lancova, K., Hajslova, J., Kostelanska, M., Kohoutkova, J., Nedelnik, J., Moravcova, H., Vanova, M. (2008) Fate of trichothecene mycotoxins during the processing: Milling and baking. *Food Addit. Contam.* **25**, 650–659.
15. Lancova, K., Hajslova, J., Poustka, J., Krplova, A., Zachariasova, M., Dostalek, P., Sachambula, L. (2008) Transfer of *Fusarium* mycotoxins and "masked" deoxynivalenol (deoxynivalenol-3-glucoside) from field barley through malt to beer. *Food Addit. Contam.* **25**, 732–744.
16. Seefelder, W., Knecht, A., Humpf, H. U. (2003) Bound fumonisin B1: analysis of fumonisin-B1 glyco and amino acid conjugates by liquid chromatography-electrospray ionization-tandem mass spectrometry. *J. Agric. Food Chem.* **51**, 5567–5573.
17. Voss, K. A., Poling, S. M., Meredith, F. I., Bacon, C. W., Saunders, D. S. (2001) Fate of fumonisins during the production of fried tortilla chips. *J. Agric. Food Chem.* **49**, 3120–3126
18. EC (2006) Commission Regulation (EC) No 401/2006 laying down the methods of sampling and analysis for the official control of the levels of mycotoxins in foodstuffs. *Off. J. Eur. Union*, **L 70**, 12–34.
19. Koppen, R., Koch, M., Siegel, D., Merkel, S., Maul, R., Nehls, I. (2010) Determination of mycotoxins in foods: current state of analytical methods and limitations. *Appl. Microbiol. Biotechnol.* **86**, 1595–1612.
20. Whitaker, T. B., Dickens, J. W., Monroe, R. J. (1974) Variability of aflatoxin test results. *J. Am. Oil Chem. Soc.* **51**, 214–218.
21. Whitaker, T. B., Whitten, M. E., Monroe, R. J. (1976) Variability associated with testing cottonseed for aflatoxin. *J. Am. Oil Chem. Soc.* **53**, 502–505.
22. Whitaker, T. B., Dickens, J. W, Monroe, R. J. (1979) Variability associated with testing corn for aflatoxin. *J. Am. Oil Chem. Soc.* **56**, 789–794.
23. FDA (2002) Investigative operations manual. Food and Drug Administration, Washington, USA.
24. Whitaker, T. B. (2006) Sampling food for mycotoxins. *Food Addit. Contam.* **23**, 50–61.
25. Whitaker, T. B., Trucksess, M. W., Weaver, C. M., Slate, A. (2009) Sampling and analytical variability associated with the determination of aflatoxins and ochratoxin A in bulk lots of powdered ginger marketed in 1-lb bags. *Anal. Bioanal. Chem.* **395**, 1291–1299.
26. Royer, D., Humpf, H.U., Guy, P. A. (2004) Quantitative analysis of *Fusarium* mycotoxins in maize using accelerated solvent extraction before liquid chromatography/atmospheric pressure chemical ionization tandem mass spectrometry. *Food Addit. Contam.* **21**, 678–692.
27. Kokkonen, M. K., Jestoi, M. N. (2009) A multi-compound LC-MS/MS method for the screening of mycotoxins in grains. *Food Anal. Method* **2**, 128–140.
28. Langseth, W., Rundberget, T. (1998) Instrumental methods for determination of nonmacrocyclic trichothecenes in cereals, foodstuffs and cultures. *J. Chromatogr. A.* **815**, 103–121.
29. Krska, R. (1998) Performance of modern sample preparation techniques in the analysis of *Fusarium* mycotoxins in cereals. *J. Chromatogr. A* **815**, 49–57.
30. Krska, R., Baumgartner, S., Joseph, R. (2001) The state-of-the-art in the analysis of type-A and -B trichothecene mycotoxins in cereals. *Fresenius J. Anal. Chem.* **371**, 285–299.
31. Lattanzio, V. M. T., Solfrizzo, M., Visconti, A. (2008) Determination of trichothecenes in cereals and cereal-based products by liquid chromatography-tandem mass spectrometry. *Food Addit. Contam.* **25**, 320–330.
32. Pascale, M., Visconti, A.: Immunoaffinity clean-up / fluorescence detection methods for mycotoxins. In: The Mycotoxin Factbook, D. Barug, D. Bhatnagar, H.P. van Egmond, J.W. van der Kamp, W.A. van Osenbruggen, A. Visconti (Ed.), Academic Publishers, Wageningen, The Netherlands (2006) pp. 269–283.
33. Kostelanska, M., Zachariasova, M., Lacina, O., Fenclova, M., Kollos, A., Hajslova, J. (2011) The study of deoxynivalenol and its masked metabolites fate during the brewing

process realised by UPLC–TOFMS method. *Food Chem.* **126**, 1870–1876.

34. MacDonald, S. J., Kelleher, B., Donelly, C., Hird, S. (2007) Multimycotoxin analysis using immunoaffinity column clean-up and LC-MS/MS determination. Poster 1438 at the XII IUPAC Symposium on Mycotoxins and Phycotoxins. Istanbul, Turkey; May 21–25, 2007.
35. Zachariasova, M., Lacina, O., Malachova, A., Kostelanska, M., Poustka, J., Godula, M., Hajslova, J., (2010) Novel approaches in analysis of *Fusarium* mycotoxins in cereals employing ultra performance liquid chromatography coupled with high resolution mass spectrometry. *Anal. Chim. Acta* **662**, 51–61.
36. Cunha, S. C., Faria, M. A., Fernandes, J. O. (2009) Determination of patulin in apple and quince products by GC-MS using C-13(5–7) patulin as internal standard. *Food Chem.* **115**, 352–359.
37. Cunha, S. C., Fernandes, J. O. (2010) Development and validation of a method based on a QuEChERS procedure and heart-cutting GC-MS for determination of five mycotoxins in cereal products. *J. Sep. Sci.* **33**, 600–609.
38. Larsen, T. O., Svendsen, A., Smedsgaard, J. (2001) Biochemical characterization of ochratoxin A-producing strains of the genus *Penicillium. Appl. Environ. Microbiol.* **67**, 3630–3635.
39. Schollenberger, M., Lauber, U., Jara, H., Suchy, S., Drochner, W., Mueller, H. M. (1998) Determination of eight trichothecenes by gas chromatography mass spectrometry after sample clean-up by a two-stage solid-phase extraction. *J. Chromatogr. A* **815**, 123–132.
40. Tanaka, T., Yoneda, A., Sugiura, S., Ueno, Y. (2000) Simultaneous determination of trichothecene mycotoxins and zearalenone in cereals by gas chromatography-mass spectrometry. *J. Chromatogr. A* **882**, 23–28.
41. Scott, P. M. (1995) Mycotoxin methodology. *Food Addit. Contam.* 12, 395–403.
42. Cirillo, T., Ritieni, A., Visone, M., Cocchieri, R. A. (2003) Evaluation of conventional and organic Italian foodstuffs for deoxynivalenol and fumonisins B-1 and B-2. *J. Agric. Food Chem.* **51**, 8128–8131.
43. Koch, P. (2004) State of the art of trichothecenes analysis. *Toxicol. Lett.* **153**, 109–112.
44. Croteau, S. M., Prelusky, D. B., Trenholm, H. L. (1994) Analysis of trichothecene mycotoxins by gas chromatography with electron capture detection. *J. Agric. Food Chem.* **42**, 928–933.
45. Jelen, H. H., Wasowicz, E. (2008) Determination of trichothecenes in wheat grain without sample cleanup using comprehensive two-dimensional gas chromatography-time-of-flight mass spectrometry. *J. Chromatogr. A* **1215**, 203–207.
46. Belajova, E., Rauova, D. (2010) Single laboratory-validated HPLC methods for determination of ochratoxin A, fumonisin B1 and B2, zearalenone and deoxynivalenol in cereals and cereal-based foods. *J. Food Nutr. Res.* **49**, 57–68.
47. Reiter, E. V., Vouk, F., Bohm, J., Razzazi-Fazeli, E. (2010) Aflatoxins in rice - A limited survey of products marketed in Austria. *Food Control.* **21**, 988–991.
48. Yang, L., Wang, L. A., Pan, J. Y., Xiang, L., Yang, M., Logrieco, A. F. (2010) Determination of ochratoxin A in traditional Chinese medicinal plants by HPLC-FLD. *Food Addit. Contam.* **27**, 989–997.
49. Bononi, M., Gallone, F., Tateo, F. (2010) Validation data for HPLC/FLD determinations of ochratoxin A in red paprika and black pepper adopting a one-step clean-up procedure. *Food Addit. Contam.* **27**, 249–254.
50. Moazami, E. F., Jinap, S. (2009) Optimisation of the determination of deoxynivalenol in wheat flour by HPLC and a comparison of four clean-up procedures. *Food Addit. Contam.* **26**, 1290–1297.
51. Shundo, L., Navas, S.A., Lamardo, L. C. S., Ruvieri, V., Sabino, M. (2009) Estimate of aflatoxin M-1 exposure in milk and occurrence in Brazil. *Food Control* **20**, 655–657.
52. Shundo, L., de Almeida, A. P., Alaburda, J., Lamardo, Leda C. A., Navas, S. A., Ruvieri, V., Sabino, M. (2009) Aflatoxins and ochratoxin A in Brazilian paprika. *Food Control* **20**, 1099–1102.
53. Gaspar, E. M. S. M., Lucena, A. F. F. (2009) Improved HPLC methodology for food control - furfurals and patulin as markers of quality. *Food Chem.* **114**, 1576–1582.
54. Muller, C., Kemmlein, S., Klaffke, H., Krauthause, W., Preiss-Weigert, A., Wittkowski, R. (2009) A basic tool for risk assessment: A new method for the analysis of ergot alkaloids in rye and selected rye products. *Mol. Nutr. Food Res.* **53**, 500–507.
55. Katerere, D. R., Stockenstrom, S., Shephard, G. S. (2008) HPLC-DAD method for the determination of patulin in dried apple rings. *Food Control*, **19**, 389–392.
56. Moukas, A., Panagiotopoulou, V., Markaki, P. (2008) Determination of patulin in fruit juices using HPLC DAD and GC MSD techniques. *Food Chem.* **109**, 860–867.
57. Trebstein, A., Seefelder, W., Lauber, U., Humpf, H. U. (2008) Determination of T-2

and HT-2 toxins in cereals including oats after immunoaffinity cleanup by liquid chromatography and fluorescence detection. *J. Agr. Food Chem.* **56**, 4968–4975.

58. La Pera, L., Avellone, G., Lo Turco, V., Di Bella, G., Agozzino, P, Dugo, G. (2008) Influence of roasting and different brewing processes on the ochratoxin A content in coffee determined by high-performance liquid chromatography-fluorescence detection (HPLC-FLD). *Food Addit. Contam.* **25**, 1257–1263.
59. Dall'Asta, C., Lindner, J. D., Galaverna, G., Dossena, A., Neviani, E., Marchelli, R. (2008) The occurrence of ochratoxin A in blue cheese. *Food Chem.* **106**, 729–734.
60. Document No. SANCO/10684/2009, Method validation and quality control procedures for pesticide residues analysis in food and feed. Available: http://ec.europa.eu/food/plant/protection/resources/qualcontrol_en.pdf via the Internet. Accessed on August 17 2010.
61. Rundberget, T., Wilkins, A.L. (2002) Determination of Penicillium mycotoxins in foods and feeds using liquid chromatography-mass spectrometry. *J. Chromatogr. A* **964**, 189–197.
62. Razzazi-Fazeli, E., Rabus, B., Cecon, B., Böhm, J. (2002) Simultaneous quantification of A-trichothecene mycotoxins in grains using liquid chromatography atmospheric pressure chemical ionisation mass spectrometry. *J. Chromatogr. A* **968**, 129–142.
63. Lagana, A., Curini, R., D'Ascenzo, G., De Leva, I., Faberi, A., Pastorini, E. (2003) Liquid chromatography/tandem mass spectrometry for the identification and determination of trichothecenes in maize. *Rapid Commun. Mass Spectrom.* **17**, 1037–1043.
64. Jestoi, M., Rokka, M., Yli-Mattila, T., Parikka, P., Rizzo, A., Peltonen, K. (2004) Presence and concentrations of the *Fusarium*-related mycotoxins beauvericin, enniatins and moniliformin in Finnish grain samples. *Food Addit. Contam.* **21**, 794–802.
65. Biselli, S., Hummert, C. (2005) Development of a multicomponent method for *Fusarium* toxins using LC-MS/MS and its application during a survey for the content of T-2 toxin and deoxynivalenol in various feed and food samples. *Food Addit. Contam.* **22**, 752–760.
66. Berthiller, F., Schuhmacher, R., Buttinger, G., Krska, R. (2005) Rapid simultaneous determination of major type A- and B-trichothecenes as well as zearalenone in maize by high performance liquid chromatography-tandem mass spectrometry. *J. Chromatogr. A* **1062**, 209–216.
67. Tanaka, H., Takino, M., Sugita-Konishi, Y., Tanaka, T. (2006) Development of a liquid chromatography/time-of-flight mass spectrometric method for the simultaneous determination of trichothecenes, zearalenone and aflatoxins in foodstuffs. *Rapid Commun. Mass Spectrom.* **20**, 1422–1428.
68. Cavaliere, C., Foglia, P., Pastorini, E., Samperi, R., Lagana, A. (2005) Development of a multiresidue method for analysis of major *Fusarium* mycotoxins in corn meal using liquid chromatography/tandem mass spectrometry. *Rapid Commun. Mass Spectrom.* **19**, 2085–2093.
69. Sorensen, L. K., Elbaek, T. H. (2005) Determination of mycotoxins in bovine milk by liquid chromatography tandem mass spectrometry. *J. Chromatogr. B Analyt. Technol. Biomed. Life Sci.* **820**, 183–196.
70. Kokkonen, M., Jestoi, M., Rizzo, A. (2005) Determination of selected mycotoxins in mould cheeses with liquid chromatography coupled to tandem with mass spectrometry. *Food Addit. Contam.* **22**, 449–456.
71. Spanjer, M., Rensen, P., Scholten, J. (2006) Multi-mycotoxin analysis by LCMS/MS in a single sample extract. Proceedings of the XIth International IUPAC Symposium on Mycotoxins and Phycotoxins, Bethesda, Maryland USA; May 17–21, 2004. Wageningen Academic Publishers, pp. 117–124.
72. Sulyok, M., Berthiller, F., Krska, R., Schuhmacher, R. (2006) Development and validation of a liquid chromatography/tandem mass spectrometric method for the determination of 39 mycotoxins in wheat and maize. *Rapid Commun. Mass Spectrom.* **20**, 2649–2659.
73. Krska, R., Molinelli, A. (2007) Mycotoxin analysis: state-of-the-art and future trends. *Anal. Bioanal. Chem.* **387**, 145–148.
74. Sulyok, M., Krska, R., Schuhmacher, R. (2007) A liquid chromatography/tandem mass spectrometric multi-mycotoxin method for the quantification of 87 analytes and its application to semi-quantitative screening of moldy food samples. *Anal. Bioanal. Chem.* **389**, 1505–1523.
75. Mol, H. G. J., Plaza-Bolanos, P., Zomer, P., de Rijk, T. C., Stolker, A. A. M., Mulder, P. P. J. (2008) Toward a generic extraction method for simultaneous determination of pesticides mycotoxins, plant toxins, and veterinary drugs in feed and food matrixes. *Anal. Chem.* **80**, 9450–9459.
76. Spanjer, M. C., Rensen, P. M., Scholten, J. M. (2008) LC-MS/MS multi-method for mycotoxins after single extraction, with validation

data for peanut, pistachio, wheat, maize, cornflakes, raisins and figs. *Food Addit. Contam.* **25**, 472–489.

77. Herebian, D., Zuhlke, S., Lamshöft, M., Spiteller, M. (2009) Multi-mycotoxin analysis in complex biological matrices using LC-ESI/MS: Experimental study using triple stage quadrupole and LTQ-Orbitrap. *J. Sep. Sci.* **32**, 939–948.
78. Frenich, A. G., Vidal, J. L. M., Romero-Gonzales, R., Del Mar Aguilera-Luiz, M. (2009) Simple and high-throughput method for the multimycotoxin analysis in cereals and related foods by ultra-high performance liquid chromatography/tandem mass spectrometry. *Food Chem.* **117**, 705–712.
79. Mavungu, J. D. D., Monbaliu, S., Scippo, M., Maghuin-Rogister, G., Schneider, Y., Larondelle, Y., Callebaut, A., Robbens, J., Van Peteghem, C., De Saeger, S. (2009) LC-MS/MS multi-analyte method for mycotoxin determination in food supplements. *Food Addit. Contam.* **26**, 885–895.
80. Romero-Gonzalez, R., Vidal, J. L. M., Aguilera-Luiz, M. M., Frenich, A. G. (2009) Application of conventional solid-phase extraction for multimycotoxin analysis in beers by ultrahigh-performance liquid chromatography-tandem mass spectrometry. *J. Agric. Food Chem.* **57**, 9385–9392.
81. Zachariasova, M., Cajka, T., Godula, M., Malachova, A., Veprikova, Z., Hajslova, J. (2010) Analysis of multiple mycotoxins in beer employing (ultra)-high resolution mass spectrometry. *Rappid Commun. Mass Spectrom.* **24**, 3357–3367.
82. Kelleman, M., Muenster, H., Zomer, P., Mol, H. (2009) Full scan MS in comprehensive qualitative and quantitative residue analysis in food and feed matrices: How much resolving power is required? *J. Am. Soc. Mass Spectrom.* **20**, 1464–1476.
83. Berthiller, F., Sulyok, M., Krska, R., Schuhmacher, R. (2007) Chromatographic methods for the simultaneous determination of mycotoxins and their conjugates in cereals. *Int. J. Food Microb.* **119**, 33–37.
84. Hajslova, J., Cajka, T., Vaclavik, L. (2008) Challenging applications offered by direct analysis in real time (DART) in food-quality and safety analysis. *TrAC-Trend Anal. Chem.* **30**, 204–218.
85. Maragos, C. M., Busman, M. (2010) Rapid and advanced tools for mycotoxin analysis: a review. *Food Addit. Contam.* **27**, 688–700.
86. Cody, R. B., Laramee, J. A., Durst, H. D. (2005) Versatile new ion source for the analysis of materials in open air under ambient conditions. *Anal. Chem.* **77**, 2297–2302.
87. Vaclavik, L., Zachariasova, M., Hrbek, V., Hajslova, J. Analysis of multiple mycotoxins in cereals under ambient conditions using direct analysis in real time (DART) ionization coupled to high resolution mass spectrometry. *Talanta.* **82**, 1950–1957.
88. Zheng, M. Z., Richard, J. L., Binder, J. (2006) A review of rapid methods for the analysis of mycotoxins. *Mycopathologia.* **161**, 261–273.
89. Krska, R., Molinelli A. (2009) Rapid test strips for analysis of mycotoxins in food and feed. *Anal. Bioanal. Chem.* **393**, 67–71.
90. van der Gaag, B., Spath, S., Dietrich, H., Stigter, E., Boonzaaijer, G., van Osenb-ruggen, T., Koopal, K. (2003) Biosensors and multiple mycotoxin analysis. *Food Control,* **14**, 251–254.
91. BioCop Download Page. Available: www.biocop.org via the Internet. Accessed on August 17 2010.
92. Conffidence Download Page. Available: www.conffidence.eu via the Internet. Accessed on August 17 2010.

Chapter 11

Multi Mycotoxin Analysis in Food Products Using Immunoaffinity Extraction

Masahiko Takino, Hiroki Tanaka, and Toshitsugu Tanaka

Abstract

We developed a method for the simultaneous determination of deoxynivalenol, T-2 toxin, HT-2 toxin and zearalenone in wheat and biscuit by liquid chromatography/electrospray ionization/tandem mass spectrometry coupled with immunoaffinity extraction. This chapter describes a method to extract, clean-up, and quantitate these mycotoxins and the effect of the ion suppression of multifunctional column and IAC in the clean-up were compared.

Key words: Immunoaffinity column, Liquid-chromatography, Tandem mass-spectrometry

1. Introduction

The Fusarium species (e.g. *Fusarium graminearum*, *F. culmorum* and *F. sporotrichioides*), one of the plant pathogenic fungi in wheat and other cereals, produces toxic metabolites such as deoxynivalenol (DON), T-2 toxin (T-2), HT-2 toxin (HT-2) and zearalenone (ZEN). These mycotoxins frequently contaminate food commodities simultaneously. Consequently, humans are exposed to serious danger when ingesting these mycotoxins directly or as residues in animal tissues (1). The determination of these toxins in foods and animal feedstuffs is therefore important for the protection of human health. DON, T-2 and HT-2 are called trichothecenes (TRs), a group of mycotoxins. DON is a type-B TRs, while T-2 and HT-2 belong to the group of type-A TRs. Both subgroups vary at C-8, where type-B TRs are characterized by a keto group and type-A TRs are esterified, hydroxylated, or non-substituted. T-2 and HT-2, which differ only at C-4, show a higher acute toxicity in comparison to DON (2). ZEN is an estrogenic metabolite produced by Fusarium species such as

Jerry Zweigenbaum (ed.), *Mass Spectrometry in Food Safety: Methods and Protocols*,
Methods in Molecular Biology, vol. 747, DOI 10.1007/978-1-61779-136-9_11, © Springer Science+Business Media, LLC 2011

F. culmorum, *F. crookwellense* (*F. cerealis*) and *F. graminearum*, and causes hyperestrogenism in livestock (3, 4). Oral exposure to a high dose of ZEN has been reported to cause hepatocellular adenomas in B6C3F1 mice 14 and apoptosis in vitro (5, 6). The Joint FAO/WHO Expert Committee on Food Additives (JECFA) evaluated the risk of Fusarium mycotoxins and established a level of 1 μg/kg of body weight per day as the provisional maximum tolerable daily intake (PMTDI) for DON 10. In addition, JECFA has established a level of 0.5 μg/kg of body weight per day as the PMTDI for ZEN and a level of 0.06 μg/kg for T-2 and HT-2 combined (7). Therefore, to grasp their various toxic effects and to ensure food safety, the accurate and convenient determination of the foods contaminated with these toxins is important for the supply of safe foods. There are numerous methods for the determination of DON, T-2, HT-2 and ZEN in food-stuffs, such as gas chromatography with mass spectrometry (8, 9), liquid chromatography with mass spectrometry (LC/MS) (10) either with tandem mass spectrometry (LC-MS/MS) (11, 12) or liquid chromatography/time of flight mass spectrometry (LC/TOF-MS) (13). With regards to clean-up methods, the demand for a multifunctional column (MFC) and a immunoaffinity column (IAC) has increased for mycotoxin analysis in recent years (14–17). MFCs are rapid and simple, and a clean-up can do more than one toxins simultaneously. However, because clean-up of the food matrix is insufficient, the examination of the ionization to perform quantitation by matrix matched standards is necessary (10, 13). On the other hand, the use of IAC in the purification step provides a number of advantages over the conventional methods, such as clean extracts due to the high specificity of the antibodies for mycotoxins, rapidity of the purification step, and reduction in the use of toxic solvents. However, the column for the simultaneous clean-up of several mycotoxins is limited. In this chapter, we describe a high sensitivity and reliable method for DON, T-2, HT-2 and ZEN, by utilizing multi-mycotoxin IAC and liquid chromatography/electrospray ionization/tandem mass spectrometry (LC/ESI-MS/MS).

2. Materials

2.1. Instrument

1. An Agilent 1200 series LC (Agilent Technologies, Waldbronn, Germany), consisting of a vacuum solvent degassing unit, a binary high-pressure gradient pump, an automatic sample injector, and a column compartment.
2. 100 mm × 2.1 mm I.D. column packed with 1.8 μm ZORBAX Extend C18 (Agilent Technologies, Santa Clara, CA, USA).

3. Agilent 6410 triple quadrupole mass spectrometer equipped with the ESI (Agilent Technologies, Santa Clara, CA, USA).
4. Nitrogen as a nebulizer gas and a drying gas of the ion source was generated from pressurized air by a AIR-TECH model AT-10ND-CS (Tokyo, Japan).
5. The immunoaffinity column of DZT Multi Myco IAC was purchased from R-BIOPHARM RHONE LTD. (Glasgow, UK).
6. The multifunctional cartridge column of MultiSep® #226 (Romer Labs, Inc., Union, MO, USA) was purchased from Showa Denko Ltd. (Tokyo, Japan).

2.2. Chemicals

1. The mycotoxins, DON, T-2, HT-2, and ZEN were purchased from Sigma-Aldrich Japan (Tokyo, Japan).
2. LC/MS-grade methanol, acetonitrile, and reagent-grade ammonium acetate were purchased from Wako Chemicals (Osaka, Japan).
3. 10 mM Ammonium acetate: dissolve 0.77 g ammonium acetate in 1,000 mL of Milli-Q water.
4. Pure water was purified with a Milli-Q system (Millipore, Tokyo, Japan).
5. Phosphate-buffered solution at pH 7.4 (PBS): dissolve one commercial phosphate-buffered saline tablets (Sigma-Aldrich) in 1,000 mL of Milli-Q water.
6. The immunoaffinity column of DZT Multi Myco IAC was purchased from R-BIOPHARM RHONE LTD. (Glasgow, UK).
7. The multifunctional cartridge column of MultiSep® #226 (Romer Labs, Inc., Union, MO, USA) was purchased from Showa Denko Ltd. (Tokyo, Japan).

3. Methods

3.1. Sample Preparation Steps

3.1.1. Standard Solution

1. Stock solution
   The 10 mg of DON, T-2, HT-2 and ZEN was dissolved in acetonitrile. Then they are stored at 4°C in the dark until use
2. Working solution
   An appropriate amount of individual stock standard solution is evaporated to dryness at 40°C under a gentle stream of nitrogen. The residue is reconstituted with 1 mL of aqueous 10 mM ammonium acetate/methanol (90/10).

3.1.2. Extraction

1. Shaking
   25 g of each sample is weighed in a 200 mL Erlenmeyer flask. After adding 100 mL methanol/water (75/25), the flask is shaken for 30 min on an automatic shaker.
2. Centrifugation
   The mixed solution is centrifuged for 5 min at 1,410 × *g*.
3. Dilution
   10 mL of the supernatant is diluted with 40 mL of PBS
4. Filtration
   The diluted extract is filtered through glass microfiber filter

3.1.3. Purification and Dryness

1. IAC purification

   The DZT Multi Myco IAC is conditioned by washing it twice with 5 mL of PBS. 5 mL of the diluted extract was applied to the DZT Multi Myco IAC.

   After sample application the column is washed with 10 mL of water and eluted using 2 mL methanol, at a flow rate of one to two drops per second. Furthermore, a backflushing through the column is carried out to obtain complete elution of all mycotoxins.

   The eluate is evaporated to dryness under a stream of nitrogen in a heating block or electric water bath at 60°C. The residue is dissolved in 0.5 mL of aqueous 10 mM ammonium acetate/methanol (90/10).

3.2. Methods for LC-MS/MS

3.2.1. LC Separation

1. LC separation is performed on a 100 mm × 2.1 mm I.D. column packed with 1.8 μm ZORBAX Extend C18 using linear gradient from A/B (90/10) to A/B (30/70) in 20 min. A solvent A is water containing 10 mM ammonium acetate and a solvent B is acetonitrile. The flow rate is set at 0.2 mL/min.

3.2.2. MS Condition

1. Detection is performed by LC-MS/MS. In Table 1 relevant LC-MS/MS parameters for mycotoxins are given
2. The MS/MS operating parameters are given in Table 1.
3. The MS/MS multi-reaction monitoring (MRM) transitions for the mycotoxins are given in Table 2.

4. Notes

Typical Performance of the Method

In this section, performance characteristics of optimized method are described for the analysis of mycotoxins in food products.

Table 1
LC-MSMS parameters

Mycotoxin	DON	HT-2	T-2	ZEN
Ionization mode	ESI-positive	ESI-positive	ESI-positive	ESI-positive
Capillary (kV)	4	4	4	4
Fragmentor (V)	100	100	100	100
Desolvation temperature (°C)	350	350	350	350
Desolvation gas flow (L/min)	10	10	10	10
Nebulizer gas pressure (kPa)	345	345	345	345

Table 2
MRM parameters 484 > 215

Mycotoxins	Transition 1	Transition 2	Collision energy (1)	Collision energy (2)
DON	297 > 249	297 > 203	10	10
HT-2	442 > 215	442 > 265	10	20
T-2	484 > 305	484 > 215	10	20
ZEN	319 > 187	319 > 283	20	10

4.1. Linearity and Sensitivity

To test the linearity of the calibration curve, various concentrations of mycotoxins in the range from 1 to 250 ng/mL were analyzed. The instrument detection limits (IDLs) were calculated as the peak to peak signal-to-noise (S/N) ratio was 3 by using the mixture standard solution of each mycotoxin (0.2 ng/mL). These results were shown in Table 3

4.2. Matrix Effects

Matrix effects are one of the major problems for LC-MS/MS quantification. The matrix could either enhance or suppress ionization of mycotoxins and ultimately affect the LC-MS/MS quantitative performance. In particular, the ion suppression by ESI is greater and the main cause of the ion suppression is matrix compounds in the sample. Consequently, if satisfactory clean-up can be done, the ion suppression is reduced or eliminated. To evaluate the clean-up ability of IAC, it was compared with the results of MFC with respect to the ion suppression.

The experiments to measure ion suppression is carried out by taking 5.0 ng of each mycotoxin standard and reconstituting in

Table 3
Linearity of calibration curves in a standard solutions and Instrument detection limits (IDL)

Mycotoxins	Calibration equation	r^{2a}	IDL (pg)[b]
DON	Y=1,013x+250	0.9999	0.28
HT-2	Y=6,022x+4,809	0.9999	0.15
T-2	Y=43,872x+28,610	0.9999	0.04
ZEN	Y=5,333x+4,524	0.9999	0.56

[a]2 is correlation coefficient
[b]Instrument detection limits defined as S/N ratio=3

Table 4
Relative abundance[a] (ion suppression) of each mycotoxin in wheat and biscuit

Mycotoxins	#226(wheat)	DZT(wheat)	#226(biscuit)	DZT(biscuit)
DON	19	81	12	119
HT-2	47	95	79	95
T-2	67	92	60	92
ZEN	51	92	58	104

[a]Relative abundance; Relative abundance (%)=abundance of the standard added to the final solution of the blank sample (10 ng/mL)/abundance of the solvent standard (10 ng/mL)

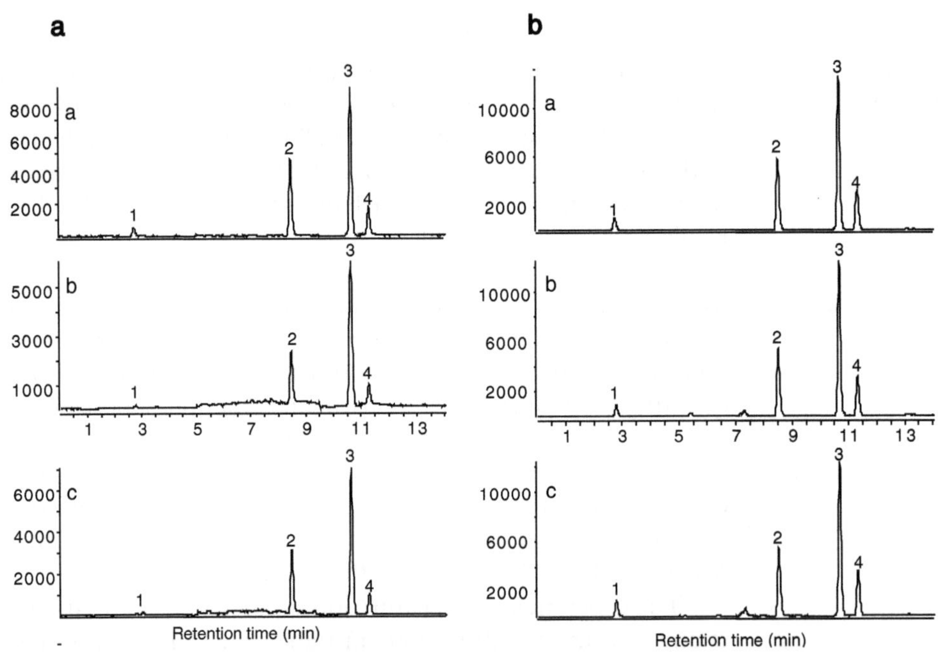

Fig. 1. MRM chromatograms of each mycotoxin in wheat and biscuit cleaned-up with MultiSep® #226 MFC and DZT Multi Myco IAC. (**a**) MultiSep® #226 MFC; (**b**) DZT Multi Myco IAC; a: standard solution (10 ng/mL); b: reconstituted wheat extract (10 ng/mL); c: reconstituted biscuit extract (10 ng/mL); 1: DON; 2: HT-2; 3: T-2; 4: ZEN.

0.5 mL of the final solution of a control sample of food products purified by MFC and IAC. The relative abundance of each mycotoxin in wheat and biscuit for the MFC and IAC purification was calculated, respectively. These results are shown in Table 4 and typical chromatograms are shown in Fig. 1.

The relative abundance of each mycotoxin in wheat and biscuit for the MFC and IAC purification ranged from 12 to 79% and from 81 to 119%, respectively. Ion suppression occurred in the result of MFC with DON, T-2, HT-2 and ZEN in comparison with the result of IAC. Especially, DON show much more ion suppression, and this confirmed that the matrix compounds frequently had an influence with the quantitation of DON. This shows that the clean-up ability of IAC is higher than MFC and the IAC clean-up step may be able to minimize the need to prepare matrix matched standards.

4.3. Recovery

To evaluate recoveries, the proposed method was applied to the analysis of spiked control samples of wheat and biscuit. Three samples of each food were each spiked with all mycotoxins at 10 ng/g. The mean recovery of each in wheat and biscuit obtained by the proposed method are presented in Table 5. They ranged from 82 to 109% with an RSD of 3.2–6.1%, which was satisfactory. The limits of detection (LODs) of each mycotoxin in wheat and biscuit were determined by the signal corresponding to three times the background noise on each MRM chromatogram for the sample spiked at 10 ng/g. The LODs of DON, T-2, HT-2 and ZEN were 0.13, 0.03–0.04, 0.08–0.17, and 0.24–0.33 ng/g, respectively (Table 4).

Table 5
Recovery of mycotoxins and limits of detection (LODs) for spiked wheat and biscuit when it was purified with IAC

Mycotoxins	Food	Recovery (%)[a]	RSD (%)[a]	LODs (ng/g)[b]
DON	Wheat	82	5.4	0.13
	Biscuit	82	5.8	0.13
HT-2	Wheat	93	3.2	0.08
	Biscuit	90	4.7	0.17
T-2	Wheat	103	5.1	0.03
	Biscuit	93	6.1	0.04
ZEN	Wheat	108	4.8	0.24
	Biscuit	109	5.7	0.33

[a] Recovery and RSD were calculated on the basis of three replicates at 10 ng/g

[b] Limits of detection calculated by the wheat-based product matrix matched standard defined as S/N ratio = 3

References

1. Ueno, Y. General toxicology. In: Trichothecenes- Chemical, Biological and Toxicological Aspects, edited by Y. Ueno (Amsterdam: Elsevier), 1983; 135–146.
2. SCF. Opinion of the scientific committee on food on Fusarium Toxins Part 6: Group of T-2 toxin, HT-2 toxin, nivalenol and deoxynivalenol (SCF/CS/CNTM/MYC/27 Final), Brussel, 2002.
3. Miller, J.D. and Trenholm, H.L. (Eds.). Mycotoxins in Grain-Compounds Other Than Aflatoxin, Eagan Press, St. Paul, 1994.
4. Mirocha, C.J., Pathre, S.V. and Christensen, M. In: J.V. Rodricks, C.W. Hesseltine, M.A. Mehlman (Eds.), Mycotoxins in Human and Animal Health, Pathotox, IL. 1977; p. 345.
5. Abid-Essefi, A., Baudrimont, I., Hassen, W., Quanes, Z., Monlo, T. A., Anane, R., Creppy, E. E. and Bacha, H. DNA fragmentation, apoptosis and cell cycle arrest induced by zearalenone in cultured DOK, Vero and Caco-2 cells: prevention by Vitamin E. Toxicology. 2003; 192: 237–248.
6. Ouanes, Z., Abid, S., Ayed, I., Anane, R. E., Mobio, T., Creppy, E. E. and Bacha, H. Induction of micronuclei by Zearalenone in Vero monkey kidney cells and in bone marrow cells of mice: protective effect of Vitamin E. Mut. Res. 2003; 538: 63–70.
7. WHO. Evaluation of Certain Food Additives and Contaminants. WHO Technical Report Series 884. World Health Organization, Geneva, 1999.
8. Tanaka, T., Yoneda, A., Inoue, S., Sugiura, Y. and Ueno, Y. Simultaneous determination of trichothecene mycotoxins and zearalenone in cereals by gas chromatography-mass spectrometry. J. Chromatogr. A. 2000; 882: 23–28.
9. Nielsen, K. F. and Thrane, U. 2001. Fast methods for screening of trichothecenes in fungal cultures using gas chromatography-tandem mass spectrometry. J. Chromatogr. A. 2001; 929: 75–87.
10. Tanaka, H., Takino, M., Sugita-Konishi, Y., Tanaka, T., Toriba, A. and Hayakawa, K. 2009. Determination of nivalenol and deoxynivalenol by liquid chromatography/atmospheric pressure photo ionizationmass spectrometry. Rapid Commun. Mass Spectrom. 2009; 23: 3119–3124.
11. Berthiller, F., Schumacher, R., Buttinger, G. and Krska, R. Rapid simultaneous determination of major type A- and B-trichothecenes as well as zearalenone in maize by high-performance liquid chromatography-tandem mass spectrometry. J. Chromatogr. A. 2005; 1062: 209–216.
12. Spanjer, M. C., Rensen P.M. and Scholten, J.M. LC-MS/MS multi-method for mycotoxins after single extraction, with validation data for peanut, pistachio, wheat, maize, cornflakes, raisins and figs. Food Add. Contam. 2008; 25: 472–489.
13. Tanaka, H., Takino, M., Sugita-Konishi, Y. and Tanaka, T. Development of a liquid chromatography/time-of-flight mass spectrometric method for the simultaneous determination of trichothecenes, zearalenone and aflatoxins in foodstuffs. Rapid Commun. Mass Spectrom. 2006; 20: 1422–1428.
14. Poapolathep, A., Poapolathep, S., Klangkaew, N., Sugita-Konishi, Y. and Kumagai, S. 2008. Detection of deoxynivalenol contamination in wheat products in Thailand. J.Food Prot. 71:1931–1933.
15. Sugiyama, K., Tanaka, H., Kamata, Y., Tanaka, T. and Sugita-Konishi, Y. 2009. A reduced rate of deoxynivalenol and nivalenol during bread production from wheat flour in Japan. Mycotoxins. 2009; 59: 1–6.
16. Campbell, H. M. and Armstrong, J. F. Determination of zearalenone in cereal grains, animal feed, and feed ingredients using immunoaffinity column chromatography and liquid chromatography: interlaboratory study. J. AOAC Int. 2007; 90: 1610–1622.
17. Trebstein, A., Seefelder, W., Lauber, U. and Humpf, H.-U. Determination of T-2 and HT-2 toxins in cereals including oats after immunoaffinity cleanup by liquid chromatography and fluorescence detection. J. Agric. Food Chem. 2008; 56: 4968–4975.

Chapter 12

Multiresidue Analysis of Antibiotics in Food of Animal Origin Using Liquid Chromatography–Mass Spectrometry

Katerina Mastovska

Abstract

Antibiotics are the most important drugs administered in veterinary medicine. Their use in food-producing animals may result in antibiotic residues in edible tissues, which are monitored to protect human and animal health, support the enforcement of regulations, provide toxicological assessment data, and resolve international trade issues. This chapter provides basic characterization of the most important classes of antibiotics used in food-producing animals (aminoglycosides, amphenicols, β-lactams, macrolides and lincosamides, nitrofurans, quinolones, sulfonamides, and tetracyclines), along with examples of practical liquid chromatographic-(tandem) mass spectrometric methods for analysis of their residues in food matrices of animal origin. The focus is on multiresidue methods that are favored by regulatory and other food testing laboratories for their ability to analyze residues of multiple compounds in a time- and cost-effective way.

Key words: Antibiotics, Liquid chromatography–mass spectrometry, Multiresidue methods, Aminoglycosides, Amphenicols, β-lactams, Macrolides, Nitrofurans, Quinolones, Sulfonamides, Tetracyclines

1. Introduction

Antibiotics are the most important drugs used in veterinary medicine to treat mainly bacterial diseases (1). They are also often administered as growth-promoting substances. Their use in food-producing animals may result in antibiotic residues in edible tissues, which are monitored to protect human and animal health, support the enforcement of regulations, provide toxicological assessment data, and resolve international trade issues. Antibiotic residues in food may cause allergic reactions in sensitive individuals. Also, the ingestion of subtherapeutic doses of antibiotics may lead

Jerry Zweigenbaum (ed.), *Mass Spectrometry in Food Safety: Methods and Protocols*,
Methods in Molecular Biology, vol. 747, DOI 10.1007/978-1-61779-136-9_12, © Springer Science+Business Media, LLC 2011

to the development of resistance strains of bacteria, which would no longer respond to drug treatment.

For monitoring and enforcement purposes, multiresidue methods, i.e., methods capable of analyzing residues of multiple compounds, are favored by regulatory and other food testing laboratories because of their time- and cost-effectiveness. Unlike pesticide residue analysis, in which multiclass, multiresidue methods represent a well-established approach, veterinary drug analysis is still performed mainly by using single-class or even single-residue methods, traditionally employing techniques such as microbial inhibition assays, immunoassays, or liquid chromatography (LC) with UV or fluorescence detection.

As opposed to the traditional techniques, LC coupled to mass spectrometry (LC-MS) offers many beneficial features for multiclass, multiresidue analysis, including mainly detection of a wide range of compounds independent of their biological function or chemical structure. Also, it provides simultaneous quantitation and structure-based identification of individual analytes. Even with LC–MS, however, the development of a multiclass method for antibiotics is a challenge because of the chemical diversity of the compounds.

2. LC–MS Instrumentation and Techniques in Antibiotic Analysis

Antibiotics are polar compounds that are readily ionized using the electrospray ionization (ESI) technique, which dominates in the published LC–MS-based methods for antibiotics over much less employed atmospheric pressure ionization (APCI) (2–7). Most antibiotics give a stronger signal in positive ionization mode, typically resulting in a protonated pseudomolecular ion $(M+H]^+$ (see details in the next section that discusses LC–MS analysis of different classes of antibiotics).

In unit-mass resolution MS instruments, the pseudomolecular ion usually does not provide enough selectivity for determination of trace levels of analytes in complex biological matrices. Therefore, modern laboratories performing antibiotic residue analysis in food are commonly equipped with LC–MS instrumentation with tandem MS (MS/MS) capabilities. MS/MS not only improves selectivity and consequently signal-to-noise ratio (analyte detectability) in real-world samples, but also provides structural information for identification/confirmation purposes. In this respect, triple quadrupole (QqQ) MS/MS instruments have become workhorses for antibiotic and other veterinary drug residue analysis in routine laboratories world-wide (2–7). Ion trap (IT) MS instruments offer MS^n capabilities, which are highly valuable for analyte identification, but QqQ-MS/MS is generally preferred over IT-MS/MS in quantitative residue methods.

Recently, accurate-mass, high-resolution (HR) time-of-flight (TOF) MS bench-top instruments became available from different vendors, either as single TOF-MS instruments or as hybrid instruments combined with a quadrupole (Q-TOF-MS) and a collision cell for MS/MS analysis. Accurate-mass HRMS measurements offer increased selectivity by eliminating potential interferences originating from the matrix that have very similar mass-to-charge ratio (m/z) as analytes (e.g., the same nominal mass) but different elemental composition, thus different accurate mass. As opposed to Q and IT instruments, TOF-MS is a non-scanning mass analyzer, which can acquire full spectral ("full-scan") data in a fast and efficient (sensitive) fashion. Therefore, TOF-MS instruments are very suitable for rapid non-targeted screening, especially when combined with a fast LC separation using for instance shorter, narrower columns packed with sub-2 μm particles (8).

3. LC–MS Analysis of Different Antibiotic Classes

The following sections provide basic characterization of the most important classes of antibiotics used in the veterinary medicine, along with examples of practical LC–MS(MS) methods for analysis of their residues in food matrices of animal origin. The focus is on methods that can analyze multiple analytes, i.e., on multiresidue methods.

For each class (except for the special case of nitrofurans, which are presented differently), a table with the most important representatives is given, which provides their molecular weight (MW), formula (elemental composition), CAS number, information on regulatory limits in the European Union (EU) and the USA, and basic MS/MS conditions, including ionization mode (POS for positive and NEG for negative), precursor ion and typical product ions. For the MW, monoisotopic molecular weight (rounded to three decimal points) is given using the most abundant isotopes present in the nature. This information together with the elemental composition can be used in accurate mass HRMS analysis.

The regulatory limits, maximum residue limits (MRLs) in the EU and tolerances in the US, are provided (if established) in various animal species and matrices as a range in the text and in more details (for each matrix) for bovine tissues in the tables. The information on regulatory limits and banned substances is important in order to establish target detection and quantitation limits of the analytical methods.

3.1. Aminoglycosides

Aminoglycosides are broad-spectrum antibiotics isolated from *Streptomyces* and *Micromonospora* bacteria that exert their antibacterial effect by targeting the bacterial ribosome, thus inhibiting

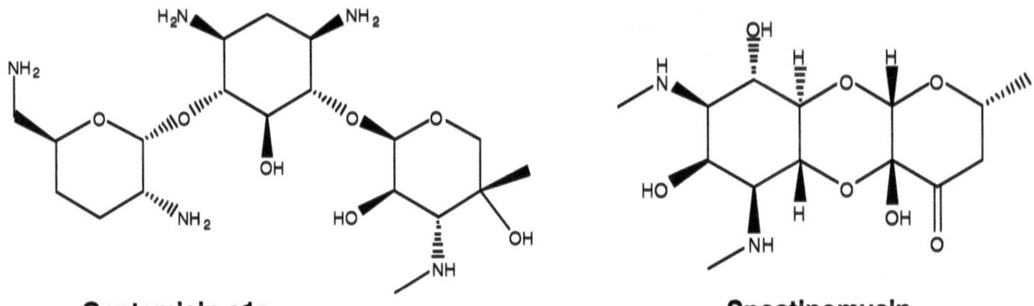

Fig. 1. Examples of aminoglycoside (gentamicin c1a) and aminocyclitol (spectinomycin) structures.

protein synthesis (1). Their structure contains two or more aminosugars linked by a glycosidic bond to an aminocyclitol group, which is 2-deoxystreptamine in most aminoglycosides or streptidine in streptomycin and dehydrostreptomycin. Most aminoglycosides are mixtures of several very similar components differing only in degree of methylation or stereochemistry of the sugar units. Closely related aminocyclitols, such as spectinomycin or apramycin, that also contain an aminocyclitol group but slightly differ in structure (see Fig. 1 comparing structures of gentamicin c1a and spectinomycin), are generally considered part of the aminoglycoside class of antibiotics.

Improper use of aminoglycosides may generate residues that are potentially harmful due to their oto-, neuro- and nephrotoxicity. The EU set MRLs for eight aminoglycosides ranging from 50 μg/kg for gentamicin in bovine and porcine muscle, and fat to 20,000 μg/kg for apramycin in bovine kidney. In the US, tolerances were established for seven aminoglycosides in the range from 100 μg/kg (apramycin in porcine kidney, gentamicin in poultry edible tissues, and porcine muscle and spectinomycin in poultry edible tissues) to 7,200 μg/kg for neomycin in kidney and fat of various species. There is a zero tolerance for hygromycin B in porcine and poultry edible tissues and eggs in the US. Also, gentamicin is not registered for use in cattle. A "safe level" of 30 μg/kg is used by the US Food and Drug Administration (FDA) as a non-binding, prosecutional guideline for gentamicin in milk.

Table 1 provides basic information on the most important aminoglycosides, their regulatory limits in bovine tissues, and MS/MS transitions used in their LC–MS/MS analysis. Aminoglycosides are ionized in positive mode, usually forming a singly charged protonated molecular ion. Spectinomycin typically gives a stronger precursor ion as the protonated ketone hydrate ($[M+H_2O+H]^+$).

Aminoglycosides are very polar compounds that show practically no retention in reversed-phase LC (RPLC), unless an ion-pairing agent, such as heptafluorobutyric acid (HFBA), is added to the

Table 1
Aminoglycosides: compound information, regulatory limits in bovine tissues, and typical MS/MS conditions

Compound information				Regulatory limits (μg/kg) in bovine tissues[a]		MS/MS conditions			
Name	MW	Formula	CAS No.	EU	USA	Mode	Precursor ion (m/z)	Product ions (m/z)	Reference
Amikacin	585.286	$C_{22}H_{43}N_5O_{13}$	37517-28-5			POS	586	163, 425	(10, 11, 14)
Apramycin	539.280	$C_{21}H_{41}N_5O_{11}$	37321-09-8	1,000 (M/F) 10,000 (L) 20,000 (K)		POS	378 540	217 217, 378	(10–12)
Dihydrostreptomycin	583.281	$C_{21}H_{41}N_7O_{12}$	128-46-1	500 (M/F/L) 1,000 (K) 200 (milk)	500 (M/F/L) 2,000 (K) 125 (milk)	POS	584	204, 221, 246, 263	(10–13)
Gentamicin c1[b]	477.316	$C_{21}H_{43}N_5O_7$	25876-10-2	50 (M/F) 200 (L)		POS	478	157, 160, 163, 322	(10–14)
Gentamicin c1a[b]	449.285	$C_{19}H_{39}N_5O_7$	26098-04-4	750 (K)		POS	450	160, 322	(10, 12)
Gentamicin c2+c2a[b]	463.301	$C_{20}H_{41}N_5O_7$	25876-11-3	100 (milk)		POS	464	160, 322	(10, 12)
Hygromycin B	527.233	$C_{20}H_{37}N_3O_{13}$	31282-04-9			POS	528	177, 352	(10)
Kanamycin A	484.238	$C_{18}H_{36}N_4O_{11}$	8063-07-8	100 (M/F) 600 (L) 2,500 (K) 150 (milk)		POS	485	163, 205, 324	(10–12, 14)

(continued)

Table 1 (continued)

Compound information				Regulatory limits (μg/kg) in bovine tissues[a]		MS/MS conditions			
Name	MW	Formula	CAS No.	EU	USA	Mode	Precursor ion (m/z)	Product ions (m/z)	Reference
Neomycin B	614.312	$C_{23}H_{46}N_6O_{13}$	1404-04-2	500 (M/F/L) 5,000 (K) 1,500 (milk)	1,200 (M) 3,600 (L) 7,200 (K/F) 150 (milk)	POS	615	161, 163, 293, 323	(10–14)
Paromomycin	615.296	$C_{23}H_{45}N_5O_{14}$	1263-89-4	500 (M) 1,500 (L/K)		POS	616	161, 163, 293	(10–12, 14)
Sisomicin	447.269	$C_{19}H_{37}N_5O_7$	32385-11-8			POS	448	160, 254	(11)
Spectinomycin	332.158	$C_{14}H_{24}N_2O_7$	1695-77-8	500 (M) 300 (F) 1,000 (L) 5,000 (K) 200 (milk)	250 (M) 4,000 (K)	POS	351[c]	98, 100, 189, 333, 207	(10–12)
Streptomycin	581.266	$C_{21}H_{39}N_7O_{12}$	57-92-1	500 (M/F/L) 1,000 (K) 200 (milk)	500 (M/F/L) 2,000 (K)	POS	582	204, 246, 263	(10–13)
Tobramycin	467.259	$C_{18}H_{37}N_5O_9$	32986-56-4			POS	468	163, 324	(10, 11, 13, 14)

[a] Bovine tissues: M=muscle, F=fat, L=liver, K=kidney
[b] In the EU, residue definition for gentamicin is sum of gentamicin c1, c1a, c2 and c2a
[c] Protonated molecule of spectinomycin hydrate $[M+H_2O+H]^+$

mobile phase (9–13). It is also possible to use hydrophilic interaction chromatography (HILIC) for retention of aminoglycosides under aqueous normal LC conditions (14) or perform derivatization using phenyl isocyanate that attaches to each amino group in aminoglycoside molecules, resulting in less polar derivatives that can be analyzed in RPLC (15).

In practice, most multiresidue methods for analysis of aminoglycoside residues in animal tissues (liver, kidney, and muscle) are based on a LC–MS confirmation method used by the Food Safety and Inspection Service (FSIS) of the US Department of Agriculture (9), which employs a buffer containing trichloroacetic acid (TCA) as a protein precipitant, followed by a solid-phase extraction (SPE) step, and an ion-pair RPLC with (20 mM) HFBA as a volatile ion pairing agent compatible with ESI. The aqueous extraction buffer solution (pH 4) in the FSIS method contains 10 mM KH_2PO_4, 0.4 mM ethylenediaminetetraacetic acid (EDTA) and 2% (*w/v*) TCA (9, 12, 13). Alternatively, 5% (*w/v*) TCA in water can be used for the extraction of aminoglycosides from tissues (10, 11). Depending on the loading and elution conditions, the SPE cleanup step may employ a weak cation exchange (WCX) sorbent (9, 12), a hydrophilic–lipophilic balance (HLB) sorbent (10), a C_{18} sorbent (13) or a combination of an anion-exchanger, HLB, and WCX (11).

3.2. Amphenicols

Amphenicols (chloramphenicol, florfenicol, and thiamphenicol) are broad-spectrum antibiotics active against a variety of pathogens through interaction with their ribosomes, resulting in inhibition of protein synthesis (1). Chloramphenicol has been used in food-producing animals for over 50 years. However, due to the reports of serious side effects (mainly aplastic anemia) in humans, chloramphenicol was banned in the EU, the USA and Canada in the 1990s. Structurally similar thiamphenicol and florfenicol, in which the nitro group of chloramphenicol is replaced by a methyl sulphonyl group (in florfenicol, a hydroxyl group is also replaced by a fluorine, see Fig. 2), have been permitted as chloramphenicol substitutes.

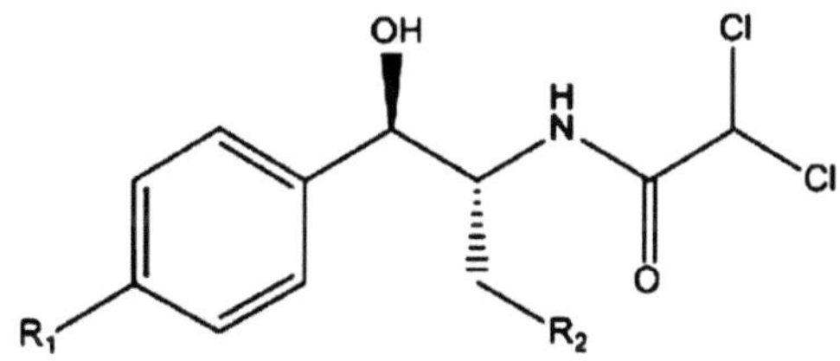

Chloramphenicol:	$R_1 = NO_2$	$R_2 = OH$
Thiamphenicol:	$R_1 = CH_3SO_2$	$R_2 = OH$
Florfenicol:	$R_1 = CH_3SO_2$	$R_2 = F$

Fig. 2. Structures of amphenicols.

For thiamphenicol, the EU established an MRL of 50 μg/kg in bovine and chicken tissues (the same provisional limit in porcine expired in 2007, in ovine and fish in 2001). Regulatory limits for florfenicol in various animal species and matrices are in the range of 100–3,000 μg/kg in the EU and 200–3,700 μg/kg in the US. In the EU, the residue definition for florfenicol is the sum of florfenicol and its metabolites measured as florfenicol amine. In the US, florfenicol amine is the marker residue for florfenicol in bovine and fish tissues, whereas the parent florfenicol is the marker residue in porcine matrices. There is no permitted limit of chloramphenicol in food; therefore it is important to make sure that an adequately sensitive method is used for its analysis in the practice. For this reason, the EU established a minimum required performance limit (MRPL) of 0.3 μg/kg in meat, eggs, milk, urine, aquaculture products, and honey in 2003 (16).

Table 2 provides basic information on the three important amphenicols (together with florfenicol amine), their regulatory limits in bovine tissues, and MS/MS transitions used in their LC–MS/MS analysis. Chloramphenicol, florfenicol, and thiamphenicol typically give a stronger signal in the negative mode (17–24), whereas florfenicol amine is analyzed in positive mode (24, 25).

Numerous LC–MS-based methods have been published for chloramphenicol since 2003 that can meet the EU requirement of MRPL of 0.3 μg/kg in various matrices (17–23). The sample preparation is based on different techniques, including extraction using ethyl acetate or acetonitrile (combined with removal of fat using hexane) and/or SPE with C18 or molecular imprinted polymer (MIP) sorbents. An isotopically labeled internal standard (d_5-chloramphenicol) is generally used to compensate for volume changes during the sample preparation and matrix effects in the LC–ESI-MS.

Recently, Zhang et al. (24) published a multiresidue method for chloramphenicol, florfenicol, florfenicol amine, and thiamphenicol in chicken muscle, using LC–QqQ-MS/MS with a C18 column and water and acetonitrile as mobile phases for a gradient elution. Their sample preparation procedure consists of multiple steps, including addition of d_5-chloramphenicol as an internal standard, extraction with ethyl acetate-ammonium hydroxide (98:2, *v/v*), concentration in a small volume of 5% acetic acid, de-fatting with hexane, SPE clean-up using Oasis MCX sorbent, evaporation to dryness, reconstitution in acetonitrile–water (30:70, *v/v*), and filtration.

3.3. β-Lactams: Penicillins and Cephalosporins

β-Lactam antibiotics are probably the most widely applied antimicrobial drugs in current veterinary practice. The class of β-lactams includes penicillins and cephalosporins that both have a β-lactam ring in their structures, but this ring is fused to a five-membered

Table 2
Amphenicols: compound information, regulatory limits in bovine tissues, and typical MS/MS conditions

Compound information				Regulatory limits (μg/kg) in bovine tissues[a]		MS/MS conditions			
Name	MW	Formula	CAS No.	EU	USA	Mode	Precursor ion (m/z)	Product ions (m/z)	Reference
Chloramphenicol	322.012	$C_{11}H_{12}Cl_2N_2O_5$	85666-84-8	0.3 (MRPL)[b]		NEG	321	152, 194, 257	(17–24)
Florfenicol amine[c]	247.068	$C_{10}H_{14}FNO_3S$	76639-93-5	200 (M)	300 (M)	POS	248	130, 230	(24, 25)
Florfenicol[c]	357.000	$C_{12}H_{14}Cl_2FNO_4S$	76639-94-6	3,000 (L) 300 (K)	3,700 (L)	NEG	356	185, 336	(24)
Thiamphenicol	355.005	$C_{12}H_{15}Cl_2NO_5S$	15318-45-3	50 (M/F/L/K/ milk)		POS	354	185, 290	(24)

[a] Bovine tissues: M=muscle, F=fat, L=liver, K=kidney

[b] Minimum required performance limit (MRPL) of 0.3 μg/kg in meat, eggs, milk, urine, aquaculture products, and honey

[c] In the EU, residue definition for florfenicol is sum of florfenicol and its metabolites measured as florfenicol amine. In the US, florfenicol amine is the marker residue for florfenicol in bovine tissues

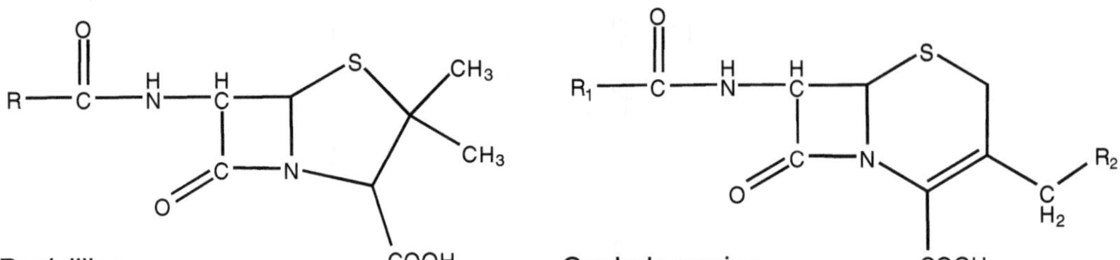

Fig. 3. Basic structures of β-lactams (penicillins and cephalosporins).

thiazolidine or a six-membered dihydrothiazine ring, respectively (see Fig. 3). The β-lactam ring is responsible for the antimicrobial activity and also for a reduced stability of β-lactams. They are thermolabile, unstable in alcohols and acidic conditions (4, 26).

In the EU, the MRLs for β-lactams range from 4 μg/kg (for amoxicillin, ampicillin, and penicillin G in milk) to 6,000 μg/kg (for ceftiofur, as desfuroylceftiofur, in bovine kidney). In the US, the tolerances for β-lactams range from 10 μg/kg (for amoxicillin, ampicillin, and cloxacillin in milk and bovine edible tissues) to 3,000 μg/kg (for ceftiofur, as desfuroylceftiofur, in porcine liver). There is a zero tolerance for penicillin G in milk in the US, with a "safe level" of 5 μg/kg used by the US FDA as a non-binding, prosecutional guideline.

Table 3 gives basic information on important β-lactams, their regulatory limits in bovine tissues, and MS/MS transitions used in their LC–MS/MS analysis. For the multiresidue analysis, positive ionization mode can be used for all β-lactams (26–28) or negative ionization can be employed for the majority of penicillins (except for the amphoteric analytes amoxicillin and ampicillin) in methods utilizing both positive and negative modes for early and late eluting β-lactams, respectively (28, 29). Both QqQ (26–31) and IT (32, 33) instruments have been used in the LC–MS/MS analysis of β-lactams, with QqQ generally giving better quantitation results (27). In the positive ionization mode, characteristic MS/MS product ions for penicillins include *m/z* 160, formed due to the cleavage of the β-lactam ring, and *m/z* 114 resulting from a further loss of COOH (34). Product ions $(M-H-CO_2]^-$ and $(M-H-141]^-$ can be predominantly seen in Table 3 for penicillins ionizing in the negative mode (31).

Based on the published studies, the addition of formic acid to the mobile phase seems to be the best choice for the LC separation and MS sensitivity (4). Mastovska and Lightfield (26) evaluated different mobile phase composition for an optimal analysis of 14 β-lactams, varying the percentage of formic acid (0–0.4%, *v/v*) in both parts of the mobile phase, which consisted of (A) water and (B) acetonitrile, methanol, or an acetonitrile–methanol (50:50, *v/v*) mixture. Independent of the amount of formic acid

Table 3
β-Lactams: compound information, regulatory limits in bovine tissues, and typical MS/MS conditions

Compound information				Regulatory limits (μg/kg) in bovine tissues[a]		MS/MS conditions			
Name	MW	Formula	CAS No.	EU	USA	Mode	Precursor ion (m/z)	Product ions (m/z)	Reference
Amoxicillin	365.105	$C_{16}H_{19}N_3O_5S$	26787-78-0	50 (M/F/L/K) 4 (milk)	10 (M/F/L/K) 10 (milk)	pos	366	114, 208, 349	(26–30)
Ampicillin	349.110	$C_{16}H_{19}N_3O_4S$	69-53-4	50 (M/F/L/K) 4 (milk)	10 (M/F/L/K) 10 (milk)	POS	350	106, 160, 174, 192	(26–30)
Cefadroxil[b]	363.089	$C_{16}H_{17}N_3O_5S$	50370-12-2			POS	364	114, 208	(26, 28)
Cefazolin	454.030	$C_{14}H_{14}N_8O_4S_3$	25953-19-9	50 (milk)		POS	455	156, 323	(26–28)
Cefoperazone	645.142	$C_{25}H_{27}N_9O_8S_2$	62893-19-0	50 (milk)		NEG	644	188, 528	(29)
Cefquinome	528.125	$C_{23}H_{24}N_6O_5S_2$	84957-30-2	50 (M/F) 100 (L) 200 (K) 20 (milk)		POS	529	113, 324	(29)
Cephalexin	347.094	$C_{16}H_{17}N_3O_4S$	15686-71-2	200 (M/F/L) 1,000 (K) 100 (milk)		POS	348	158, 174	(26, 28, 29)
Cephalonium	458.072	$C_{20}H_{18}N_4O_5S_2$	5575-21-3	20 (milk)		POS	459	152, 337	(29)
Cephapirin[c]	423.056	$C_{17}H_{17}N_3O_6S_2$	21593-23-7	50 (M/F) 100 (K) 60 (milk)	100 (M/F/L/K) 20 (milk)	POS	424	152, 292	(26, 29, 30)
Cloxacillin	435.066	$C_{19}H_{18}ClN_3O_5S$	61-72-3	300 (M/F/L/K) 30 (milk)	10 (M/F/L/K) 10 (milk)	POS NEG	436 434	160, 277 293, 390	(26–28, 30) (28, 29, 31)

(continued)

Table 3
(continued)

Compound information				Regulatory limits (μg/kg) in bovine tissues[a]		MS/MS conditions			
Name	MW	Formula	CAS No.	EU	USA	Mode	Precursor ion (m/z)	Product ions (m/z)	Reference
DCCD[d]	548.028	$C_{17}H_{20}N_6O_7S_4$		1,000 (M) 2,000 (F/L) 6,000 (K) 100 (milk)	1,000 (M) 2,000 (L) 400 (K) 100 (milk)	POS NEG	549 547	183, 241, 366, 397 239, 382	(26–28) (28)
Desacetyl cephapirin[c]	381.045	$C_{15}H_{15}N_3O_5S_2$		50 (M/F) 100 (K) 60 (milk)		POS	382	112, 152, 226, 292	(26–29)
Dicloxacillin	469.027	$C_{19}H_{17}Cl_2N_3O_5S$	3116-76-5	300 (M/F/L/K) 30 (milk)		POS NEG	470 468	160, 311 267, 327, 424	(26–28) (28, 29)
Nafcillin	414.125	$C_{21}H_{22}N_2O_5S$	147-52-4	300 (M/F/L/K) 30 (milk)		POS NEG	415 413	171, 199, 256 272, 369	(26–28) (28, 29, 31)
Oxacillin	401.105	$C_{19}H_{19}N_3O_5S$	66-79-5	300 (M/F/L/K) 30 (milk)		POS NEG	402 400	160, 243 259, 356	(26, 27, 30) (28, 31)
Penicillin G	334.099	$C_{16}H_{18}N_2O_4S$	61-33-6	50 (M/F/L/K) 4 (milk)	50 (M/F/L/K) 0 (milk)	POS NEG	335 333	160, 176 192, 289	(26–28, 30) (28, 29, 31)
Penicillin V[b]	350.094	$C_{16}H_{18}N_2O_5S$	87-08-1			POS NEG	351 349	114, 160, 192 114, 208, 305	(26–28) (28, 29, 31)

[a] Bovine tissues: M=muscle, F=fat, L=liver, K=kidney

[b] Cefadroxil and penicillin V are not used in cattle, thus can be employed as QC standards for analysis of bovine tissues (26)

[c] In the EU, residue definition for cephapirin is sum of cephapirin and desacetylcephapirin

[d] DCCD (desfuroylceftiofur cysteine disulfide) is the most abundant free metabolite of ceftiofur suitable for screening using a multiresidue method. To determine total residues of ceftiofur in a single-residue method, desfuroylceftiofur has to be released from the disulfide bonds by their reduction, followed by stabilization of the thiol group by derivatization (e.g., acetylation). In the EU, residue definition for ceftiofur is sum of all residues retaining β-lactam structure expressed as desfuroylceftiofur. In the US, desfuroylceftiofur is the marker residue for all bovine and porcine tissues, with kidney being the target tissue for enforcement and monitoring purposes

in the mobile phase, acetonitrile provided overall better sensitivity for the tested β-lactams than methanol. For LC–ESI(+)–MS/MS multiresidue analysis, 0.1% formic acid as an additive in the mobile phase (in water and in acetonitrile) provided the best overall performance (sensitivity, speed, and ruggedness). However due to the stability issues, formic acid should not be added to the final extracts prior to the LC–MS/MS analysis to match the mobile phase composition. For multiresidue analysis using both ESI+ and ESI−, Becker et al. (29) recommended using a very small amount of formic acid (0.005%).

For monitoring and enforcement purposes, kidney is the target tissue for β-lactam analysis. Fagerquist et al. (27) published a relatively simple preparation method for a multiresidue analysis in bovine kidney, which was further improved and overall streamlined by Mastovska and Lightfield (26). The streamlined method is based on a simple extraction using acetonitrile-water (4:1, *v/v*), followed by dispersive solid-phase extraction cleanup with C_{18} sorbent, concentration of an extract aliquot, and filtration of the final extracts using syringeless filter vials, which are used for the sample introduction in the LC–MS/MS analysis. The stable isotope [$^{13}C_6$]sulfamethazine is added to the homogenized sample as an internal standard (to correct for volume changes) before the extraction step together with two β-lactams (penicillin-V and cefadroxil), which serve for method performance controls. This method can be used not only for β-lactam analysis, but also for a multiresidue screening of multiple classes of antibiotics in various bovine matrices (25).

For milk, the sample preparation procedure usually involves extraction with acetonitrile, solvent exchange (with addition of saturated sodium chloride solution to avoid foaming) into a phosphate buffer (pH 8.5), SPE cleanup using Oasis HLB cartridges, elution of analytes with 40 or 50% acetonitrile in water and solvent exchange to water or ammonium acetate buffer (pH 6.7), and filtration or centrifugation prior to the LC–MS/MS analysis (29, 33).

3.4. Macrolides and Lincosamides

Macrolides are basic macrocyclic antibiotics that have a common 14-, 16-, or 17-membered ring in their structure, which is linked by glycoside bonding to one or more molecules of deoxy sugars. They are widely used in veterinary practice to treat respiratory diseases and to promote growth (1). Lincosamides (lincomycin, clindamycin, and pirlimycin) are monoglycosides with an amino acid side chain. They are highly effective against a broad spectrum of gram-positive and anaerobic bacteria. Both macrolides and lincosamides target the bacterial ribosome and inhibit protein synthesis. Figure 4 shows examples of macrolide and lincosamide structures.

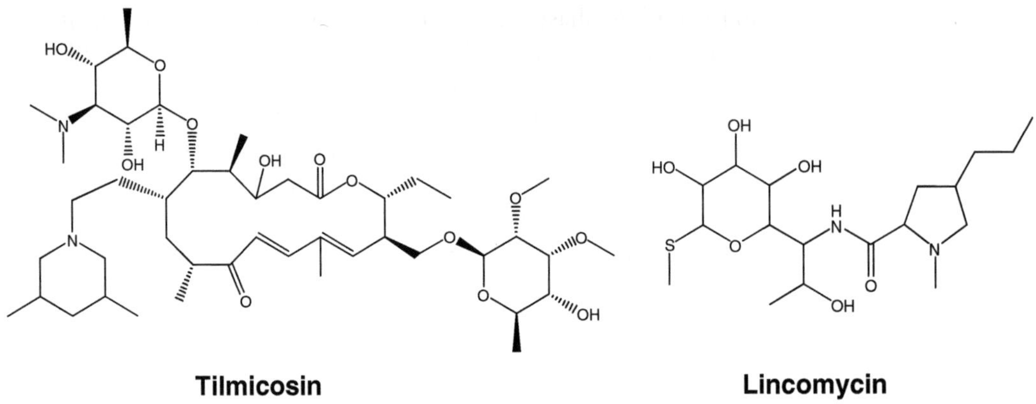

Fig. 4. Examples of macrolide (tilmicosin) and lincosamide (lincomycin) structures.

In the EU, the MRLs for macrolides and lincosamides range from 40 μg/kg (for erythromycin in milk) to 3,000 μg/kg (for tulathromycin in bovine and porcine kidney and liver). In the US, the tolerances for macrolides and lincosamides range from 25 μg/kg (for erythromycin in eggs) to 15,000 μg/kg (for tulathromycin in porcine kidney). There is a zero tolerance for erythromycin in milk in the US, with a "safe level" of 50 μg/kg used by the US FDA as a non-binding, prosecutional guideline.

Table 4 gives basic information on important macrolides and lincosamides, their regulatory limits in bovine tissues, and MS/MS transitions used in their LC–MS/MS analysis. Macrolides and lincosamides are ionized in the positive mode. In some cases, such as spiramycin, tilmicosin or tulathromycin, doubly charged protonated pseudomolecular ions $[M+2H]^{2+}$ may be more abundant than the singly charged species $[M+H]^{+}$ (35, 38–41).

The recently published multiresidue methods for analysis in animal-derived matrices employ mostly LC–QqQ-MS/MS (25, 35–39). Wang and Leung (37) compared LC–QqQ-MS/MS with ultra-performance (UP)LC–TOFMS (using a Q-TOF instrument) for the analysis of six macrolides in eggs, milk, and honey. The former technique provided somewhat better repeatability and lower detection limits (0.01–0.5 vs. 0.2–1 μg/kg for UPLC–TOFMS). However, they demonstrated that a TOFMS instrument could be highly beneficial for targeted and non-targeted screening and/or confirmation of LC–QqQ-MS/MS results based on the accurate mass measurement.

Berrada et al. used LC–MS–SIM with a single quadrupole to analyze seven macrolides in bovine kidney and liver (40) or in fish and bovine, porcine, and poultry meat (41). The extraction of lyophilized meat and fish samples was done using a pressurized liquid extraction (PLE) with methanol (41).

In terms of sample preparation, a very simple method for analysis of six macrolides, (including tulathromycin A as a parent

Table 4
Macrolides and lincosamides: compound information, regulatory limits in bovine tissues, and typical MS/MS conditions

Compound information				Regulatory limits (μg/kg) in bovine tissues[a]		MS/MS conditions			
Name	MW	Formula	CAS No.	EU	USA	Mode	Precursor ion (m/z)	Product ions (m/z)	Reference
Clindamycin	424.180	$C_{18}H_{33}ClN_2O_5S$	18323-44-9			POS	425, 427	126	(39)
Erythromycin A	733.461	$C_{37}H_{67}NO_{13}$	114-07-8	200 (M/F/L/K) 40 (milk)	100 (M/F/L/K) 0 (milk)	POS	734 (735)	158, 522, 540, 558, 576 (577)	(25, 35–38)
Josamycin	827.467	$C_{42}H_{69}NO_{15}$	16846-24-5			POS	829	109, 174, 229, 600	(36, 39)
Lincomycin	406.214	$C_{18}H_{34}N_2O_6S$	154-21-2	100 (M) 50 (F) 500 (L) 1,500 (K) 150 (milk)		POS	407	126, 359	(25, 38, 39)
Neospiramycin[b]	698.435	$C_{36}H_{62}N_2O_{11}$	70253-62-2	200 (M) 300 (F/L/K) 200 (milk)		POS	699	142, 174	(37)
Oleandomycin	687.419	$C_{35}H_{61}NO_{12}$	3922-90-5			POS	688	158, 544	(37, 38)
Roxithromycin[c]	836.525	$C_{41}H_{76}N_2O_{15}$	80214-83-1			POS	837	158, 679	(35–38)
Pirlimycin	410.164	$C_{17}H_{31}ClN_2O_5S$	79548-73-5	100 (M/F) 1,000 (L) 400 (K) 100 (milk)	300 (M) 500 (L) 400 (milk)	POS	411, 413	112, 363	(25, 39)

(continued)

Table 4
(continued)

Compound information				Regulatory limits (μg/kg) in bovine tissues[a]		MS/MS conditions			
Name	MW	Formula	CAS No.	EU	USA	Mode	Precursor ion (m/z)	Product ions (m/z)	Reference
Spiramycin[b]	842.514	$C_{43}H_{74}N_2O_{14}$	8025-81-8	200 (M) 300 (F/L/K) 200 (milk)		POS	844 (843) 438 (422)[d]	101, 142, 174, 318, 522, 540, 700 101, 174	(36, 37, 39) (38)
Tilmicosin	868.566	$C_{46}H_{80}N_2O_{13}$	050-54-0	50 (M/F) 1,000 (L/K) 50 (milk)	100 (M) 1,200 (L) 50 (milk)	POS	870 435	132, 174, 506, 522, 679, 697 88, 99, 174, 696	(25, 36–38) (35, 39)
Tulathromycin A[c]	805.566	$C_{41}H_{79}N_3O_{12}$	217500-96-4	100 (F) 3,000 (L/K)	5,500 (L)	POS	807 404	578 158	(39)
Tylosin A	915.519	$C_{46}H_{77}NO_{17}$	1405-54-5	100 (M/F/L/K) 50 (milk)	200 (M/F/L/K) 50 (milk)	POS	917 (916)	145, 174, 318, 407, 598, 773 (772)	(25, 35, 37–39)

[a] Bovine tissues: M=muscle, F=fat, L=liver, K=kidney

[b] In the EU, residue definition for spiramycin is the sum of spiramycin and neospiramycin

[c] Roxithromycin is often used as an internal standard in macrolide multiresidue methods (35–38)

[d] The doubly charged precursor ion m/z 438 is a methanol adduct $[M+CH_3OH+2H]^{2+}$ that can be seen for spiramycin in the presence of methanol. In the presence of acetonitrile, doubly charged protonated pseudomolecular ion $[M+2H]^{2+}$ at m/z 422 is formed

[e] In the EU and the US, the marker residue for tulathromycin is (2R,3S,4R,5R,8R,10R,11R,12S,13S,14R)-2-ethyl-3,4,10,13-tetrahydroxy-3,5,8,10,12,14-hexamethyl-11-[[3,4,6-trideoxy-3-(dimethylamino)-β-D-xylo-hexopyranosyl]oxy]-1-oxa-6-azacyclopentadecan-15-one (also known as CP-60,300, the acid hydrolysis product of tulathromycin and selected metabolites using 2N HCl), expressed as tulathromycin equivalents. Tulathromycin cannot be used in animals from which milk is produced for human consumption

compound) and three lincosamides in bovine (beef and calf), porcine, and poultry meat was recently introduced by Martos et al. (39). The method involves extraction with acetonitrile, followed by 3:7 (*v/v*) dilution with water and extract defatting with hexane. Other published methods for macrolides generally include a laborious SPE cleanup step. Also, the LC separation employed a short (20 mm × 3.9 mm, 3 μm) C_{18} guard column and 1 mL/min mobile phase flow rate for a fast LC–ESI–MS/MS analysis. The mobile phase for a gradient elution consisted of (A) 0.1% formic acid and (B) acetonitrile.

3.5. Nitrofurans

Nitrofurans are synthetic antibacterial compounds, which contain a characteristic 5-membered nitrofuran ring in their structure (see Fig. 5). They are very effective for treatment of infections caused by bacteria and protozoa (by inhibiting glucose metabolism and ribosomal function) and do not contribute to the development of antimicrobial resistance (1). However due to their toxicological effects (carcinogenity and mutagenicity), nitrofurans (nitrofurazone, nitrofurantoin, furaltadone, furazolidone and later also nifursol) were banned in many countries, including the US, the EU, Japan and Australia, starting in mid-1990s to early 2000s.

Nitrofurans are rapidly metabolized to smaller molecules that are detectable in eggs, milk, muscle, liver, and kidney as protein-bound residues. Therefore, the detection of the parent drugs is

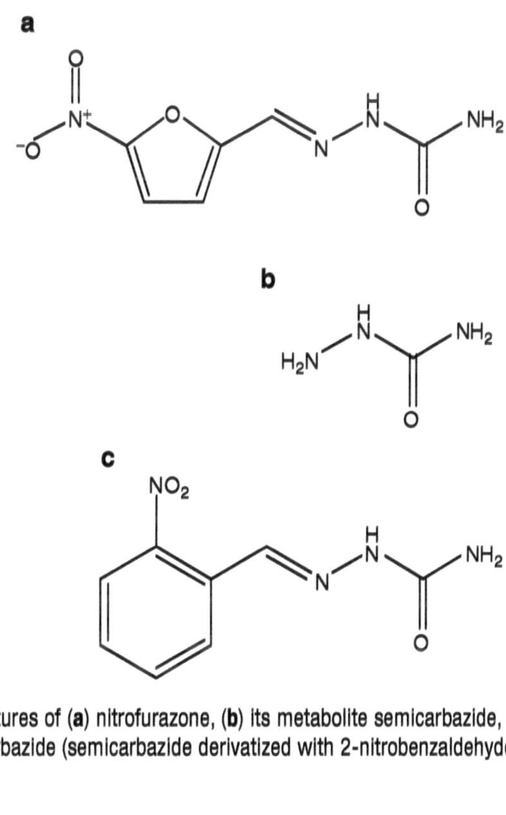

Fig. 5. Structures of (**a**) nitrofurazone, (**b**) its metabolite semicarbazide, and (**c**) nitrophenylsemicarbazide (semicarbazide derivatized with 2-nitrobenzaldehyde).

generally not very likely (except for their accumulation in avian eyes (42)), and methods detecting nitrofuran residues in food should be aimed at their metabolites. A well-established sample preparation procedure for LC–MS analysis involves acid hydrolysis and overnight derivatization of the released metabolites with 2-nitrobenzaldehyde, followed by extraction with ethyl acetate (in SPE or LLE, liquid–liquid extraction, format) and LC–MS(MS) analysis (43–50). As an example, Fig. 5 shows structures of nitrofurazone along with its metabolite semicarbazide and derivatized semicarbazide (the analytical form). Table 5 gives basic information on the five important nitrofurans, including MW, formula, and CAS number for the parent drugs and metabolites, and, for the derivatized metabolites, MW, formulas and MS/MS transitions used in their LC–MS/MS analysis.

In 2003, The EU established an MRPL of 1 μg/kg for the nitrofuran metabolites in poultry meat and aquaculture products (16). Since then, several multiresidue methods (based on the above sample preparation procedure) were published that can analyze nitrofuran metabolites at levels below 1 μg/kg in poultry muscle (44–46) and pork muscle (44, 47), eggs (45, 48), milk (49), and shrimp (50). It should be noted that the metabolite of nitrofurazone, semicarbazide, can occur in food also from other sources, such as from food packaging and carrageenan (51). This is one of the reasons why laboratories generally set their reporting limits for all nitrofurans at the EU MRPL of 1 μg/kg, even though their analytical methods can often detect and quantify lower concentrations of nitrofuran metabolites (51).

3.6. Quinolones

Quinolones are synthetic antibiotics (derived from 3-quinolenecarboxylic acid) that show very high activity against a wide range of diseases in livestock and aquaculture. They are also highly important human drugs, and their widespread use in food-producing animals is of great concern due to the recent evidence of development of bacterial resistance to these antibiotics (1). The first generation of quinolones includes mainly oxolinic acid and nalidixic acid that are effective only against Gram-negative bacteria. The second-generation quinolones are fluoroquinolones, such as enrofloxacin, danofloxacin, ciprofloxacin etc., which contain a fluorine atom at the C-3 position and a piperazinyl group at the C-7 position (see Fig. 6 for structures of oxolinic acid and enrofloxacin), which increases the activity against Gram-positive and Gram-negative bacteria, respectively. The mode of action of quinolones most likely involves inhibition of bacterial DNA gyrase enzymes.

Quinolones have amphoteric and zwitterionic properties due to the presence of a carboxylic acid group (pK_a about 5) and one or more amine functional groups (pK_a about 8–9) in their molecules (4, 5). At pH 6–8, they have poor water solubility, but are soluble in lipids, thus can penetrate tissues.

Table 5
Nitrofurans, their metabolites and derivatization products: compound information and typical MS/MS conditions for the metabolites derivatized with 2-nitrobenzaldehyde (nitrophenyl derivatives)

Nitrofuran parent compounds					
Name	Furaltadone	Furazolidone	Nifursol	Nitrofurantoin	Nitrofurazone
MW	324.107	225.039	365.024	238.034	198.039
Formula	$C_{13}H_{16}N_4O_6$	$C_8H_7N_3O_5$	$C_{12}H_7N_5O_9$	$C_8H_6N_4O_5$	$C_6H_6N_4O_4$
CAS	139-91-3	67-45-8	16915-70-1	67-20-9	59-87-0
Nitrofuran metabolites					
Name	3-Amino-5-morpholinomethyl-2-oxazolidinone	3-Amino-2-oxazolidinone	3,5-Dinitrosalicylic acid hydrazide	1-Aminohydantoin	Semicarbazide
Acronym	AMOZ	AOZ	DNSAH	AHD	SEM
MW	201.111	102.043	242.029	115.038	75.043
Formula	$C_8H_{15}N_3O_3$	$C_3H_6N_2O_2$	$C_7H_6N_4O_6$	$C_3H_5N_3O_2$	CH_5N_3O
CAS	43056-63-9	80-65-9		2827-56-7	563-41-7
Nitrophenyl (NP) derivatives of nitrofuran metabolites					
Acronym	NP-AMOZ	NP-AOZ	NP-DNSAH	NP-AHD	NP-SEM
MW	334.128	235.059	375.045	248.055	208.060
Formula	$C_{15}H_{18}N_4O_5$	$C_{10}H_9N_3O_4$	$C_{14}H_9N_5O_8$	$C_{10}H_8N_4O_4$	$C_8H_8N_4O_3$
MS ionization mode	POS	POS	POS	POS	POS
Precursor ion (m/z)	335	236	376	249	209
Product ions (m/z)	128, 262, 291	78, 101, 104, 134, 149	166, 211	104, 134, 178	134, 166, 192
Reference	(43–50)	(43–50)	(46)	(43–50)	(43–50)

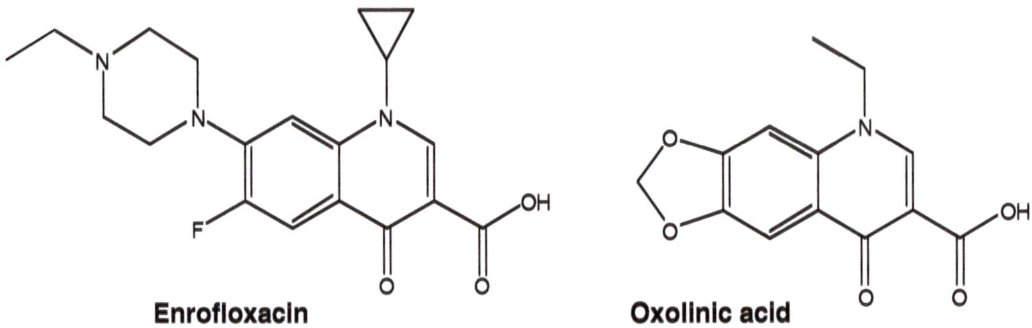

Fig. 6. Examples of quinolone structures.

In the EU, the use of seven quinolones (danofloxacin, difloxacin, enrofloxacin, flumequine, marbofloxacin, oxolinic acid, and sarafloxacin) is approved in food-producing animals (except for animals from which eggs are produced for human consumption) and aquaculture, with the MRLs ranging from 10 μg/kg (for sarafloxacin in chicken skin and fat) to 1,900 μg/kg (for difloxacin in poultry liver). In the USA, only two fluoroquinolones are currently approved: enrofloxacin for swine, dairy cows (under 20 months of age) and beef cattle (excluding calves) and danofloxacin for beef cattle (excluding calves). The US tolerances for enrofloxacin are 100 μg/kg in bovine liver (as desethylene ciprofloxacin) and 500 μg/kg in porcine liver (as enrofloxacin). Danofloxacin's tolerance in bovine liver (the target tissue) and muscle is 200 μg/kg. Table 6 gives basic information on important quinolones, their regulatory limits in bovine tissues, and MS/MS transitions used in their LC–MS/MS analysis.

There have been many methods published for quinolone residue analysis, using different sample preparation and determination strategies (52, 53). Among LC–MS-based multiresidue methods, a notable example includes analysis of 16 quinolones in honey using automated extraction by turbulent flow chromatography coupled to LC–MS/MS (54). Other methods that analyze a larger number of quinolones are described in recently published papers that involve analysis of milk (55), porcine kidney (56, 57), porcine liver (58), eggs (59, 60), chicken muscle (61) or fish (62). Most of these methods employ LC–QqQ-MS/MS for quantitation and identification of target quinolones (54–59, 62). Schneider and Donoghue (60, 61) used fluorescence detection for quantitation of eight fluoroquinolones in chicken muscle and eggs and performed confirmation using IT-MS^3 for most of the analytes (except for desethylene ciprofloxacin, in which case only MS^2 was used). Hermo et al. (58) compared performance of LC–QqQ-MS/MS with LC–TOFMS for the analysis of nine quinolones in porcine liver and demonstrated suitability of TOFMS for residue quantitation and high selectivity for confirmation using accurate mass measurements.

Table 6
Quinolones: compound information, regulatory limits in bovine tissues, and typical MS/MS conditions

Compound information				Regulatory limits (μg/kg) in bovine tissues[a]		MS/MS conditions			
Name	MW	Formula	CAS No.	EU	USA	Mode	Precursor ion (m/z)	Product ions (m/z)	Reference
Cinoxacin	262.060	$C_{12}H_{10}N_2O_5$	28657-80-9			POS	263	189, 217, 245	(54, 56, 57)
Ciprofloxacin[b]	331.133	$C_{17}H_{18}FN_3O_3$	85721-33-1	100 (M/F) 300 (L) 200 (K) 100 (milk)		POS	332	245, 288, 314	(25, 54–59, 62)
Danofloxacin	357.149	$C_{19}H_{20}FN_3O_3$	112398-08-0	200 (M) 100 (F) 400 (L/K) 30 (milk)	200 (M/L)	POS	358	96, 283, 314, 340	(25, 54–59, 62)
Desethylene ciprofloxacin[b]	305.118	$C_{15}H_{16}FN_3O_3$	103222-12-4		100 (L)	POS	306	268, 286	(25, 61)
Difloxacin	399.139	$C_{21}H_{19}F_2N_3O_3$	98106-17-3	400 (M) 100 (F) 1,400 (L) 800 (K)		POS	400	299, 356, 382	(25, 54, 58, 59)
Enoxacin	320.128	$C_{15}H_{17}FN_4O_3$	74011-58-8			POS	321	206, 257, 277, 303	(54, 56, 57)
Enrofloxacin[b,c]	359.165	$C_{19}H_{22}FN_3O_3$	93106-60-6	100 (M/F) 300 (L) 200 (K) 100 (milk)	100 (L)	POS	360	245, 316, 342	(25, 54–59, 62)
Fleroxacin	369.130	$C_{17}H_{18}F_3N_3O_3$	79660-72-3			POS	370	269, 326, 352	(54)

(continued)

Table 6 (continued)

Compound information				Regulatory limits (μg/kg) in bovine tissues[a]		MS/MS conditions			
Name	MW	Formula	CAS No.	EU	USA	Mode	Precursor ion (m/z)	Product ions (m/z)	Reference
Flumequine	261.080	$C_{14}H_{12}FNO_3$	42835-25-6	200 (M) 300 (F) 500 (L) 1,500 (K) 50 (milk)		POS	262	202, 220, 244	(54–59, 62)
Lomefloxacin	351.139	$C_{17}H_{19}F_2N_3O_3$	98079-51-7			POS	352	237, 251, 265, 308, 334	(25, 54, 57, 59)
Marbofloxacin	362.139	$C_{17}H_{19}FN_4O_4$	115550-35-1	150 (M/L/K) 50 (F) 75 (milk)		POS	363	72, 122, 277, 320, 345	(54–59)
Nalidixic acid	232.085	$C_{12}H_{12}N_2O_3$	389-08-2			POS	233	159, 187, 215	(54, 56, 57, 62)
Norfloxacin	319.133	$C_{16}H_{18}FN_3O_3$	70458-96-7			POS	320	233, 276, 302	(25, 54, 56–59)
Ofloxacin	361.144	$C_{18}H_{20}FN_3O_4$	82419-36-1			POS	362	261, 318, 344	(25, 54, 56, 57)
Orbifloxacin	395.146	$C_{19}H_{20}F_3N_3O_3$	113617-63-3			POS	396	295, 352	(25, 61, 62)
Oxolinic acid	261.064	$C_{13}H_{11}NO_5$	14698-29-4	100 (M) 50 (F) 150 (L/K)		POS	262	160, 216, 244	(25, 54, 56–59, 62)
Pipemidic acid	303.133	$C_{14}H_{17}N_5O_3$	51940-44-4			POS	304	189, 217, 286	(54)
Piromidic acid	288.122	$C_{14}H_{16}N_4O_3$	19562-30-2			POS	289	173, 201, 243	(62)
Sarafloxacin	385.124	$C_{20}H_{17}F_2N_3O_3$	98105-99-8			POS	386	299, 342, 368	(25, 54, 58, 62)

[a] Bovine tissues: M=muscle, F=fat, L=liver, K=kidney
[b] In the EU, residue definition for enrofloxacin is sum of enrofloxacin and ciprofloxacin
[c] In the US, desethylene ciprofloxacin is the marker residue for enrofloxacin in bovine liver

3.7. Sulfonamides

Sulfonamides are synthetic antibiotics that are used for prophylactic and therapeutic treatment of bacterial and protozoal infections and also as growth-promoting substances. They share a common chemical nucleus that comes from sulfanilamide (see Fig. 7) and is responsible for the exhibited antimicrobial activity. In practice, several sulfonamides may be combined in one preparation to reduce toxicity and cover a wider activity range. Also, sulfonamides are often administered together with synthetic diaminopyrimidines, such as baquiloprim, ormetoprim or trimethoprim, which act as potentiators of sulfonamides, targeting in synergy bacterial DNA synthesis. Sulfonamides compete with *p*-aminobenzoic acid and block its enzymatic conversion to dihydropholic acid. Diaminopyrimidines inhibit the subsequent conversion of dihydropholic acid to tetrahydropholic acid (1).

Parent sulfonamides are relatively insoluble in water. Their sodium salts have better water solubility, thus are commonly used in drug preparations. Sulfonamides have amphoteric properties given by the weak basic anilinic nitrogen (a protonation site in ESI or APCI) and a weakly acidic sulfonamidic group.

Some sulfonamides are potential carcinogens. They can also cause hypersensitive allergic reactions. Thus, the presence of sulfonamide residues in food is of significant toxicological and regulatory concern. In the EU, the combined total residues of all substances within the sulfonamide group should not exceed 100 μg/kg. This MRL is set for all tissues (muscle, fat, liver, and kidney) and milk of all food-producing species. In the US, the tolerances for approved sulfonamides are set at zero for sulfaethoxypyridazine in milk, sulfamerazine in trout tissues, sulfachloropyrazine, sulfanitran and sulfomyxin in poultry edible tissues, and sulfathiazole and sulfaethoxypyridazine in porcine edible tissues; at 10 μg/kg for sulfabromomethazine and sulfadimethoxine in milk; and at 100 μg/kg for seven sulfonamides in bovine, porcine, poultry and/or fish edible tissues. Extra-label use of sulfonamides in lactating dairy cows (except for the approved use of sulfabromomethazine, sulfadimethoxine and sulfaethoxypyridazine) is strictly prohibited in the US. A "safe level" of 10 μg/kg is used by the US FDA as a non-binding, prosecutional guideline for sulfonamides in milk.

Table 7 gives basic information on important sulfonamides, their regulatory limits in bovine tissues, and MS/MS transitions

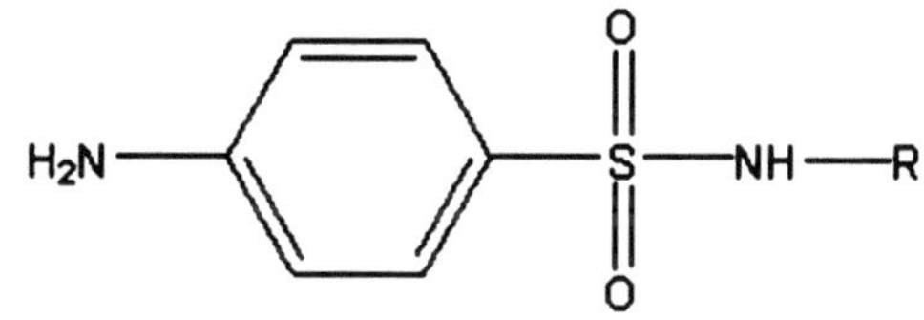

Fig. 7. Basic structure of sulfonamides.

Table 7
Sulfonamides: compound information, regulatory limits in bovine tissues, and typical MS/MS conditions

Compound information				Regulatory limits (μg/kg) in bovine tissues[a]		MS/MS conditions			
Name	MW	Formula	CAS No.	EU[b]	USA	Mode	Precursor ion (m/z)	Product ions (m/z)	Reference
Dapsone	248.062	$C_{12}H_{12}N_2O_2S$	80-08-0	All sulfonamides: 100 (M/F/L/K) 100 (milk)		POS	249	92, 156	(66)
Sulfabenzamide	276.057	$C_{13}H_{12}N_2O_3S$	127-71-9			POS	277	92, 108, 156	(63, 66)
Sulfabromomethazine	355.994	$C_{12}H_{13}BrN_4O_2S$	116-45-0		100 (M/F/L/K) 10 (milk)	POS	359	108, 156	(25)
Sulfacetamide	214.041	$C_8H_{10}N_2O_3S$	144-80-9			POS	215	108, 156	(25, 63)
Sulfachloropyrazine	284.013	$C_{10}H_9ClN_4O_2S$	1672-91-9						
Sulfachloropyridazine	284.013	$C_{10}H_9ClN_4O_2S$	80-32-0		100 (M/F/L/K)	POS	285	92, 108, 156, 207	(25, 63–67)
Sulfadiazine	250.052	$C_{10}H_{10}N_4O_2S$	68-35-9			POS	251	92, 108, 156	(25, 63–67)
Sulfadimethoxine	310.074	$C_{12}H_{14}N_4O_4S$	122-11-2		100 (M/F/L/K) 10 (milk)	POS	311	92, 108, 156	(25, 63–67)
Sulfadoxine	310.074	$C_{12}H_{14}N_4O_4S$	2447-57-6			POS	311	92, 108, 156	(25, 63, 66, 67)
Sulfaethoxypyridazine	294.079	$C_{12}H_{14}N_4O_3S$	963-14-4		100 (M/F/L/K) 0 (milk)	POS	295	108, 156	(25)
Sulfaguanidine	214.052	$C_7H_{10}N_4O_2S$	57-67-0			POS	215	108, 156	(25, 63)

Sulfamerazine	264.068	$C_{11}H_{12}N_4O_2S$	127-79-7		POS	265	92, 108, 110, 156, 172	(25, 63–67)
Sulfameter	280.063	$C_{11}H_{12}N_4O_3S$	651-06-9		POS	281	92, 108, 126, 156, 215	(63, 64, 66)
Sulfamethazine	278.084	$C_{12}H_{14}N_4O_2S$	57-68-1	100 (M/F/L/K)	POS	279	124, 156, 186	(25, 63–67)
Sulfamethizole	270.025	$C_9H_{10}N_4O_2S_2$	144-82-1		POS	271	92, 108, 156	(25, 63–65)
Sulfamethoxazole	253.052	$C_{10}H_{11}N_3O_3S$	723-46-6		POS	254	92, 108, 156	(25, 63–66)
Sulfamethoxypyridazine	280.063	$C_{11}H_{12}N_4O_3S$	80-35-3		POS	281	92, 108, 126, 156	(25, 63–66)
Sulfamonomethoxine	280.063	$C_{11}H_{12}N_4O_3S$			POS	281	92, 156, 215	(63, 65)
Sulfamoxole	267.068	$C_{11}H_{13}N_3O_3S$	729-99-7		POS	268	92, 108, 113, 156	(63–66)
Sulfanilamide	172.031	$C_6H_8N_2O_2S$	63-74-1		POS	173	92, 132, 156	(25, 63, 64)
Sulfanitran	335.058	$C_{14}H_{13}N_3O_5S$	122-16-7		POS	336	156, 198, 294	(63, 64)
Sulfaphenazole	314.084	$C_{15}H_{14}N_4O_2S$	526-08-9		POS	315	156, 159, 160, 222	(25, 63)
Sulfapyridine	249.057	$C_{11}H_{11}N_3O_2S$	144-83-2		POS	250	92, 156, 184	(25, 63–67)
Sulfaquinoxaline	300.068	$C_{14}H_{12}N_4O_2S$	59-40-5	100 (M/F/L/K)	POS	301	92, 108, 156	(25, 63–66)
Sulfasalazine	398.068	$C_{18}H_{14}N_4O_5S$	599-79-1		POS	399	119, 381	(25)
Sulfathiazole	255.014	$C_9H_9N_3O_2S_2$	72-14-0		POS	256	92, 108, 156	(25, 63, 64, 66, 67)
Sulfisomidine	278.084	$C_{12}H_{14}N_4O_2S$	515-64-0		POS	279	124, 156, 186	(63, 64, 66)
Sulfisoxazole	267.068	$C_{11}H_{13}N_3O_3S$	127-69-5		POS	268	108, 113, 156	(25, 63, 64)

[a] Bovine tissues: M=muscle, F=fat, L=liver, K=kidney

[b] In the EU, the combined total residues of all substances within the sulfonamide group should not exceed 100 μg/kg in milk and edible tissues of all food-producing species

used in their LC–MS/MS analysis. Sulfonamides are typically ionized in positive mode (25, 63–68), resulting in protonated pseudomolecular ions as respective precursor ions. Product ions common to the majority of sulfonamides include the *p*-aminobenzene sulphonic acid moiety $[M\text{-}RNH_2]^+$ at *m/z* 156, $(M\text{-}RNH_2\text{-}SO]^+$ at *m/z* 108, $[M\text{-}RNH_2\text{-}SO_2]^+$ at *m/z* 92, and ions formed from the various amino substituents RNH_3^+ after loss of the *p*-aminobenzene sulphonic acid moiety at *m/z* M+H-155. Sulfonamides can be also ionized in negative mode, which was employed by Sheridan et al. (69) in a simultaneous analysis of 14 sulfonamides and chloramphenicol in honey using an LC–ESI–MS/MS instrument incapable of fast positive/negative mode switching.

Most quantitative methods for sulfonamides employ LC–QqQ–MS/MS (63–67, 69). Heller et al. (68) reported on a confirmation method for sulfonamides in eggs using LC–IT–MS/MS. They also used this instrumentation for identification of N_4-acetyl metabolites of sulfonamides, which give $(M+H]^+$ ions 42 Da higher than the parent drugs.

For multiresidue analysis of 24 sulfonamides in meat samples, Cai et al. (63) developed a very simple sample preparation procedure based on acetonitrile extraction, followed by defatting with hexane, addition of water to the remaining acetonitrile layer, liquid–liquid extraction into ethyl acetate, solvent exchange into initial mobile phase (0.2% formic acid in 96:4 water-methanol), and filtration prior UPLC–MS/MS analysis. A little bit more elaborate method (also based on acetonitrile extraction and defatting with hexane but involving an SPE cleanup step) for LC–MS/MS multiresidue analysis of 17 sulfonamides in porcine meat, kidney, and liver was reported by Shao et al. (64). Sergi et al. (65) employed matrix solid phase dispersion (MSPD) technique for analysis of 13 sulfonamides in meat. In their method, homogenized meat samples are dispersed with C18 sorbent, packed into a glass cartridge and eluted with methanol at 0°C. A freeze-out at −18°C is then used to remove co-extracted lipids, followed by extract evaporation and solvent exchange into methanol-water (50:50, *v/v*) prior to the LC–MS/MS analysis.

Several multiresidue methods were published for the LC–MS/MS analysis of sulfonamides in honey (66, 67, 69). It is important to note that sulfonamides bind to sugars, thus it is necessary to perform an acid hydrolysis to release the bound residues. Mohamed et al. (66) evaluated ESI, APCI, and atmospheric pressure photoionization (APPI) in positive mode for analysis of 16 sulfonamides in honey and validated their method using APPI because this ionization technique provided the highest signal-to-noise ratios for all tested sulfonamides and minimal matrix effects in honey extracts. Sheridan et al. (69) used their method for analysis of 14 sulfonamides and chloramphenicol to perform a survey of 116 honey samples from various countries. In addition to chloramphenicol (in nine samples), they found sulfathiazole (19 samples), sulfamethoxazole (six samples), sulfamethaxine (four samples), and sulfamethoxypyridazine (one sample).

3.8. Tetracyclines

Tetracyclines are broad-spectrum antibiotics active against both Gram-positive and Gram-negative bacteria by inhibiting their protein biosynthesis. They are widely used in veterinary medicine for cost-effective prophylactic and therapeutic treatment and also as growth-promoting substances in cattle and poultry (1). Chlortetracycline, oxytetracycline, and demeclocycline are three naturally occurring tetracyclines that were isolated from fungi. Other members of the group, such as tetracycline, doxycycline, or minocycline, were prepared by modification of the basic hydronaphthacene skeleton containing four fused rings (see Fig. 8).

In the EU, the MRLs for chlortetracycline, doxycycline, oxytetracycline, and tetracycline range from 100 μg/kg in milk and muscle to 600 μg/kg in kidney. Doxycycline should not be used in animals, from which milk and eggs are produced for human consumption. In the US, tolerances were established for the sum of residues of the tetracyclines including chlortetracycline, oxytetracycline, and tetracycline, at 300 μg/kg in milk, 2,000 μg/kg in muscle, 6,000 μg/kg in liver and 12,000 μg/kg in fat and kidney of the main food-producing species.

Table 8 gives basic information on important tetracyclines, their regulatory limits in bovine tissues, and MS/MS transitions used in their LC–MS/MS analysis. Tetracyclines produce ions in positive mode. In MS/MS, typical product ions of tetracyclines

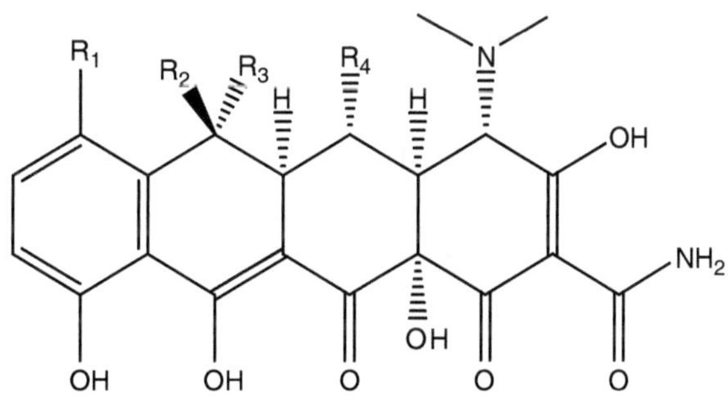

	R_1	**R_2**	**R_3**	**R_4**
Chlortetracycline	Cl	CH_3	OH	H
Demeclocycline	Cl	H	OH	H
Doxycycline	H	H	CH_3	OH
Minocycline	$N(CH_3)_2$	H	H	H
Oxytetracycline	H	CH_3	OH	OH
Tetracycline	H	CH_3	OH	H

Fig. 8. Structures of important tetracyclines.

Table 8
Tetracyclines: compound information, regulatory limits in bovine tissues, and typical MS/MS conditions

Compound information				Regulatory limits (µg/kg) in bovine tissues[a]		MS/MS conditions			
Name	MW	Formula	CAS No.	EU[b]	USA[c]	Mode	Precursor ion (m/z)	Product ions (m/z)	Reference
Chlortetracycline	478.114	$C_{22}H_{23}ClN_2O_8$	57-62-5	100 (M) 300 (L) 600 (K) 100 (milk)	2,000 (M) 6,000 (L) 12,000 (F/K) 300 (milk)	POS	479	444, 461, 462	(70–74)
Demeclocycline	464.099	$C_{21}H_{21}ClN_2O_8$	127-33-3			POS	465	430, 448	(70–74)
Doxycycline	444.153	$C_{22}H_{24}N_2O_8$	564-25-0	100 (M) 300 (L) 600 (K)		POS	445	154, 321, 428	(70–74)
Minocycline	457.185	$C_{23}H_{27}N_3O_7$	10118-90-8			POS	458	352, 441	(70, 73)
Oxytetracycline	460.148	$C_{22}H_{24}N_2O_9$	6153-64-6	100 (M) 300 (L) 600 (K) 100 (milk)	2,000 (M) 6,000 (L) 12,000 (F/K) 300 (milk)	POS	461	426, 443, 444	(70–74)
Tetracycline	444.153	$C_{22}H_{24}N_2O_8$	60-54-8	100 (M) 300 (L) 600 (K) 100 (milk)	2,000 (M) 6,000 (L) 12,000 (F/K) 300 (milk)	POS	445	410, 427, 428	(70–74)

[a] Bovine tissues: M=muscle, F=fat, L=liver, K=kidney
[b] In the EU, the residue definition for chlortetracycline, oxytetracycline, and tetracycline is sum of parent drug and its 4-epimer (for doxycycline, it is only the parent drug)
[c] In the US, tolerances are established for the sum of residues of the tetracyclines, including chlortetracycline, oxytetracycline, and tetracycline

include $[M+H-NH_3]^+$, $[M+H-H_2O]^+$, and $[M+H-H_2O-NH_3]^+$ (70–74). The loss of NH_3 comes from the carboxyamide group and the loss of water occurs for tetracyclines with a hydroxyl group in position C6 (R_2 or R_3 in Fig. 8).

Chemical properties of tetracyclines may complicate their extraction from biological matrices and chromatographic separation (2, 75). The presence of two ketone groups in positions C1 and C11 in their molecules gives tetracyclines the ability to chelate with metal ions and interact with silanol groups. To avoid losses during sample preparation and prevent tailing on silica-based columns, it is recommended to use chelating agents, such as oxalic acid and EDTA salts, during the extraction and as additives in the mobile phase. In LC–ESI–MS, however, these non-volatile agents should be avoided due to significant reduction of signal intensities and source contamination. The use of end-capped, high-purity stationary phases without residual metal ions (76) and/or extensive column pre-washing with a chelating solution of Na_2EDTA (77) can provide acceptable peak shapes even only with the addition of formic acid into the mobile phase.

Another problem in tetracycline chromatographic separation involves rapid isomerization of chlortetracycline and doxycycline to their 4-epimers in aqueous solutions at pH 2–6. In addition, keto tautomers may be also formed in aqueous conditions, with both isomerization products eluting before peaks of the original compounds. This results in broad peak fronting and complicates quantitation. It should be noted that, in the EU, 4-epimers are included in the residue definitions for chlortetracycline, oxytetracycline and tetracycline, but not for doxycycline. Bruno et al. (77) minimized partial conversion of chlortetracycline and doxycycline to their isomers during the LC separation by keeping the column temperature at 15°C.

Most sample preparation methods for tetracyclines use EDTA–McIlvaine buffer (citric acid with disodium hydrogen phosphate) or sodium succinate buffer (both at pH 4) for extraction of tetracyclines and their 4-epimers from tissues, eggs and milk, typically followed by an SPE step prior to the LC–MS/MS analysis (70–72). Recently, the use of heated water as an extraction medium in PLE or MSPD has been reported for isolation of tetracyclines from muscle samples dispersed with EDTA-treated sand (73, 74).

4. Multiclass, Multiresidue LC–MS Analysis of Antibiotics

As mentioned in the introduction, the use of multiclass, multiresidue methods is the most effective approach to the analysis of chemical residues in food. In the case of antibiotics, their diverse chemical properties pose a true challenge for development of a fully quantitative, wide-scope multiclass method. Table 9 reviews

Table 9
Examples of multiclass, multiresidue methods for antibiotic residues in animal-derived food

Analytes	Matrix	Sample preparation	LC column and mobile phase conditions	MS detection	Reference
29 antibiotics: β-lactams (5) Fluoroquinolones (4) Sulfonamides (15) Tetracyclines (5)	Eggs	Extraction with sodium succinate buffer (pH 3.5) Cleanup with HLB SPE Elution with MeOH Solvent exchange to water	Phenyl column (50×4.0 mm, 3 μm) A: 0.1% FA in water B: MeCN Gradient: 3–85% B Flow=0.5 mL/min Time=23 or 29 min	ESI-IT-MS/MS	(78)
22 antibiotics: β-lactams (4) Macrolides (2) Lincosamides (1) Quinolones (2) Sulfonamides (8) Tetracyclines (4) Others (1)	Milk	Extraction with acetonitrile Centrifugation Dilution (1:9, *v/v*) with 0.1% FA in water HLB SPE clean-up Elution with MeCN Solvent exchange to 0.1% FA in water	C18 column (100×2.0, 3 μm) A: 0.1% FA in water B: MeCN Gradient: 5–100% B Flow=0.25 mL/min T=35°C Time=21 min	ESI-QqQ-MS/MS	(79)
42 antibiotics: Aminoglycosides (3) Amphenicols (2) β-lactams (8) Macrolides (7) Sulfonamides (17) Tetracyclines (5)	Honey	Four subsequent LLE steps of honey sample dissolved in 0.1 M disodium hydrogen phosphate buffer (pH 8): (1) Extraction with MeCN, followed by centrifugation and evaporation of the supernatant to dryness (2) Extraction with 10% TCA and MeCN, followed by centrifugation, neutralization of the supernatant with ammonium hydroxide and evaporation to dryness	C18 column (50×2.1, 1.8 μm) A: 0.5% FA and 1 mM NFPA in water B: 0.5% FA in MeCN–MeOH (50:50, *v/v*) Gradient: 0–99% B Flow=0.3 mL/min Time=25 min	ESI-QqQ-MS/MS (4 stacked injections=1 run)	(80)

		(3) Extraction with 4% NFPA in MeCN, followed by centrifugation, neutralization of the supernatant with ammonium hydroxide and evaporation to dryness (4) Hydrolysis (65 °C, 1 h) of sugar-bound sulfonamides, followed by sample neutralization, extraction with MeCN, centrifugation and evaporation of the supernatant to dryness- all four dry extracts individually re-suspended in MeOH–water (20:80, *v/v*)			
19 antibiotics: β-lactams (3) Macrolides (4) Quinolones (4) Sulfonamides (4) Tetracyclines (4)	Muscle and kidney from various species	Addition of 0.1 M EDTA, followed by extraction with MeOH–water (70:30, *v/v*) Centrifugation Dilution (1:4, *v/v*) with water	C18 column (50 × 2.1, 4 μm) A: 0.2% FA and 0.1 mM oxalic acid in water B: MeCN Gradient: 0–75% B Flow = 0.3 mL/min Time = 7 min	ESI–QqQ–MS/MS	(81)
31 antibiotics: β-lactams (3) Macrolides (4) Lincosamides (1) Quinolones (8) Sulfonamides (10) Tetracyclines (3) Others (2)	Bovine and porcine muscle	PLE with water of samples dispersed with EDTA-treated sand	C18 column (100 × 2.1, 3.5 μm) A: 0.1% FA in water B: 0.1% FA in MeOH Gradient: 10–90% B Flow = 0.2 mL/min Time = 32 min	ESI–QqQ–MS/MS	(82)

(continued)

Table 9 (continued)

Analytes	Matrix	Sample preparation	LC column and mobile phase conditions	MS detection	Reference
39 antibiotics: β-lactams (7) Macrolides (4) Quinolones (9) Sulfonamides (14) Tetracyclines (4) Others (1)	Chicken muscle	Extraction with MeOH–water (70:30, *v/v*) with addition of EDTA Centrifugation Dilution (1:3, *v/v*) with water	UPLC C18 column (100×2.1, 1.7 μm) A: 0.2% FA and 1 mM oxalic acid in water B: 0.1% FA in MeCN Gradient: 5–90% B Flow=0.3 mL/min T=40°C Time=13 min	ESI–QqQ–MS/MS	(83)
41 veterinary drugs, including 28 antibiotics Quinolones (12) Sulfonamides (16)	Chicken muscle	Extraction with 1% acetic acid in MeCN with addition of anh. sodium sulphate Centrifugation Dispersive SPE with NH_2 sorbent	C18 column (100×2.0, 2.5 μm) A: 0.1% FA in water B: 0.1% FA in MeCN Gradient: 0–75% B Flow=0.25 mL/min Time=28 min	ESI–QqQ–MS/MS	(84)
56 antibiotics: Amphenicols (1) β-lactams (12) Macrolides (3) Lincosamides (2) Quinolones (11) Sulfonamides (20) Tetracyclines (5) Others (2)	Bovine kidney juice and serum	Extraction with MeCN–water (4:1, *v/v*) Centrifugation Dispersive C18 SPE clean-up Solvent exchange to water	C18 column (150×3.0, 5 μm) A: 0.1% FA in water B: 0.1% FA in MeCN Gradient: 2–100% B Flow=0.3 mL/min Time=25 min	ESI–QqQ–MS/MS	(25)

130 veterinary drugs, including 46 antibiotics Amphenicols (3) Macrolides (9) Lincosamides (2) Quinolones (10) Sulfonamides (18) Others (4)	Bovine, porcine, and chicken muscle	Extraction with MeCN–MeOH (95:5, *v/v*) with addition of sodium sulphate Centrifugation Defatting with hexane Solvent exchange to MeOH	C18 column (100 × 2.1, 3.0 μm) A: 0.3% acetic acid in 10 mM ammonium acetate B: MeCN–MeOH (1:4, *v/v*) Gradient: 20–95% B Flow = 0.22 mL/min Time = 25 min	ESI–QqQ–MS/MS (2 injections)	(85)
100 veterinary drugs, including 67 antibiotics β-lactams (12) Macrolides (8) Lincosamides (3) Quinolones (13) Sulfonamides (23) Tetracyclines (6) Others (2)	Bovine muscle, liver and kidney	Extraction with MeCN Followed by addition of ammonium sulphate and extraction with EDTA in succinate buffer (pH 5) Centrifugation Evaporation of MeCN pH of the remaining aqueous solution adjusted to 6.5 Evaporation vessel rinsed with succinate buffer-dimethylsulfoxide (1:1, *v/v*) HLB SPE clean-up of the vessel rinse, followed by water and the extraction solution (pH 6.5) Elution with MeCN and succinate buffer Solvent exchange into water (mostly)	UPLC C18 (100 × 2.1, 1.8 μm) A: 0.3% FA in water–MeCN (95:5, *v/v*) B: 0.3% FA in MeCN–water (95:5, *v/v*) Gradient: 0–100% B Flow = 0.4 mL/min Time = 15 min	ESI–TOFMS	(86)

(continued)

Table 9 (continued)

Analytes	Matrix	Sample preparation	LC column and mobile phase conditions	MS detection	Reference
100 veterinary drugs, including 51 antibiotics: Amphenicols (2) β-lactams (7) Macrolides (8) Lincosamides (2) Quinolones (11) Sulfonamides (16) Tetracyclines (4) Others (1)	Bovine, porcine, and chicken muscle Fish Eggs	Extraction with MeCN–water (3:2, *v/v*) Centrifugation Dilution of supernatant with water StrataX SPE clean-up Fish and meat: elution with MeOH:MeCN (1:1, *v/v*) Eggs: elution with MeOH:ethyl acetate (1:1, *v/v*) Solvent exchange to 0.1% FA in water–MeCN (9:1, *v/v*)	UPLC C18 (100 × 2.1, 1.7 μm) A: 0.1% FA in water B: 0.1% FA in MeCN–water (9:1, v/v) Gradient: 0–100% B Flow = 0.4 mL/min Time = 12 min	ESI–TOFMS screening	(87)
150 veterinary drugs, including 80 antibiotics: β-lactams (25) Macrolides (10) Quinolones (14) Sulfonamides (25) Tetracyclines (6)	Milk	Protein precipitation with MeCN Ultracentrifugation (cut-off at 3 kDa) Evaporation of MeCN	UPLC C18 column (100 × 2.1, 1.7 μm) A: 0.1% FA in water B: 0.1% FA in MeCN Gradient: 5–95% B Flow = 0.4 mL/min Time = 9 min	ESI–TOFMS screening	(88)

recently published multiclass methods (25, 78–88) that include different number of analytes from various classes of antibiotics (and other veterinary drugs in some cases). It should be noted that acceptable performance characteristics for quantitative methods, such as analyte recoveries, repeatability/reproducibility and/or desirable quantitation limits, were usually not achieved for all of the analytes that were on the target list (numbers listed in the table) but the methods could still serve for screening purposes in most of those cases. Notable examples of wide-scope screening methods include those for 100–150 different veterinary drugs (46–80 antibiotics) that either employs two LC–QqQ–MS/MS runs (85) or a single run using UPLC–TOFMS (86–88).

Certain antibiotics are very difficult to include in a single screening approach. In the case of nitrofurans, their tissue-bound residues (nitrofuran metabolites) have to be first released by strong acid hydrolysis, followed by a derivatization and cleanup steps prior to the LC–MS analysis. This procedure would destroy labile analytes, such as β-lactams.

Aminoglycosides represent another example of a group that is not compatible with the other antibiotics. Due to their high polarity, aminoglycosides are not retained in RPLC. They can be retained by HILIC columns, which however do not provide retention for most of the other antibiotic classes. Ion-pairing agents are needed for separation of aminoglycosides in RPLC (9–13), but these additives significantly suppress signal of other antibiotics in a multiclass LC–ESI–MS approach (76, 80). Mastovska and Lightfield (76) tried to develop a LC–MS methodology, in which aminoglycosides could be analyzed together with the other antibiotic classes using a single LC column and mobile phase system consisting of (A) 0.1% formic acid in water and (B) 0.1% formic acid in acetonitrile. They evaluated a hydride-based silica bonded C_{18} column, which offers unique retention characteristics and can be employed in RP, normal-phase, or aqueous normal-phase (ANP) mode, depending on the mobile phase composition (89). As opposed to ordinary silica, hydride-based silica surface is predominantly populated by nonpolar silicon–hydride (Si–H) groups instead of the polar silanol groups (Si–OH), which reduces adsorption of water on the surface and provides other unique features. Bonded chemical moieties (such as the hydrophobic C_{18}) are attached to the hydride silica support surface by stable Si–C. For multiclass antibiotic analysis, alternate ANP- and RP-based runs can be employed for separation of aminoglycosides and other antibiotics, respectively (76). The alternating RP and ANP runs save time required for column equilibration because the gradient of the RP method basically starts at final conditions of the ANP method and vice versa. Unfortunately, the evaluated hydride-based silica bonded C_{18} column did not provide good peak shapes for tetracyclines, which are known to chelate with metal ions (extensive washing of the stationary phase with an EDTA solution had only a very little effect on tetracycline peak shapes). Therefore, the authors opted for a dual-column approach, employing a conventional C_{18} column for the RPLC runs.

5. Future Trends

Current and future trends in multiresidue antibiotic analysis are closely connected to the advances in LC–MS instrumentation. Modern QqQ–MS/MS instruments can accommodate a large number of analytes (MS/MS transitions) due to their speed (short dwell times and interscan delays) and other features, such as fast positive/negative mode switching that enables simultaneous analysis of positively and negatively charged ions or time-scheduled multiple reaction monitoring that simplifies the MS/MS method development for a large number of analytes. The introduction of affordable accurate mass HR-TOFMS bench-top instruments opened a door for fast, non-targeted screening of potentially unlimited number of compounds. Another accurate mass HRMS technology, orbitrap MS, has just started getting the attention among the residue community (90).

These advances on the MS side have been complemented with developments in the LC instrumentation and column technology. For instance, the commercial introduction of sub-2 μm particle LC columns and compatible LC instruments (with high pressure limits and low dead volumes) brought fast LC separations from academic environment to routine laboratories, resulting in higher sample throughput and reduced solvent consumption.

As the modern LC–MS instruments become more and more sensitive, selective and rugged, they can tolerate dirtier samples. This enables minimum sample preparation prior to the LC-MS run, thus introduction of very generic methods that can determine not only multiple veterinary drug residues, but also look for other analytes of interest relevant to the given matrix, such as pesticides, mycotoxins, or plant toxins (91).

References

1. Botsoglou, N.A. and Fletouris, D.J. (2001) *Drug Residues in Foods: Pharmacology, Food Safety, and Analysis*, Marcel Dekker, Inc., New York, NY, USA.
2. Di Corcia, A. and Nazzari, M. (2002) Liquid chromatographic-mass spectrometric methods for analyzing antibiotic and antibacterial agents in animal food products. *J. Chromatogr. A* **974**, 53–89.
3. Balizs, G. and Hewitt A. (2003) Determination of veterinary drug residues by liquid chromatography and tandem mass spectrometry. *Anal. Chim. Acta* **492**, 105–131.
4. Gentili, A., Perret, D., Marchese, S. (2005) Liquid chromatography-tandem mass spectrometry for performing confirmatory analysis of veterinary drugs in animal-food products. *Trends Anal. Chem.* **24**, 704–733.

5. Stolker, A.A.M. and Brinkman, U.A.Th. (2005) Analytical strategies for residue analysis of veterinary drugs and growth-promoting agents in food-producing animals - a review. *J. Chromatogr. A* **1067**, 15–33.

6. Blasco, C., Torres, C.M., Pico, Y. (2007) Progress in analysis of residual antibacterials in food. *Trends Anal. Chem.* **26**, 895–913.

7. Stolker, A.A.M., Zuidema, T., Nielen, M.W.F. (2007) Residue analysis of veterinary drugs and growthpromoting agents. *Trends Anal. Chem.* **26**, 967–979

8. Mastovska, K. (2008) Recent developments in chromatographic techniques, in *Comprehensive Analytical Chemistry*, Volume 51: Food Contaminants and Residue Analysis (Pico, Y., ed.), Elsevier, Oxford, UK, pp. 175–200.

9. SOP No. CLG-AMG1.02 (2005) US Department of Agriculture, Food Safety and Inspection Service, Office of Public Health Science (www.fsis.usda.gov/PDF/CLG_AMG_1_02.pdf)

10. Zhu, W.-X., Yang, J.-Z., Wei, W., Liu, Y.-F., Zhang, S.-S. (2008) Simultaneous determination of 13 aminoglycoside residues in foods of animal origin by liquid chromatography_electrospray ionization tandem mass spectrometry with two consecutive solid-phase extraction steps. *J. Chromatogr. A* **1207**, 29–37

11. Kaufmann, A. and Maden, K. (2005) Determination of 11 aminoglycosides in meat and liver by liquid chromatography with tandem mass spectrometry. *J. AOAC Int.* **88**, 1118–1125.

12. van Holthoon, F.L., Essers, M.L., Mulder, P.J., Stead, S.L., Caldow, M., Ashwin, H.M., Sharman, M. (2009) A generic method for the quantitative analysis of aminoglycosides (and spectinomycin) in animal tissue using methylated internal standards and liquid chromatography tandem mass spectrometry. *Anal. Chim. Acta* **637**, 135–143.

13. Babin, Y. and Fortier, S. (2007) A high-throughput analytical method for determination of aminoglycosides in veal tissues by liquid chromatography/tandem mass spectrometry with automated cleanup. *J AOAC Int.* **90**, 1418–1426

14. Oertel, R., Neumeister, V., Kirch, W. (2004) Hydrophilic interaction chromatography combined with tandem-mass spectrometry to determine six aminoglycosides in serum. *J. Chromatogr. A* **1058**, 197–201.

15. Turnipseed, S.B., Clark, S.B., Karbiwnyk, C.M., Andersen, W.C., Miller, K.E., Madson, M.R. (2009) Analysis of aminoglycoside residues in bovine milk by liquid chromatography electrospray ion trap mass spectrometry after derivatization with phenyl isocyanate. *J. Chromatogr. B* **877**, 1487–1493.

16. Commission Decision 2003/181/EC of 13 March 2003 amending Decision 2002/657/EC as regards setting of minimum required performance limits (MRPLs) for certain residues in food of animal origin, Brussels, Belgium, 2003. *Official Journal of the European Union* **L71**, 17–18.

17. Ashwin, H.M., Stead, S.L., Taylor, J.C., Startin, J.R., Richmond, S.F., Homer, V., Bigwood, T., Sharman M. (2005) Development and validation of screening and confirmatory methods for the detection of chloramphenicol and chloramphenicol glucuronide using SPR biosensor and liquid chromatography–tandem mass spectrometry. *Anal. Chim. Acta* **529**, 103–108.

18. Vinci, F., Guadagnuolo, G., Danese, V., Salini, M., Serpe, L., Gallo, P. (2005) In-house validation of a liquid chromatography/electrospray tandem mass spectrometry method for confirmation of chloramphenicol residues in muscle according to Decision 2002/657/EC. *Rapid Commun. Mass. Spectrom.* **19**, 3349–3355.

19. Vivekanandan, K., Swamy, M.G., Prasad, S., Mukherjee, R. (2005) A simple method of isolation of chloramphenicol in honey and its estimation by liquid chromatographycoupled to electrospray ionization tandem mass spectrometry. *Rapid Commun. Mass Spectrom.* 19, 3025–3030.

20. Ronning, H.T., Einarsen, K., Asp, T.N. (2006) Determination of chloramphenicol residues in meat, seafood, egg, honey, milk, plasma and urine with liquid chromatography–tandem mass spectrometry, and the validation of the method based on 2002/657/EC. *J. Chromatogr. A* **1118**, 226–233.

21. Nicolich, R.S., Werneck-Barroso, E., Sipoli Marques, S.A. (2006). Food safety evaluation: detection and confirmation of chloramphenicol in milk by high performance liquid chromatography-tandem mass spectrometry. *Anal. Chim. Acta* **565**, 97–102.

22. Boyd, B., Bjork, H., Billing, J., Shimelis, O., Axelsson, S., Leonora, M., Yilmaz, E. (2007) Development of an improved method for trace analysis of chloramphenicol using molecularly imprinted polymers. *J. Chromatogr. A* **1174**, 63–71.

23. Rodziewicz, L., Zawadzka, I. (2008) Rapid determination of chloramphenicol residues in milk powder by liquidchromatography-elektrospray ionization tandem mass spectrometry. *Talanta* **75**, 846–850.

24. Zhang, S., Liu, Z., Guo, X., Cheng, L., Wang, Z., Shen, Z. (2008) Simultaneous determination and confirmation of chloramphenicol, thiamphenicol, florfenicol and florfenicol amine in chicken muscle by liquid chromatography-tandem mass spectrometry. *J. Chromatogr. B* **875**, 399–404.
25. Schneider, M.J., Mastovska, K., Lehotay, S.J., Lightfield, A.R., Kinsella, B., Shultz, C. (2009) Comparison of screening methods for antibiotics in beef kidney juice and serum. *Anal. Chim. Acta* **637**, 290–297.
26. Mastovska, K. and Lightfield, A.R. (2008) Streamlining methodology for the multiresidue analysis of beta-lactam antibiotics in bovine kidney using liquid chromatography-tandem mass spectrometry. *J. Chromatogr. A* **1202**, 118–123.
27. Fagerquist, C.K, Lightfield, A.R., Lehotay, S.J. (2005) Confirmatory and quantitative analysis of β-lactam antibiotics in bovine kidney tissue by dispersive solid-phase extraction and liquid chromatography-tandem mass spectrometry. *Anal. Chem.* 77, 1473–1482.
28. Mastovska, K. and Lightfield, A.R. (2007) Use of the echo peak technique to compensate for matrix effects in the LC-MS/MS analysis of β-lactam antibiotics, in *Proceedings of 55th ASMS Conference on Mass Spectrometry*, Indianapolis, IN, USA, June 3–7, 2007.
29. Becker M, Zittlau E, Petz M (2004) Residue analysis of 15 penicillins and cephalosporins in bovine muscle, kidney and milk by liquid chromatography–tandem mass spectrometry. *Anal. Chim. Acta* **520**, 19–32.
30. Riediker, S. and Stadler, R.H. (2001) Simultaneous determination of five β-lactam antibiotics in bovine milk using liquid chromatography coupled with electrospray ionization tandem mass spectrometry. *Anal. Chem.* **73**, 1614–1621.
31. Ito, Y., Ikai, Y., Oka, H., Matsumoto, H., Miyazaki, Y., Takeba, K., Nagase, H. (2001) Application of ion-exchange cartridge clean-up in food analysis IV. Confirmatory assay of benzylpenicillin, phenoxymethylpenicillin, oxacillin, cloxacillin, nafcillin and dicloxacillin in bovine tissues by liquid chromatography–electrospray ionization tandem mass spectrometry. *J. Chromatogr. A* **911**, 217–223.
32. Fagerquist, C.K and Lightfield, A.R. (2003) Confirmatory analysis of β-lactam antibiotics in kidney tissue by liquid chromatography/electrospray ionization selective reaction monitoring ion trap tandem mass spectrometry. *Rapid Commun. Mass Spectrom.* **17**, 660–671.
33. Holstege, D.M., Puschner, B., Whitehead, G., Galey, F.D. (2002) Screening and mass spectral confirmation of β-lactam antibiotic residues in milk using LC-MS/MS. *J. Agric. Food Chem.* **50**, 406–411.
34. Straub, R.F. and Voyksner, R.D. (1993) Determination of penicillin-g, ampicillin, amoxicillin, cloxacillin and cephapirin by high-performance liquid-chromatography electrospray massspectrometry. *J. Chromatogr.* **647**, 167–181.
35. Draisci, R., Palleschi, L., Ferretti, E., Achene, L., Cecilia, A. (2001) Confirmatory method for macrolide residues in bovine tissues by micro-liquid chromatography–tandem mass spectrometry. *J. Chromatogr. A* **926**, 97–104.
36. Dubois, M., Fluchard, D., Sior, E., Delahaut, P. (2001) Identification and quantification of five macrolide antibiotics in several tissues, eggs and milk by liquid chromatography–electrospray tandem mass spectrometry. *J. Chromatogr. B* **735**, 189–202.
37. Wang, J. and Leung D. (2007) Analyses of macrolide antibiotic residues in eggs, raw milk, and honey using both ultra-performance liquid chromatography/quadrupole time-of-flight mass spectrometry and high-performance liquid chromatography/tandem mass spectrometry. *Rapid Commun. Mass. Spectrom.* **21**, 3213–3222.
38. Bogialli, S., Di Corcia, A., Lagana, A., Mastrantoni, V., Sergi, M. (2007) A simple and rapid confirmatory assay for analyzing antibiotic residues of the macrolide class and lincomycin in bovine milk and yoghurt: hot water extraction followed by liquid chromatography/tandem mass spectrometry. *Rapid Commun. Mass Spectrom.* **21**, 237–246.
39. Martos, P.A, Lehotay S.J., Shurmer, B. (2008) Ultratrace analysis of nine macrolides, including tulathromycin a (draxxin), in edible animal tissues with minicolumn liquid chromatography tandem mass spectrometry. *J. Agric. Food Chem.* **56**, 8844–8850.
40. Berrada, H., Borrull, F., Font, G., Moltó, J.C., Marcé, R.M. (2007) Validation of a confirmatory method for the determination of macrolides in liver and kidney animal tissues in accordance with the European Union regulation 2002/657/EC. *J. Chromatogr. A* **1157**, 281–288.
41. Berrada, H., Borrull, F., Font, G., Marcé, R.M. (2008) Determination of macrolide antibiotics in meat and fish using pressurized liquid extraction and liquid chromatography–mass spectrometry *J. Chromatogr. A* **1208**, 83–89.

42. Cooper, K.M., McCracken, R.J., Kennedy, D.G. (2005) Nitrofurazone accumulates in avian eyes - a replacement for semicarbazide as a marker of abuse. *Analyst* **130**, 824–827.
43. Leitner, A., Zollner, P., Lindner, W.J. (2001) Determination of the metabolites of nitrofuran antibiotics in animal tissue by high-performance liquid chromatography–tandem mass spectrometry. *J. Chromatogr A* **939**, 49–58.
44. Mottier, P., Khong, S.-P., Gremaud, E., Richoz, J., Delatour, T., Goldmann, T., Guy, P.A. (2005) Quantitative determination of four nitrofuran metabolites in meat by isotope dilution liquid chromatography–electrospray ionisation–tandem mass spectrometry. *J. Chromatogr A* **1067**, 85–91.
45. Finzi, J.K., Donato, C.L., Sucupira, M., De Nucci, G. (2005) Determination of nitrofuran metabolites in poultry muscle and eggs by liquid chromatography-tandem mass spectrometry. *J. Chromatogr B* **824**, 30–35.
46. Verdon, E., Couedor, P., Sanders, P. (2007) Multi-residue monitoring for the simultaneous determination of five nitrofurans (furazolidone, furaltadone, nitrofurazone, nitrofurantoine, nifursol) in poultry muscle tissue through the detection of their five major metabolites (AOZ, AMOZ, SEM, AHD, DNSAH) by liquid chromatography coupled to electrospray tandem mass spectrometry - In-house validation in line with Commission Decision 657/2002/EC. *Anal. Chim. Acta* **586**, 336–347.
47. Xia, X., Li, X., Zhang, S., Ding, S., Jiang, H., Li, J., Shen, J. (2008) Simultaneous determination of 5-nitroimidazoles and nitrofurans in pork by high-performance liquid chromatography–tandem mass spectrometry. *J. Chromatogr A* **1208**, 101–108.
48. Bock, C., Stachel, C., Gowik, P. (2007) Validation of a confirmatory method for the determination of residues of four nitrofurans in egg by liquid chromatography–tandem mass spectrometry with the software InterVal. *Anal. Chim. Acta* **586**, 348–358.
49. Rodziewicz, L. (2008) Determination of nitrofuran metabolites in milk by liquid chromatography-electrospray ionization tandem mass spectrometry. *J. Chromatogr B* **864**, 156–160.
50. Chu, P.-S., and Lopez, M.I. (2005) Liquid chromatography-tandem mass spectrometry for the determination of protein-bound residues in shrimp dosed with nitrofurans. *J. Agric. Food Chem.* **53**, 8934–8939.
51. S. Phongvivat (2004) Nitrofurans Case Study: Thailand's Experience, in report from Joint FAO/WHO Technical Workshop on Residues of Substances without ADI/MRL in Food, Bangkok, Thailand, 125–149 (ftp://ftp.fao.org/docrep/fao/008/y5723e/y5723e00.pdf).
52. Hernandez-Arteseros, J.A., Barbosa Compano, J.R., Prat, M.D. (2002) Analysis of quinolone residues in edible animal products. *J. Chromatogr. A* **945**, 1–24.
53. Andrei, V., Blasco, C., Pico, Y. (2007) Analytical strategies to determine quinolone residues in food and the environment, *Trends Anal. Chem.* **26**, 534–556.
54. Mottier, P., Hammel, Y.-A., Gremaud, E., Guy, P.A. (2008) Quantitative high-throughput analysis of 16 (fluoro)quinolones in honey using automated extraction by turbulent flow chromatography coupled to liquid chromatograph-tandem mass spectrometry. *J. Agric. Food Chem.* **56**, 35–43.
55. Hermo, M.P., Nemutlu, E., Kir, S., Barron, D., Barbosa, J. (2008) Improved determination of quinolones in milk at their MRL levels using LC–UV, LC–FD, LC–MS and LC–MS/MS and validation in line with regulation 2002/657/EC. *Anal. Chim. Acta* **613**, 98–107.
56. van Vyncht, G., Janosi, A., Bordin, G., Toussaint, B., Maghuin-Rogister, G., De Pauw, E., Rodriguez, A.R. (2002) Multiresidue determination of (fluoro)quinolone antibiotics in swine kidney using liquid chromatography–tandem mass spectrometry. *J. Chromatogr. A* **952**, 121–129.
57. Toussaint, B., Chedin, M., Bordin, G., Rodriguez, A.R. (2005) Determination of (fluoro)quinolone antibiotic residues in pig kidney using liquid chromatography–tandem mass spectrometry: I. Laboratory-validated method. *J. Chromatogr. A* **1088**, 32–39.
58. Hermo, M.P., Barron, D., Barbosa, J. (2008) Determination of multiresidue quinolones regulated by the European Union in pig liver samples. High-resolution time-of-flight mass spectrometry versus tandem mass spectrometry detection. *J. Chromatogr. A* **1201**, 1–14.
59. Bogialli, S., D'Ascenzo, G., Di Corcia, A., Lagana, A., Tramontana, G. (2009) Simple assay for monitoring seven quinolone antibacterials in eggs: Extraction with hot water and liquid chromatography coupled to tandem mass spectrometry Laboratory validation in line with the European Union Commission Decision 657/2002/EC. *J. Chromatogr. A* **1216**, 794–780.
60. Schneider, M.J. and Donoghue, D.J. (2003) Multiresidue determination of fluoroquinolone antibiotics in eggs using liquid chromatography–fluorescence–mass spectrometry". *Anal. Chim. Acta* **483**, 39–49.

61. Schneider, M.J. and Donoghue, D.J. (2002) Multiresidue analysis of fluoroquinolone antibiotics in chicken tissue using liquid chromatography-fluorescence-multiple mass spectrometry. *J. Chromatogr. B* **780**, 83–92.
62. Johnston, L., Mackay, L., Croft, M. (2002) Determination of quinolones and fluoroquinolones in fish tissue and seafood by high-performance liquid chromatography with electrospray ionisation tandem mass spectrometric detection. *J. Chromatogr. A* **982**, 97–109.
63. Cai, Z., Zhang, Y., Pan, H., Tie, X., Ren, Y. (2008) Simultaneous determination of 24 sulfonamide residues in meat by ultra-performance liquid chromatography tandem mass spectrometry. *J. Chromatogr. A* **1200**, 144–155.
64. Shao, B., Dong, D., Wu, Y., Hu, J., Meng, J., Tu, X., Xu, S. (2005) Simultaneous determination of 17 sulfonamide residues in porcine meat, kidney and liver by solid-phase extraction and liquid chromatography–tandem mass spectrometry. *Anal. Chim. Acta* **546**, 174–181.
65. Sergi, M., Gentil, A., Perret, D., Marchese, S., Materazzi, S., Curini, R. (2007) MSPD extraction of sulphonamides from meat followed by LC tandem MS determination. *Chromatographia* **65**, 757–761.
66. Mohamed, R., Hammel, Y.-A., LeBreton, M.-H., Tabet, J.-C., Jullien, L., Guy, P.A. (2007) Evaluation of atmospheric pressure ionization interfaces for quantitative measurement of sulfonamides in honey using isotope dilution liquid chromatography coupled with tandem mass spectrometry techniques. *J. Chromatogr. A* **1160**, 194–205.
67. Thompson, T.S. and Noot, D.K. (2005) Determination of sulfonamides in honey by liquid chromatography–tandem mass spectrometry. *Anal. Chim. Acta* **551**, 168–176.
68. Heller, D.N., Ngoh, M.A., Donoghue, D., Podhorniak, L., Righter, H., Thomas, M.H. (2002) Identification of incurred sulfonamide residues in eggs: methods for confirmation by liquid chromatography–tandem mass spectrometry and quantitation by liquid chromatography with ultraviolet detection. *J. Chromatogr. B* **774**, 39–52.
69. Sheridan, R., Policastro, B., Thomas, S., Rice, D. (2008) Analysis and occurrence of 14 sulfonamide antibacterials and chloramphenicol in honey by solid-phase extraction followed by LC/MS/MS analysis. *J. Agric. Food Chem.* **56**, 3509–3516.
70. Zhenfeng, Y., Yueming, Q., Xiuyun, L., Caini, J. (2006) Determination of multi-residues of tetracyclines and their metabolites in milk by high performance liquid chromatography-tandem positive-ion electrospray ionization mass spectrometry. *Chin. J. Anal. Chem.* **34**, 1255–1259.
71. Nakazawa, H., Ino, S., Kato, K., Watanabe, T., Ito, Y., Oka, H. (1999) Simultaneous determination of residual tetracyclines in foods by high-performance liquid chromatography with atmospheric pressure chemical ionization tandem mass spectrometry. *J. Chromatogr. B* **732**, 55–64.
72. Cherlet, M., Schelkens, M., Croubels, S., De Backer, P. (2003) Quantitative multi-resideu analysis of tetracycline and their 4-epimers in pig tissues by high-performance liquid chromatography combined with positive-ion electrospray ionization mass spectrometry. *Anal. Chim. Acta* **492**, 199–213.
73. Bogialli, S., Curini, R., Di Corcia, A., Lagana, A., Rizzuti, G. (2006) A rapid confirmatory method for analyzing tetracycline antibiotics in bovine, swine, and poultry muscle tissues: matrix solid-phase dispersion with heated water as extractant followed by liquid chromatography-tandem mass spectrometry. *J. Agric. Food Chem.* **54**, 1564–1570.
74. Blasco, C., Di Corcia, A., Pico, Y. (2009) Determination of tetracyclines in multi-specie animal tissues by pressurized liquid extraction and liquid chromatography-tandem mass spectrometry. *Food Chem.*, **116**, 1005–1012.
75. Anderson, C.R., Rupp, H.R., Wu, W.-H. (2005) Complexities in tetracycline analysis – chemistry, matrix extraction, cleanup, and liquid chromatography. *J. Chromatogr. A* **1075**, 23–32.
76. Mastovska, K. and Lightfield, A.R. (2008) Reversed phase and aqueous normal phase retention in multiclass LC-MS analysis of antibiotics. *Am. Lab.* (on-line edition) 6–7, 37–40.
77. Bruno, F., Curini, R., Di Corcia, A., Nazzari, M., Pallagrosi, M. (2002) An original approach to determining traces of tetracycline antibiotics in milk and eggs by solid-phase extraction and liquid chromatography/mass spectrometry. *Rapid Commun.Mass Spectrom.* **16**, 1365–1376.
78. Heller, D.N., Nochetto, C.B., Rummel, N.B., Thomas, M.H. (2006) Development of multiclass methods for drug residues in eggs: hydrophilic solid-phase extraction cleanup and liquid chromatography/tandem mass spectrometry of tetracycline, fluoroquinolone, sulfonamide, and β-lactam residues. *J. Agric. Food Chem.* **54**, 5267–5278.
79. Turnipseed, S.B., Andersen, W.C., Karbiwnyk, C.M., Madson, M.R., Miller, K.E. (2008)

Multi-class, multi-residue liquid chromatography-tandem mass spectrometry screening and confirmation methods for drug residues in milk. *Rapid Commun. Mass Spectrom.* **22**, 1467–1480.

80. Hammel, Y.-A., Mohamed, R., Gremaud, E., LeBreton, M.-H., Guy, P.A. (2008) Multi-screening approach to monitor and quantify 42 antibiotic residues in honey by liquid chromatography–tandem mass spectrometry. *J. Chromatogr. A* **1177**, 58–76.
81. Granelli, K. and Branzell, C. (2007) Rapid multi-residue screening of antibiotics in muscle and kidney by liquid chromatography-electrospray ionization–tandem mass spectrometry. *Anal. Chim. Acta* **586**, 289–295.
82. Carretero, V., Blasco, C., Pico, Y. (2008) Multi-class determination of antimicrobials in meat by pressurized liquid extraction and liquid chromatography-tandem mass spectrometry. *J. Chromatogr. A* **1209**, 162–173.
83. Chico, J., Rubies, A., Centrich, F., Companyo, R., Prat, M.D., Granados, M. (2008) High-throughput multiclass method for antibiotic residue analysis by liquid chromatography-tandem mass spectrometry. *J. Chromatogr. A* **1213**, 189–199.
84. Stubbings, G. and Bigwood, T. (2009) The development and validation of a multiclass liquid chromatography tandem mass spectrometry (LC–MS/MS) procedure for the determination of veterinary drug residues in animal tissue using a QuEChERS (Quick, Easy, Cheap, Effective, Rugged and Safe) approach. *Anal. Chim. Acta* **637**, 68–78.
85. Yamada, R., Kozono, M., Ohmori, T., Morimatsu, F., Kitayama, M. (2006) Simultaneous determination of residual veterinary drugs in bovine, porcine, and chicken muscle using liquid chromatography coupled with electrospray ionization tandem mass spectrometry. *Biosci. Biotechnol. Biochem.* **70**, 54–65.
86. Kaufmann, A., Butcher, P., Maden, K., Widmer, M. (2008) Quantitative multiresidue method for about 100 veterinary drugs in different meat matrices by sub 2-μm particulate high-performance liquid chromatography coupled to time of flight mass spectrometry. *J. Chromatogr. A* **1194**, 66–79.
87. Peters, R.J.B., Bolck, Y.J.C., Rutgers, P., Stolker, A.A.M., Nielen, M.W.F. (2009) Multi-residue screening of veterinary drugs in egg, fish and meat using liquid chromatography accurate mass time-of-flight mass spectrometry. *J. Chromatogr. A,* **1216**, 8206–8216.
88. Ortelli, D., Cognard, E., Jan, P., Edder, P. (2009) Comprehensive fast multiresidue screening of 150 veterinary drugs in milk by ultra-performance liquid chromatography coupled to time of flight mass spectrometry. *J. Chromatogr. B,* **877**, 2363–2374.
89. Pesek, J.J., Matyska, M.T., Larrabee, S. (2007) HPLC retention behavior on hydride-based stationary phases. *J. Sep. Sci.* **30**, 637–647.
90. van der Heeft, E., Bolck, Y.J.C., Beumer, B., Nijrolder, A.W.J.M., Stolker, A.A.M., Nielen, M.W.F. (2009) Full-scan accurate mass selectivity of ultra-performance liquid chromatography combined with time-of-flight and orbitrap mass spectrometry in hormone and veterinary drug residue analysis. *J. Am. Soc. Mass Spectrom.* **20**, 451–463.
91. Mol, H.G.J., Plaza-Bolanos, P., Zomer, P., de Rijk, T.C., Stolker, A.A.M., Mulder, P.P.J. (2008) Toward a generic extraction method for simultaneous determination of pesticides, mycotoxins, plant toxins, and veterinary drugs in feed and food matrixes. *Anal. Chem.* **80**, 9450–9459.

Chapter 13

The LC-MS/MS Methods for the Determination of Specific Antibiotics Residues in Food Matrices

Gui-Liang Chen and Yan-Yan Fang

Abstract

This chapter describes the LC-MS/MS methods for the determination of antibiotics residues in food matrices. The types of antibiotics include β-lactam antibiotics, sulfonamides, tetracyclines, fluoroquinolones, nitrofurans, and chloramphenicol (CAP). The food matrices are mainly from animal origin, such as animal tissues, fishes (marine products), eggs, milk, honey, and so on. The methods and procedures are covered, including three parts: (1) Liquid chromatographic conditions, (2) mass spectrometer conditions, including ionization source, analyzer, and acquisition, and (3) extraction and clean-up methods. In each case, the standard operating procedures (SOPs) for analysis are given with sensitivity, linearity, precision, and recovery. Some criteria of maximum residue limits (MRLs) from the legislation are listed.

Key words: Antibiotics, Residue analysis, LC-MS/MS, Food matrices

1. General Consideration for the Analysis of Antibiotics in Food

1.1. Introduction (1, 2)

Antibiotics are the substances produced by fungi and bacteria at low concentrations for inhibiting the growth of other microorganisms. Traditionally, antibiotics should include only five classes, i.e., penicillins, tetracyclines, macrolides, aminoglycosides, and amphenicols. However, currently, the term antibiotic is considered synonymous with antibacterial, so synthetic drugs (e.g., sulfonamides, quinolones, or nitrofurans) and substances of high molecular weight (e.g., peptide antibiotics) also belong to this group. This group of compounds comprises the most detected veterinary drug residues in food through their dual use for preventing and treating diseases or promoting growth in food-producing animals. In modern agricultural practice, the use of antibiotics in veterinary medicine, which began in the 1950s as feed additives, is common. In addition, as veterinary drugs antibiotics are given

Jerry Zweigenbaum (ed.), *Mass Spectrometry in Food Safety: Methods and Protocols*,
Methods in Molecular Biology, vol. 747, DOI 10.1007/978-1-61779-136-9_13, © Springer Science+Business Media, LLC 2011

in cases of disease, for dehydration, or to prevent losses during transportation.

Residual antibiotics in food constitute a risk to human health. Their presence in food can provoke allergic reactions in some hypersensitive individuals and may compromise the human immune system. Even more important, the presence of subtherapeutic doses of the above drugs in foodstuffs for long periods has led to the problem of drug-resistant pathogenic bacterial strains. The more antibiotics are used, the more rapidly resistance develops. The occurrence and the fate of antibacterial compounds in the environment have also been recognized as an emerging problem as well as a prevailing problem.

To ensure the safety of food for consumers, the World Health Organization (WHO) and the Food and Agriculture Organization (FAO) have proposed standards of residual antibiotics to animal food since early 1969, and the US Food and Drug Administration (FDA), the European Union (EU), and the State Food and Drug Administration (SFDA) in People's Republic of China have set maximum residue limits (MRLs) for antibiotics in food. Analysis of these residues plays not only a key role, but also a challenging task in ensuring food safety because there is a large number of antibiotics. In addition, in many cases, antibacterial residues comprise parent drugs and metabolites because most of the antibacterials administered to food-producing animals are oxidized, reduced, or hydrolyzed in phase I of the metabolic process and biotransformed during phase II metabolism to water-soluble conjugates, primarily by glucuronidation, sulfation, or conjugation with glycine. The extremely low part per billion (PPM) levels at which an antibacterial residue need to be analyzed further complicates the analysis. MRLs are fixed at the parts per million level (ppm or mg/kg) or even at the ppb (μg/kg) level depending on the antibiotic. Residues below the set limits are considered safe. Legislation may differ considerably in different countries and, for many food commodity residue combinations, there are no set MRLs or clear guidance of the levels of residues permitted. Finally, the complexity of the food matrix should also be taken into account.

Recently, many technologies have been developed for the analysis of antibiotics residues in food from rapid screening to confirmatory methods. The rapid screening methods include the microbiological assay (inexpensive, easy to perform on a large scale and possessing a wide, nonspecific spectrum of sensitivity), immunological techniques, such as fluorescent immunoassay (FIA), enzyme-linked immunosorbent assay (ELISA), radioimmunoassay (RIA), multi-antibiotic-ELISA, chemiluminescent immunosensor, fluorescent biosensor immune assay (BIA), and surface plasmon resonance biomolecular interaction analysis (SPR-BIA). Confirmatory analysis methods are mainly LC-MS,

GC-MS, and CE-MS. Although HPLC-UV or fluorescence detection (FLD) was used in the analysis of antibiotic residues in food, they were primarily a screening method followed by MS detection. Nowadays, confirmation of antibiotic residues in food is performed mainly by LC-MS/MS. When mass fragments are measured using techniques other than full-scan, the system of identification points (IPs) is applied. Of the methods reported in the literature, LC-MS/MS using triple quadrupole (QqQ) mass spectrometers in selected reaction monitoring (SRM) mode is the number one analytical methodology now selected for simultaneous, unambiguous identification or confirmation and quantification of antibiotic residues. However, LC can be combined with different mass analyzers (e.g., quadrupole ion trap (IT), time-of-flight (TOF), quadrupole time-of-flight (QqTOF), quadrupole linear ion-trap (QTRAP) or the Orbitrap) through different atmospheric pressure ionization (API) sources. However, the application of LC-TOF-MS for the elucidation of unknown immunoactive compounds is presently being developed.

This chapter basically describes the LC-MS/MS methods for the determination of β-lactam antibiotics, sulfonamides, tetracyclines, fluoroquinolones, nitrofurans, and chloramphenicol (CAP) residues in food matrices. Presently, these methods are applied to routine detection of samples from food markets in our laboratory.

1.2. General Methods

1.2.1. Extraction and Clean-Up (3, 4)

Antibiotics residues in food exist not only in a complex matrix, but also at low concentrations for qualitative and quantitative analysis. The procedures usually include sample preparation, enrichment, and purification. Extraction methods commonly used for purification are Soxhlet extraction, oscillation extraction, liquid–liquid extraction, column layer chromatography analysis (florisil silica gel, diatomite, and alumina column), solid-phase extraction (SPE), solid-phase microextraction (SPME), matrix solid-phase dispersion (MSPD), and supercritical fluid extraction (SFE). There also appears a simple and rapid sample treatment procedure that couples positive features of the MSPD technique, i.e., simplicity and intimate contact between the extractant and the matrix, to those offered by heated water as an extractant (5). The ideal method should meet the requirements of less solvent consumption, improved extraction throughput (in some instances linked to automation), higher recoveries, and better reproducibility. The usual techniques recently employed for extraction and cleanup of antibiotics from food matrices include protein precipitation, liquid–liquid extraction (LLE), and SPE.

1.2.2. Protein Precipitation

Deproteinization is commonly used in the extraction of antibiotics from biological matrices, where the removal of interferences is necessary while retaining good recoveries of the analytes of interest. It is a simple off-line procedure. Deproteinization solvents usually use acetonitrile. Acids, such as trichloroacetic acid or perchloric acid, can also be used for protein precipitation prior to the analysis of food samples. Phosphate buffer and acetonitrile are used simultaneously to precipitate the proteins.

1.2.3. Liquid–Liquid Extraction

LLE has been exploited as an extraction procedure for antibiotics from complex matrices. A number of methods employ extraction with acetonitrile prior to clean-up of the extracts by LLE with hexane in each case. Sometimes, this procedure is followed by solid-phase extraction. Besides acetonitrile, solvent extraction can be carried out using different solvent or mixtures (e.g., ethyl acetate, dichloromethane, methanol, or water), e.g., methanol and acetonitrile simplify this part of the process because they can simultaneously precipitate the proteins and extract the antibacterial agent.

Supported liquid membrane (SLM) extraction and/or enrichment are similar to LLE and dialysis combined. In SLM, an organic liquid is embedded in small pores of a polymer support and is held there by capillary forces. If the organic liquid is immiscible with the aqueous feed and strip streams, SLM can be used to separate the two aqueous phases. It may also contain a diluent, which is generally an inert organic solvent to adjust viscosity, and sometimes also a modifier to avoid the so-called third phase formation. One of the advantages of SLM is that the relatively small volume of organic components in the membrane and simultaneous extraction and re-extraction in one technological step results in high separation factors, easy scale-up, lower energy requirements, and thus lower overall running costs.

1.2.4. Solid-Phase Extraction

SPE is always employed to clean-up and to preconcentrate a sample. SPE involves liquid–solid partition, where the extracting phase is a solid sorbent. This technique and versions thereof have been used extensively to extract and concentrate trace organic materials from samples. A wide choice of sorbents, which rely on different mechanisms for extraction/retention of analytes, is available. While there are drawbacks associated with SPE, such as the importance of packing uniformity to avoid poor efficiency, this technique can be used to extract veterinary residues from even the most challenging matrices, such as shrimp, soil, or sewage sludge.

Clean-up is frequently carried out by SPE. Alumina, amino, or strong cation exchangers (SCX) have been proposed for ionic antibiotics while C18 or polymeric sorbents, especially hydrophilic–lipophilic balance (HLB) polymeric reversed phases, are used for neutral or ionizable compounds working at a pH lower than the pKa of the analytes. For compounds with varied chemical properties, mixed-mode sorbents are recommended (e.g., Bond Elut SCX cartridges for multiresidue of basic drugs). SPE can be directly used for the extraction of antibacterials from liquid food only (e.g., milk or honey, which can be dissolved in water).

LLE and/or SPE for sample pretreatment are used in most employed methods for the determination of veterinary drug residues. A specific combination of LLE and SPE can be very selective for a specific class of veterinary drugs.

1.3. LC Methods (6–8)

Briefly, building an LC method includes selecting both a stationary phase and a mobile phase to perform separation procedure and suitable for the MS detection.

1.3.1. Chromatograph Column

Generally, the RP-LC stationary phases consist of a silica support to which the hydrophobic octadecyl group is attached. This group can be either monomerically or polymerically bonded by chlorosilanes, but normally does not occupy all silanols on the silica support. To avoid interactions of basic antibiotics with these underivatized silanol groups, resulting in peak tailing, end-capped materials are normally used. Another type of RP-LC stationary phase consists of polymeric support. An advantage of polymeric stationary phases, such as polystyrene-divinylbenzene (PS-DVB), is that they allow a larger pH range than silica-based phases. Normal-phase conditions with flammable mobile solvents (mixtures of alkanes and alcohols) at a high flow rate are incompatible with Mass [electrospray ionization (ESI) and atmospheric pressure chemical ionization (APCI)] detection, due to concern about the explosion hazards associated with APCI (corona discharge) and ESI (high voltage discharge).

Particle sizes of stationary phases are generally between 3 and 5 μm. Smaller packing particles can increase column efficiency and speed of analysis, but also cause higher pressures. The pore sizes of the column beds vary between 80 and 300 Å. The LC particles with large pore sizes (~300 Å) have been especially developed for the separation of large molecules, as these compounds are too large to enter smaller pores. For peptide antibiotics separations, the smaller pore sizes have therefore only been used for the smaller peptides (<2,000 Da), whereas the larger pores have been particularly applied for the larger antibiotics.

Narrow-bore (±2 mm ID) columns are most frequently used with column lengths varying from 50 to 150 mm. Miniaturization of the column's inner diameter tends to increase ionization efficiency of electrospray for MS detection, by providing a lower volumetric flow rate and sample preconcentration. Although column efficiency is not improved and analysis time not reduced, it can facilitate the coupling of LC to MS, saves solvents, and is advantageous for the analysis of small amounts of biological samples.

1.3.2. Mobile Phases

The choice of the correct mobile phase and mobile phase additives in RP-LC is very important to achieve good chromatographic separation as well as mass spectrometric performance. When LC is coupled to MS with electrospray ionization (ESI) the mobile phase has to be compatible with the ESI source and, furthermore, the additives have to enhance or at least not to suppress ionization of the analyte. Salts and nonvolatile organic compounds can lead to decreased signal intensities and signal-to-noise ratios as well as contamination of the ion source. A suitable mobile phase for ESI-MS contains an organic modifier (methanol or acetonitrile) and should not contain nonvolatile buffers or mobile phase additives. Modification of the mobile phase has been accomplished with volatile substances (e.g., ammonium acetate, oxalic acid, or formic acid) to avoid clogging at the interface and a build-up of deposits in the ion source. Sometimes, the use of ion-pair reagents is needed, as the analytes are present as ionic forms. In these cases, perfluorinated pairing agents [e.g., heptafluorobutyric anhydride (HFBA)] have been used. Separations are performed with an isocraticor with gradient elution mode.

The composition of the mobile phase is also an important aspect on the analyte's ESI response. Solvent pH affects the protonation or deprotonation of the analyte. Acidic solvents are therefore ideally used for the analysis in the positive ion mode of compounds with basic functional groups. However, protonated ions can also be observed when ESI-MS analysis is performed with basic solvents and deprotonated ions can be observed in the ESI-MS analysis of acidic solutions. A possible explanation can be gas phase-ion-molecule reactions or collision-induced dissociation of base or acid adduct ions in basic or acidic solutions, respectively.

1.3.3. LC Trends (9, 10)

In the multiresidue method using full-scan MS, the LC part of the system sometimes causes certain limitations. Due to the high number of compounds to be separated, the run times become relatively long. LC runs can be shortened by utilizing short columns and/or fast gradients. The price to be paid is low(er) chromatographic resolution, which necessitates a further reduction of the already limiting dwell times if MRM transitions are to be monitored. The introduction of pressure-stable sub-2-μm particulate packing materials and novel low-dead volume, high-pressure (1,000 bar) LC equipment (i.e., ultra-high pressure liquid chromatography or UHPLC), provides strategies to improve resolution while maintaining or even shortening run times. An essential aspect of UHPLC is the use of sub-2-μm particulate packing materials while maintaining other aspects of the column geometry (e.g., column length and very low dead volumes). UHPLC-ToF-MS provides significant advantages in selectivity, sensitivity, and speed. The most outstandingly novel approach to chromatographic separation is the development of UHPLC, which has sensitivity two to three times greater than high-performance liquid chromatography (HPLC) by using columns packed with particles of <2 μm, resulting in better chromatographic efficiency (narrower peaks), increased peak capacity and reduced run time.

1.4. Mass Spectrometry Methods (1, 9–13)

In the past, HPLC analysis was employed for the antibiotics residues analysis widely using two main detectors: UV and FLD using derivatization agents. The development of API interfaces (e.g., ESI and APCI) enabled mass spectrometric detection's increased utilization. Now, mass spectrometry has become popular for residue analysis due to its sensitivity, speed, and specificity. Tandem mass spectrometry (MS/MS) has emerged as the most advisable technique for regulatory purposes not only because it constitutes the most powerful detection mode in the analysis of compounds at the required residue level, but also because it gives the capability of being used as a confirmatory technique as it provides unequivocal structural information. Mass spectrometric methods include some parameters associated with ionization sources, analyzers, and acquisition.

1.4.1. Ionization Sources

As MS detection is only possible when the antibiotics exist as ions in the gaseous state, coupling of LC to MS necessitates the volatilization of these biomolecules from the LC eluent. A large number of different LC-MS interfaces have been developed to achieve this goal, of which nowadays API is generally accepted as the optimum interface for online LC-MS coupling. Two main API sources are ESI and APCI.

In ESI, charged droplets are formed by spraying the sample solution through a high voltage (2–5 kV) capillary in the presence of a strong electric field. The charged droplets move toward the mass spectrometer inlet, generating analyte ions during evaporation and droplet fission. As electrospray efficiency depends on eluent composition, the optimal mobile-phase composition for adequate chromatographic separation is sometimes unsuitable for obtaining maximal electrospray response. However, the correct choice of a postcolumn solvent to add to modify pH, or to increase the percentage of organic solvent, can enhance ionic evaporation and improve the signal.

In APCI, a heated pneumatic probe is used for nebulization and a high voltage needle is used to produce a corona discharge for ionization of the evaporated solvent, reacting with the analyte. APCI is very sensitive for weakly basic compounds that can be easily protonated by gas-phase, mobile-phase ions, according to their proton affinity. APCI is a more energetic source than electrospray, since effluent is nebulized by a coaxial nitrogen flow and rapidly evaporated using high temperatures (350–500°C). These high temperatures must be taken into account when working with thermally degradable compounds.

Regarding interfaces, ESI sources have been by far the most frequently applied in antimicrobial residue determination, since it is particularly suitable for polar and moderately nonpolar analytes and for thermally labile substances, although it is well known to be more prone to ion suppression caused by matrix than APCI. Quantitative LC-MS analysis of residues is mostly performed by ESI, although some methods have been used by APCI. The main reason for using APCI was either to overcome complications like matrix effects and ionization suppression, or better sensitivity for a simultaneously quantified compound. ESI and APCI performances have been compared resulting in the selection of ESI as it was easier in use or showed significantly lower limits of detection (LODs).

To fulfill the increasing expectations of the ESI response, novel advances in ESI techniques have been introduced. Electrospray sources that allow low flow rates, e.g. micro-ESI and nanospray, have been developed to improve ESI response. Miniaturization of the ESI interface should result in a larger fraction of ions entering the mass detector and thus in increased sensitivity. Also different types of ESI sources, all claiming more efficient ion transfer into the mass spectrometer, are available.

Another API technique is atmospheric pressure photoionization (APPI). With APPI, the analyte is ionized by photons emitted from a vacuum-ultraviolet lamp.

1.4.2. Mass Analyzers

The three most common mass analyzers are quadrupole, TOF, and ion trap (IT). Mass analyzers separate ions by their mass-to-charge ratio (m/z). In quadrupole mass spectrometers specific radio frequency (RF) and direct current (DC) voltages are applied on four parallel rods, allowing only ions with a specific m/z values to move between the rods and to reach the detector. TOF analyzers accelerate the ions by a short voltage gradient and measure the flight time of the ions which is proportional to the square root of the m/z. Ion trap mass spectrometers trap the ions in an electric field, where specific ions can be activated and ejected by manipulation of this electric field. Tandem mass spectrometers are made of two successive mass analyzers with a collision cell in between that can select specific ions, induce their fragmentation in the collision cell, and measure the m/z of the fragment ions. The most common tandem mass spectrometer is the triple quadrupole, where the second quadrupole is the collision cell and the first and third quadrupole are the actual mass analyzers. Quadrupoles are very suitable for quantitative measurements because of their relative high ion transmission efficiency, large dynamic range, and high sensitivity when used in the static mode, i.e., no full scan (FS) mode. Ion traps can also perform multiple MS (MS^n) as the trapped ions can be fragmented inside the trap and is commonly known as MS/MS in time rather than space.

LC-MS/MS is typically required to quantify the low concentration levels of antibiotics in real-world samples. Analyzers used most in the detection or quantification of antimicrobials ion trap (IT) and most especially QqQ. This is principally due to their easier operating performance, their better robustness for routine analysis, and their very high sensitivity and selectivity in complex matrices. Triple-quadrupole analyzers display high sensitivity when working in multiple reaction monitoring (MRM) mode, and are thus best suited to obtain the strict MRLs regulated for various residues in different food matrices. Recently, the QqTOF hybrid analyzer has been explored for identification purposes, as it offers higher resolution than QqQ and elevated mass accuracy working in both MS and MS^2 modes.

When performing MS/MS, ion-trap instruments are generally less sensitive than triple-quadrupole analyzers, but they have the advantage of working in product-ion scan without losses in sensitivity. Moreover, ion traps offer the possibility of performing multiple-stage fragmentation (MS^n). Such advantages are important tools for characterizing and correctly identifying residues in complex food matrices, and have frequently been used for screening purposes. For this application, both product-ion full-scan MS^n and SRM are used.

An important parameter in mass analysis is the mass resolution of the mass spectrometer. The resolution defines the ability of the mass spectrometer to distinguish different masses and is commonly described as the width of a mass peak halfway up the height of the peak (full width at half maximum or FWHM). Baseline separation of two peaks with 1 amu mass difference is approximately achieved at 0.7 FWHM or unit resolution. Measuring multiply charged ions, a lower FWHM value is required when distinction between the different isotope peaks is desired. However, the applied resolution for the quantitative assay is not often mentioned.

1.4.3. Acquisition

Mass spectrometry can be performed in either the positive or negative ionization mode, measuring either positively or negatively charged ions.

The measurement of ions in MS and MS/MS can occur in different scan modes. In the MS full scan mode a complete mass spectrum is generated, showing the molecular ion in different charge states and the different adduct ions. Collection of all mass spectral information can be useful when specific information is required, for example to monitor possible interfering ions. For quantitative assays, both sensitivity and specificity of the method are usually increased when only one or few specific m/z ions are selected in the selected ion monitoring (SIM) mode. To overcome the problem of irreproducible quantification in the SIM mode due to variable intensities of the differently charged molecular ions, a summation of all charged molecular ions can be used.

Also in MS/MS, a full scan of the fragments of the selected ion can be recorded, but for very selective and sensitive MS/MS quantification the selection of one or few fragment ions is favored. This is called MRM and eliminates the interference of co-eluting contaminants more efficiently than SIM, especially in the case of antibiotics with high molecular masses. However, fragmentation of these larger compounds during collision-induced dissociation usually requires higher collision energies, thereby more commonly producing multiple fragments, providing lower sensitivity. Therefore, sometimes SIM is preferred despite the ability to perform MRM.

In general, ions originating from APCI sources correspond to protonated molecules $[M+H]^+$ in positive ionization mode and deprotonated molecules $[M-H]^-$ in negative ionization mode. However, in some instances, additional signals for *m/z* corresponding to Na^+, K^+ or NH_4^+ adducts appeared in the spectra. Indeed, these cations are usually present as impurities in organic solvents of mobile phases in LC. Sometimes, these adducts appear as the base peak in the spectra.

In the past, MS methods using single-quadrupole MS produce low fragmentation, so abundant stable pseudomolecular ions, $[M+H]^+$ or $[M+H]^-$, are obtained. However, this technique can prevent reliable analysis due to the co-fragmentation of matrix components other than those targeted. Nowadays, by using LC-tandem MS (LC-MS^2), it is possible to distinguish individual compounds having the same molecular mass (isobaric) by the different fragments obtained after an induced collision with an inert gas. So it is preferable to use MS/MS detection for better selectivity and sensitivity in complex matrices.

2. Specific Methods for Antibiotic Analysis

2.1. The Residue Analysis of 7 Fluoroquinolones in Meats, Marine Products, and Eggs: LC-MS/MS

2.1.1. Introduction

The method can be used for the analysis of ofloxacin, norfloxacin, ciprofloxacin, enrofloxacin, danofloxacin, sarafloxacin, and difloxacin residues in poultry meat, seafood (marine product), and eggs.

2.1.2. Materials

2.1.2.1. Reagents

1. High purity water: obtained from a Milli-Q water system (Millipore, Billerica, MA, USA).
2. Acetonitrile: HPLC grade.
3. Glacial acetic acid: liquid chromatography grade.
4. *n*-Hexane: liquid chromatography grade.
5. Formic acid: excellent pure class.
6. *n*-Hexane saturated with acetonitrile: Place 80 ml of *n*-hexane to a 100 ml liquid funnel, add appropriate amount of

acetonitrile, shake thoroughly, standing for balance, dispose acetonitrile layer, use the *n*-hexane.

7. A 0.2% formic acid solution: dilute 2 ml of formic acid with water to 1,000 ml, mix well.
8. Formic acid–acetonitrile: Add 2 ml of formic acid into 98 ml of acetonitrile, mix well.

2.1.2.2. Standards

Ofloxacin, norfloxacin, ciprofloxacin, enrofloxacin, danofloxacin , sarafloxacin, and difloxacin standard stock solution: Separately weigh accurately each quinolone standards above at 10 mg each to the same 100 ml flask, dissolve with acetonitrile and dilute to volume, mix well. The standard stock solution of each at 0.1 mg/ml is obtained (store at 4°C for 3 months).

1. A solution at 10 μg/ml each of ofloxacin, norfloxacin, ciprofloxacin, enrofloxacin, danofloxacin , sarafloxacin, and difloxacin: Measure accurately 5.0 ml of the quinolone standards stock solution into 50 ml flask, dilute with acetonitrile to volume, mix well, and the middle standard solution is obtained (store at 4°C for months).
2. Ofloxacin, norfloxacin, ciprofloxacin, enrofloxacin, danofloxacin, sarafloxacin, and difloxacin-mixed standard solution: Measure accurately an appropriate amount of quinolone middle standard solution, dilute with 0.2% formic acid solution–acetonitrile (75:25) solution to prepare a series of the standard working solutions with the concentration of 2.5, 5.0, 10.0, 25.0, 50.0, 100.0, 250.0, and 500.0 ng/ml of mixed standard quinolone (store at 4°C, use within 1 week): Measure accurately the middle standard solution (10 μg/ml) 1.0, 2.5, and 5.0 ml each into 100 ml flasks, dilute with 0.2% formic acid solution–acetonitrile (75:25) solution to volume, mix well, and label as S6, S7, and S8; Measure accurately S8 0.25, 0.5, 1.0, 2.5, and 5.0 ml each into 50 ml flasks, dilute with 0.2% formic acid solution–acetonitrile (75:25) solution to volume, mix well, and label as S1, S2, S3, S4, and S5. The concentration of standard solutions was listed in the Table 1.

2.1.2.3. Equipment

1. Pipette: 20–200 μl and 1 ml.
2. Homogenizor.
3. Rotary evaporator.
4. Oscillator.
5. Centrifuge (temperature control and the minimum speed at 4,000 rpm (3,488 × *g*) or more).
6. Centennial bottle: 100 ml.
7. Filters Teflon and Nylon membranes 0.45 μm.
8. Vials with PTFE screw-cap 2 ml.
9. Analytical Balance: accurate to 0.1 mg.

Table 1
Concentration of ofloxacin, norfloxacin, ciprofloxacin, enrofloxacin, danofloxacin, sarafloxacin, and difloxacin mixed standard solutions:

No.	W (mg)	Dilute factor from stocks	The concentration of stock (μg/ml)	The concentration of medium (μg/ml)	The amounts for linear test (ml)		Volume (ml)	Concentration of solution (ng/ml)	The amounts (ng/g)
S1	10	100	100	10	S8	0.25	50	2.5	1
S2					S8	0.5	50	5	2
S3					S8	1.0	50	10	4
S4					S8	2.5	50	25	10
S5					S8	5.0	50	50	20
S6					Medium 1	1.0	100	100	40
S7					Medium 2	2.5	100	250	100
S8					Medium 3	5.0	100	500	200

10. Balance: accurate to 0.01 g.
11. Polypropylene centrifuge tubes 50 ml.

2.1.3. Methods

2.1.3.1. Specimen Treatment

Samples were taken out from all the representative edible parts of about 1 kg, fully broken up, and mix thoroughly, transfer into a clean container. Stored at −20°C until use.

2.1.3.2. Sample Preparation Procedure

1. Extraction
   Place 2.5 g of the sample (accurate to 0.01 g), in a 50 ml centrifuge tube, add 10 ml of formic acid–acetonitrile solution, mix for 1 min using homogenizer, wash mixture with about 2 ml acetonitrile, combine the washings to centrifuge tube (to avoid cross-contamination homogenize after washing aliquot with acetonitrile two times). Centrifuge at the rate of 4,000 rpm (3,488 × *g*) for 10 min, transfer the supernatant into another 50 ml of centrifuge tube. Add 10 ml of formic acid–acetonitrile to the residue and then extract once, combine the supernatants.
2. Purification
   Transfer the supernatant to another 50 ml centrifuge tube, add 12 ml of *n*-hexane saturated with acetonitrile, vortex for 2 min, centrifuge at the rate of 4,000 rpm (3,488 × *g*) for 10 min, discharge the upper solution, transfer the lower solution to a 100 ml evaporator flask, and rotary evaporate in a water bath at 40°C to near dry. Accurately add 1.0 ml of 0.2% formic acid solution–acetonitrile (75:25) solution, vortex for 1 min to dissolve the residue, transfer the solution to a 2 ml centrifuge tube, centrifuge at 13,000 rpm (11,337 × *g*) for 5 min, and filter the supernatant through 0.45 μm filter for LC-MS/MS analysis.

2.1.3.3. HPLC Conditions

1. HPLC: Agilent liquid chromatography (HP1100) is equipped with an automatic degasser, a quaternary pump, an autosampler, and a column thermostat.
2. Column: WATERS XTerra MS C18, 3.5 μm, 100 mm × 4.6 mm.
3. Flow rate: 0.35 ml/min.
4. Injection volume: 40 μl (for MS 3200Qtrap); 20 μl (for MS API4000).
5. Mobile phase: acetonitrile −0.2% formic acid aqueous solution (25:75).
6. Column temperature: 30°C.

2.1.3.4. MS Conditions (Applied Biosysytem API4000 or 3200QTrap)

1. Ion Source: ESI ion source.
2. Scan Mode: Positive Ion Scan.
3. Detection mode: multiple reaction monitor.

Table 2
MS conditions for the analysis of fluoroquinolones

Components	Transition ions precursor/ product (*m/z*)	Collision energy/V	Declustering potential/V
Ofloxacin	362.4/344.3 362.4/318.2 362.4/261.2	31 29 37	70
Norfloxacin	320.4/302.2 320.4/276.3 320.4/233.2	29 26 37	70
Ciprofloxacin	332.2/314.3 332.2/245.3 332.2/231.2	30 37 52	70
Enrofloxacin	360.2/342.2 360.2/316.2 360.2/245.3	32 28 39	70
Danofloxacin	358.1/340.3 358.1/283.3 358.1/255.1	34 34 57	70
Sarafloxacin	386.0/368.2 386.0/342.2 386.0/299.2	33 28 40	80
Difloxacin	400.1/382.1 400.1/356.4 400.1/299.2	35 32 40	75

4. ESI voltage: 5,000 V.
5. Ion source temperature: 500°C.
6. Quantitative and qualitative ions on Collision Energy (CE) and Declustering Potential (DP) values were listed in Table 2.

2.1.4. Comments

1. The retention time are listed in Table 3 for reference.
2. The results are calculated using following equation:

$$X = c \times V \times 1{,}000/(m \times 1{,}000),$$

   X is the component residues in sample (μg/kg), c is calculated in accordance with the standard curve concentration (ng/ml), V is the sample volume (ml), and m is the sample solution represented by the quantity of the specimen (g).
3. Quality Control Samples: Place separately three 2.5 g of blank samples in three 50 ml centrifuge tube, add separately 500 μl of S2 and S5 in 2 of the three tubes, respectively, add 10 ml

Table 3
Retention times for the fluoroquinolones

Components	Retention time (min)
Ofloxacin	3.2
Norfloxacin	3.2
Ciprofloxacin	3.3
Enrofloxacin	3.5
Danofloxacin	3.3
Sarafloxacin	3.8
Difloxacin	3.9

of formic acid–acetonitrile solution into three tubes, mix well, respectively, as a blank quality control, QC-LOD and QC-10LOD samples. Carry out the procedure described under Subheading "Sample Preparation Procedure," and then analyze samples in sequence before, during, and afterward. Calculate the average recoveries of samples in accordance with the standard curve for the QC-LOD and QC-10LOD. The recoveries are between 86.7% and 90.4%.

4. In the end of the analysis, inject accurately a constant volume of S4 linear solution, calculate the recovery in accordance with the standard curve. The recovery rate should be 90–105%.
5. The detection limits of all components are near 1.0 μg/kg.
6. The reference chromatograms are shown in Fig. 1.

2.2. The Residue Analysis of 6β-Lactam Antibiotics in Milk by LC-MS/MS

2.2.1. Introduction

This method can be used for the analysis of six kinds of penicillin drugs in milk (ampicillin, amoxicillin, penicillin V, penicillin G, oxacillin, and dicloxacillin). The penicillin-type drug residues in milk samples were extracted with acetonitrile, the fat removed with *n*-hexane, then condensed after filtration. The extract was determined by LC-MS/MS method with external standard for quantification.

2.2.2. Materials

2.2.2.1. Reagents

1. Acetonitrile: HPLC grade.
2. *n*-Hexane: HPLC grade.
3. *n*-Hexane saturated with acetonitrile: Place 80 ml of *n*-hexane to a 100 ml liquid funnel, add appropriate amount of acetonitrile, shake thoroughly, let stand for separation, dispose acetonitrile layer, use the acetonitrile saturated *n*-hexane.

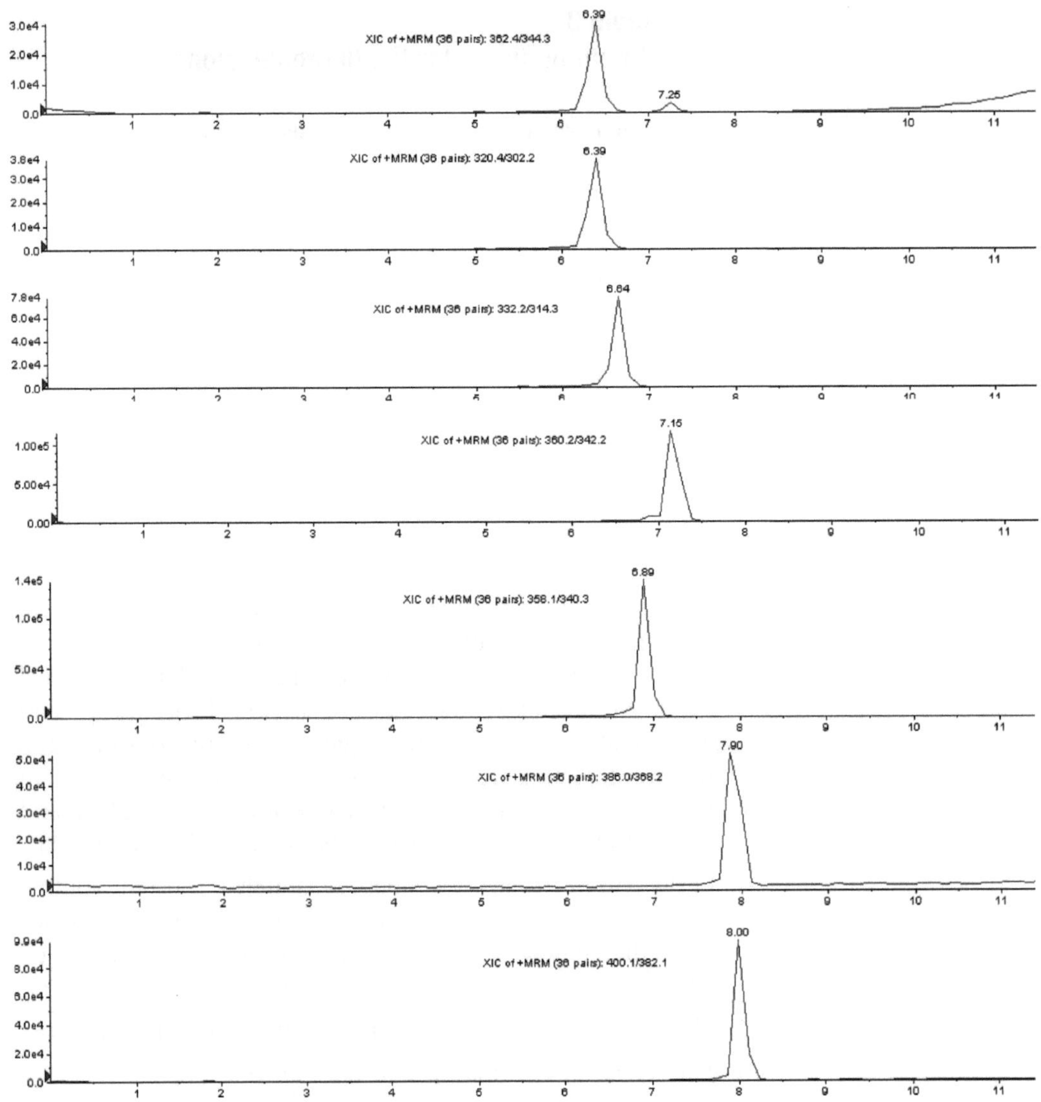

Fig. 1. Chromatograms of fluoroquinolone transitions.

4. Penicillin standard stock solutions (prepare the solution before use): weigh accurately each 10 mg of ampicillin, amoxicillin, penicillin V, penicillin G, oxacillin, and dicloxacillin RS to separately 100 ml flasks, dissolve with water, and diluted to volume. The standard stock solutions each contains 0.1 mg/ml. (Penicillin G should be dissolved with an appropriate volume of acetonitrile before adding water to volume).
5. Penicillin standard storage solutions: measure accurately 1.0 ml of penicillin G standard stock solution, and 2.0 ml of each of the other penicillin standard stock solution to separate 100 ml flask, dilute with water to volume, mix well.

The standard storage solutions with the concentration of 1.0 or 2.0 μg/ml are obtained.

6. Penicillin standard working solutions: measure accurately 1.0 ml of penicillin standard storage solutions, dilute with water to 20 ml, respectively. The concentration of the solutions was 50 ng/ml and 100 ng/ml, respectively.
7. Mixed standard solution: Add 10, 20, 40, 80, 100, 200, 300, and 400 μl, accurately measured penicillin standard working solutions to 1.0 ml of blank milk, mix well. Carry out the procedure described under Subheading "Preparation of Sample Solution," a series of the mixed standard solutions with the concentration of penicillin G at 0.5, 1.0, 2.0, 4.0, 5.0, 10.0, 15.0, 20.0 ng/ml and of other penicillins 1.0, 2.0, 4.0, 5.0, 10.0, 15.0, 20.0, 40.0 ng/ml, respectively, are obtained and labeled as S1–S8, respectively.

2.2.2.2. Instrument

1. LC-MS/MS detector equipped with a heated capillary interface, an ESI source, and a QqQ mass analyzer.
2. Analytical balance: accurate to 0.1 mg.
3. Balance: accurate to 0.01 g.
4. Oscillator.
5. Microsyringe: 1 ml.
6. Centrifuges: 15 ml centrifuge tube PTFE screw-cap.
7. Nitrogen drying instrument
8. High-speed desktop centrifuge: 15,000 rpm (15,093 × *g*).

2.2.3. Methods

2.2.3.1. Preparation of Sample Solution

Place 1.0 ml, measured accurately, of sample in centrifuge tube, add 3.0 ml of acetonitrile, shake, and centrifuge at 3,000 rpm (1,962 × *g*) for 5 min, take the supernatant, and repeat the operation using the residue. Add 2.0 ml of *n*-hexane saturated acetonitrile into the both combined supernatants, oscillate at 2,000 rpm (872 × *g*) for 1 min, shake, centrifuge at 4,000 rpm (3,488 × *g*) for 5 min, take the layer of *n*-hexane, dry the supernatant at 35°C under nitrogen to about 0.5 ml, add water to 1 ml. After vortexing, the supernatant is filtered through 0.45 μm filter for LC-MS/MS analysis.

2.2.3.2. HPLC conditions

1. Column: SunFireTMC18, 3.5 μm, 150 mm × 2.1 mm (diameter) or a considerable column.
2. Mobile phase: operated on gradient program in Table 4.
3. Column temperature: 30°C.
4. Injection volume: 20 μL.

2.2.3.3. MS Conditions

1. Ion source: ESI ion source.
2. Scan mode: positive ion scan.

Table 4
Mobile phase gradient program and flow rate for penicillins

Time/min	Flow rate (μL/min)	Water (containing 0.3% acetic acid)/(%)	Acetonitrile (containing 0.3% acetic acid)/(%)
0.0	300	90.0	10.0
2.0	300	90.0	10.0
2.1	300	50.0	50.0
8.0	300	50.0	50.0
8.1	300	15.0	85.0
16.0	300	15.0	85.0
16.1	300	90.0	10.0
22.0	300	90.0	10.0

3. Detect mode: multiple reaction monitoring.
4. ESI voltage: 5,500 V.
5. Atomization air pressure: 0.055 MPa.
6. Air-curtain air pressure: 0.079 MPa.
7. Auxiliary air speed: 6 L/min.
8. Ion source temperature: 500°C.
9. Ion pairs for qualitative, quantitative, and decluster voltage (DP), focusing voltage (FP), collision gas energy (CE), and collision cell export voltage (CXP) see Table 5 below.

2.2.3.4. Liquid Chromatography-Tandem Mass Spectrometry

1. Qualitative Analysis
   The penicillin residues should be judged as positive when both of the following conditions are met: (1) at the same test conditions, the deviation of the retention time from test samples are within ±2.5% of the retention time from the standard substances, and (2) the deviation does not exceed the scope of the provisions of Table 5 in the sample chromatogram in the characterization of ion relative abundance and concentration, compared to the nearest concentration of mixed standard solution.
2. Quantitative Analysis
   Record the chromatographic peak areas of the mixed standard working solution, mapping the standard working curve using area for *Y* aisle and concentration for *X* aisle, calculate the concentration of residues by the external standard method, respectively. Each of the six penicillins detected in the samples should be in the linear range of the calibrated standards.

Table 5
MS parameters for the six penicillins

Compounds	Ion-pairs for quantitation (*m*/*z*)	Ion-pairs for qualifier (*m*/*z*)	Collision gas energy/V	Decluster voltage/V	Highlights voltage/V	Collision chamber voltage/V
Ampicillin	350.0/192.1 350.0/160.1 350.0/106.0	350.0/160.1	23 20 20	20	90	10
Amoxicillin	366.1/113.9 366.1/208.0 366.1/349.1	366.1/208.0	30 19 19	21	90	10
Penicillin V	351.0/160.2 351.0/192.1	351.0/160.2	20 15	40	90	10
Penicillin G	335.0/160.1 335.0/176.1	335.0/160.1	20 20	23	90	10
Oxacillin	402.1/160.0 402.1/243.1 402.1/374.2	402.1/160.0	20 20 20	23	90	10
Dicloxacillin	470.3/160.0 470.3/311.2 470.3/442.2	470.3/160.0	20 22 20	20	90	10

2.2.4. Comments

1. The results are calculated using following equation:

$$X = c \times \frac{V}{m} \times \frac{1{,}000}{1{,}000}, \quad (1)$$

where X sample test component residues, units in micrograms per liter (μg/L); c is the concentration from standard curves obtained in the sample solution measured the concentration of components, units in ng per milliliter (ng/ml); V is the sample solution volume size, units in milliliters (ml); m is the sampling volume, units in milliliters (ml).

2. Quality Control Samples: Place 1.0 ml of blank sample into 15 ml centrifuge tube, add 10 μl and 100 μl measured accurately and carry out the procedure under Subheading "Preparation of Sample Solution." The QC samples with the concentration at the LOD and ten times of the LOD (10× LOD) are obtained. The recovery of the added concentrations should be within 60–120%.

3. Correlation coefficient of linearity for each component should be greater than 0.99. The LOD of the method: penicillin G is 0.5 μg/L; penicillin V and oxacillin is 1.0 μg/L; amoxicillin, dicloxacillin, and ampicillin are 2.0 μg/L.

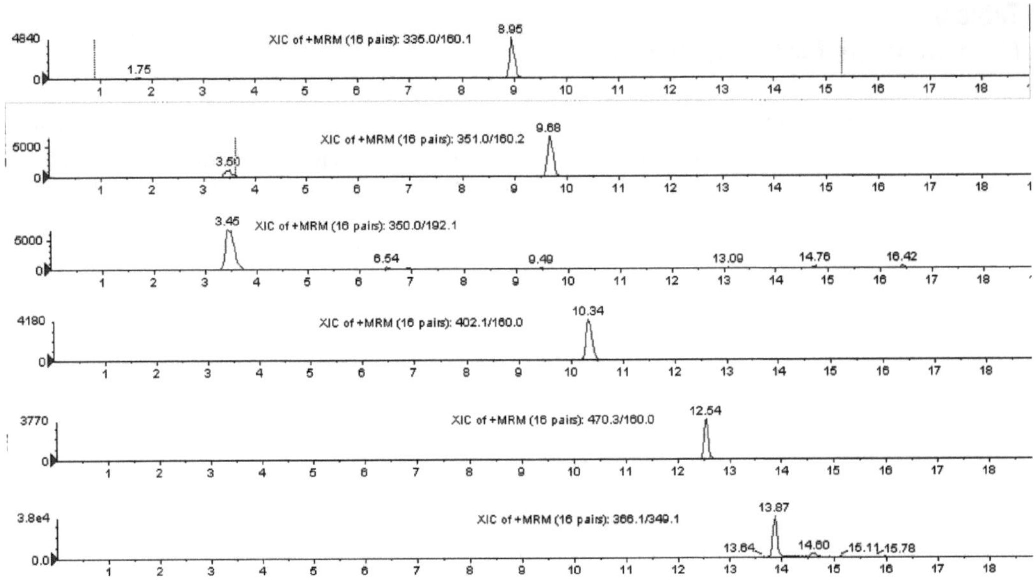

Fig. 2. MRM chromatograms of the six penicillins.

4. In Subheading "Preparation of Sample Solution," cannot be taken to dryness when evaporating with nitrogen or the recovery will be less than required. The whole procedure should be done as soon as possible to prevent the hydrolysis of penicillins. Also if an emulsion occurs after stabilization of the two phases for the LLE, the sample solution should be centrifuged again at 15,000 rpm (15,093 × *g*) and then separated and filtered.
5. The references chromatogram is given in Fig. 2.
6. Both the quantitative transition ion pair (precursor and product ion) and two qualitative transition ion pairs for the qualifier are shown for ampicillin in Fig. 3. The relative abundances of each ion pair must match the standard for the compound with the tolerances shown in Table 6.

2.3. The Residue Analysis of 17 Sulfonamides in Pork, Fishes, and Eggs by LC-MS/MS

2.3.1. Introduction

Sulfonamides are broad spectrum antibiotics frequently used as veterinary medicines. They are commonly used for the prevention and treatment of dairy cattle for several infectious diseases, prophylactic, or as feed additives to promote growth in farm animals. The presence of these drug residues, regardless of their minute amounts, can trigger potential adverse side effects in humans, such as allergic reactions in hypersensitive individuals, other long-term health effects, or they can be potentially carcinogenic. Prolonged exposure to residue antibiotics can result in an increase of drug-resistant bacteria. Therefore, monitoring of their residues is very important in controlling the safety of food for human consumption. In order to ensure correct identification and quantitative analysis of the compounds, a reliable and sensitive analytical method is

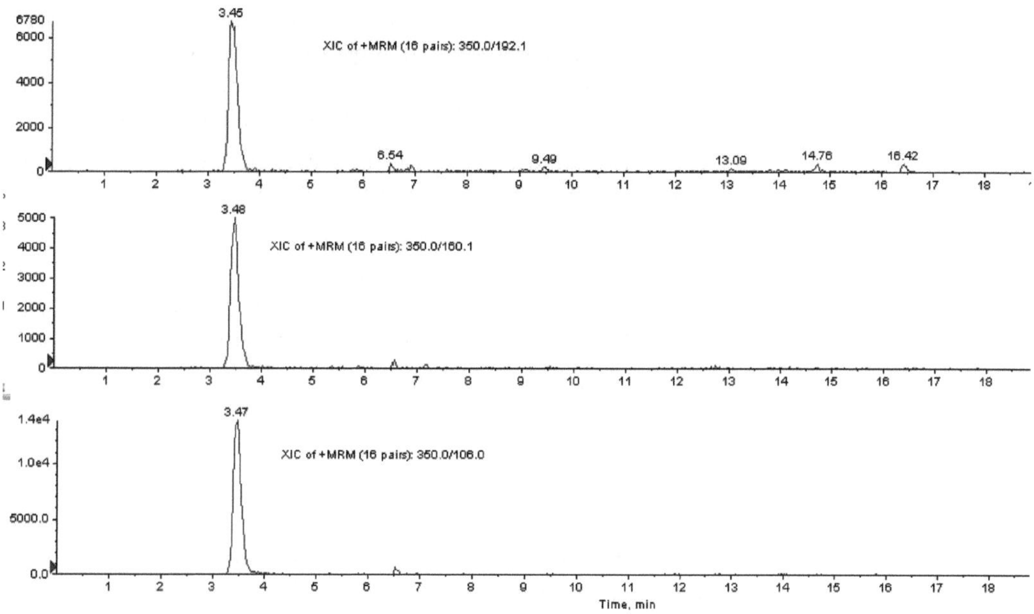

Fig. 3. The qualitative pairs for ampicillin figure are shown.

Table 6
Qualitative relative ion abundance of the maximum allowable deviation (%)

Relative ion abundance	>50	>20~50	>10–20	≤10
Permissible maximum deviation	±20	±25	±30	±50

needed. This method can be used to simultaneous detect 17 sulfonamides residues in pork, fishes, and eggs.

2.3.2. Materials

2.3.2.1. Reagents

1. Acetonitrile: HPLC grade.
2. Isopropanol: HPLC grade.
3. *n*-hexane: HPLC grade.
4. Ammonium acetate: AA.
5. Anhydrous sodium sulfate: AA.
6. Reference substances: sulfacetamide, sulfadiazine, sulfathiazole, sulfasalazine, sulfamethyldiazine, sulfamethazine, sulfamethoxazole thiadiazole, sulfamethoxypyridazine, sulfametoxydiazine, sulfa-6-methoxy pyrimidine, sulfachloropyridazine, o-sulfadimethoxine, sulfamethoxazole, dimethyl sulfisoxazole, M-suladimethoxypyrimidine, sulfaphenazole, $^{13}C_6$-sulfamethoxazole (as is). All should have a specific purity, and meet requirements applied to chromatographic analysis.

7. The standard stock solution of the 17 sulfonamides (0.1 mg/ml): weigh accurately about 10 mg (accurate to 0.1 mg) of each sulfonamide in 100 ml flask, dissolve and dilute with methanol to volume, mix well. Store at 4°C for less than 2 months.
8. Internal standard stock solution: prepare a solution containing $^{13}C_6$-sulfamethoxazole 100 μg/ml with acetonitrile, store at 4°C, the valid period is 24 months.
9. Standard working solution (5 μg/ml): measure accurately 1.0 ml of standard stock solution in 20 ml flask, dilute with methanol to volume. Prepare directly before use.
10. Standard working solution (0.5 μg/ml): measure accurately 1.0 ml of standard working solution (5 μg/ml) in 10 ml flask, dilute with methanol to volume. Prepare directly before use.
11. Internal standard working solution: Measure accurately 1.0 ml of internal standard stock solution, dilute with acetonitrile to 100 ml. The solution contains 1 μg/ml $^{13}C_6$-sulfamethoxazole. Store at 4°C, the valid period is 3 months.
12. Standard working solution mixed with matrix: measure accurately 20, 50, and 100 μl of standard working solution (0.5 μg/ml) and 20, 50, 100, and 200 μl of standard working solution (5 μg/ml) separately, dilute with blank sample extract (prepare the extracts separately using pork, fish, and eggs) to 1.0 ml and mix well. The concentrations are 10, 25, 50, 100, 250, 500 and 1,000 ng/ml (labeled as S1–S7). Store at 4°C for less than a week.

2.3.2.2. Equipment

1. Pipette: 2–20 μl, 20–200 μl.
2. Homogenizer.
3. Rotary evaporator.
4. Oscillator.
5. Centrifuge (temperature control and the minimum speed at 3,000 rpm or more).
6. Centennial bottle: 100 ml.
7. Membrane filters: 0.22 μm.
8. Vials: 2 ml, with PTFE screw-cap.
9. Analytical balance: accurate to 0.1 mg.
10. Balance: accurate to 0.01 g.
11. Centrifuge tubes, 50 ml polypropylene.

2.3.3. Methods

2.3.3.1. Specimen Treatment

Samples are taken out from all the representative edible parts of about 1 kg, fully broken up, and mixed thoroughly, and transferred into a clean container. Store at −18°C till use.

2.3.3.2. Sample Preparation Procedure

Place 5 g sample (accurate to 0.01 g) in 50 ml centrifuge tubes, add 100 μl, measured accurately, of internal standard working solution, then add 20 g of anhydrous sodium sulfate and 20 ml of acetonitrile (for egg samples, add 25 g of anhydrous sodium sulfate and acetonitrile before homogenizing), homogenize for 2 min, rinse aliquot with about 2 ml of acetonitrile, combine mixture (to avoid cross-contamination, rinse aliquot with 0.1 M hydrochloric acid solution, followed by methanol solution and the water after homogeneous), centrifuge at 3,000 rpm (1,962 × *g*) for 3 min. Repeat the extraction procedure for the residue with 20 ml of acetonitrile and combine both supernatant extracts. Transfer the combination into a 100 ml Centennial bottle, and then add 10 ml of isopropanol, mix well, dry at 30°C in a water bath using rotary evaporator to dryness. Add 1.0 ml of acetonitrile, 0.01 M ammonium acetate (12:88), and 0.8 ml of *n*-hexane. Vortex for 1 min to dissolve the residue, transfer to 2 ml centrifuge tube, centrifuge at 12,000 rpm (9,660 × *g*) for 5 min. Discharge the upper layer *n*-hexane, and then add 0.8 ml of *n*-hexane. Vortex for 1 min, repeat the above procedure till the lower layer water phase is transparent. The lower aqueous phase is filtered through 0.22 μm filter for LC-MS/MS analysis. Prepare the blank sample as the above steps.

2.3.3.3. HPLC Conditions

1. HPLC: Agilent liquid chromatography (HP1100).
2. Column: WATERS XTerra MS C18, 3.5 μm, 100 mm×4.6 mm.
3. Column temperature: 25°C.
4. Injection volume: 10 μl.
5. The mobile phase operated with gradient program in Table 7. A: Acetonitrile –0.02% formic acid aqueous solution (18:82). B: Acetonitrile.

Table 7
Mobile phase gradient program and flow rate for sulfonamides

Time (min)	A (%)	B (%)	Flow rate (μl/min)
0	100	0	250.0
2	100	0	250.0
5	95	5	250.0
7	60	40	250.0
9	60	40	250.0
9.5	100	0	300.0
11.8	100	0	300.0
12	100	0	250.0

2.3.3.4. MS Conditions (Applied Biosysytem API4000 or 3200QTrap)

1. Ion Source: ESI ion source.
2. Scan Mode: Positive Ion Scan.
3. Detect mode: multiple reaction monitoring.
4. ESI voltages: 5,000 V.
5. Ion source temperature: 500°C.
6. Quantitative and qualitative ions, CE and DP values were listed in Table 8.

Table 8
MS/MS parameters for sulfonamide antibiotics

Components	Ion-pairs for quantitation (*m*/*z*)	Ion-pairs for qualifier (*m*/*z*)	Collision gas energy/V	Tube lens
Sulfanilamide	173/92	173/156	19	77
	173/108		16	77
	173/156		6	73
Sulfacetamide	215/92	215/156	21	97
	215/108		21	97
	215/156		10	97
Sulfadiazine	251/92	251/156	27	83
	251/108		29	83
	251/156		16	83
Sulfathiazole	256/92	256/156	26	87
	256/108		24	87
	256/156		15	87
Sulfasalazine	250/92	250/156	29	91
	250/156		17	91
	250/184		18	91
Sulfamethyldiazine	265/92	265/156	29	93
	265/156		17	93
	265/172		16	93
Sulfamethazine	279/124	279/156	27	94
	279/156		19	94
	279/186		18	94
Sulfamethoxazole thiadiazole	271/92	271/156	29	83
	271/108		22	83
	271/156		16	83
Sulfamethoxypyridazine	281/92	281/156	31	94
	281/108		29	94
	281/156		18	94
Sulfametoxydiazine	281/92	281/156	31	94
	281/108		29	94
	281/156		18	94

(continued)

Table 8 (continued)

Components	Ion-pairs for quantitation (*m/z*)	Ion-pairs for qualifier (*m/z*)	Collision gas energy/V	Tube lens
Sulfa-6-methoxy pyrimidine	281/92 281/108 281/156	281/156	31 29 18	94 94 94
sulfachloropyridazine	285/92 285/108 285/156	285/156	31 28 15	90 90 90
o-Sulfadimethoxine	311/92 311/108 311/156	311/156	28 27 18	97 97 97
Sulfamethoxazole	254/92 254/108 254/156	254/156	33 23 16	83 83 83
Dimethyl sulfisoxazole	268/108 268/113 268/156	268/156	22 17 15	93 93 93
M-sulfadimethoxypyrimidine	311/92 311/108 311/156	311/156	34 29 21	97 97 97
Sulfaphenazole	315/156 315/158 315/160	315/156	20 30 20	97 97 97

2.3.4. Comments

1. The retention time are listed in Table 9 for reference. For the sample solution and QC solution, the retention times should be within ±5%.
2. The results are calculated by internal standards (IS) or by using following equation:

$$X = c \times V \times 1{,}000/(m \times 1{,}000),$$

where X is the component residues (μg/kg) in sample, c is the concentration (ng/ml) calculated in accordance with the standard curve, V is the sample volume (ml), and m is the quantity of the specimen (g).

3. Quality Control samples: place 5.0 g of blank sample 3 each in separate 50 ml centrifuge tubes, add 50 μl of standard working solution (0.5 μg/ml) one tube and standard working solution (5 μg/ml) in the other two tubes. Carry out the procedure under Subheading "Sample Preparation Procedure." The blank sample and QC samples with the concentration at the LOD and ten times the LOD (10× LOD are obtained.

Table 9
Retention times of sulfonamide antibiotics

Components	The retention time (min)	Components	The retention time (min)
Sulfanilamide	2.17	Sulfacetamide	2.94
Sulfadiazine	3.14	Sulfathiazole	3.26
Sulfasalazine	3.52	Sulfamethyldiazine	4.01
Sulfamethazine	4.88	Sulfamethoxazole thiadiazole	5.20
Sulfamethoxypyridazine	5.28	Sulfametoxydiazine	5.59
Sulfa-6-methoxy pyrimidine	6.87	sulfachloropyridazine	7.69
o-Sulfadimethoxine	8.03	Sulfamethoxazole	8.36
Dimethylsulfisoxazole	8.63	M-sulfadimethoxypyrimidine	9.07
Sulfaphenazole	9.23		

The extract recovery rate of added concentration should be more than 20% for sulfadiazine, sulfachloropyridazine, sulfamethoxazole, and dimethylsulfisoxazole and more than 40% for the other components.

4. The LODs are 2.5 μg/kg for sulfamethoxazole thiadiazole; 5.0 μg/kg for sulfanilamide, sulfacetamide, sulfadiazine, sulfasalazine, sulfamethyldiazine, sulfa-6-methoxy pyrimidine, sulfachloropyridazine, o-sulfadimethoxine, sulfamethoxazole, and dimethylsulfisoxazole; 10.0 μg/kg for sulfathiazole, sulfamethoxypyridazine, and M-suladimethoxypyrimidine; 20.0 μg/kg for sulfamethazine and sulfametoxydiazine; and 40.0 μg/kg for sulfaphenazole.
5. The chromatograms of 16 sulfanilamides (except sulfanilamide) are shown in Fig. 4 for reference.

2.4. The Residue Analysis of Nitrofuran Metabolisms in Animal Tissues, Marine Products and Eggs: LC-MS/MS

2.4.1. Introduction

Nitrofuran antibiotics, which include furazolidone, nitrofurantoin, furaltadone, and nitrofurazone are rapidly metabolized and have in vivo half-lives of only a few hours. Their metabolites, however, are highly stable and have significant potential for genotoxicity. These compounds include 3-amino-2-oxazolidinone (AOZ), 3-amino-5-morpholinomethyl-2-oxazolidinone (AMOZ), semicarbazide (SEM), and 1-aminohydantoin (AHD) (13). The structures of the antibiotics and their metabolites are given in Table 10.

This method is suitable for food of animal origin, such as aquatic products, muscles, tissues, and eggs. Considering the complex of the matrix, the LC-MS/MS is adopted for this analysis.

Sample preparation requires hydrolysis under acidic conditions, derivatization with 2-nitrobenzaldehyde (2-NBA) liquid–liquid extract, and then drying the extract and reconstituting in initial mobile phase. Internal standards are used for quantitation.

2.4.2. Materials

2.4.2.1. Reagents

1. Pure water: abide by GB/T6682.
2. Methanol: HPLC grade.
3. Acetonitrile: HPLC grade.

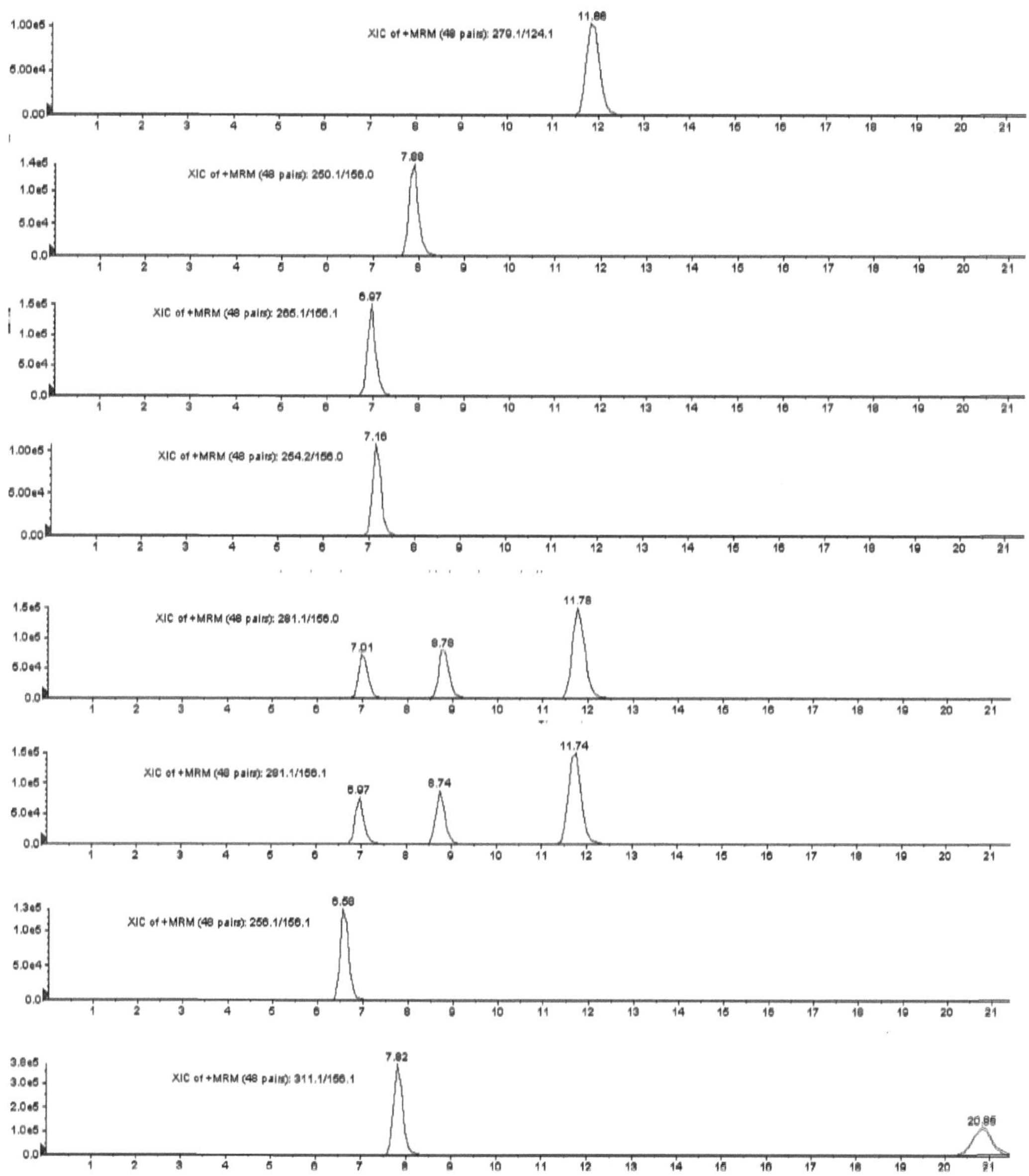

Fig. 4. MRM chromatograms of sulfonamide antibiotics.

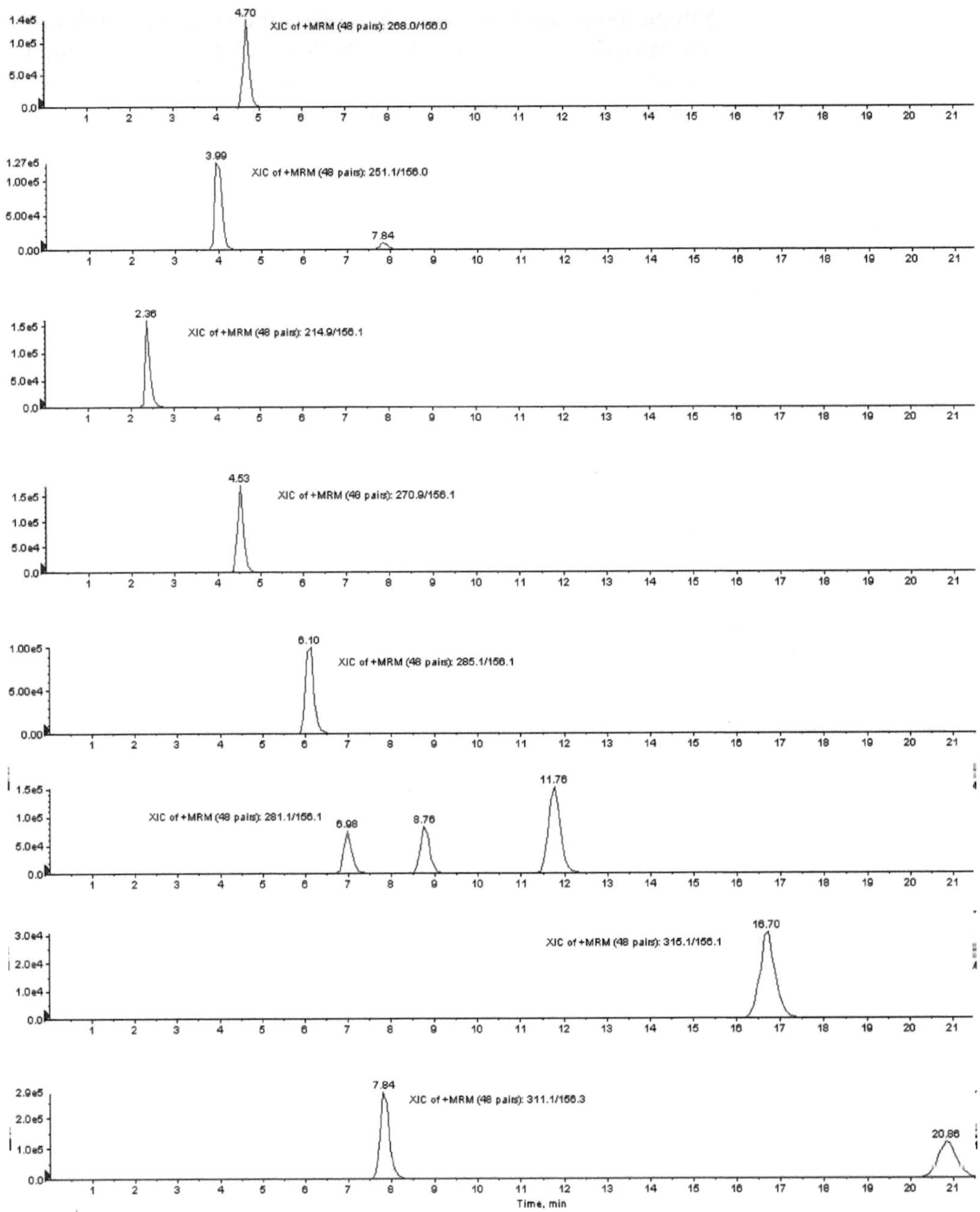

Fig. 4. (continued)

4. Ethyl acetate: HPLC grade.
5. Dipotassium hydrogen phosphate trihydrate.
6. Formic acid: AA.
7. DMSO: AA.

Table 10
Structure of the nitrofuran antibiotics and their metabolites

Patent drugs	Metabolites
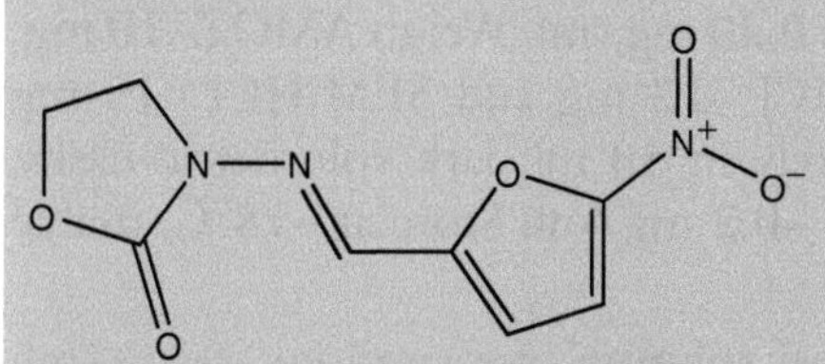	AOZ 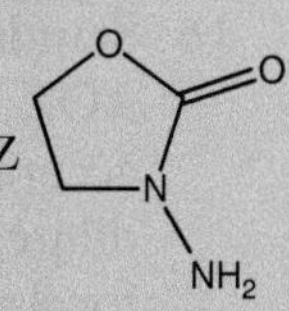
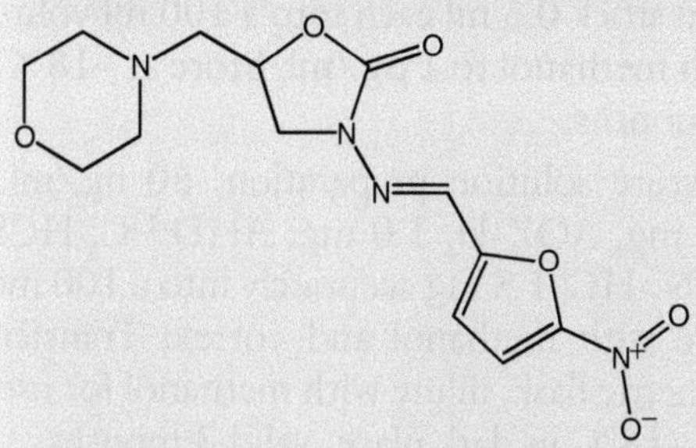	AMOZ 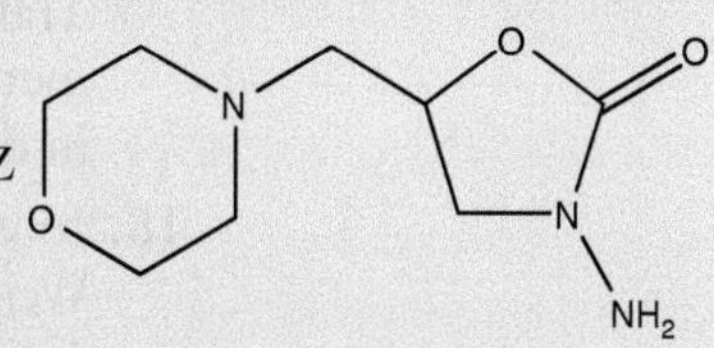
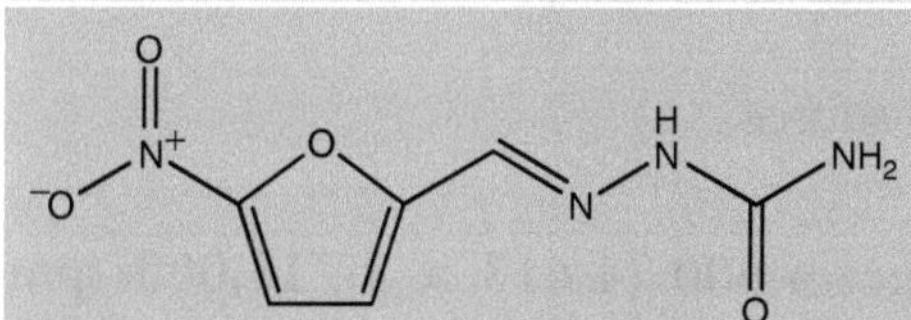	SEM 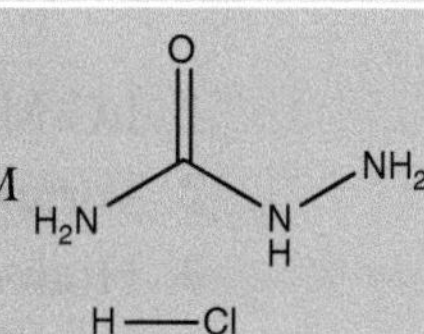
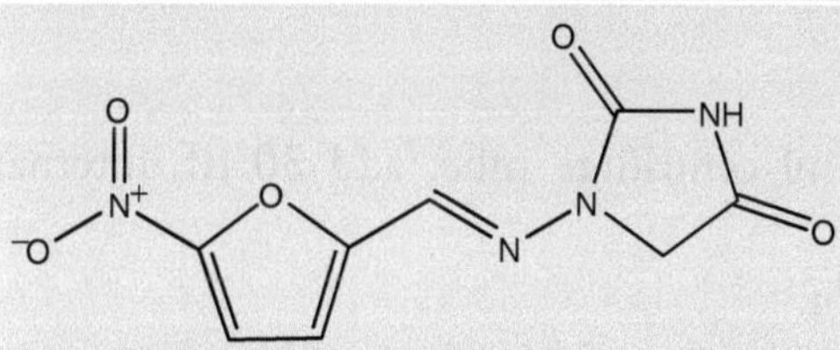	AHD 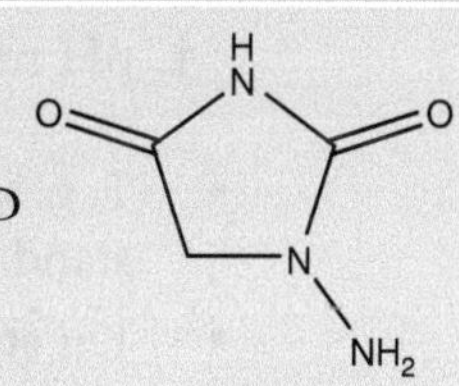

8. Hydrochloric acid: AA.
9. Sodium hydroxide, NaOH: AA.
10. 2-NBA: Purity≥99%.
11. Potassium phosphate dibasic solution: 0.1 mol/L. Weigh 5.7 g potassium phosphate dibasic, dissolve with water to 250 ml.
12. HCl solution: 0.1 mol/L. Transfer 9.0 ml concentrated HCl, dilute with water to 1,000 ml.
13. NaOH solution: 1 mol/L. Weigh 40 g sodium hydroxide, dissolve with water to 1,000 ml.

14. 2-NBA solution: 20 g/L. Weigh 100 mg 2-NBA with 5 ml DMSO before use.
15. Standards: AMOZ (FW201.11), AOZ (FW102.04), AHD (FW115.04/), SEM (FW75.043) and the internal standards: AMOZ-D_5, AOZ-D_4, AHD-$^{13}C_3$ · SEM-$^{13}C_{15}N_2$: purity ≥98%.
16. Standard stock solution: 0.20 mg/ml. Weigh AMOZ~10 mg, AOZ ~10 mg, AHD·HCl ~13 mg and SEM·HCl ~15 mg accurately, place separately in 50 ml dark volumetric flasks, dilute with methanol to ~0.2 mg/ml. Store at −18°C, usable for 3 months.
17. Standard stock mixture solution preparation: 1 μg/ml. Transfer above solution stock 0.5 ml each into a 100 ml volumetric flask, dilute with methanol to 1 μg/ml. Store at −18°C in dark place, valid 3 months.
18. Internal standards mixture solution preparation: 50 ng/ml. Weigh AMOZ-D_5 1.0 mg, AOZ-D_4 1.0 mg, AHD-$^{13}C_3$·HCl 1.4 mg and SEM-$^{13}C_{15}N_2$·HCl 1.5 mg accurately into a 100 ml volumetric flask, dilute with methanol and vortex; Transfer 1 ml into 200 ml volumetric flask, dilute with methanol for use (~50 ng/ml). Store at −18°C in dark place, valid 3 months.

2.4.2.2. Instruments and Equipments

1. LC-MS/MS with ESI source.
2. Vortexer.
3. High-speed centrifuge: 4,600 (4,613 × *g*)/15,000 rpm (15,093 × *g*).
4. pH meter.

2.4.3. Methods

2.4.3.1. Sample Preparation

- 1 g sample in 50 ml centrifuge tube, add 50 μL internal standards.
- Let stand for 15 min.
- Add 10 ml 0.125 M HCl sol., 100 μL 0.05 M 2-NBA sol. mix gently.
- Incubate on water bath (37°C for 16 h).
- Cool at room temperature.
- Adjust to pH 7.4 with 10 ml 0.1 M potassium phosphate and 500 μL 0.8 M NaOH.
- Centrifuge at 4,000 rpm (3,488 × *g*) for 5 min and transfer the supernatant into new tube.
- Add 10 ml hexane, mix and centrifuge at 4,000 rpm (3,488 × *g*) for 5 min.
- Transfer the aqueous lower phase into new tube.
- Add 7 ml ethyl acetate and centrifuge at 4,400 rpm (4,221 × *g*) for 10 min.
- Transfer the supernatant into glass tube, repeat two times.
- Evaporate to dryness under N_2 (40°C).

- Dissolve with 0.5 ml 50% methanol:water.
- Filter with 0.2 μm nylon syringe filter.
- Inject into LC-MS or LC-MS/MS.

Note: Adjust the same pH of both samples solution and standards solutions.

2.4.3.2. LC Setting

(a) Column: C18 (100 mm × 4.6 mm × 3.5 μm) or equivalent.
(b) Column temp: 35°C.
(c) Injection volume: 20 μL.
(d) Mobile phase: methanol-ammonium acetate (NH_4OAc) buffer (1 mmol/L NH_4OAc and 1 mmol/L formic acid solution) (45:55).
(e) Flow rate: 0.4 ml/min.
(f) Stop time: 10 min.

2.4.3.3. MS/MS Settings

(a) Ion Source: ESI.
(b) Scan mode: Positive.

The MRM transitions and collision energies are given in Table 11.

2.4.4. Comments

1. Identification:
   Using RT as the identification of quantifier and qualifier ions, the ratio of the two should be <30% of standard for each analyte.
   Blank solution: same procedure with no standards added (internal standards added)
   Double blank solution: same procedure with no standards and internal standards added.

Table 11
MRM setting for metabolites of nitrofurans

Name	Quantitation ion (*m/z*)	Qualifier ion (*m/z*)
AMOZ	335.3/291.2	335.3/262.2
AOZ	236.1/134.1	236.1/104.0
AHD	249.1/134.1	249.1/178.1
SEM	209.1/166.2	209.1/192.1
AMOZ-D_5	340.2/296.4	
AOZ-D_4	240.1/134.0	
AHD-C_3	252.1/134.1	
SCA-^{13}C-$^{15}N_2$	212.1/168.2	

2. Calculate the results using following equation,

$$X = \frac{C \times C_i \times A \times A_{si} \times V}{C_{si} \times A_i \times A_s \times W}, \quad (2)$$

where X is the sample result (ng/g); C AMOZ, SEM, AHD, or AOZ standard calibration solution (ng/ml); C_{si} is the concentration of the standard solution (ng/ml); Ci is the concentration in the samples (ng/ml); A is the response of AMOZ, SEM, AHD, or AOZ in the real samples; A_s is the response of AMOZ, SEM, AHD, or AOZ in the standard solution; A_{si} is the internal standard response in the calibration solution; Ai is the response of the IS in the samples; V is the volume of the samples (ml); and W is the weight of samples (g).

3. LOD, Accuracy, and precisions
   LOD: AMOZ, AOZ, SEM, and AHD LOD is 0.25 ng/g in aquatic products.
   At the concentration of LOD, recovery varies from 50 to 120%
   At the concentration of 10× LOD, recovery varies from 60 to 120%.
   In general, the recovery is between 85 and 115%
   Coefficient of linearity, r 0.99 for each compound.
4. Limits
   Using this method, the total of AMOZ, AOZ, SEM, and AHD are lower than 1 μg/kg.
5. The chromatogram of the analytes is shown in Fig. 5.

2.5. The Residue Analysis of 3 Tetracyclines in Meats and Eggs

2.5.1. Introduction

Oxytetracycline, tetracycline, chlorteytractcline, and doxycycline are determined in chicken muscle.

2.5.2. Materials

2.5.2.1. Reagents

1. Standards: oxytetracycline, tetracycline, chlortetracycline, and doxycycline.
2. Mcllvaine-EDTA Buffer: dissolve Na_2HPO_4 54.93 g, monohydrate citric acid 23.60 g, and EDTA-2Na 82.36 g with water into 2,000 ml, mix well.
3. Metaphosphate solution: dissolve 2.0 g with 80 ml of water, then add 20 ml of acetonitrile, mix well.
4. Methanol + water (1 + 19).
5. Mobile phase: acetonitrile + methanol + 0.01 mol/L oxalic acid solution (2 + 1 + 7).
6. Oasis® HLB (500 mg, 6 ml waters) SPE column, or equivalent.

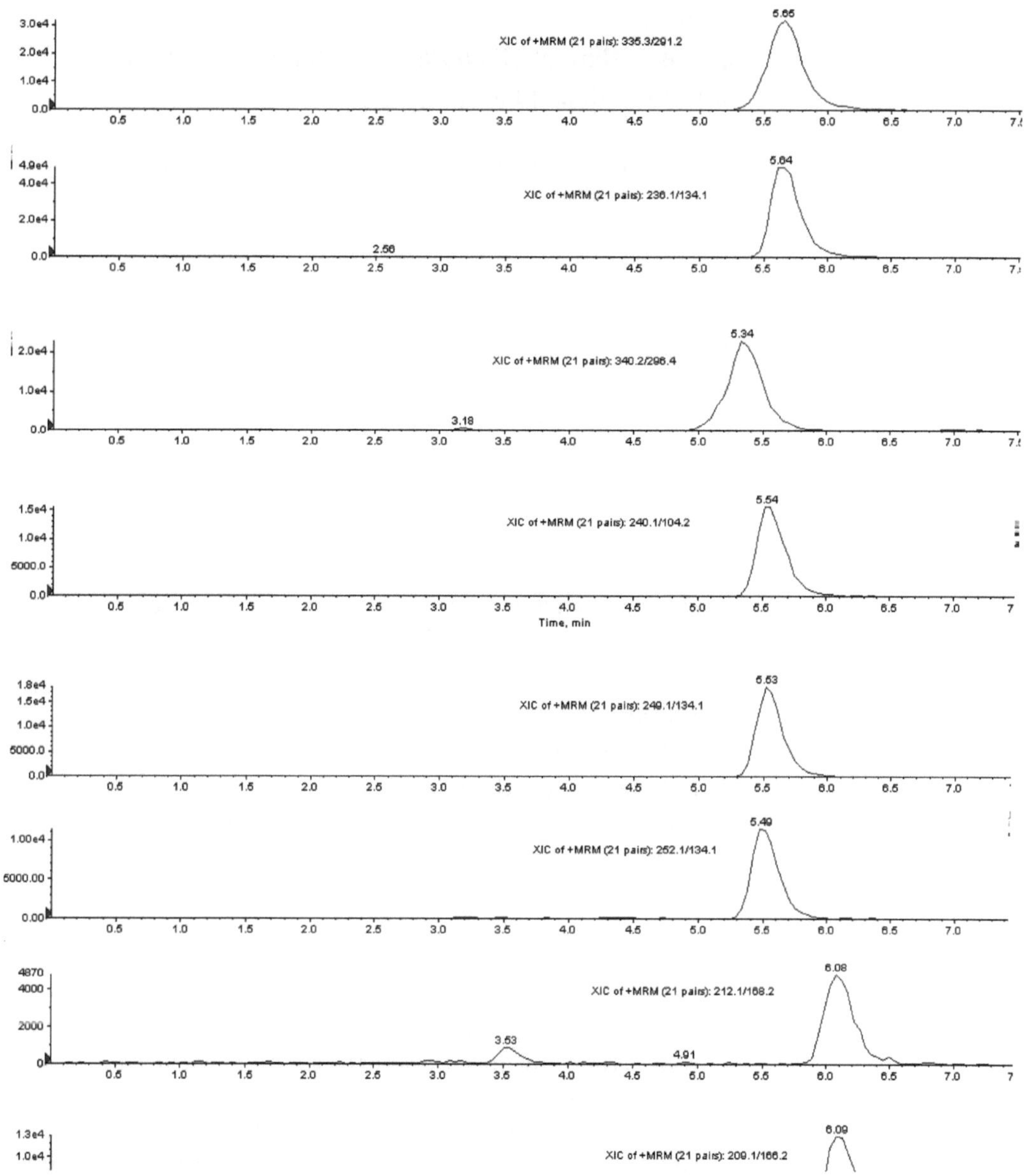

Fig. 5. MRM chromatogram of each of the derivatized nitrofuran metabolites.

2.5.2.2. Instruments

1. LC with DAD detector.
2. Balance.

2.5.3. Methods

2.5.3.1. Standards Solutions

1. Oxytetracycline, tetracycline, and chlortetractcyline standard stock solution (100 μg/ml): weigh oxytetracycline, tetracycline, and chlortetractcyline about 10 mg, dissolve to 100 ml with methanol.
2. Oxytetracycline, tetracycline, and chlorteytractcline mixture standard solution: transfer oxytetracycline, tetracycline, and

Table 12
Concentration of standards prepared for the analysis of tetracyclines [unit: (mg/kg)]

	Oxytetracycline residue	Tetracycline residue	Chlortetracyline residue
S1	0.01	0.01	0.01
S2	0.02	0.02	0.02
S3	0.05	0.05	0.05
S4	0.10	0.10	0.10
S5	0.20	0.20	0.20
S6	0.30	0.30	0.30

chlorteytractcline standard stock solution above 0.5 ml each into a 10 ml volumetric flask, dilute with mobile phase (fresh preparation).

3. Internal standards solution: weigh 10 mg doxycycline, dissolve with methanol to 100 ml.
4. Calibration standard solution: Transfer the above mixture standard solution 0, 12, 24, 60, 120, 240, 360 μl, and add 80 μl IS solution, then to 4 ml to obtain the concentrations for the standards as shown in Table 12.

2.5.3.2. Sample Preparation

1. Homogenize the sample for use.
2. Extraction and Clean-up
   Extraction: weigh 6.00 g of samples, into 50 ml centrifuge tube, add 7 ml metaphosphate solution, vortex for 1 min, then add 10 ml Mcllvaine-EDTA Buffer, vortex for 1 min, at < 10°C, centrifuge at 4,600 rpm for 15 min, and transfer the supernatant to another 50 ml centrifuge tube; repeat above steps and combine the extracts, add 80 μl IS solution, then add 10 ml *n*-hexane vortex for 3 min, at <10°C centrifuge at 4,600 rpm for 15 min, discard the hexane layer.
   Clean-up: take Oasis® HLB (200 mg, 6 ml waters) SPE cartridges, using 6 ml methanol, 6 ml water to active cartridges, wash with methanol + water (1 + 19) 3 ml, dry for 10 min, then using methanol to wash the column 15 ml, collect eluant. Then, using (500 mg, 3 ml Agela) cartridges for further clean-up, collect the eluant of 4 ml mobile phase.
3. Quality control sample preparation: three blank samples of 6.0 g each, put into 50 ml centrifuge tubes, add two of mixture

standard solution 24 μl and 240 μl as LOD and 10 LOD quality control samples, for the rest add the metaphosphate solution, 7 ml, for the same procedure.

4. Blank solution: same procedure without samples.

2.5.3.3. LC Setting

1. Column: C_{18} 250 mm × 4.6 mm, 5 μm.
2. ACN + methanol + 0.01 mol/L oxalic acid solution (2 + 1 + 7).
3. Flow rate: 1.0 ml/min.
4. Injection volume: 100 μl.
5. Temp: 35°C.
6. Detection wavelength: 350 nm.

2.5.4. Comments

1. LOD is 0.02 mg/kg.
2. Recovery
   In muscles, recovery of oxytetracycline, tetracycline, and chlortetracycline varies in the range of 40–110% at 0.1–0.6 mg/kg.
   In egg, recovery of oxytetracycline, tetracycline, and chlortetracycline varies in the range of 40–110% at 0.1–0.6 mg/kg.
   In aquatic samples, recovery of oxytetracycline, tetracycline, and chlortetracycline varies in the range of 40–110% at 0.1–0.6 mg/kg.
3. Confirmation by LC-MS
   (a) LC condition:
       Column: Venusil XBP C_8: 5 μm 2.1 mm × 15.0 cm
       Same with the above LC method.
   (b) MS parameters:
       Ion Source Type: ESI.
   (c) Scan mode: Positive.
   (d) MRM *m/z*:
       Oxytetracycline: 461.1/425.9 (quantitation ion) 461.1/381.0 (qualifier ion).
       Tetracycline: 445.2/410.0 (quantitation ion) 445.2/427.0 (qualifier ion).
       Chlortetracycline: 479.1/444.1 (quantitation ion) 479.1/461.9 (qualifier ion).
4. The MS chromatogram for confirmations is shown in Fig. 6 for reference. Note that only the quantitation ion is shown but for confirmation both quantitation and qualifier ion must be detected at the ratio of the standard for confirmation and the retention time must be within tolerance of the standard.

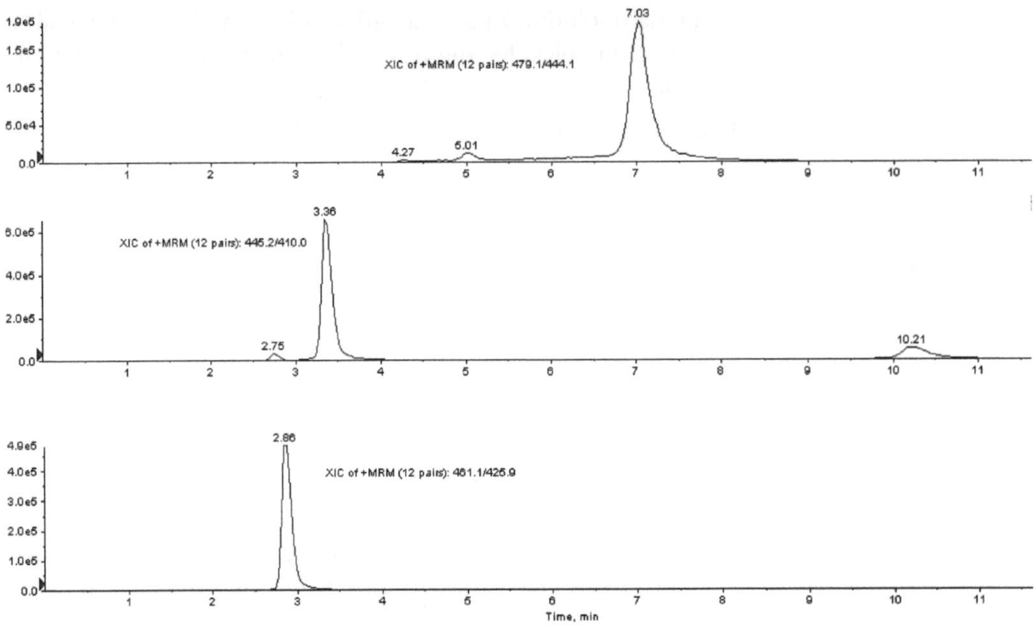

Fig. 6. MRM chromatograms of quantitation ions for the three tetracyclines.

2.6. The Residue Analysis of Chloramphenicol in Meats, Marine Products, Honey, and Eggs

2.6.1. Introduction

This is the LC-MS/MS method to detect CAP in the animal products.

The target matrix covers meat, egg, aqueous, etc. animal original.

2.6.2. Materials

2.6.2.1. Reagents

1. Water: LC grade.
2. Methanol, acetonitrile, and ethyl acetate: LC grade.
3. NaCl, *n*-hexane: ammonium acetate.
4. A solution of 4% sodium chloride: weigh NaCl 4.0 g, disslove with water to 100 ml. Prepare before use.
5. CAP: purity more than 99%.
6. d5-CAP: internal standards with purity more than 98%.
7. Standard stock solution: weigh CAP to about 10 mg, put into 100 ml volumetric flask, diluted to 100 μg/ml with methanol. Store at −20°C, valid for 1 year.
8. Internal standard stock solution: weigh d5-CAP 10 mg, put into 100 ml volumetric flask, diluted to 100 μg/ml with methanol. Store at −20°C and valid for 1 year.
9. Calibration curve solutions: Transfer above standard stock solution, 0.5 ml accurately, put into 100 ml volumetric flask, dilute with mobile phase to make the concentration at

500 ng/ml standard calibration curve solution, transfer this solution at 5 ml and 1 ml to two 50 ml volumetric flasks, prepare the concentration at 50 ng/ml and 10 ng/ml standard calibration curve solution diluted with mobile phase. Store at 4°C, valid for 3 months.

d5-CAP standard calibration solution: transfer above IS stock solution, 0.5 ml, into a 500 ml volumetric flask, make the concentration at 100 ng/ml IS calibration solution with mobile phase. Store at 4°C, valid for 3 months.

2.6.2.2. Instruments

1. LC-MS with Electrospray source.
2. Micropipet.
3. Balance (limitation 0.01 mg and 0.01 g each).
4. N_2 evaporator.
5. Low temperature centrifuge.
6. High-speed centrifuge.
7. Vortexer.
8. Plastic tube with caps.

2.6.3. Methods

2.6.3.1. Guideline

After adding the IS, the samples are extracted by ACN with 4% NaCl to remove the protein, using *n*-hexane to remove the grease, then extract with ethyl acetate, filter and ready for use.
Use the internal standards method for the quantitation.

2.6.3.2. Extraction

Weigh about 5 ± 0.02 g homogenized samples into 50 ml centrifuge tubes, add 50 μL IS calibration solution, then 10 ml ACN, 10 ml of 4% NaCl solution, vortex for 20 min to extract thoroughly. Centrifuge at 4,600 rpm (4,613 × *g*) for 10 min, and transfer the supernatant into another 50 ml centrifuge tube. Then, add 5 ml *n*-hexane, vortex for 1 min, centrifuge at 4,000 rpm (3,488 × *g*) for 10 min, and discard the supernatant. Add 6 ml ethyl acetate, vortex 1 min, centrifuge at 4,600 rpm (4,613 × *g*) for 10 min, transfer the supernatant to a 15 ml centrifuge tube, dry under N_2, dissolve with 1 ml mobile phase, and injection at 10 μl.

2.6.3.3. Calibration Curve

Transfer CAP standard calibration solution and d5-CAP calibration solution as shown in the Table 13, diluted with mobile phase to 1.0 ml, make the CAP solution at 0.5, 1.0, 2.0, 5.0, 10, and 50 ng/ml, d5-CAP concentration at 5.0 ng/ml.

2.6.3.4. LC Setting

1. Column: Xterra C18 10 cm × 4.6 mm (or equivalent).
2. Mobile phase: methanol-water (50:50).
3. Flow rate: 0.4 ml/min.
4. Injection volume: 10 μl.

Table 13
Volume of solutions to make working standards for chloramphenicol quantitation

Solutions	0.5 ng/ml	1.0 ng/ml	2.0 ng/ml	5.0 ng/ml	10.0 ng/ml	50.0 ng/ml
10 ng/ml CAP	50 μL	100 μL	–	–	–	–
50 ng/ml CAP	–	–	40 μL	100 μL	–	–
500 ng/ml CAP	–	–	–	–	20 μL	100 μL
IS	50 μL	50 μL	50 μL	50 μL	50 μL	50 μL
Diluent[a]	900 μL	850 μL	910 μL	850 μL	930 μL	850 μL

[a]Use initial mobile phase

Table 14
MRM setting for chloramphenicol (CAP)

Compounds	Precursor ion (*m/z*)	Ions for quantitation and qualifier (*m/z*)	CE/V
CAP	320.9 320.9	152.1 257.1	23 15
D5-CAP	325.9	157.1	23

2.6.3.5. MSD Setting

1. Ion Source: ESI.
2. Scan mode: Negative.

MRM transitions and collision energies are given in Table 14.

2.6.4. Comments

1. Quality control samples preparation
   Weigh 3 of 5 ± 0.02 g samples, put into three 50 ml centrifuge tubes, the first add none, the next two add 10 ng/ml and 50 ng/ml CAP calibration solution 50 μl and 100 μl, respectively, to make the quality control samples at 0 μg/kg (blank), 0.1 μg/kg (LOQ)and 1.0 μg/kg, add IS calibration solution 50 μL, test as below.
2. The LOD of the method is 0.1 μg/kg.
3. Recovery: at 0.1 μg/kg (LOQ), recovery varies between 60 and 120%; At the concentration of 1 μg/kg (10× LOD), the recovery is between 80 and 120%.

4. Double blank: prepare as the procedure above without samples or IS added.
5. Quality control sample: recovery is between 80 and 120%.
6. Calculation

$$\frac{A_s}{A_{is}} = a \times \frac{C_s}{C_{is}} + b, \quad (3)$$

$$C = \frac{C_{is}}{a}\left(\frac{A}{A_{is}} - b\right), \quad (4)$$

The CAP concentration is:

$$X = \frac{CV}{m}, \quad (5)$$

where A_s is the response of CAP standard solution; A_{si} is the response of IS standard solution; C_s is the CAP standard concentration (ng/ml); C_{is} is the CAP IS concentration (ng/ml); C is the CAP concentration in the sample (ng/ml); C_{is} is the IS concentration in the sample (ng/ml); A is the response of the CAP of sample; A_{is} is the response of the IS in the sample; X is the CAP residue (μg/kg); V is the dissolve solvent amount (ml); m is the sample weight (g).

7. The chromatogram for CAP is shown in Fig. 7 as reference.

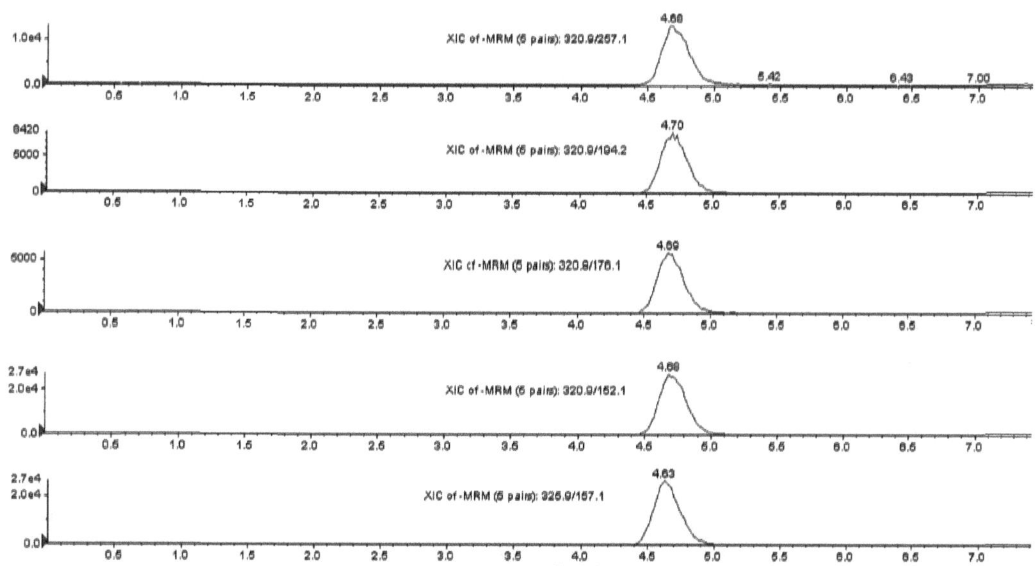

Fig. 7. MRM chromatograms of four transitions of chloramphenicol. Bottom chromatogram is that of d5-CAP used as internal standards.

3. General Notes on Regulations

There is increasing awareness of food safety by the consumer with respect to antimicrobial resistance due to the discovery of new resistant strains of bacteria and others that are becoming increasingly resistant over time. As a result, legislation with more and stricter criteria for monitoring antimicrobial residues has been laid down by governments all over the world. This is a very important area for the monitoring of veterinary residues in agriculture, as there are so many different compounds and matrices required to be monitored under the strict legislation. The requirement to be able to measure even lower concentration levels is a great analytical challenge. Despite the activity in this area of research, there still exist many gaps for certain matrices and species that residue laboratories are required to monitor in their national residue plans.

Residues are divided into two groups, A and B, for the purposes of monitoring and defining the legislation relating to them. Group A refers to substances having an anabolic or antibiotic effect and are unauthorized substances while group B refers to veterinary drugs and contaminants. The nitrofurans and CAP mentioned in this chapter are both listed under Group A, and β-lactam antibiotics, sulfonamides, tetracyclines, and fluoroquinolones are listed under Group B. These antibacterial substances listed under Group B mentioned in this chapter with established MRLs are shown in Table 15, which was laid down in China (14), the USA, the EU Council, and Japan in 2002. The definition of MRLs in the Table is according to the EU Council (15) regulation is given as: "'maximum residue limit' and means the maximum concentration allowed of residue resulting from the use of a veterinary medicinal product (expressed in mg/kg or μg/kg on a fresh weight basis) which may be accepted by the Community to be legally permitted or recognized as acceptable in or on a food." For Group A, the limits are completely dependent on the sensitivity of analytical method as they are banned substances.

4. Conclusions

LC-MS is a suitable technique for the quantification of antibiotics residues in food samples and is still under further development. Due to the complexity of the food matrices, the sample pretreatment is not only a crucial, but also a time-consuming step in the analytical process, and the most important part of the method development is the generic extraction of the compounds of interest from the matrix. In a large majority of studies, liquid partitioning

Table 15
Some antibacterial substances residues mentioned in this chapter with established MRLs required in China, the USA, the EU Council, and Japan

Compounds	Controlled/monitored compounds	Source	Target tissues	MRLs in China (μg/kg)	MRLs in the USA (μg/kg)	MRLs in the EU (μg/kg)	MRLs in Japan (μg/kg)
Amoxicillin	Amoxicillin	Food from animal origin	Muscle	50	10	50	
			Fat	50	10	50	
			Liver	50	10	50	
			Kidney	50	10	50	
			Milk	10	10	10	
Ampicillin	Ampicillin	Food from animal origin	Muscle	50	10	50	
			Fat	50	10	50	
			Liver	50	10	50	
			Kidney	50	10	50	
			Milk	10	10	10	
Benzylpenicillin/procainebenzylpenicillin	Benzylpenicillin	Food from animal origin	Muscle	50		50	50
			Fat	50		50	50
			Liver	50		50	50
			Kidney	50		50	50
			Milk	4		4	4
Cefalexin	Cefalexin	Cattle	Muscle	200		200	
			Fat	200		200	
			Liver	200		200	
			Kidney	1,000		1,000	
			Milk	100		100	

(continued)

Table 15 (continued)

Compounds	Controlled/monitored compounds	Source	Target tissues	MRLs in China (μg/kg)	MRLs in the USA (μg/kg)	MRLs in the EU (μg/kg)	MRLs in Japan (μg/kg)
Cefquinome	Cefquinome	Cattle	Muscle	50		50	
			Fat	50		50	
			Liver	100		100	
			Kidney	200		200	
			Milk	20		20	
		Pigs	Muscle	50		50	
			Skin + fat	50		50	
			Liver	100		100	
			Kidney	200		200	
Ceftiofur	Desfuroylceftiofur	Cattle/pigs	Muscle	1,000	1,000	1,000	1,000
			Fat	2,000		2,000	2,000
			Liver	2,000	2,000	2,000	2,000
			Kidney	6,000	8,000	6,000	6,000
		Cattle	Milk	100		100	100
Cloxacillin	Cloxacillin	Food from animal origin	Muscle	300		300	
			Fat	300		300	
			Liver	300		300	
			Kidney	300		300	
			Milk	30		30	
Danofloxacin	Danofloxacin	Cattle/sheep	Muscle	200		200	
			Fat	100		100	
			Liver	400		400	
			Kidney	400		400	
			Milk	30		30	
		Poultry	Muscle	200		200	
			Skin + fat	100		100	

			Liver	400		400	
			Kidney	400		400	
		Other animals	Muscle	100		100	
			Fat	50		50	
			Liver	200		200	
			Kidney	200		200	
Difloxacin	Difloxacin	Cattle/sheep	Muscle	400		400	
			Fat	100		100	
			Liver	1,400		1,400	
			Kidney	800		800	
		Pigs	Muscle	400		400	
			Skin + fat	100		100	
			Liver	800		800	
			Kidney	800		800	
		Poultry	Muscle	300		300	
			Skin + fat	400		400	
			Liver	1,900		1,900	
			Kidney	600		600	
		Other animals	Muscle	300		300	
			Fat	100		100	
			Liver	800		800	
			Kidney	600		600	
Doxycycline	Doxycycline	Cattle (forbidden for lactiferous cattle)	Muscle	100	2,000	100	100
			Liver	300	6,000	300	600
			Kidney	600	12,000	600	12,000
		Other animals (producing species)	Muscle	100	2,000	100	
			Skin + fat	300	12,000	300	
			Liver	300	6,000	300	
			Kidney	600	12,000	600	

(continued)

Table 15 (continued)

Compounds	Controlled/ monitored compounds	Source	Target tissues	MRLs in China (μg/kg)	MRLs in the USA (μg/kg)	MRLs in the EU (μg/kg)	MRLs in Japan (μg/kg)
		Poultry (forbidden for layer chicken)	Muscle	100	2,000	100	
			Skin + fat	300	12,000	300	
			Liver	300	6,000	300	
			Kidney	600	12,000	600	
Enrofloxacin	Enrofloxacin+ Ciprofloxacin	Cattle/sheep	Muscle	100	300	100	
			Fat	100		100	
			Liver	300	100	300	
			Kidney	200		200	
		Cattle/sheep	Milk	100		100	
		Pigs/rabbit	Muscle	100		100	
			Fat	100		100	
			Liver	200		200	
			Kidney	300		300	
		Poultry (forbidden for layer chicken)	Muscle	100		100	
			Skin + fat	100		100	
			Liver	200		200	
			Kidney	300		300	
		Other animals (producing species)	Muscle	100		100	
			Fat	100		100	
			Liver	200		200	
			Kidney	200		200	

Oxacillin	Oxacillin	Food from animal origin	Muscle	300		300	
			Fat	300		300	
			Liver	300		300	
			Kidney	300		300	
			Milk	30		30	
Oxytetracycline/ Chlortetracycline/ Tetracycline	Parent drug, single or complex	Food from animal origin	Muscle	100	2,000	100	
			Liver	300	6,000	300	
			Kidney	600	12,000	600	
		Cattle/sheep	Milk	100		100	
		Poultry	Eggs	200		200	
		Fish/shrink	Meat	100	2,000	100	
Sarafloxacin	Sarafloxacin	Chicken/turkey	Muscle	10		10	
			Fat	20		20	
			Liver	80		80	
			Kidney	80		80	
		Fishes	Muscle + skin	30		30	
Sulfonamides	Parent drug (total residues)	Food from animal origin	Muscle	100	100	100	100
			Fat	100	100	100	100
			Liver	100	100	100	100
			Kidney	100	100	100	100
		Cattle/sheep	Milk	100		100	
Sulfadimidine	Sulfadimidine	Cattle	Milk	25		25	25

(or LLE) is used for analyte isolation, with subsequent clean-up and analyte enrichment by means of dryness or SPE. This will no doubt cause increased interest in a variety of modern sample pretreatment techniques.

LC is necessary to separate antibiotics from interfering components, although selective MS detection diminishes the chromatographic requirements. LC columns have been developed in ways to improve analysis speed and chromatographic performance. The LC effluent usually enters the MS with ESI. However, during ionization some sensitivity of the MS detection can be lost as the signal can be deviated over multiple charges, isotopes, or adduct ions.

To improve sensitivity, the ionization has to be sufficiently optimized. MS/MS can improve the selectivity by selecting a specific product of the residues after dissociation. The different kinds of the MS remain susceptible to developments and improvements, likely to make LC-MS and/or LC-MS/MS the most versatile technique for residues quantification in food matrices.

To improve quantification of the analytical methods, there is a need for reference materials, standardized procedures, and certified reference standards for the interpretation of results.

In the future, it will be most important for the analysis of antibiotics in food to develop methods for analyzing nontarget antibiotics, unexpected residues and/or metabolites other than those included in the residues already defined. As a good indicator of the perspectives and areas for future progress, there need to be the development of (semi-)automated sample pretreatment techniques, the improvement of the new LC column for better separate interests from complexity matrixes, and the introduction of new MS techniques.

Acknowledgment

The authors are very grateful to Qin Feng and the other staff of Shanghai Institute for Food and Drug Control for using the LC QqQ mass spectrometers and supplying chromatograms.

References

1. Cristina Blasco, Carmen Maria Torres, Yolanda Pico. Progress in analysis of residual antibacterials in food. Trends in Analytical Chemistry, 26(No. 9, 2007):895–913
2. A.A.M. Stolker, U.A.Th. Brinkman. Analytical strategies for residue analysis of veterinary drugs and growth-promoting agents in food-producing animals—a review. Journal of Chromatography A, 1067 (2005): 15–53
3. Urairat Koesukwiwat, Siripastr Jayanta, Natchanun Leepipatpiboon. Validation of a liquid chromatography–mass spectrometry multi-residue method for the simultaneous determination of sulfonamides, tetracyclines, and pyrimethamine in milk. Journal of Chromatography A, 1140 (2007): 147–156
4. Tara A. McGlincheya, Paul A. Raftera, Fiona Reganb, Gillian P. McMahonb. A review of

analytical methods for the determination of aminoglycoside and macrolide residues in food matrices. analytica chimica acta 624 (2008): 1–15

5. Sara Bogialli , Giuseppe D'Ascenzo, Antonio Di Corcia, Aldo Lagana`, Simone Nicolardi. A simple and rapid assay based on hot water extraction and liquid chromatography–tandem mass spectrometry for monitoring quinolone residues in bovine milk. Food Chemistry 108 (2008): 354–360
6. M.J. Gonz'alez, de la Huebra, U. Vincent. Analysis of macrolide antibiotics by liquid chromatography. Journal of Pharmaceutical and Biomedical Analysis 39 (2005): 376–398
7. Sandra Pe'rez, Damia` Barcelo'. Applications of LC-MS to quantitation and evaluation of the environmental fate of chiral drugs and their metabolites. Trends in Analytical Chemistry, 27 (No. 10, 2008): 836~846
8. Linling Wang, Hai Yang, Chunwei Zhang, Yulin Mo, Xiaohua Lu. Determination of oxytetracycline, tetracycline and chloramphenicol antibiotics in animal feeds using subcritical water extraction and high performance liquid chromatography. analytica chimica acta 619 (2008): 54–58
9. Oscar Nu'n˜ ez, Encarnacio'n Moyano, Maria Teresa Galceran. LC–MS/MS analysis of organic toxics in food. Trends in Analytical Chemistry, 24 (No. 7, 2005): 683~703
10. Irene van den Broeka,*, RolfW. Sparidansa, Jan H.M. Schellensa, Jos H. Beijnena. Quantitative bioanalysis of peptides by liquid chromatography coupled to (tandem) mass spectrometry. Journal of Chromatography B, 872 (2008): 1–22
11. M. Silvia Dı'az-Cruz, Damia` Barcelo. Recent advances in LC-MS residue analysis of veterinary medicines in the terrestrial environment. Trends in Analytical Chemistry, 26(No. 6, 2007):637~646
12. M. Espinosa Bosch, A.J. Ruiz Sánchez, F. Sánchez Rojas, C. Bosch Ojeda. Recent developments in analytical determination of furosemide. Journal of Pharmaceutical and Biomedical Analysis 48 (2008): 519–532
13. A. Conneely, A. Nugent and M. O' Keeffe. Use of solid phase extraction for the isolation and clean-up of a derivatised furazolidone metabolite from animal tissues Analyst 127 (6, 2002): 705–709
14. The Maximum Residue Limits of Residues of Veterinary Drugs in Animal Derived Food, issued No. 235 bulletin (2002) by Ministry of Agriculture, People's Republic of China
15. European Commission, Commission Decision (2002/657/EC) of 12 August 2002

Chapter 14

Identification of Unknown Migrants from Food Contact Materials

Malcolm Driffield, Emma L. Bradley, Laurence Castle, and Leon Coulier

Abstract

Materials that come into contact with foodstuffs can transfer components that may cause odour or taint problems or in the worse case cause the foodstuff to be unsafe to eat. The identities of some of these are easily predicted from the chemistry of known components but others are not. In this respect, it is important to be able to identify and quantify these chemicals. This chapter describes the need for methods of identification of unknown chemicals that may migrate. Mass spectrometric analytical methods are described, including headspace-gas chromatography with mass spectrometry (HS-GC-MS), liquid injection gas chromatography with MS, and liquid chromatography with time-of-flight MS (LC-TOF-MS).

Key words: Food contact materials, Unknowns, Identification, GC-MS, LC-TOF-MS

1. General Introduction

1.1. What Are Food Contact Materials?[1]

Virtually, all of the food and drink that we buy and then consume is packaged in some way (1, 2). It is also nearly certain that it comes into contact with other materials during harvesting, production, transport, storage, and cooking. A food contact material (FCM) is any material or article intended to be placed in contact with foodstuffs. Food packaging materials are not only the most notable example, but also include cutlery, dishes and plates, containers, parts of food processing equipment, etc. These can be made from a range of different materials, such as plastics, paper and board, metals, glass, ceramics, cork and rubber, along with the adhesives,

[1] This chapter describes some aspects of testing food contact materials. The same considerations apply for other contact materials such as packaging for pharmaceuticals, medical devices and tobacco products

Jerry Zweigenbaum (ed.), *Mass Spectrometry in Food Safety: Methods and Protocols*,
Methods in Molecular Biology, vol. 747, DOI 10.1007/978-1-61779-136-9_14, © Springer Science+Business Media, LLC 2011

inks, varnishes, and coatings needed to make finished materials and articles. There are literally thousands of different starting substances (chemicals) needed to make this variety of materials and give them the characteristics and properties that are desired.

1.2. Legislation

When a foodstuff – meaning a food or beverage – comes into contact with a FCM, there is the potential for migration of any chemicals from the material into the foodstuff. Depending on the chemical substance(s) involved, this can compromise the safety and/or the quality of food and so most countries have legislation in place to keep any chemical migration within acceptable limits. The legislation takes different forms in different countries but the principles, to help ensure consumer protection and avoid contamination of foodstuffs, are universal. FCMs should be made and used in such ways that they do not transfer constituents to food in quantities that could:

1. Endanger human health.
2. Bring about an unacceptable change in composition.
3. Bring about deterioration in organoleptic characteristics.

1.3. Testing for Specified Chemicals

Keeping a check on the migration of all the starting substances that may be used to make FCMs is a huge undertaking. It involves the chemical analysis of either the material itself or testing for migration of chemicals into foods or into model foods that are called food simulants. For this job – targeted analysis – mass spectrometric methods and especially gas chromatography with mass spectrometric detection (GC-MS) and liquid chromatography with mass spectrometric detection (LC-MS) are used widely. However, this chapter focuses on the use of MS methods for the identification and quantification of hitherto unknown substances, not for targeted substances which, by definition, are known.

1.4. Testing for the Unexpected and for the Unknown

There is a need to assess the safety of all potential migrants and not just known starting substances. These have become known in some circles as the NIAS – the non-intentionally added substances. Though not used deliberately, if they migrate then these substances can raise safety and quality concerns and so there is a need to test for them. So where do the NIAS come from?

1.4.1. Impurities

The NIAS can be present as a consequence of impurities in the starting substances used to make FCMs. Even minor impurities can have a disproportionate impact with respect to migration. First, because they themselves may not participate in the chemistry intended of the starting substance (e.g. polymerisation) and so the impurity may appear relatively enriched in the finished material. Second, minor impurities can migrate disproportionately if they are of a lower molecular weight (and so size) because migration is usually a diffusion-controlled process.

1.4.2. Reaction and Breakdown Products

The NIAS may also be decomposition and degradation products formed from the starting substances. These chemical transformations can be especially important under the often severe reaction conditions and processing conditions that can be used, such as:

- Stoving of can coatings at high temperature then subsequent sterilisation of the canned foods at elevated temperature and pressure.
- Using ionising radiation (gamma ray or electron beam) to sterilise packaging materials or to irradiate pre-packaged foods to kill insects and microorganisms so as to prolong shelf-life.
- Breakdown of antioxidants, UV stabilisers, and other similar chemicals that are intended to degrade as part of their function.

1.5. The Analytical Requirement

1.5.1. The Scope of the Analysis

If any substance can migrate, then there is a need to identify and quantify it. This simple maxim applies equally for the producer of the packaging, the food packer, the regulator, or the consumer of the food. The requirements place emphasis on the information-rich separation techniques using mass spectrometry (MS) as the detection system, i.e. GC-MS and LC-MS (/MS).

As migration is largely a diffusion-controlled phenomenon, then low molecular weight (size) substances migrate fastest to higher concentration, and so they are of first concern. These substances are accommodated best using GC-MS. The reproducibility of the electron ionisation (EI) and chemical ionisation (CI) MS detection allows the production of mass spectral libraries containing many thousands of substances. Software available with most GC-MS instruments allows deconvolution and comparison of the peaks detected with the MS libraries to provide a confident identification. In recent times, the accurate mass data provided by GC-time-of-flight (TOF)-MS adds further information to aid in the identification of unknown substances and adds greater confidence in that assignment.

Larger and/or polar molecules are not well suited for analysis by GC-MS unless derivatisation is carried out. This is not ideal if the class of chemicals, that may or may not be present, is unknown as chemically different substances can require different derivatisation reagents and conditions. Also, as producers turn to higher molecular weight additives to give reduced migration levels, the molecular weight range of the NIAS of interest increases too. These polar and higher molecular weight substances are best tested for using LC-MS. For the identification of unknown substances, the accurate mass data derived by LC coupled to high resolution MS, for example LC-TOF-MS, aids in identification and can give mass accuracy values of 3 parts per million (ppm) or less, with a wide dynamic range with full scan capability down to picogram levels. Few mass spectral libraries exist for LC-MS, due to the atmospheric ionisation process needed, for example atmospheric pressure chemical ionisation (APCI), electrospray ionisation (ESI), or atmospheric pressure photo ionisation (APPI). Although MS libraries cannot aid in identification, the use of the LC-TOF-MS accurate mass data produced for the unknown substances can be compared to a database of accurate masses for possible structures (and retention times if authentic standards are available). This database can contain thousands of substances and should be populated using experience and knowledge about the samples.

1.5.2. The Detection Limits Needed

There can be an impasse in deciding on the level of detection needed when testing for the NIAS. Analytical chemists ask toxicologists – how low do we need to go? At what concentration is a substance of no safety concern to you? But when it comes to unknowns, with no chemical structure available to guide the toxicologist by structural alerts for example, the question can often be turned around to – how low can you go? If it cannot be detected, then we cannot worry about it – well, not productively at least!

To escape this impasse, it is very useful to consider the threshold of toxicological concern (TTC) or the threshold of regulation (ToR) limits used by some expert bodies and regulatory agencies, such as the US Food and Drug Administration (US-FDA). In brief, the so-called cohort of concern chemicals is excluded first. These chemicals, such as dioxins, N-nitroso compounds, azoxy compounds, aflatoxin-type substances, and steroids, are so uniquely toxic or potent that they require separate consideration. Fortunately, with some minor exceptions, these types of chemicals have no place in FCMs and so can effectively be ruled-out. Then, even for unidentified substances, the lowest level of interest is about 1.5 μg in the daily diet – meaning 1.5 μg per person per day. Considering that a daily diet may be simplified as about 1 kg of food plus 2 L of drink, an amount of 1.5 μg per person per day would be a concentration of 0.5 parts per billion (ppb) in the whole diet.

This need not mean, of course, that we have to test complex foods for unknown chemicals at concentrations of 0.5 ppb. There are simplification and amplification possibilities available. For example, if the packaging under evaluation is used for 10% of the diet (e.g. used to pack meat and fish, say), then the concentrations of interest becomes ca. 5 ppb in that portion of the daily diet that may be affected. A further simplification of the analytical task is to resort to using food simulants. These are simple media, such as solvents, that are intended to mimic the migration behaviour of foodstuffs but which are themselves easier to analyse than complex foods. An extension of this is to use other solvents that are more aggressive than foodstuffs and test for extractables. The concentrations of any substances identified in the extract can then be scaled-back using reduction factors or migration modelling. Not only do simulants and solvents present a simplified matrix for testing, but if the right choice is made they can be used to test an exaggerated quantity (area) of packaging and can also be evaporated to concentrate the sample further.

The remainder of this chapter describes analytical methods that can be used to identify previously unknown substances and estimate their concentration. It should be noted that the experimental conditions, including instrumental parameters, were optimised for use in the author's laboratory, and should be optimised in individual laboratories before use. Other chromatographic columns and instrument types may be used to produce similar results.

The examples given are taken from a study funded by the UK Food Standards Agency (FSA) with the major aim to identify and catalogue the NIAS (reaction and breakdown products) derived from starting materials and known additives used in the manufacture of polymers used as FCMs (3). In this study, six polymers (polypropylene, polyethylene, polystyrene, polyvinyl chloride, polyethylene terephthalate, and polyamide) were prepared containing additives chosen to be representative of those typically used in plastic materials and articles intended for contact with food. Test plastics containing the selected additives at concentrations typically used in the manufacture of food contact plastics were moulded into sheets. Control plastics were also prepared in which no additives were incorporated. By comparing the chromatograms obtained for the analysis of the plastic and additive samples with those obtained from the analysis of the plastics only, any additional substances (impurities, reaction products, or breakdown products) present could be identified. This approach has been reported using other FCM types too (4, 5).

2. Specific Methods

2.1. Headspace GC-MS

2.1.1. Introduction

HS-GC-MS is used to detect volatile substances that may migrate from FCMs into foodstuffs. Sample preparation is minimal as the specimen is placed in a vial and warmed to allow equilibrium of the volatile substances between the sample and the headspace gas in the vial. The vapour in the headspace is then injected directly onto the GC column for separation and MS detection and identification. In this example, the packaging sample itself is analysed for volatiles. A similar analytical approach can be used for the analysis of food simulant samples.

2.1.2. Materials and Sample Preparation

1. Cut up the packaging sample (approximately 250 cm^2 if available to get a representative specimen) into small pieces (approximately 0.5 cm^2) and mix well.
2. Prepare a solution of fluorobenzene (1 μg/mL) to act as internal standard.
3. Prepare headspace vials in two ways:
   (a) Sample (0.25 cm^2 taking into account the actual contact area with the food and making note of the mass of sample used).
   (b) Sample (25 cm^2 taking into account the actual contact area with the food and making note of the mass of sample used) and internal standard (100 μL).
4. Prepare each vial type in duplicate and cap tightly prior to analysis. Prepare corresponding blanks in the same way but without the addition of the sample.

2.1.3. Methods

Incubate each sample at 100°C for 30 min and inject the resulting volatile substances onto the GC column. The following conditions have been found to be suitable on an Agilent 6980 gas chromatograph (Agilent, Palo Alto, California, USA) coupled with an Agilent 5973 inert mass selective detector. Splitless injection of 1 mL of headspace gas onto a DB-VRX (low polarity phase with selectivity for the optimum resolution of volatiles analysis. Similar phases include: VOCOL, NON-PAKD T-624, Rtx-Volatiles, Rtx-502.2, PE-624, PE-502, 007-624, 007-524, HP-624, CP-624) capillary column 30 m long, 0.25 mm diameter with a film thickness of 1.2 μm (J&W Scientific, Folsom, California, USA). The injector port is held at 280°C with helium as the carrier gas at a constant flow of 1 mL/min. The GC oven is held at 40°C for 1 min and then raised at 10°C/min to 250°C, where it is held for 5 min. The MS is operated in electron ionisation mode with scanning between 40 and 440 amu.

1. Examine each chromatogram in turn, first eliminating those peaks present in procedural blanks and then assign the identity of any remaining peaks using available MS libraries.
2. Estimate the concentration of each peak by comparing the area with the area of the internal standard peak.

2.1.4. Method notes

1. Substances other than fluorobenzene could be used as internal standard, for example deuterated volatile compounds, such as d_{10}-ethylbenzene. Quantification is only approximate since a uniform response factor of 1 is used at first. If accurate quantification is required, then authentic standards of the identified peaks can be used for calibration purposes.
2. Care should be taken when carrying out library searches as the first hit from the software may not be the correct one,

especially if low mass fragments are present in the library mass spectrum but not the sample (as a consequence of the electron impact scanning mass range).

3. In some cases, the presence of water in the headspace vials can change the equilibrium of the volatile compounds between the solid sample and headspace and this technique may be used to detect other substances.

2.1.5. Example: Polyvinyl Chloride

Three peaks were detected in the total ion chromatograms of the polyvinyl chloride (PVC) sample that were not detected in a sample of PVC not containing the three additives of interest (Fig. 1). The mass spectra of each of these additional peaks were derived and compared with the NIST library. The best library matches of these peaks were: ethanol (retention time 7.0 min), 1-butanol (11.2 min), and 2-ethyl-1-hexanol (19.1 min, Fig. 2, shown as an example). The other two peaks detected in the PVC + additive chromatograms were also detected in the PVC only chromatograms. The best library matches for these substances were methyl acetate and acetic acid. Additional peaks were detected in the PVC only samples but were not in the PVC + additive samples. This suggests that these volatile substances were lost and/or reacted with other substances present when the PVC was mixed with the additives. The three substances detected in the headspace above the PVC + additive samples that were not in the control were attributed to be breakdown products from the known additives: 2-ethyl-1-hexanol derived from di-*n*-octyltin-bis(ethylhexylthioglycolate) (a stabiliser), ethanol derived from di-*n*-octyltin bis(ethyl maleate) (a stabiliser), and 1-butanol from acetyltributyl citrate (a plasticiser).

2.2. Solvent Extraction Followed by GC-MS

2.2.1. Introduction

Solvent extraction followed by GC-MS analysis can be used to determine any semi-volatile substances with the potential to migrate from food contact materials into foodstuffs. Again, in this example, the packaging sample itself is analysed. A similar analytical approach can be used for solvent extracts of food simulant samples although sample clean-up to remove coextractives, such as oils and fats, can often be required.

2.2.2. Materials and Sample Preparation

1. Cut up the sample (100 cm^2 if available) into small pieces (approximately 1 cm^2) and transfer to a glass vial.
2. Prepare a solution of d_{10}-benzophenone (1 mg/mL) to act as internal standard. Add 100 μL of internal standard solution and 20 mL of ethanol to the vial ensuring that the sample is totally emerged. Cap the vial and incubate at 60°C for 24 h with occasional shaking.
3. After cooling, transfer 1 mL of extract to a vial suitable for GC-MS analysis and cap.

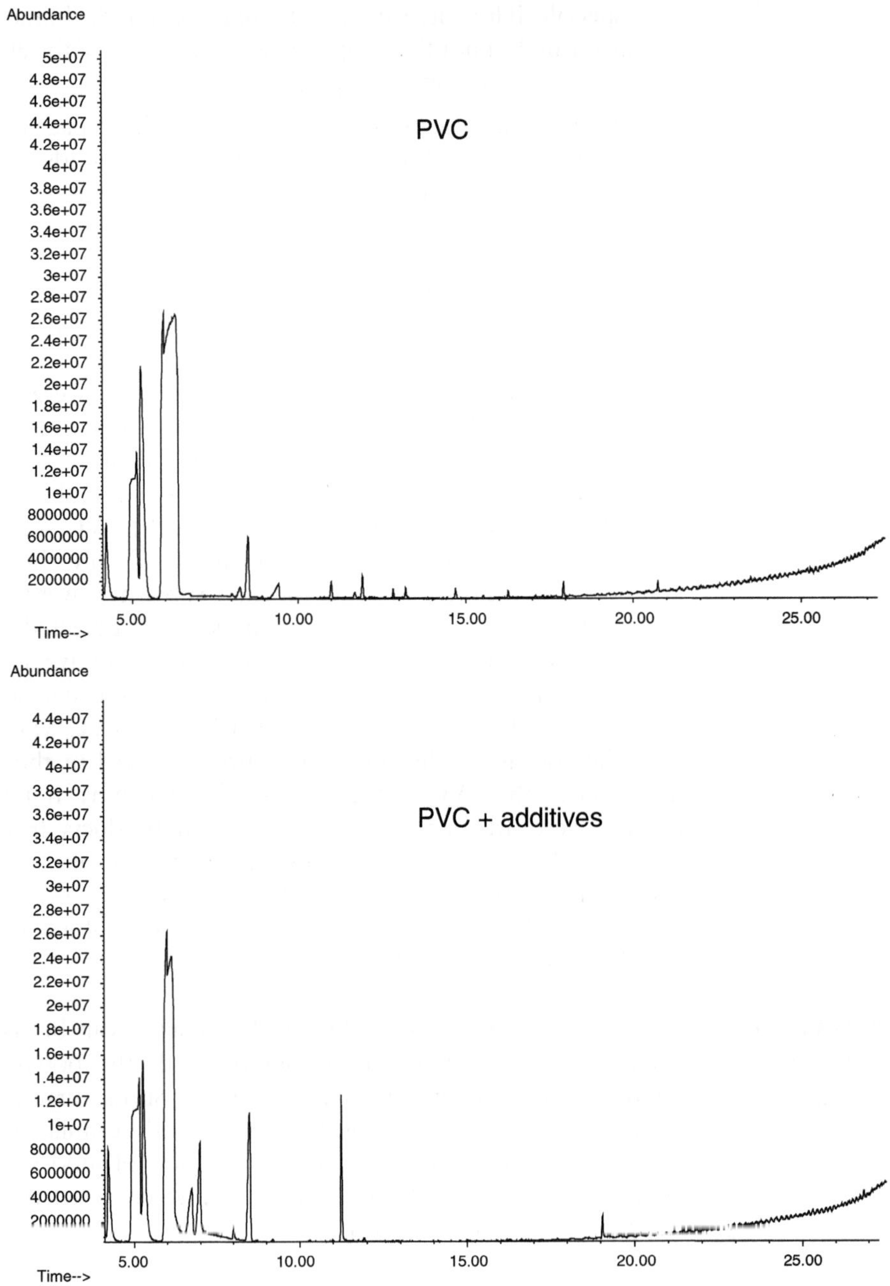

Fig. 1. HS-GC-MS chromatograms of PVC and PVC + additives.

4. Concentrate the remainder of the solvent to 2 mL under a gentle stream of nitrogen gas and transfer to a vial suitable for GC-MS analysis.
5. Repeat steps 1–4 but using isooctane rather than ethanol.
6. Prepare suitable procedural blanks as described in steps 1–5, but without the addition of sample.

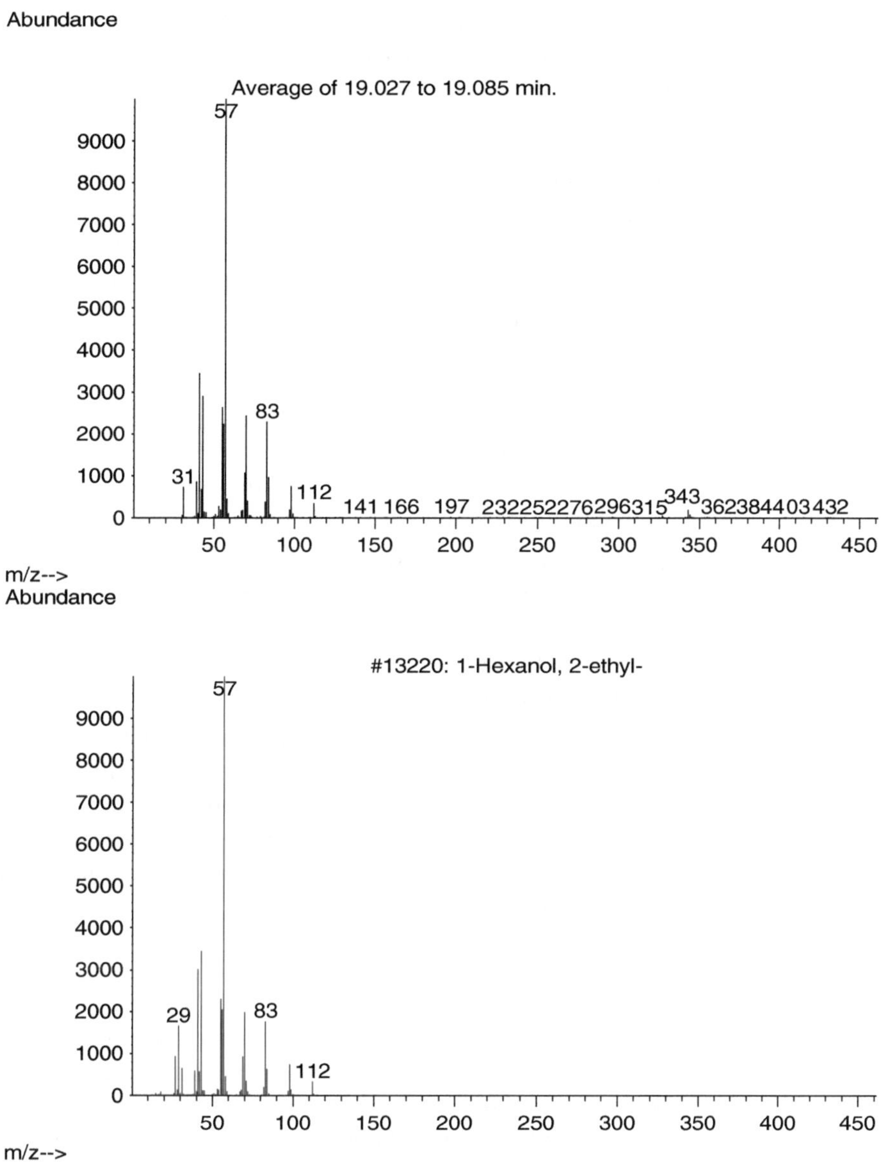

Fig. 2. Best library match of the peak eluting at retention time 19.1 min in the HS-GC-MS chromatogram of the PVC + additives sample.

2.2.3. Methods

Inject each extract into the GC-MS. The following conditions have been found to be suitable on an Agilent 6980 gas chromatograph (Agilent, Palo Alto, California, USA) coupled with an Agilent 5973 inert mass selective detector. Splitless injection of 1 μL of extract onto a DB-5MS ((5% Phenyl)-methylpolysiloxane capillary column 30 m long, 0.25 mm diameter with a film thickness of 0.25 μm (J&W Scientific, Folsom, California, USA). Similar phases include Ultra-2, Rtx-5MS, HP-5MS, and PTE-5. The injector port is held at 250°C with helium as the carrier gas at a constant flow of 1 mL/min. The GC oven is held at 60°C for 5 min and then raised at 10°C/min to 320°C, where it is held for 5 min. The MS is operated in electron ionisation mode with scanning between 50 and 550 amu.

1. Examine each chromatogram in turn, first eliminating those peaks present in procedural blanks and then assign the identity of any remaining peaks using available MS libraries.
2. Estimate the concentration of each peak by comparing the area with the area of the internal standard peak.

2.2.4. Method notes

1. If only a small number of peaks or no peaks are detected in the extract injected directly, then the concentrated extract should be analysed too. Care should be taken not to overheat the extract during the concentration step to avoid any volatile compounds being driven off and lost.
2. Isooctane and ethanol are chosen as extraction solvents as they are also recognised solvents used as extraction media for FCMs. As such they allow worst-case estimations of migration to be calculated. They differ in polarity meaning different substances are extracted into each solvent.
3. d_{10}-Benzophenone was chosen as the internal standard. Other compounds could be used instead. Quantification is only approximate since a uniform response factor of 1 is used at first. If accurate quantitation is needed, then quantification using authentic standards of the identified peaks can be carried out.

2.2.5. Example: PVC

The chromatograms obtained from the analysis of ethanol and isooctane extracts of the PVC and PVC + additive samples are shown in Figs. 3 and 4, respectively. Due to the high concentration of additives detected in the ethanol extracts and the resulting problems with contamination of the GC-MS instrumentation, then the concentrated isooctane extracts were analysed but not the concentrated ethanol extracts. Few peaks are detected in the extracts of the control PVC samples; however, the PVC + additive extracts contained in excess of 144 peaks that can be assigned either to the additives themselves, their impurities or their reaction and/or breakdown products. Of the 144 peaks listed (145 substances), just one was a known additive. Of the remaining 144 substances, 24 were considered to be impurities of the additives since they were also found in the analysis of the additives themselves. The remaining 115 substances were assigned as reaction/breakdown products since they were not detected in the analysis of the additives or the PVC alone but were detected only once the additives were subjected to thermal processing of the PVC plastic. Problems can

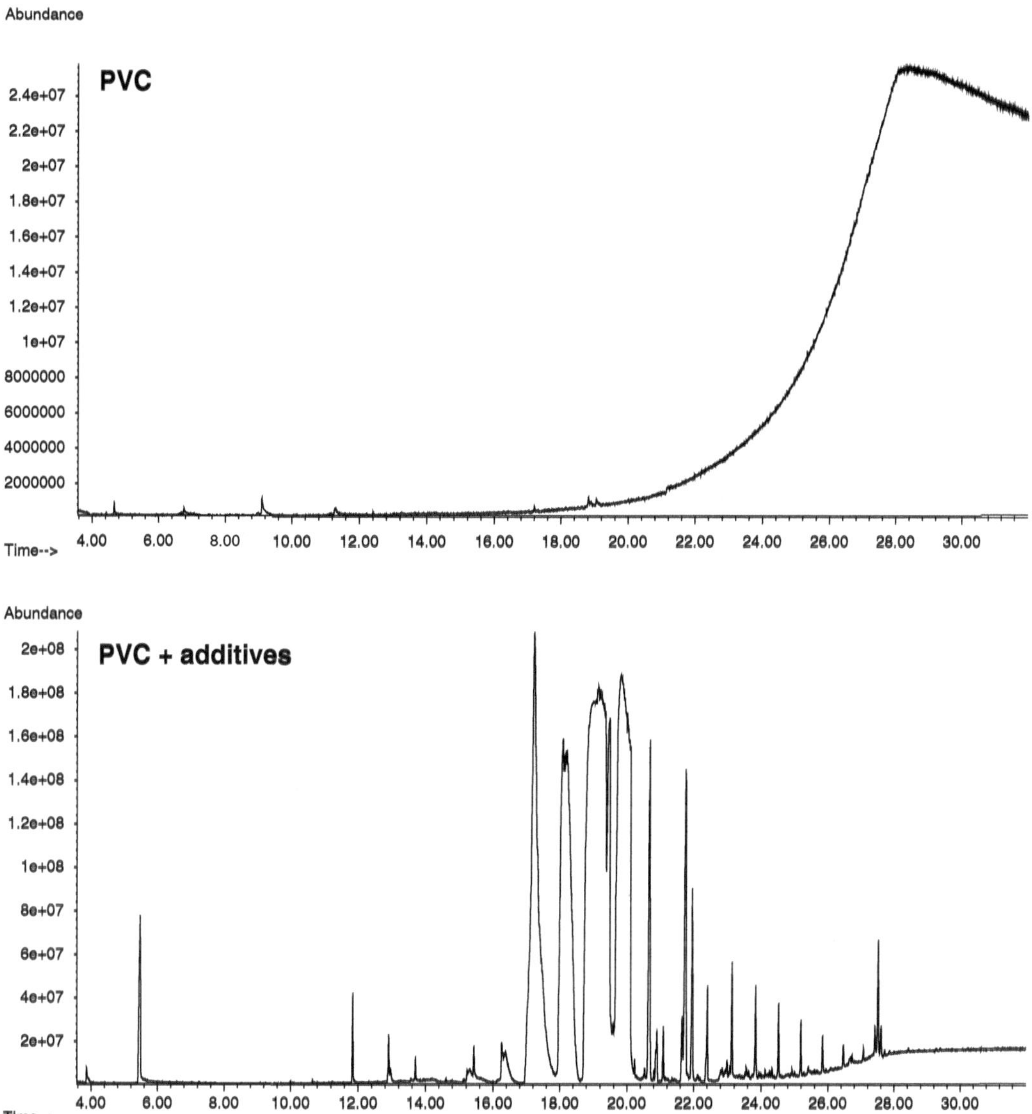

Fig. 3. GC-MS chromatograms of ethanol extracts of PVC and PVC + additives.

occur when testing for unknown substances as the GC separation cannot be optimised and so co-elution can give mixed spectra that cannot be assigned.

2.3. Solvent Extraction Followed by LC-TOF-MS

2.3.1. Introduction

Polar or non-volatile substances can be determined using LC-MS. As already mentioned, the use of TOF-MS allows determination of the accurate mass for each substance, and this adds confidence and aids in the identification of unknowns.

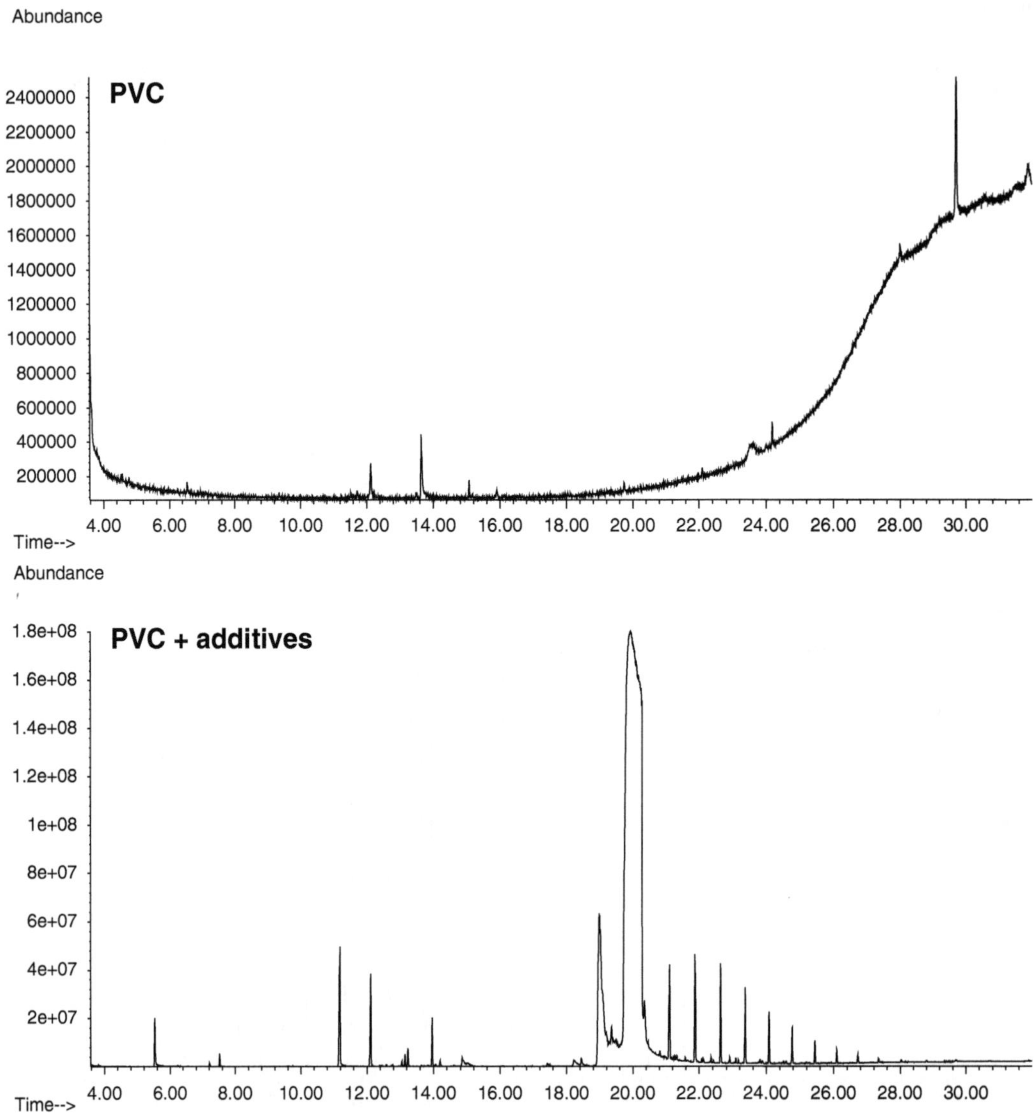

Fig. 4. GC-MS chromatograms of isooctane extracts of PVC and PVC + additives.

2.3.2. Materials and Sample Preparation

1. Prepare ethanol and isooctane extracts of the sample as described in Subheading 2.2.2.
2. Take 1 mL of the each extract, evaporate at 40°C under a gentle stream of nitrogen and redissolve the residue in 1 mL of acetonitrile. Transfer this to a vial suitable for injection on the LC-TOF-MS system and cap tightly.
3. Prepare suitable procedural blanks but without the addition of sample.

2.3.3. Methods

Inject each sample into the LC-TOF-MS. The following conditions have been found to be suitable on an Agilent LC/MSD TOF (Agilent, Santa Clara, California, USA) consisting of a 1200 series

LC and 6220 TOF MS. Two separate methods can be used to increase the coverage of substances that can be detected. In both cases, separation is carried out on an Agilent ZORBAX Eclipse XDB-C18 column 100 mm long, 2.2 mm diameter, and particle size 3.5 μm. For positive mode electrospray, the mobile phase consists of 0.1% (w/v) aqueous acetic acid (Channel A) and acetonitrile (Channel B). For negative mode electrospray, the mobile phase is 5 mM aqueous ammonium formate at pH 5.5 (Channel A) and 0.1% (v/v) 5 mM ammonium formate at pH 5.5 in acetonitrile (Channel B). In both cases, the mobile phase is initially at 80% Channel A changing linearly to 50% A by 25 min and held for 20 min. This is then changed linearly to 0% A by 60 min and held for a further 10 min before returning to the starting conditions for column equilibration for 5 min. The flow rate is 0.2 mL/min and injection volume is 5 μL. TOF-MS analysis was carried out with a nebulizer pressure of 45 psi, capillary of 4,000 V, gas temperature of 325°C, drying gas flow of 10 L/min, skimmer of 60 V, fragmentor of 150 V, and octopole RF voltage of 250 V. The mass range measured is 100–1,000 m/z.

1. Examine each chromatogram in turn; first eliminating those peaks present in procedural blanks. Assign the identity of any remaining peaks using accurate mass data and any databases available.
2. Estimate the concentration of each peak by comparing the area with the area of the internal standard peak.

2.3.4. Method notes

1. d_{10}-Benzophenone was chosen as internal standard. Other substances could be used instead. Quantification is only estimated due to comparison with the internal standard and not authentic standards of each substance present. If substantial quantities of a substance are detected, then full quantification using authentic standards should be carried out.
2. Following LC-TOF-MS analysis, the data should be interrogated using the data processing software of the system used for analysis. In our laboratory, Agilent MassHunter Qualitative software is used. This employs algorithms to automatically identify all the detectable substances or molecular features in accurate mass data even when analyzing very complex mixtures. Key among these are the Molecular Feature Extractor and Empirical Formula Generation algorithms. Molecular Feature Extractor is a data-mining tool that generates a list of molecular features with retention time, neutral mass, and ion abundance. All of the related ions of a molecular feature (isotopes, charge states, adducts, and multimers) are grouped together, and areas of noise are removed. As the name suggests, the Empirical Formula Generator calculates potential empirical formulae for TOF-MS peaks. It uses accurate mass

MS, isotope spacing, and mass peak abundance information to decrease the number of potential formulae generated and then lists them in order of likeliness using a unique scoring system. The software generates a report describing the peaks detected, including retention time, accurate mass, predicted empirical formulae and score.

3. In some cases, the chromatographic peaks detected in the sample extracts can be very small, but are often elucidated nevertheless, highlighting the power of TOF-MS. Many of the peaks are not visible from the total ion chromatogram but are extracted from the raw data by the data processing software.
4. In LC-TOF-MS, data molecular adducts are seen in the mass spectra and the characteristic differences in accurate mass between the $[M+H]^+$, $[M+NH_4]^+$, $[M+Na]^+$, and $[M+K]^+$ adducts can be used to identify the molecular ion.
5. Another useful tool in LC-TOF-MS for assisting in the identification of unknown substances is in-source fragmentation. This can be induced by changing voltages in the source and some substances will then break apart. The accurate mass of these fragments can be used to determine structural information about the substances detected.
6. It should be noted that accurate mass data and the molecular formulae predicted do not allow differentiation between isomers or other substances with the same formula.

2.3.5. Example: High Density Polyethylene

The chromatograms obtained from the analysis of isooctane extracts of the high density polyethylene (HDPE) and HDPE + additive samples are shown in Fig. 5. Of the eight peaks detected, four were known additives (*N,N*-bis-(2-hydroxyethyl)alkyl(C13) amine, *N,N*-bis-(2-hydroxyethyl)alkyl(C15)amine, oleamide and glycerol monooleate) and three others were reaction or breakdown products. The remaining peak was dibutyl phthalate a common contaminant present at low concentrations in many samples.

3. General Summary

In order to assess the safety of all potential migrants from FCMs, not just the known starting substances, but also the NIAS whose identity may not be known, it is necessary to use analytical techniques involving mass spectrometry. This chapter has described GC-MS methodology, including headspace GC-MS to detect volatile substances and solvent extraction followed by GC-MS analysis to determine semi-volatile substances. A wide range of large mass spectral libraries exist for GC-MS analyses, and these can be used to aid in identification. Other substances that are

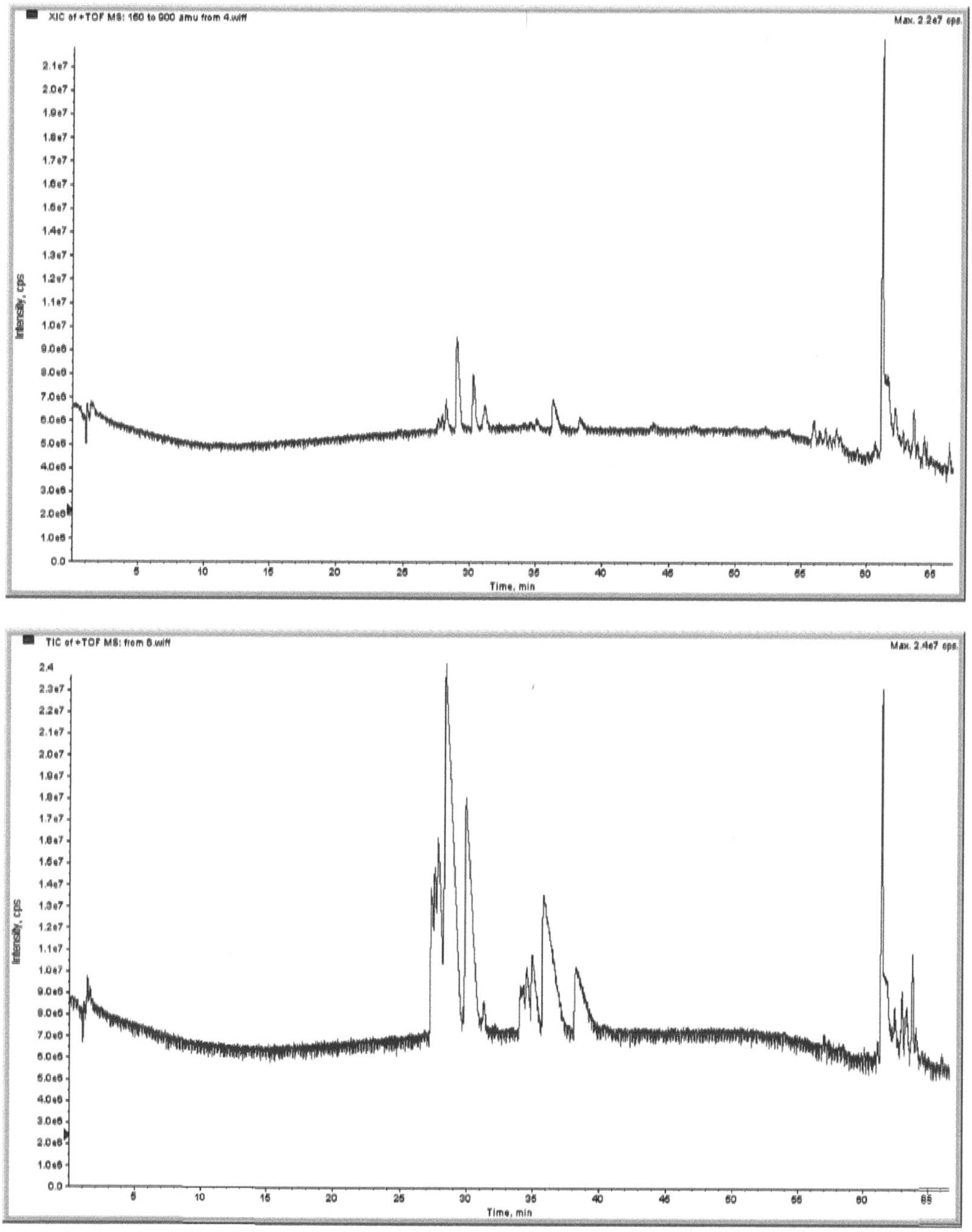

Fig. 5. Total ion chromatogram for HDPE and HDPE + additives (isooctane extraction, final solvent acetonitrile) in positive mode electrospray TOF-MS.

polar or non-volatile can be determined LC-TOF-MS. These techniques are suitable for the identification of potential migrants from food contact materials into foodstuffs across a wide range of molecular weights and different polarities and also provide an estimation of the concentrations present.

Acknowledgements

We gratefully acknowledge financial support from the UK FSA for funding work used as examples in this chapter: Project A03054 – An investigation into the reaction and breakdown products from starting substances used to produce food contact plastics. The statements and conclusions in this paper are the responsibility of the authors alone, and they should not be taken to represent the opinion of the FSA.

References

1. Chemical migration and food contact materials. K.A. Barnes, R. Sinclair and D. Watson, (eds). Woodhead Publishing, 2007. ISBN-13:978-1-84569-029-8.
2. Migration From Food Contact Materials. L.L. Katan (Ed) Springer, 1996. ISBN-13:978-0751402377.
3. FSA report A03054 - An investigation into the reaction and breakdown products from starting substances used to produce food contact plastics. Available at www.food.gov.uk
4. E.L. Bradley, M. Driffield, N. Harmer, P.K. Oldring, L. Castle (2008) Identification of potential migrants in epoxy phenolic can coatings. *International Journal of Polymer Analysis and Characterisation*, 13, 200–213.
5. E.L. Bradley, C. Jiang, J.T. Guthrie, M. Driffield, N. Harmer, P.K. Oldring, L. Castle (2009) Analytical approaches to identify potential migrants in can coatings. *Food Additives and Contaminants Part A*. 26, 1602–1610.

Chapter 15

Halogenated Persistent Organic Pollutants and Polycyclic Aromatic Hydrocarbons in Food

Tomas Cajka and Jana Hajslova

Abstract

During recent years, mass spectrometry (MS) and hyphenated chromatographic instrumentation and techniques have been a subject of dramatic developments, resulting in the introduction of various useful tools for the analysis of halogenated persistent organic pollutants (POPs) and polycyclic aromatic hydrocarbons (PAHs) in food and environmental matrices. This chapter describes state-of-the-art in the field of MS as a primary detection tool for the halogenated POPs and PAHs previously separated using either gas chromatography (GC) or liquid chromatography (LC). Since sample preparation practice plays a crucial role for obtaining optimal performance characteristics of a particular analytical method, a brief overview of sample extraction and clean-up procedures in the POPs/PAHs analysis is also briefly outlined.

Key words: Persistent organic pollutants, Polycyclic aromatic hydrocarbons, Food, Mass spectrometry, Gas chromatography, Liquid chromatography, Sample preparation

1. Introduction

Persistent organic pollutants (POPs) represent chemicals with long half-lives in all compartments of the environment including biota. Based on the Stockholm convention on POPs (last up-date in May 2009), the following groups of compounds are of main interest (Table 1) (1):

- Organochlorine pesticides (OCPs)
- Polychlorinated biphenyls (PCBs)
- Polychlorinated dibenzo-*p*-dioxins (PCDDs) and polychlorinated dibenzofurans (PCDFs)
- Brominated flame retardants (BFRs) including hexabromobiphenyl (HBB), tetra-, penta-, hexa-, and heptabromodiphenyl ethers (a group of compounds commonly known as polybromodiphenyl ethers, PBDEs)

Jerry Zweigenbaum (ed.), *Mass Spectrometry in Food Safety: Methods and Protocols*,
Methods in Molecular Biology, vol. 747, DOI 10.1007/978-1-61779-136-9_15, © Springer Science+Business Media, LLC 2011

Table 1
An overview of POPs (based on the Stockholm convention, 2009) and a group of so-called "PAH4" (EFSA, 2008) (1, 2)

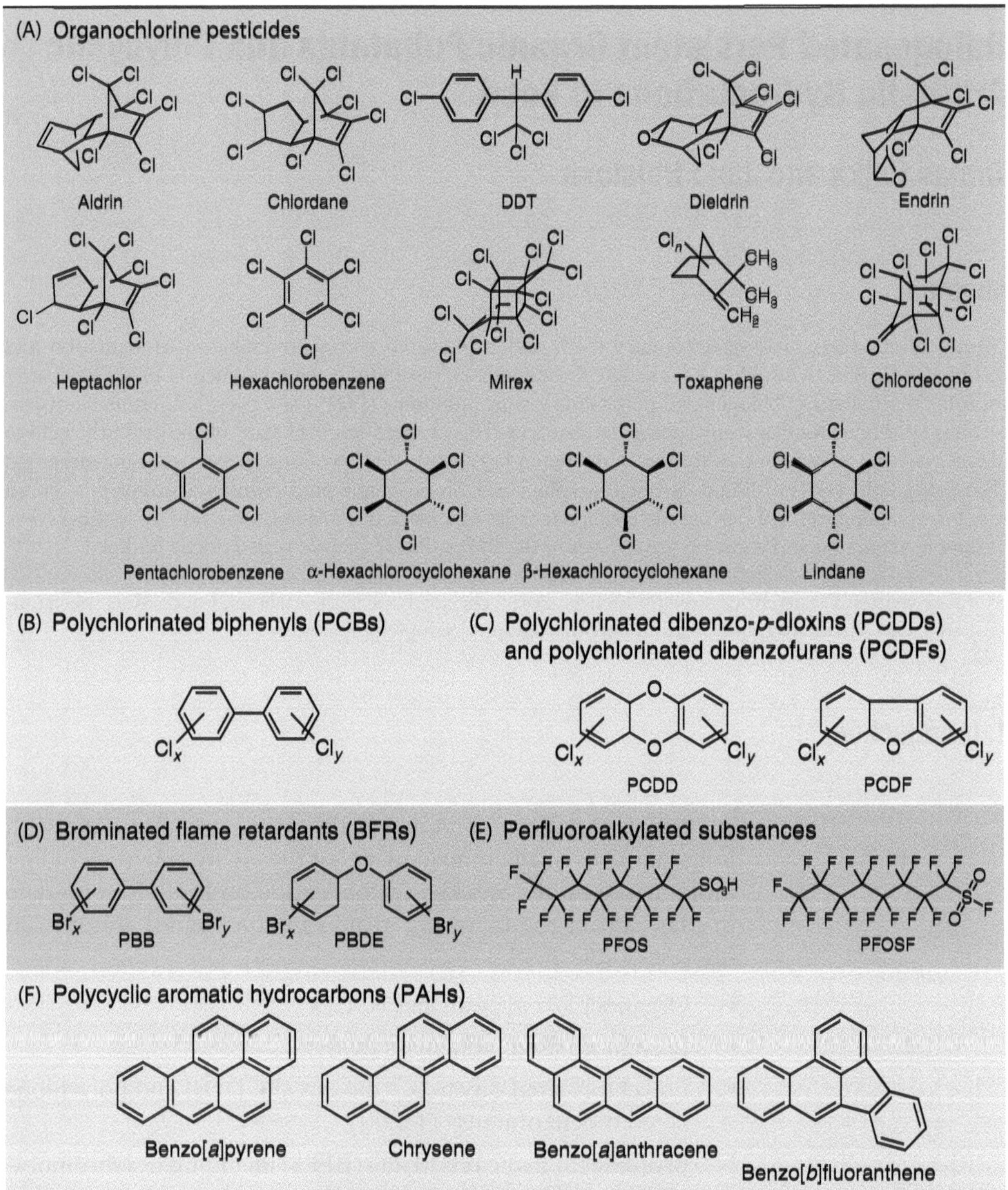

- Perfluoroalkylated substances represented by perfluorooctane sulfonic acid (PFOS), its salts and perfluorooctane sulfonyl fluoride (PFOSF)

While the group of OCPs, PCBs, PCDDs/PCDFs, and BFRs is accumulated mainly in lipid tissue (2), the perfluoroalkylated substances are, on the other hand, bound to proteins (3).

In addition to these halogenated POPs, polycyclic aromatic hydrocarbons (PAHs) are often included in the monitoring programs as indicators of environmental pollution. Although PAHs do not meet POPs definition (their bioaccumulative potential, due to a relatively rapid metabolism in biota, is low), they are also included in this chapter under this term. With regards to typical physicochemical properties, analytical strategies applicable for their determination are similar to those employed for POPs.

During recent years, several analytical approaches have been successfully developed not only for the "older" POPs such as OCPs, PCBs, PCDDs/PCDFs, and PAHs, but also for "emerging" contaminants such as BFRs and perfluoroalkylated substances. The determination of these analytes in complex matrices represents a challenging task since the concentrations in food samples are typically at ultra-trace levels, thus, requiring advanced analytical strategies for their accurate determination.

In practice, the methods used for the analysis of POPs/PAHs in food typically consist of the following basic steps: (1) *sampling and homogenisation*; (2) isolation of target analytes from a representative sample (*extraction step*); (3) separation of POPs/PAHs from bulk co-extracted matrix components (*clean-up step*); in this step also further *fractionation* of some groups of POPs might be required (typically to enable pre-concentration of minor analytes); (4) *separation* of the compounds of interest employing relevant chromatographic technique; (5) *identification and quantification* – nowadays mainly using mass spectrometric techniques. If the need is important enough, this is followed by (6) *confirmation* of results by an additional analysis. In the following sections, an overview of current state-of-the-art in the field of POPs/PAHs analysis in food will be presented with the attention to the applications employing mass spectrometry.

Although not discussed in this chapter, effect-based bio-analysis methods employing transcriptomics, proteomics, and biosensor-based technologies are of growing use as an efficient tool for hazard screening. For instance, CALUX (Chemically-Activated Luciferase eXpression) bioassay represents a very popular tool for rapid and easy screening of dioxins and dioxin-like PCBs (4).

2. Extraction

In general, extraction techniques rely on a favourable partition of POPs/PAHs from the sample matrix into the extraction matrix. However, the extraction procedures are typically not selective enough for the isolation of POPs/PAHs from complex food matrices, thus, additional clean-up and further fractionation steps are included in respective analytical procedures. In most cases, the samples are homogenised with sodium sulphate or other desiccant (e.g. hydromatrix, magnesium sulphate) causing rupture of cell walls and binding water present in the sample. The flowing powder is then extracted (in some cases after overnight drying) with a suitable solvent or their mixture. The principles and the use of the most extraction techniques are summarised bellow and in Table 2 (5–8).

Liquid–liquid extraction (LLE) is applicable only to liquid matrices (e.g. milk, oils). In most cases, LLE uses about 100 mL of solvent per 5–50 g of sample. The major drawbacks of this technique are low sample throughput resulting from the need for manual concentration step, and using of large amounts of organic solvents. As far as stable emulsions are formed, centrifugation is needed to assist the separation of phases.

Solid-phase extraction (SPE) represents a feasible alternative for isolation/pre-concentration of POPs from aqueous and other liquid samples. Non-polar analytes are adsorbed by stationary phase such as octadecyl silica (C_{18}) in an extraction cartridge or disc. This approach is more advantageous compared to LLE because of reduced sample preparation time, decreased solvent usage, and improved sensitivity. SPE-based sample processing can be fully automated; and a wide range of SPE extractors are available in the market. However, problems such as clogging of the SPE cartridge can be encountered when solid particles are dispersed in the sample.

Soxhlet extraction represents the most frequently used extraction technique for isolation of lipophilic POPs from solid low moisture matrices or flowing powder obtained by desiccation of a sample (1–100 g). In common practice, extraction with polar and non-polar organic solvents such as dichloromethane, hexane–acetone, hexane–dichloromethane takes 4–18 h. To increase sample throughput, semi-automated extractor batteries are employed. Due to co-extraction of lipids and other sample components, "dirty" extracts obtained by this liquid–solid extraction, need subsequent extensive clean-up.

Accelerated solvent extraction (ASE) also known as *pressurised solvent extraction* (PLE) uses organic solvent/solvent mixtures at increased pressure during the extraction. This allows to keep the solvent(s) in liquid phase even at higher temperatures. Higher speed extraction of POPs/PAHs under these conditions is a result of their increased solubilities, better desorptions, and enhanced diffusion. The PLE system consists from a stainless-steel extraction cell, where temperature and pressure are controlled by electronic heaters and pumps. Extraction steps in the static mode involve: (1) loading the sample into the extraction cell; (2) filling the cell with an organic solvent; (3) heating and pressuring the cell to adjusted values; (4) transfer of the extract to the collection bottle and rinsing the sample with an additional solvent; and (5) purging the remaining solvent from the sample to the collection bottle using a suitable gas. Compared to Soxhlet extraction, only minutes

Table 2
Characterisation of extraction techniques employed in food analysis

	Extraction technique				
	Soxhlet	**Sonication**	**PLE**	**MAE**	**SFE**
Extraction time	3–48 h	10–60 min	5–30 min	3–30 min	10–60 min
Sample size	1–30 g	1–30 g	1–30 g	1–10 g	1–5 g
Solvent demands	100–500 mL	30–200 mL	10–100 mL	10–40 mL	2–5 mL (solid trap); 5–20 mL (liquid trap)
Investment	Low	Low	High	Moderate	High
Advantages	No filtration required	Multiple extraction	Fast extractions; low solvent volumes; elevated temperatures; no filtration required; automated systems	Fast and multiple extractions; low solvent volumes; elevated temperatures	Minimal solvent volumes; elevated temperatures; relatively selective towards matrix interferences; no clean-up or filtration needed; concentrated extracts; automated systems
Limitations	Long extraction times; large solvent volumes; clean-up step needed	Large solvent volumes; repeated extractions may be required; clean-up step needed	Clean-up step needed	Extraction solvent must be able to absorb microwaves; clean-up step needed; waiting time for the vessels to cool down	Many parameters to optimise, especially analyte collection

Source: From (5) with permission

are needed for the automated extraction process, but sample clean-up is still necessary, unless sorbents such as florisil, aluminium oxide are employed as fat, pigments, or other components retainers. To achieve good selectivity of the POPs/PAHs isolation, careful tuning of extraction conditions is necessary in such a case.

Ultrasonic extraction is a simple extraction technique, in which the sample is suspended in an organic solvent in a vessel and placed in an ultrasonic bath. The main parameters influencing the extraction efficiency are the polarity of the solvent, the homogeneity of the matrix, and the ultrasonication time. After extraction, the mixture of the sample and organic solvent is separated by filtration and rinsing with the solvent. Although this extraction procedure does not require expensive instruments and is not laborious, large consumption of the solvent (30–200 mL per 1–30 g of sample) is the main drawback of this technique.

Matrix solid-phase dispersion (MSPD) allows extraction of various POPs/PAHs from homogenously dispersed food samples with a sorbent phase (e.g. C_{18} silica). The homogenised sample is placed in a glass-syringe-barrel column and the POPs/PAHs are selectively eluted with suitable organic solvent (e.g. hexane), followed by the immediate instrumental analysis since the sample extraction and clean-up are conducted in one step. Compared to "conventional" extraction procedures, this technique requires a smaller sample size, has a shorter analysis time, and uses less organic solvent.

Supercritical fluid extraction (SFE) offers short sample processing times and use of a cheap environment-friendly extraction agent. In SFE, the sample is loaded in a high-pressure vessel and extracted with low viscosity supercritical fluid (in most cases carbon dioxide at pressures of 150–450 bar and temperatures of 40–150°C). The analytes are collected in a small volume of solvent or onto a solid-phase trap, from which they are rinsed with organic solvent in a subsequent step. Fat retainers (e.g. basic alumina, neutral alumina, florisil, and/or silica) can be introduced into the extraction thimble to obtain a fat-free extract. The use of SFE in POPs analysis has partially vanished during recent years because of operation problems such as a need to optimise many parameters, problems for matrices with high water content, and the high cost of automated instrumentation.

Microwave-assisted extraction (MAE) allows rapid extraction of POPs/PAHs from solid matrices by employing microwave energy as a source of heat. The partitioning of the analytes from the sample to the extractant depends upon the temperature and the nature of the extractant. Since the microwave device heats the entire sample simultaneously without heating the vessel, the solution reaches its boiling point rapidly, leading to a very short extraction time. The attraction of this technique includes also somewhat easy optimisation and it is cheaper than other modern extraction techniques (SFE, PLE).

Solid-phase microextraction (SPME) represents solvent-free isolation/pre-concentration technique employing a fused-silica fibre that is coated with an appropriate stationary phase. Analytes present in the sample are directly extracted (from the headspace or by direct immersion) and concentrated onto the fibre coating. The SPME sampling procedure is then followed by the transfer of pre-concentrated analytes into the chromatographic system using either a GC injector (thermal desorption) or an SPME–HPLC interface (desorption by the solvent). The main features of SPME include unattended operation via robotics (if a fully automated option is available) and in the case of GC-amenable analytes, elimination of maintenance of the liner and column (contamination by non-volatiles does not occur). This sample extraction technique, however, is susceptible to strong matrix effects, which can produce complications in quantification. In addition, variability of limits of detection for different analytes depends on the equilibrium between the coating material and the matrix.

Stir-bar sorptive extraction (SBDE) can be used as an effective tool for sample enrichment in aqueous solutions. A glass-lined magnetic bar is covered with a thick layer of sorbent (similar to that in SPME). By magnetically stirring the bar in the sample solution, the analytes are enriched in the sorbent phase. After this pre-concentration, the compounds are thermally desorbed from the bar with GC–MS.

In addition to these extraction techniques, sample preparation can be simplified by using a single-step organic solvent extraction and salting out effect to enhance liquid–liquid partitioning from water in the sample. This strategy is the main concept of the *QuEChERS* (Quick, Easy, Cheap, Effective, Rugged, and Safe) sample extraction method (9). During the development of this method, great emphasis was placed on streamlining this sample preparation procedure whenever possible by simplifying or omitting impractical, laborious, and time-consuming steps. The "original" QuEChERS method involves initial extraction with acetonitrile (MeCN), liquid–liquid partitioning after addition of a mixture of anhydrous $MgSO_4$ and NaCl, which reduces some polar matrix components, followed by a simple clean-up step in which the extract is mixed with primary secondary amine (PSA) sorbent and anhydrous $MgSO_4$ (dispersive-SPE). After these steps the extract is ready for GC–MS and LC–MS (directly or after dilution with water containing formic acid). The QuEChERS concept has been successfully used in the analysis of various POPs in food in its original version, or after some modifications such as change of extraction solvent (MeCN → MeCN containing acetic acid, ethyl acetate, methanol), the amount and kind of salts (NaCl → sodium acetate). While the use of MeCN or ethyl acetate is suitable for POPs such as OCPs, PCBs, and PBDEs, methanol is preferred as an extraction solvent for perfluoroalkylated substances.

3. Clean-Up

As mentioned earlier, not only are the target POPs/PAHs isolated during the extraction from the sample, but also various matrix components are co-extracted and may lead to worsened method performance. Although in some cases little or no clean-up is needed, the impact of matrix effects (caused mainly by the matrix co-extracts) such as inaccurate quantification, decreased method ruggedness, poor analyte detectability, and even reporting of false positive or negative results have to be considered. Therefore, some clean-up step is typically involved in the POPs/PAHs analysis (Table 3) (5, 8–10).

Gel permeation chromatography (GPC), sometimes referred as *size exclusion chromatography* (SEC), represents a non-destructive clean-up procedure. In most cases, spherical porous styrene–divinylbenzene copolymers (commercially available as soft Bio-Beads S-X3, or rigid types e.g. PL-gel) are used for separation of lipids (>500 Da), which are the first eluting compounds from the column, followed by the smaller molecules, including the POPs. However, size of the molecules is not the only separation mechanism in this particular case since π–π interactions of this copolymer with planar compounds may cause different elution order not reflecting the "size rule." Dichloromethane, chloroform, or mixtures of dichloromethane–hexane or ethyl acetate–cyclohexane are the most often applied eluents. The GPC can be fully automated and, contrary to adsorption chromatography, it is more suitable for the isolation of "unknown" contaminants. This method can handle a relatively large amount of lipids (up to 500 mg). However, in some cases, the use of a second GPC elution or other clean-up techniques is needed to remove all lipids. In addition, this technique does not separate individual groups of POPs, thus, follow-up fractionation, if needed, is employed to obtain different classes of POPs.

Adsorption column chromatography involves passing the extracts though adsorbent columns. Various sorbents such as alumina, silica, and florisil, available in different mesh sizes, levels of activity and column size, either separately or in combination, were successfully evaluated for this purpose to reduce sample handling and analysis time. Alumina columns have a fat capacity of ~250 mg per 10–20 g, which may not be enough in ultra-trace analysis required to remove large quantities of lipids. With regards to silica gel, it allows fractionation of the extract according to the polarity of different classes of POPs.

Dispersive solid-phase extraction (d-SPE) is a very simple clean-up procedure where suitable sorbent (primary-secondary amine – PSA, C_{18} silica, or activated charcoal) is added to an extract aliquot. After mixing and centrifugation, the extract is used for subsequent

Table 3
Characterisation of lipid removal techniques

	Requirement					
Technique	**Lipid removal**	**Destruction of compounds**	**Recovery**	**Amount of time spent**	**Amount of work invested**	**Automation**
GPC	Two steps necessary	None observed	Good	High	High	Easy
Florisil, silica, alumina	Few lipids removed or large amount of absorbent used	None observed	Good	Fair	Fair	Difficult
Sulphuric acid (and/or sulphuric acid/silica)	Good	Some OCPs	Good	Low	Fair	Difficult
Saponification	Good	Some OCPs, PCBs with fully chlori-nated aryl-ring	Fair	Low	Fair	Difficult

Source: From (5) with permission

analysis employing either GC or LC. The d-SPE step reduces the amount of common matrix co-extractives typical for foods, particularly fatty acids supposing PSA is employed. As far as C_{18} silica is employed, lower recoveries of some OCPs might occur. Regarding charcoal, this non-specific sorbent was shown to be very suitable in analysis of PFOS and related perfluorinated compounds.

Destructive lipid removal includes either alkaline treatment (saponification), or oxidative dehydration by sulphuric acid treatment. In the later case, mineralisation of lipids and other bulky matrix components is realised either by direct addition of concentrated acid to the extract or by passing the crude extract through impregnated silica columns, and are the most commonly used lipid removal destructive methods. It has been shown that PBBs, PBDEs, and PCBs are stable under strong acid conditions. Basic conditions of saponification are critical as too high temperatures and too long of process time may cause degradation of highly brominated PBDEs, PBBs, and PCBs. Also, silica gel impregnated with alcoholic KOH or of a multilayer column with neutral silica, acidified silica, and basic silica can be employed.

4. Determination of POPs/PAHs by GC–MS and LC–MS

For the analysis of trace levels of lipophilic POPs (OCPs, PCBs, PCDDs/PCFDs, and BFRs), and PAHs occurring in complex matrices such as foodstuffs, high-resolution gas chromatography (GC) interfaced to mass spectrometry employing a suitable single or tandem mass analyser represents the key separation/detection technique (2, 8). For those POPs, which are either not amenable to GC due to their ionic nature (perfluoroalkylated acids) or their diastereomers are interconverted in a hot GC injector, moreover, poorly separated on conventional GC columns (e.g. HBCD), LC–MS is the method of choice for their analysis (8, 11).

Regardless of the continuously improving detection capabilities of modern GC–MS or LC–MS systems (discussed below), the sample preparation practice remains a crucial role in obtaining required performance characteristics of a particular analytical procedure (mainly limits of quantification and uncertainty of measurement might be adversely affected by sample matrix).

During the last few years, a large number of novel ambient desorption ionisation techniques, such as desorption electrospray ionisation (DESI), atmospheric-pressure solids analysis probe (ASAP), direct analysis in real time (DART), and many others have become available. Their main advantages compared to conventional techniques involve the possibility of direct sample examination in the open atmosphere, minimal or no sample preparation requirements, and, remarkably high sample throughput (chromatographic separation is not involved in this particular case) (12).

4.1. Sample Injection in GC–MS

Several of existing GC inlet systems are applicable for trace analysis of POPs/PAHs in complex food matrices; the most common being a "hot" splitless injector, a programmed temperature vaporiser (PTV), and, in the last decade, also a direct sample introduction/difficult matrix introduction injector (DSI/DMI). The choice of an optimum injection strategy depends on many factors including the concentration range of target analytes, their physicochemical properties, and, to a significant extent, on the amount and nature of matrix co-extracts present in the food sample extract (8, 13–15).

Hot *splitless injection* (250–300°C) has been in use in many laboratories concerned with routine trace analysis of POPs/PAHs. Depending on the type of injector liner and expansion volume of sample solvent, the volumes introduced onto the GC capillary are typically in the range of 1–3 μL with 0.5–2 min of splitless period. However, this inlet suffers from the potential thermal degradation (e.g. *p,p′*-DDT→*p,p′*-DDD and/or *p,p′*-DDE; BDE-209→nona-BDE congeners), rearrangement (HBCD), and/or adsorption of susceptible analytes.

To overcome, or at least partly compensate for these problems, *pulsed splitless injection* can be applied. Increased column head pressure for a short time period during the sample injection splitless period (usually 1–2 min) leads to a higher carrier gas flow rate through the injector (8–9 mL/min vs. 0.5–1 mL/min during classical splitless injection), thus faster transport of sample vapours onto the GC column. In this way, the residence time of analytes, and, consequently, their interaction with active sites in the GC inlet, is fairly reduced. In addition, the detection limits can be lowered by injection of higher sample volumes (for most liners up to 5 μL) without any risk of backflash. For splitless injections >1–2 μL, a retention gap prior to the analytical column is strongly recommended to avoid excessive contamination of front part of separation column, and consequent peaks distortion.

A *programmable temperature vaporisation* (PTV) *injector* represents the most versatile GC inlet offering significant reduction of most problems typically encountered when using hot vaporising devices (splitless inlets) in trace POPs/PAHs analysis. The most important fact is that a PTV injector chamber is cool at the moment of injection. A rapid temperature increase, following withdrawal of the syringe from the inlet, allows efficient transfer of the volatile analytes onto the GC column while leaving behind non-volatiles in the injection liner. With regard to these operational features, PTV is ideally suited for thermally labile analytes and analytes representing a wide boiling range. PTV enables introduction of large sample volumes (up to hundreds of microlitres) into the GC system. This feature makes the use of PTV for POPs/PAHs analysis particularly attractive and also enables its on-line coupling with various enrichment and/or clean-up techniques such as automated solid phase extraction (SPE) approaches. From a practical point of view, PTV is compatible with any capillary GC column diameter including microbore columns. However, to attain its optimal performance in a particular application, many parameters have to be optimised (e.g. initial and final injector temperature, inlet heating rate, venting time, flow and pressure, transfer time, injection volume, type of liner). Due to the inherent complexity of this inlet, method development might become on some occasions a rather demanding task. Despite this, the use of PTV in food analysis is becoming a gold standard.

Direct sample introduction (DSI) and its fully automated version, *difficult matrix introduction* (DMI), represent a relatively novel large volume injection (LVI) technique. The DSI approach involves adding up to 30 μL of the extract to a microvial that is placed in an adapted GC liner. After evaporating and venting sample solvent at a relatively low temperature, the injector is ballistically heated to transfer analytes at the front of a relatively cold GC column (some matrix components with similar volatility range can be pre-concentrated here). In the next phase, the column undergoes normal temperature programming to separate volatilised compounds. Then, during the cooling period, the microvial containing residues of non-volatile matrix components is removed and discarded. In the commercial DMI system, the entire liner along with the microvial is replaced after each injection. In this way, time-consuming and expensive purification step can be omitted or significantly reduced for some matrices. Since the bulk (semi)volatile matrix components introduced from the sample into the injector may influence the quantitative aspects of the injection process and/or interfere in analytes detection, instruments with MS analysers (single or tandem) providing more accurate results should be preferably used. Regardless of the sample preparation strategy, reduced demands for the GC system maintenance represents a positive feature of this technique.

4.2. Sample Separation Using GC and LC

With regard to a (typically) complex mixture of matrix components occurring in food extracts (in many cases even after purification) in fairly higher amounts as compared to concentrations of isolated toxicants, the optimisation of GC and LC separation requires careful attention to a number of important variables and their interaction. Physical (length, internal diameter, and stationary phase), parametric (temperature and flow velocity) column variables, and mobile phase composition and its additives affect the separation process. The nature of functional groups as well as the percentage of substitution of those functionalities govern the stationary phase–analytes–interferences interactions thus influencing their retentions. The choice of the separation system is closely associated with the selectivity/specificity of the detection system employed in a particular analysis. For instance, poor resolution of critical pairs (analyte–analyte or analyte–interference) by chromatography might be compensated by the resolving power of the respective mass spectrometric detector.

Organochlorine pesticides. A conventional approach to GC analysis of OCPs employs capillary columns with low-bleed stationary phases mostly consisting of (5%-phenyl)-methylpolysiloxane (or cyanopropyl, cyanopropylphenyl or increased phenyl content up to 50%). Relatively long analytical column 30–60 m of 0.25–0.32 mm inner diameters are commonly used in routine practice, with GC analysis time approaching 1 h. However, the growing number of required analyses (not only organochlorine but also other pesticides) leads to requiring decreased analysis time using fast GC techniques (mostly in combination with MS detection), thus increasing sample throughput and reducing the laboratory operating costs (16, 17). As an example, Fig. 1 illustrates the

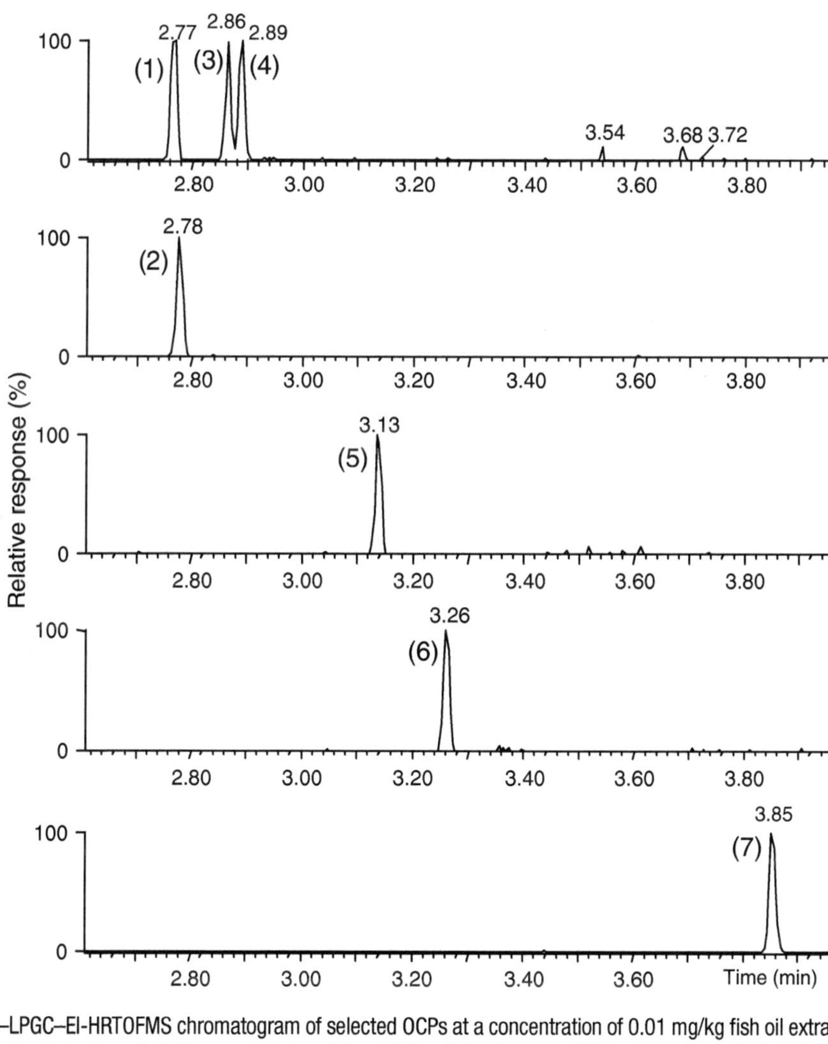

Fig. 1. PTV–LPGC–EI-HRTOFMS chromatogram of selected OCPs at a concentration of 0.01 mg/kg fish oil extract. The target ions were extracted using a 0.02 Da mass window. (1) α-HCH (*m/z* 180.938), (2) HCB (*m/z* 283.810), (3) β-HCH (*m/z* 180.938), (4) γ-HCH (*m/z* 180.938), (5) heptachlor (*m/z* 271.810), (6) aldrin (*m/z* 262.852), (7) *p,p′*-DDT (*m/z* 235.008).

rapid analysis of selected OCPs in fish oil extract at a very low level (0.01 mg/kg). In this particular case, fast GC under the conditions of a high temperature programming (60°C/min) and vacuum conditions in a megabore GC column (10 m × 0.53 mm × 0.5 μm) coupled to a restriction capillary (3 m × 0.15 mm), so called low-pressure gas chromatography (LPGC), was used. Using this chromatographic set-up, the analysis is completed within 7 min, which reduces the GC run time.

Polychlorinated biphenyls. GC combined with specific detectors, either "conventional" electron-capture detector (ECD), or currently preferred MS, are routinely used in PCBs analysis. Since even high-resolution capillaries do not allow separation of all 209 congeners, either simultaneous separation on two parallel columns differing in polarity or comprehensive two-dimensional GC (GC × GC) separation and detection with electron-capture detector (ECD) or MS is the method of choice for routine analysis. Typically, non-polar columns such as 100%-methylpolysiloxane or (5%-phenyl)-methylpolysiloxane are employed for their separation. However, because of coelution of a number of congeners (critical pairs), alternative phases such as (50%-phenyl)-methylpolysiloxane, (8%-phenyl)-polycarboranesiloxane, or (14%-cyanopropyl-phenyl)-methylpolysiloxane have to be used (5, 17).

Polychlorinated dibenzo-p-dioxins and polychlorinated dibenzofurans. Because the concentrations of interest are at the parts per trillion (ppt) level, the analytical methods for PCDDs/PCDFs require laborious and time-consuming sample clean-up and preconcentration processes. Additionally, detection techniques of high selectivity and high sensitivity are required since samples often contain matrix components (potential chemical interferences) at concentrations several orders of magnitude higher than those of target analytes. One of the key factor, which makes the analysis of dioxins so difficult, is the existence of many congeners (i.e. 75 PCDDs, 135 PCDFs). Differences in toxicities (expressed as toxic equivalency factors, TEFs) of several orders of magnitude exist between various isomers (with 2,3,7,8-tetrachlorodibenzo-*p*-dioxin, TCDD, being the most toxic), thus, the separation and reliable identification/quantification of each is a crucial task for the risk assessment. Since the monitoring of all PCDDs/PCDFs is hardly attainable, the legislation reduces the monitoring only of those compounds with the highest toxicological potential (19).

For GC separation, long narrow bore capillary columns are often used (30–60 m × 0.32–0.25 mm × 0.15–0.25 μm) with different stationary phases: (5%-phenyl)-methylpolysiloxane, (50%-cyanopropylphenyl)-dimethylpolysiloxane, or 44%-methyl–28%-phenyl–20%-cyanopropylpolysiloxane–8% Carbowax 20 M (DB-Dioxin) (2).

Brominated flame retardants. Although 209 BDE congeners are theoretically possible, only a small number of these contaminants can be found in the earlier produced technical PBDE mixtures (e.g. BDE 28, 47, 99,100, 153, 154, 183, 209), which allows using a single capillary GC column that offers sufficient resolution for a congener-specific PBDE determination. A non-polar or medium-polar column, e.g. 100%-methylpolysiloxane, (5%-phenyl)-dimethyl polysiloxane, 14%-cyanopropylphenyl–86%-dimethylpolysiloxane, with a length of 25–60 m and small diameters (<0.25 mm) are most frequently employed. The use of sufficiently long columns is important for achieving enough separation between BDE congeners and possible interferences supposing selectivity of detection is not sufficient (8, 10).

Special attention is usually paid to BDE 209, not only because of its susceptibility to thermo-degradation in the GC system, but also due to relatively very long retention times, thus reduced GC throughput. For that reason, the analysis of BDE 209 is sometimes carried out separately from the analysis of the other PBDEs. To overcome this problem, relatively short column enabling reduction of residence time of this congener, can be considered. Typically, non-polar columns, 100%-methylpolysiloxane, (5%-phenyl)-methylpolysiloxane, with a length of 10–15 m and 0.1–0.2 μm film thickness are utilised (8, 10). As an example, Fig. 2 shows a rapid separation of PBDEs (including the troublesome BDE 209 congener) under the conditions of LPGC with a high temperature programming (60°C/min). In this particular case, a combination of a restriction capillary (2.5 m × 0.15 mm) and megabore column (10 m × 0.53 mm × 0.5 μm) was employed for the separation.

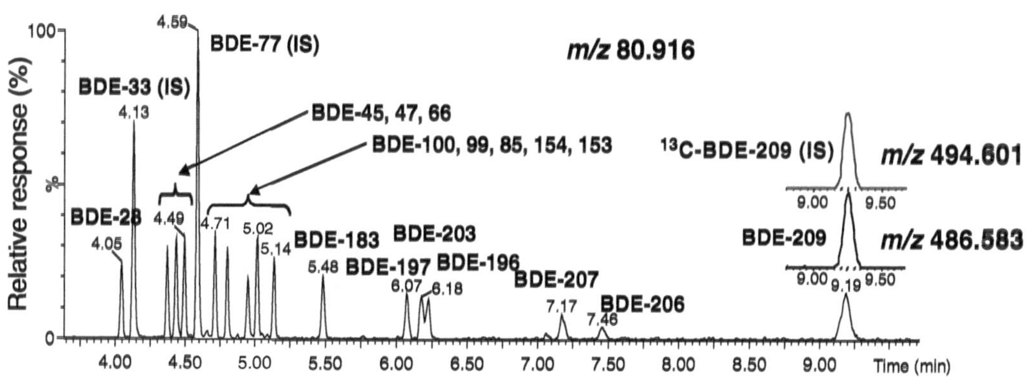

Fig. 2. LPGC–HRTOFMS chromatogram of standard solution of PBDEs detected in NCI mode (BDE-28, 47, 49, 66, 85, 99, 100, 153, 154, 183, 196, 197, 203 at a concentration of 0.5 ng/mL; BDE 206, 207, 209 at a concentration of 1 ng/mL; BDE 37 and 77 at a concentration of 5 ng/mL, and ^{13}C-BDE-209 at a concentration of 50 ng/mL; 1 μL injected). The target ions (bromine $[Br]^-$ with exception of BDE 209) were extracted using a narrow 0.05 Da mass window.

Besides PBDEs, signals of several other BFRs such as pentabromotoluene (PBT), pentabromoethylbenzene (PBEB), hexabromobenzene (HBB), bis(2,4,6-tribromophenoxy)ethane (BTBPE), octabromotrimethylphenylindane (OBIND), or decabromodiphenyl ethane (DBDPE), can be found in the chromatogram (the latter compounds elutes even after BDE 209) obtained at conditions relevant to those in Fig. 2. Determination of one of the most widely used BFR, hexabromocyclododecane (HBCD) can also be carried out using GC–MS with stationary phases similar to those typically used for PBDEs. However, such a GC set-up does not allow quantification of individual diastereomers of HBCD (α-, β-, and γ-HBCD, each having two enantiomers) since they are not separated using common GC stationary phases; moreover, they undergo interconversion at temperatures above 160°C. Contrary to GC, reversed-phase HPLC employing non-polar (C_{18}, C_{30}) or chiral columns for their separation represent a versatile tool for the isomer-specific determination of HBCD isomers (8, 10). With regards to tetrabromobisphenol A (TBBPA), acidification and derivatisation are required before the GC–MS analysis, while LC–MS allows its direct determination. For the separation, LC columns 50–250 mm × 2–4 mm × 3.5–5 μm are typically used with gradient elution employing a mixture of MeCN:H_2O, MeOH:H_2O, or MeCN:MeOH:H_2O. For optimised chromatographic separation and/or ionisation response, mobile phase additives such as formic acid, tris(hydroxymethyl) aminomethane and ammonium acetate, are often used. In the case of TBBPA, mobile phase MeOH:H_2O is preferred over MeCN:H_2O due to more stable detector baseline and thus lower LODs (8, 10).

Perfluoroalkylated substances. GC can be used for direct determination of fluorotelomer alcohols, perfluoroalkyl sulfonamidoethanols, and perfluoroalkyl sulfonamides, which have high vapour pressure. For their separation, narrow bore capillary columns (30 m × 0.25 mm × 0.20–0.25 μm) with polar stationary phases (e.g. polyethylene glycol) are employed (20). On the other hand, perfluoroalkyl sulfones, perfluorocarboxylic acids, are only amenable to LC analysis. In this case, e.g. LC columns 50–150 mm × 2–4 mm × 3.5–5 μm with C_{18} sorbent can be used with gradient elution employing MeOH:H_2O (with ammonium acetate as an additive) (11, 21).

Polycyclic aromatic hydrocarbons. For GC separation, long narrow bore capillary columns are often used (60–30 m × 0.25 mm × 0.25 μm) with stationary phases such as (5–50%-phenyl)-methylpolysiloxane (22). Separation of some isomeric PAHs (e.g. B[*a*]A/Chr; B[*b*]F/B[*j*]F/B[*k*]F; I[1,2,3-*cd*]P/DB[*ah*]A) might be a problem even on long non-polar capillaries. If comprehensive analysis of all 16 carcinogenic PAHs (2) is required, then "heavy," low volatile representatives, such as dibenzopyrenes, can be a limiting factor in GC

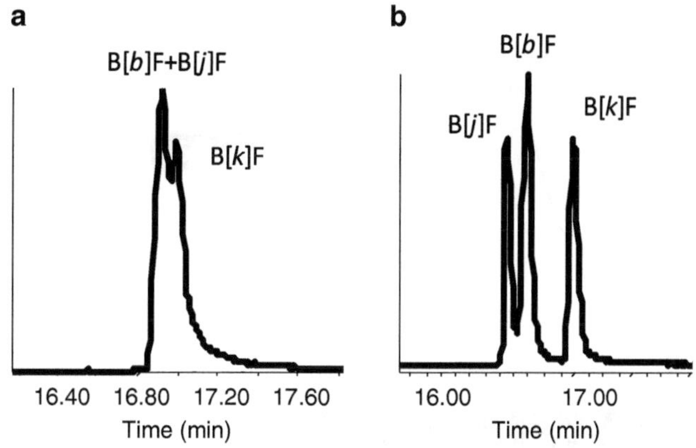

Fig. 3. GC separation of "critical" PAH isomers employing (**a**) (5%-phenyl)-methylpolysiloxane (30 m × 0.25 mm × 0.25 μm) and (**b**) 50% liquid crystalline–methylpolysiloxane (10 m × 0.18 mm × 0.1 μm). Quadrupole mass analyser used for detection (*m/z* 252 displayed).

throughput due to long retention times; moreover, band broadening might contribute to high detection limits. Under these conditions, low-bleed columns allowing programming up to high temperature limits are needed. If the analysis of these late eluting PAHs is not required, medium-polar stationary phases, 50% liquid crystalline–methylpolysiloxane stationary phase may provide a greater selectivity for some critical pairs as compared to non-polar low bleed stationary phases (Fig. 3). HPLC analysis of PAHs is very common, LC columns 100–250 mm×2–4 mm×3.5–5 μm with C_{18} sorbent (or some special sorbent developed for PAHs separation, e.g. LC-PAH, Chromspher 5PAH) can be used with gradient elution employing most often MeCN:H_2O (23).

4.2.1. Comprehensive Two-Dimensional GC (GC × GC)

The introduction of systems allowing continuous transfer of the entire sample from the first column to be separated on the second column with different selectivity has been made. Thanks to enormous increase of chromatographic resolution, significant improvements of both target and non-target screening of food components in a wide range of matrices is possible. This approach, called comprehensive two-dimensional gas chromatography (GC × GC), is introduced in the following sections in a greater detail (24, 25).

The heart of the GC × GC system is a modulator that connects the first dimension conventional-size column with a short microbore column in the second dimension.

There are three fundamental functions of this interface: (1) trapping of small adjacent fractions (typically 2–10 s) of the effluent from the first separation column; (2) refocusing these fractions (either in time or in space); and (3) injection of the refocused fractions as narrow pulses into the second-dimension column. The separation on the latter column is extremely fast and takes only 2–10 s vs. 20–120 min for the first dimension, and is, therefore, performed under essentially isothermal conditions.

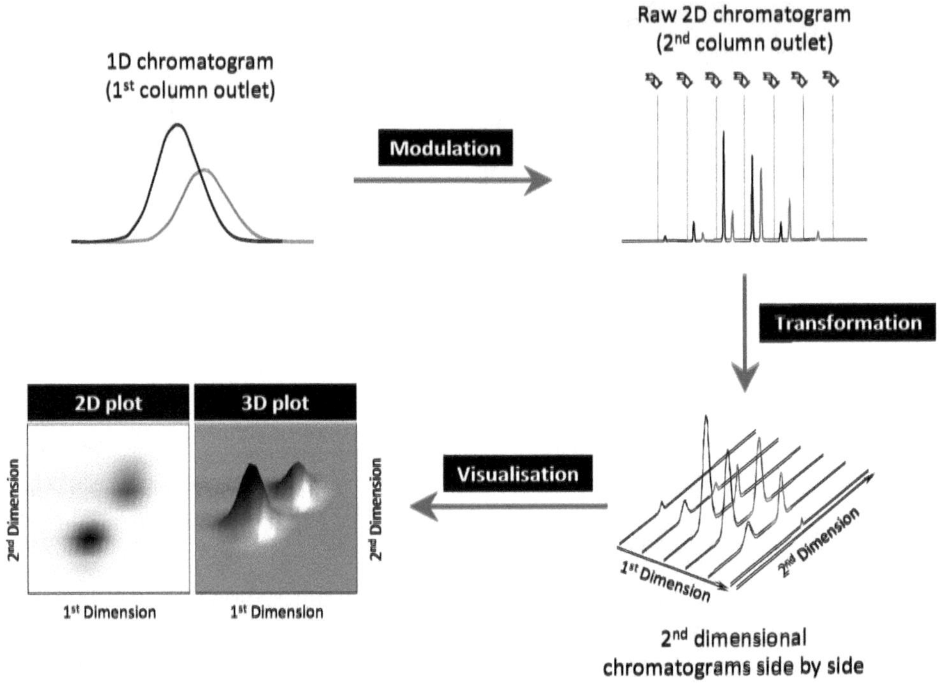

Fig. 4. Generation and visualisation of a GC × GC chromatogram.

A large series of high-speed chromatograms as the outcome of the transfer of chromatographic band from the first to the second dimension is generated during the GC×GC run. As shown in Fig. 4, these adjacent pulses are usually stacked side-by-side by special software to form a 2D chromatogram with one dimension representing the retention time on the first column (t_{R1}) and the other, the retention time on the second column (t_{R2}). The most convenient way to visualise GC×GC data is as contour plots representing the bird's eye view, where peaks are displayed as spots on a plane using colours and shading to indicate the signal intensity (Fig. 4).

Compared to conventional 1D-GC, the optimisation of GC×GC analysis requires rather more complex approach. The changes in operational parameters such as oven temperature or carrier gas flow rate have different impacts on the performance of separation columns since these differ both in their geometry and separation mechanism. Furthermore, new parameters such as modulation frequency and modulator temperature have to be optimised.

Conventional columns, typically 15–30 m length × 0.25–0.32 mm internal diameter × 0.1–1 μm film thickness, are used in the first dimension. This allows application of virtually all sample introduction techniques (split/splitless, on column, PTV, DMI/DSI). Stationary phases commonly used in first-dimension columns are typically non-polar such as 100% dimethylpolysiloxane or (5% phenylene)-dimethylpolysiloxane on which the separation is governed mainly by analyte volatility (dispersive van der Waals forces play a key role). The size of columns for the second dimension is commonly in a range of 0.5–2 m length × 0.10–0.18 mm internal diameter × 0.10–0.18 μm film thickness. More polar stationary phases such as 35–50% phenylene–65–50% dimethylpolysiloxane, polyethylene glycol, carborane, and/or cyanopropyl-phenyl–dimethylpolysiloxane are often employed. Analytes interact with these medium-polar/polar phases via various mechanisms such as π–π interactions, hydrogen bonding etc., hence the requirement for different, independent separation principle is met. In most applications orthogonality is achieved using non-polar × polar separation mechanisms. However, for the analysis of POPs/PAHs, only highly stable stationary phases are typically used for their separation. This actually limits the use of stationary phases in most applications only to non-polar and medium-polar ones (Table 4).

Effective and robust modulation is a key process in the GC × GC analysis. In practice, a fixed modulation period, typically in a range of 1–10 s is employed during the analysis. Under ideal experimental conditions, the retention time of the most retained compound in the second dimension is shorter than the modulation time. If this is not the case, i.e. analytes do not elute in their modulation cycle, so called "wrap-around," which might cause some coelutions with less retained compounds from the following cycle. Avoiding this phenomenon can be achieved e.g. by an increase of the second dimension column temperature (if an independent oven is available) thus reducing the second dimension retention time. In any case, optimal function of modulator is essential for the quality of the separation and detection process.

The fast separation on a short and microbore second dimension column results in very narrow peaks with widths of 50–1,000 ms at the baseline. Unfortunately, conventional scanning MS detectors are typically too slow and do not provide reliable spectra and peak reproduction. At present, only time-of-flight mass spectrometers can acquire 50 or more mass spectra per second, which are required for the proper reconstruction of the chromatogram and for quantification in GC × GC.

4.2.1.1. Advantages of GC × GC in the POPs Analysis

A number of characteristics of GC × GC have been reported that documents superiority of this technique over conventional 1D-GC in POPs analysis (24, 25).

High Peak Capacity. The peak capacity, characterised as a maximal number of chromatographic peaks that can be placed side by side into the available separation space (chromatogram), is significantly enhanced. As an example, the merit in dioxin analysis resulting from the separation power is shown in Fig. 5.

Table 4
Overview of GC × GC methods used for the determination of POPs in food matrices

Matrix	Analytes	Injection	GC set-up	Column flow (mL/min)	Modulation period (s)	Ionisation/ detection	Acquisition speed (Hz)	Run time (min)	Ref.
Fish	PCBs, PBDEs, PAHs	Pulsed splitless, 280°C, 1 µl	^{1}D: BPX-5 (30 m × 0.25 mm × 0.25 µm) ^{2}D: Rt-LC35 (1 m × 0.15 mm × 0.1 µm)	1.3	2	EI-TOFMS	100	32.3	(28)
Fish oil	OCPs, PCBs, PBDEs	LV-DSI, 10 µl	^{1}D: Rtx-5Sil-MS (15 m × 0.25 mm × 0.25 µm) ^{2}D: DB-17MS (2 m × 0.18 mm × 0.18 µm)	1–1.5	3.5	EI-TOFMS	100	51.5	(29)
Fish oil	PCDDs/ PCDFs	LV-DSI, 10 µl	^{1}D: Rtx-Dioxin 2 (60 m × 0.25 mm × 0.25 µm) ^{2}D: Rtx-PCB (3 m × 0.18 mm × 0.18 µm)	2–2.5	3.5	EI-TOFMS	50	56.5	(30)
Fish	PCBs, PCDDs/ PCDFs	Splitless, 270°C, 1 µl	^{1}D: Rtx-Dioxin 2 (60 m × 0.25 mm × 0.25 µm) ^{2}D: Rtx-500 (2 m × 0.18 mm × 0.18 µm)	2.5	3	EI-TOFMS	100	35.0	(31)
Fish, pork, milk	PCBs, PCDDs/ PCDFs	Splitless, 250°C, 1.2 µl	^{1}D: Rtx-500 (40 m × 0.18 mm × 0.10 µm) ^{2}D: BPX-50 (1.5 m × 0.10 mm × 0.10 µm)	0.8	4	EI-TOFMS	60	48.0	(18)
Fish	PBDEs	Splitless, 280°C, 1 µl	^{1}D: DB-1 (30 m × 0.25 mm × 0.25 µm) ^{2}D: 007-65HT (1 m × 0.10 mm × 0.10 µm)	1.2	5	NCI-qMS	24	33.0	(32)
Tea	OCPs	Splitless, 270°C, SPME	^{1}D: BPX-5 (40 m × 0.18 mm × 0.18 µm) ^{2}D: SupelcoWax (2.5 m × 0.10 mm × 0.10 µm)	1.0	5	EI-TOFMS	125	17.2	(33)
Fruit	OCPs	Pulsed splitless, 250°C, 1 µl	^{1}D: DB-XLB (30 m × 0.25 mm × 0.1 µm) ^{2}D: DB-17 (1 m × 0.10 mm × 0.10 µm)	1.2	2	EI-TOFMS	250	22.6	(34)
Fruit	OCPs	Splitless, 250°C, 1 µl	^{1}D: Rtx-5MS (10 m × 0.18 mm × 0.20 µm) ^{2}D: TR-50MS (1 m × 0.10 mm × 0.10 µm)	1.0	4	EI-TOFMS	100	24.0	(35)
Milk	PCBs, PBDEs, OCPs	Splitless, 280°C, 1.2 µl	^{1}D: DB-1 (15 m × 0.25 mm × 0.25 µm) ^{2}D: HT-8 (1.2 m × 0.10 mm × 0.10 µm)	0.8	3	EI-TOFMS	60	50.3	(36)

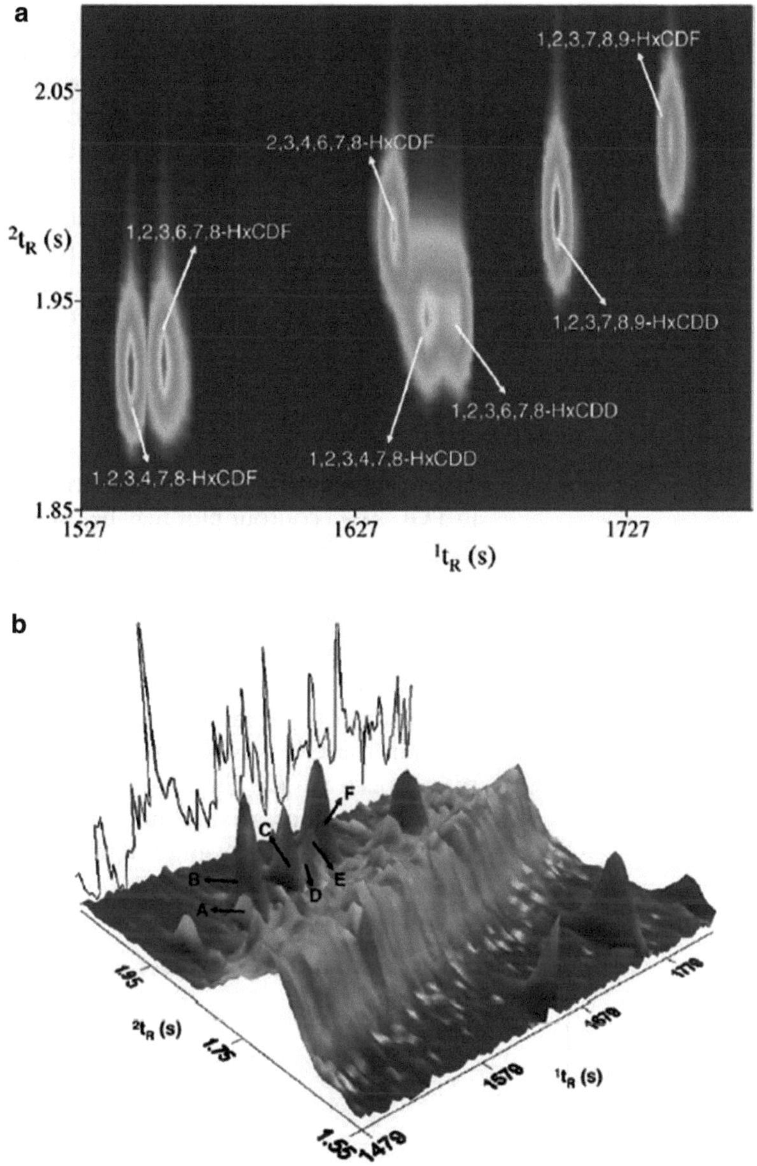

Fig. 5. (**a**) GC × GC–TOFMS contour plot of a standard solution containing 1 ng of HxCDD/Fs. The deconvoluted ion current (DIC) is based on the sum of the molecular ions corresponding to HxCDD/Fs (*m/z* 390 + 374). (**b**) Expended section of the HxCDD/F region of a GC × GC shade surface plot after injection of the clean-up fraction containing PCDD/Fs isolated from a real fish sample. DIC based on *m/z* 390 and 374. Concentrations are in the range of 2–3 pg/μL (A: 1,2,3,4,7,8-HxCDF; B: 1,2,3,7,8,9-HxCDF; C: 2,3,4,6,7,8-HxCDF; D: 1,2,3,4,7,8-Hx CDD; E: 1,2,3,6,7,8-HxCDD; F: 1,2,3,7,8,9-HxCDD). (Reproduced from (26) with permission from Elsevier).

Enhanced Detectability. Compared to 1D-GC separation, pronounced improvement of detection limits in GC×GC system is obtained thanks to compressing the peak in the modulation capillary and/or front part of the second column (following fast chromatography avoids band broadening of focused peaks). Furthermore, thanks to improved separation of analytes and matrix interferences (chemical noise) in the GC × GC system, typically the signal to noise ratio is also improved.

Structured Chromatograms. Due to complementary separation mechanisms occurring in both columns, the chromatograms resulting from particular GC × GC set-up are ordered, i.e. molecules have their definite locations in the retention space based on their structure. In the reconstructed 2D contour plots, characteristic patterns are obtained, in which the members of homological series differing in their volatility are ordered along the first dimension axis (a non-polar capillary is typically employed in first dimension), whereas the compounds differing by polarity separated on other interaction principles are spread along the second dimension axis. The formation of clusters of the various subgroups of compounds in a GC × GC contour plot may be useful for the group type analysis (Fig. 6); in this way, information supporting identification of isomeric species is obtained.

Improved Identification of Unknowns. Non-target screening allows obtaining (even retrospectively) an overview of the sample constituents. This approach consists from: (1) peak finding and deconvolution (algorithm for recognising of partly coeluting peaks in the GC–MS chromatogram and obtaining their "pure" mass spectra); (2) library searching, and (3) further post-processing. Since a large amount of data has to be processed, automated data processing is employed.

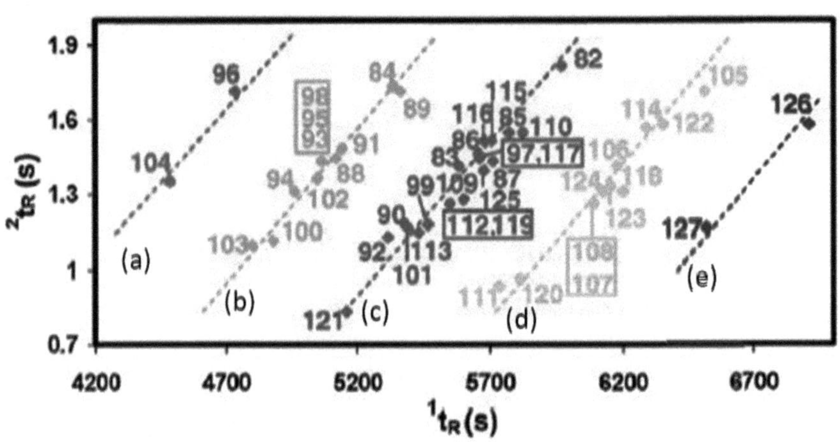

Fig. 6. GC × GC–TOFMS apex plots of the penta-CB homologue groups using the HT-8 (50 m × 0.22 mm × 0.25 μm) × BPX-50 (2.5 m × 0.1 mm × 0.10 μm) column set. A separate apex plot is shown for each homologue group. The *ortho* substitution level is described as follow: (a) tetra-*ortho*-CBs, (b) tri-*ortho*-CBs, (c) di-*ortho*-CBs, (d) mono-*ortho*-CBs, (e) non-*ortho*-CBs. Boxes represent coeluting congeners. (Reproduced from (27) with permission from Elsevier).

4.3. Sample Ionisation and Detection Using MS

Currently, mass spectrometric detectors (MS) play a dominating role in analysis of food since they are by far the most powerful and flexible tools. The advantage over "conventional" GC detectors (e.g. ECD, NPD, FPD) is the possibility to obtain, in addition to selective detection of analyte eluted at certain retention time, also structural information, enabling either confirmation of target compounds, or identification of unknown species, supposing full spectral information was acquired during the run (17). Following the introduction of relatively inexpensive single quadrupole and ion trap mass analysers in the middle of the 90s of the last century, a lot of innovative developments have occurred in the field of mass spectrometry. Atmospheric pressure ionisation in LC–MS caused extensive spreading of its use for food analysis including determination of the parent POPs polar metabolites not amenable to GC. Also, the introduction of and, at the end of the last century, the rediscovery of time-of-flight mass analysers allowed the use of this sophisticated instrumental technique in both research as well as routine applications (37). The choice of an optimal mass analyser to be coupled to respective separation systems is determined by several key operating parameters such as (1) mass range, (2) mass resolution/mass resolving power, (3) mass accuracy, (4) spectral acquisition speed, (5) acquisition mode, (6) detectability (or often expressed as sensitivity) of the instrument, (7) linear dynamic range, (8) availability of tandem MS function, (9) versatility, and (10) cost (38). The general specifications and features of selected mass analysers hyphenated to both GC and LC are shown in Tables 5 and 6.

The most common analytical techniques used in POPs/PAHs analysis are summarised in Table 7. In the text below, more details for particular techniques are given.

Organochlorine pesticides. The types of MS instruments used for pesticide residue analysis include single mass analysers (quadrupole, ion trap operated in full scan, and TOF). When employing SIM (quadrupole) and MS/MS (ion trap, triple quadrupole) modes, settings of time segments are typically needed, which may limit the number of targeted analytes that can be detected in a particular time slot at desired low levels. The trade-off with SIM relates to the difficulty of identifying analytes due to fewer ions monitored and higher chance of matrix interferences as compared to MS/MS (16, 17, 38).

Recent progress in instrumentation design as well as the use of fast recording electronics together with improvements of signal processing techniques has led to rediscovery of TOF mass analysers for the determination of a wide range of pesticide residues. As indicated earlier, GC–TOFMS has been demonstrated as a powerful tool not only for quantification of target analytes but also for identifying non-targeted compounds in complex matrices (37).

Table 5
General specifications and features of selected mass analysers coupled to gas chromatography[a]

Criteria	Scanning: Quadrupole	Triple quadrupole	Ion trap	DF magnetic sector	Non-scanning: High-speed TOF	High-resolution TOF
Mass range	Up to 1,200 Da	Up to 1,500 Da	Up to 1,000 Da	Up to 4,000 Da	Up to 1,000 Da	Up to 1,500 Da
Mass resolution/mass resolving power	Unit mass	Unit mass	Unit mass	>10,000 (10% valley definition)	Unit mass	>7,000 (FWHM)
Mass accuracy	0.1–0.2 Da	0.1–0.2 Da	0.1–0.2 Da	<5 ppm	0.1 Da	<5 ppm
Maximal spectral acquisition speed (m/z 50–550 Da)	12,500 amu/s (i.e. theoretically 25 Hz)		5 scans/s	7 scans/s	500 spectra/s	20 spectra/s
Acquisition mode	Full scan, SIM	Full scan, SIM, SRM, MRM	Full scan, SRM, MRM	Full scan, SIM	Full spectra	Full spectra
Sensitivity	>pg in full scan, fg in SIM	>pg in full scan, <pg in MS/MS	>pg in full scan	fg in SIM	pg	fg–pg
Linear dynamic range (orders of magnitude)	>5	>5	4–5	>5	4	4
Ionisation	EI, PCI, NCI	EI, PCI, NCI	EI, PCI, NCI	EI, PCI, NCI	EI	EI, PCI, NCI
MS/MS	None	MS^2	MS^n, $n = 2–10$	Only with special configuration	None	None
Cost	+	+++	+	++++	+++	+++

[a] *DF* double focussing, *EI* electron ionisation, *MS* mass spectrometry, *NCI* negative chemical ionisation, *PCI* positive chemical ionisation, *SIM* selected ion monitoring, *SRM* selected reaction monitoring, *MRM* multiple reaction monitoring, *pg* picogram, *fg* femtogram

Table 6
General specifications and features of selected mass analysers coupled to liquid chromatography[a]

	Scanning			Non-scanning		
Criteria	**Quadrupole**	**Triple quadrupole**	**Ion trap**	**High-speed TOF**	**High-resolution TOF**	**Orbitrap**
Mass range	Up to 1,000–3,000 Da	Up to 1,000–3,000 Da	Up to 2,000–6,000 Da	Up to 6,000 Da	Up to 30,000 Da	Up to 4,000 Da
Mass resolution/mass resolving power	Unit mass	Unit mass	Unit mass	>2,000 FWHM	>40,000 FWHM	Up to 100,000 FWHM
Mass accuracy	0.1–0.2 Da	0.1–0.2 Da	0.1–0.2 Da	<15 ppm	<3 ppm	<2 ppm
Acquisition mode	Full scan, SIM	Full scan, SIM, SRM, MRM	Full scan, SRM, MRM	Full spectra	Full spectra	Full spectra
Maximal spectral acquisition speed (m/z 50–1,000 Da)	≈6 scans/s		≈2 scans/s	100 spectra/s	40 spectra/s	10 spectra/s
Sensitivity	>pg in full scan, <pg in SIM	>pg in full scan, <pg in MS/MS	>pg in full scan	>pg	pg	pg
Linear dynamic range	>5	>5	4–5	>3	4–5	>4
Ionisation	ESI, APCI	ESI, APCI, APPI	ESI, APCI	ESI	ESI, APCI, APPI	ESI, APCI, APPI
MS/MS	None[b]	MS^2	MS^n, n=2–10	None[b]	None[b]	None[b]
Cost	+	+++	+	+++	+++	+++

[a] *APCI* atmospheric pressure chemical ionisation, *APPI* atmospheric pressure photoionisation, *ESI* electrospray ionisation, *FWHM* full width at half maximum, *MS* mass spectrometry, *SIM* selected ion monitoring, *TOF* time-of-flight, *SRM* selected reaction monitoring, *MRM* multiple reaction monitoring, *pg* picogram, *fg* femtogram
[b] MS/MS option only when hyphenated, such as QTOF

Table 7
Overview of the most common analytical techniques used in POPs and PAHs analysis[a]

Group	Separation	Ionisation	MS detection
Organochlorine pesticides	GC	EI	Q, IT, TOF, QqQ
Polychlorinated biphenyls	GC	EI, NCI	Q, IT, TOF
Polychlorinated dibenzo-*p*-dioxins and dibenzofurans	GC	EI	Sector, IT (MS^2), TOF (in combination with GC×GC)
Brominated flame retardants			
Polybrominated diphenyl ethers and related brominated aromatic compounds	GC	NCI	Q, IT, sector, TOF
Hexabromocyclododecane	GC LC	NCI ESI(−)	Q, IT, sector, TOF QqQ
Tetrabromobisphenol A	GC (deriv.) LC	EI ESI(−)	Q, IT, TOF QqQ
Perfluoroalkylated substances			
Fluorotelomer alcohols Perfluoroalkyl sulfonamidoethanols Perfluoroalkyl sulfonamides	GC	PCI (NCI)	Q, IT, TOF
Perfluoroalkyl sulfones Perfluorocarboxylic acids Perfluoroalkyl sulfonamides	LC	ESI(−)	Q, QqQ, QTOF
Polycyclic aromatic hydrocarbons	GC LC	EI APPI(+)	Q, IT, TOF, QqQ QqQ

[a] *APPI*(+) atmospheric pressure photoionisation in positive ion mode, *GC* gas chromatography, *GC× GC* comprehensive two-dimensional gas chromatography, *EI* electron ionisation, *ESI*(−) electrospray ionisation in negative ion mode, *IT* ion trap, *LC* liquid chromatography, *MS* mass spectrometry, *NCI* negative chemical ionisation, *PCI* positive chemical ionisation, *TOF* time-of-flight, *Q* quadrupole, *QqQ* triple quadrupole, *QTOF* quadrupole/time-of-flight

TOFMS can be performed emphasising either high speed (unit mass resolution) or high resolution, albeit at slower acquisition speed. In the case of high-speed TOFMS, GC×GC is often used for good resolution of sample components, permitting unbiased identification of sample components and achieving low LODs. The advantages of high-resolution TOF over common mass analysers (unit resolution quadrupoles and ion traps) in residue analysis, can be summarised as follows: (1) Acquisition of spectral data across a wide mass range is possible at any time during the GC run without a decrease in detection sensitivity (i.e. full spectral information searchable in spectral library is available within the elution of sample component). (2) Due to a high mass resolving power, matrix components yielding ions with the same nominal mass (isobaric interferences) as that of the target analyte can often be partially or completely resolved, and hence do not

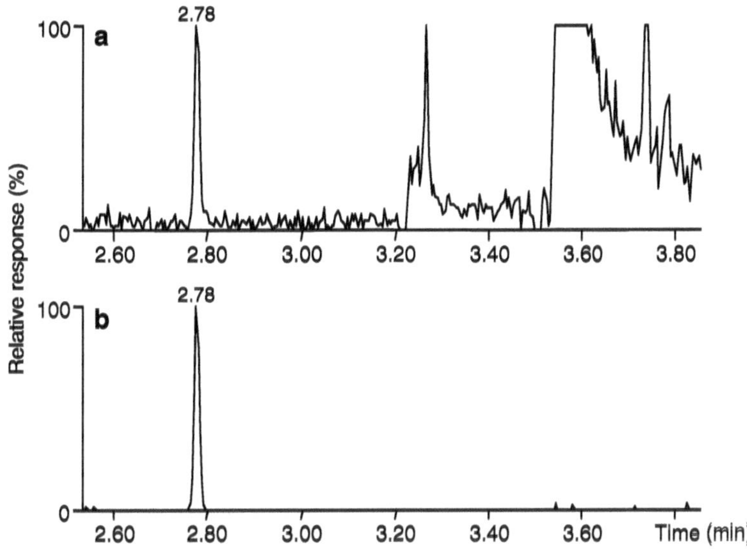

Fig. 7. PTV–LPGC–HRTOFMS chromatograms of hexachlorobenzene (t_R = 2.78 min) at a concentration of 0.003 mg/kg in apple baby food extract (2 μL injected). Target ion *m/z* 283.810 extracted using different mass windows. While by using (**a**) 1 Da mass window (setting corresponding to a unit mass resolution instrument) the peak-to-peak (PtP) signal-to-noise (S/N) ratio was only 9:1, setting the mass window as narrow as (**b**) 0.02 Da led to a distinctly improved PtP S/N value of 63:1. Mass measurement accuracy allowed determination of the mass of hexachlorobenzene's quantification ion $[C_6{}^{35}Cl_5{}^{37}Cl]$ with the error as low as +0.7 mDa.

interfere. (3) Mass measurement accuracy permits estimation of the elemental composition of the detected ions (39). An example of improved detection of HCB at trace level is demonstrated in Fig. 7. Because of high mass resolving power and monitoring of the exact mass of a target analyte, chemical noise originating from various sources (e.g. matrix coextracts, column bleed) can be significantly reduced, resulting in an improved limit of quantification (LOQs).

Various ionisation techniques are also possible in GC–MS; nevertheless, in the case of OCPs, electron ionisation (EI) is commonly preferred. Chemical ionisation (CI) in both positive (PCI) and negative (NCI) modes, as a softer ionisation technique, tends to give lower LODs depending on the pesticide, but it is not as widely applicable in OCPs methods and does not provide as much structural information about the analyte as EI (16).

A practical example demonstrating improvement of detection of heptachlor present at trace level in fish oil extract is shown in Fig. 8; in this case the use of tandem MS led to unambiguous identification and quantification of this analyte compared to single MS (SIM mode), or by using conventional detection (ECD) (40).

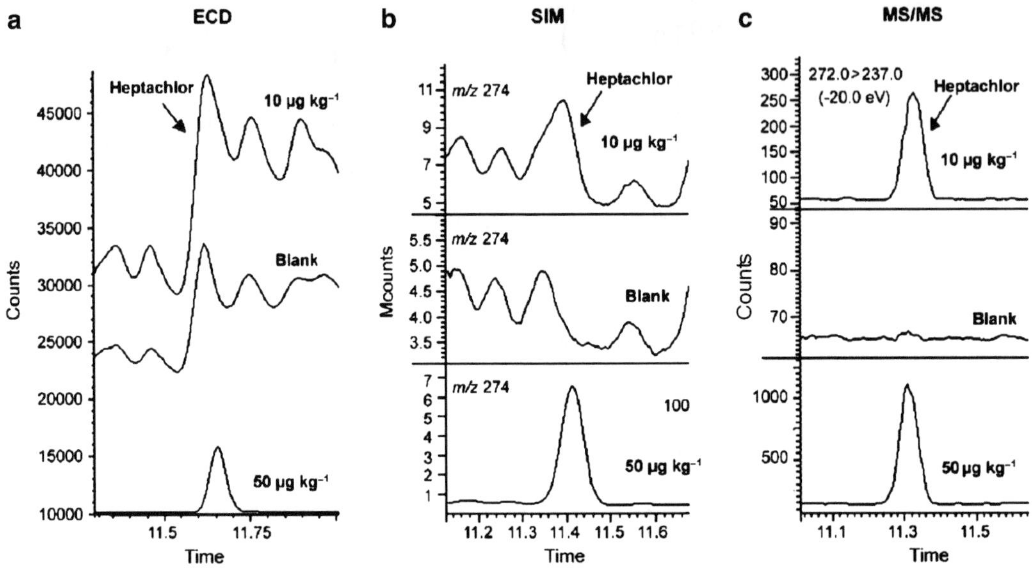

Fig. 8. Comparison of selectivity between (**a**) electron-capture detector (ECD), (**b**) selected ion monitoring (SIM) of a single quadrupole, and (**c**) tandem MS (MS/MS) of a triple quadrupole for heptachlor at 0.01 mg/kg in fish oil. Chromatograms of the corresponding blank extract and a higher matrix-matched standard are provided for contrast and identification (Reproduced from (40) with permission from Elsevier).

Polychlorinated biphenyls. GC–MS represents a reliable technique for PCBs quantification, particularly given by the availability of ^{13}C-labelled PCB standards. Although EI mode can be used, the NCI allows obtaining lower LODs of these analytes. Both single quadrupole (SIM) and ion trap (MS/MS) are frequently employed in the routine analysis of PCBs. In addition to these detection principles, triple quadrupole is becoming more and more popular in the analysis of PCBs. The use of high-resolution MS permits quantification of lower PCBs differing by two chlorines because of the high mass resolving power that allows unbiased measurement of ions. Also, the application of both high-resolution and high-speed TOFMS (the latter in combination with GC×GC replacing the "heart cut" (GC–GC) approach) represents a tool successfully applied in the analysis of PCBs (5, 38).

Polychlorinated dibenzo-p-dioxins and dibenzofurans. Quantitative determination of PCDDs/PCDFs is typically performed using GC coupled to high-resolution MS (double focusing sector analyser). Although PCI/NCI techniques can be used for their determination, a majority of laboratories employ EI. High-resolution systems (mass resolving power of >20,000 FWHM) provide higher selectivity compared to unit mass resolution instruments especially when the levels of potentially interfering compounds are too extensive (41). However, this instrumentation is very expensive, bulky, and requires operation by a highly trained specialist. Therefore, alternate analytical instruments (less expensive) have been investigated for dioxin analysis in several laboratories. GC–MS/MS employing the ion trap analyser and GC × GC–TOFMS have been reported as a valuable technique for improved selectivity in dioxin analysis. In the case of GC–MS/MS the high selectivity (minimised chemical noise → increasing of S/N → lowered LOD) is obtained due to formation of characteristic dioxin product ions produced by the collision induced dissociation from the precursor ion(s), while in GC × GC–TOFMS the improvement of selectivity is achieved employing the secondary column with different selectivity that can better separate the target compounds from co-eluting matrix components (42, 43).

Brominated flame retardants. The determination of PBDEs and related brominated aromatic compounds is performed by GC–MS operated either in EI or NCI mode. The low-resolution MS is routinely applied compared to the high-resolution MS that requires more experienced users and is much more costly and labour intensive. The high-resolution MS (sector) has several advantages over low-resolution MS (e.g. increased sensitivity and selectivity), but is almost exclusively operated in EI mode. For low-resolution MS, NCI, in addition of EI, can be applied to obtain an increased sensitivity for higher-brominated BDE congeners (8, 10). Recently, the application potential of high-resolution TOFMS under NCI conditions in the analysis of PBDEs has been demonstrated (44). EI is preferred in the analysis of PBDEs, whenever the identification of mixed organohalogenated compounds has to be carried out. Another advantage of EI mode is the possibility to use ^{13}C-labelled internal standards. This is not applicable in NCI, since generally only the $[Br]^-$ ions (*m/z* 79 and 81) are monitored. The main benefits of NCI include efficient ionisation, lower LODs, and less fragmentation compared with EI. Recently, the application of LC techniques for the analysis of PBDEs has been described. The use of atmospheric pressure chemical ionisation (APPI) in negative mode was found to be a promising tool mainly for the BDE 209 congener, seeing the difficulties encountered for this congener during GC–MS analysis (8).

Traditionally, HBCD has been analysed using GC–MS operated in NCI (similarly to PBDEs) for which the $[Br]^-$ ions are monitored because of their high selectivity (8, 10). In the case of LC employed for the isomer-specific determination, electrospray ionisation (ESI) or APCI are utilised for ionisation. Using LC–ESI-MS/MS and single reaction monitoring, the transition $[M\text{–}H]^-$ (*m/z* 640.6) → $[Br]^-$ (*m/z* 79 and 81) is monitored. The derivatised product of TBBPA is ionised typically under GC–EI-MS conditions followed by its detection using a single mass analyser (e.g. quadrupole, ion trap, TOF). In the case of direct analysis of TBBPA employing LC, ESI in negative ion mode combined with tandem MS (e.g. QqQ, IT) is most commonly used (8).

Perfluoroalkylated substances. For GC-amenable perfluoroalkylated substances (fluorotelomer alcohols, perfluoroalkyl sulfonamidoethanols, and perfluoroalkyl sulfonamides) EI is not useful because of the low intensity of molecular ions and the lack of specific fragments. However, this is not the case when PCI is employed for the ionisation. The fluorotelomer alcohols provide intensive protonated molecules ($[M+H]^+$) if methane is used as a reagent gas, but also some other useful selective (high *m/z*) fragments and/or adduct ions ($[M+C_2H_5]^+$). The perfluoroalkyl sulfonamides give also $[M+H]^+$ ions in PCI, but no suitable fragments, therefore, in this particular case, NCI can be used for their qualitative confirmation (formation of high *m/z* fragments). In the same way, the perfluoroalkyl sulfonamidoethanols provide intensive $[M+H]^+$ ions in PCI and also some fragments, but NCI can be also used for qualitative confirmation (20). Regarding the detection, all common GC–MS instruments (quadrupole, ion trap, and TOF) can be used for their analysis (the only requirement is the availability of chemical ionisation). However, the detection limits remain the limitation since these compounds occur in biotic matrices at ultra-trace levels.

For LC-amenable perfluoroalkylated substances (perfluoroalkyl sulfones, perfluorocarboxylic acids, and perfluoroalkyl sulfonamides) electrospray ionisation in negative mode coupled to either single MS or tandem MS has enabled to improve the analysis of these compounds. LC with a single MS (e.g. quadrupole), though a sensitive technique, requires more thorough clean-up of the sample in order to remove matrix interferences. Therefore, LC with tandem MS employing QqQ, IT, or QTOF can be considered as the current standard for the analysis of LC-amenable perfluoroalkylated substances (11, 21).

Polycyclic aromatic hydrocarbon. GC–EI-MS operated in SIM mode (quadrupole) represents probably the most common GC technique for determination of PAHs in food matrices. The problem encountered in the analysis of PAHs is separation of isomers and limited EI fragmentation, which does not allow reliable confirmation at ultra-trace levels. Although HPLC with a fluorescence detector (FLD) is also often routinely used, unfortunately, some of carcinogenic PAHs do not provide a fluorescence yield. LC–APCI-MS allows determination of PAHs without derivatisation (post column), which is typically required in LC–ESI-MS. The recently developed APPI enhances the ionisation of the PAH analytes, thus, lowering LODs. To improve selectivity, tandem MS (triple quadrupole) is preferred for their determination (22, 23).

5. Comprehensive POPs/PAHs Profiling

Most analytical methods for POPs focus on individual groups of targeted analytes. Therefore, analysis of multiple classes of POPs typically entails several sample preparations, fractionations, and injections, whereas other chemicals of possible interest are neglected or lost. A comprehensive POPs profiling is a novel instrumental approach employing GC×GC–TOFMS. Thanks to the recent revival of TOFMS instruments several hundreds of analytes, belonging to different classes of organic pollutants such as PCBs, PAHs, BFRs, pesticides can be theoretically measured in one run. During recent years, some effort has been spent to develop such profiling approach resulting in the introduction of GC×GC–TOFMS methods, typically in combination with large volume injection to achieve low LODs of target compounds, allowing simultaneous analysis of various groups of POPs/PAHs in food and environmental matrices (18, 28, 29, 31, 36). The main benefits of such a strategy involve: (1) more efficient monitoring of POPs, POP-like compounds, and other chemicals of interest in food; (2) possibility of non-target screening (even retrospectively) since full spectral information are acquired during the GC×GC run, and (3) significantly higher sample throughput.

6. Matrix Effects

Under real-world conditions, some residues of matrix co-extractives unavoidably remain in the purified sample prepared for examination by GC or LC analysis. Inaccurate quantification, decreased method ruggedness, poor analyte detectability, and even reporting of false positive or negative results are the most serious matrix-associated problems, which can be encountered (45, 46).

Matrix-induced chromatographic response enhancement is presumably the most discussed matrix effect adversely impacting quantification accuracy of certain, particularly more polar analytes during GC analysis. In principle, during injection of particular compounds in pure solvent, adsorption and/or thermo-degradation of susceptible analytes on the active sites (mainly free silanol groups) present in the GC injection port and in GC chromatographic column may occur. On this account, the number of analyte molecules reaching GC detector is reduced. This is, however, not the case when a real-world sample is analysed. Co-injected matrix components tend to block the active sites in GC system thus reducing the analyte losses and, consequently, enhancing their signals as compared to the injection in pure solvent. If these facts are ignored and calibration standards in solvent only are used for calculation of target analytes concentration, recoveries as high as even several hundred percent might be obtained. Repeated injections of nonvolatile matrix components, which are gradually deposited in the GC inlet and/or front part of the GC column, can give rise to successive formation of new active sites, which might be responsible for the effect, sometimes called *matrix-induced diminishment.* Gradual decrease in analyte responses associated with this phenomenon together with distorted peak shapes (broadening, tailing) and shifting the retention times towards higher values negatively impact ruggedness, i.e. long-term repeatability of analyte peak intensities, shapes, and retention times, performance characteristic of high importance in routine trace analysis.

On the other hand, LC–MS with atmospheric pressure ionisation (API) interface is considerably influenced by the composition of liquid entering the detector, i.e. the type and amount of organic mobile phase modifiers and volatile buffers, and also the type and amount of sample matrix components. These substances present in the injected sample can cause serious quantification problems when co-eluted with the analyte of interest; either by suppression or enhancement of the analyte signal. It is assumed that matrix components influence the efficiency of the ionisation processes in API interface (causing a mutual positive or negative effect in the amount of ions formed from the target analyte). Those components may also influence the ion formation in the ionisation process by altering the surface tension of electrospray droplets and by building adduct ions or ion pairs with the analytes. As a result of matrix suppression/enhancement phenomena, the response of an analyte in pure solvent standard may differ significantly from that in matrix sample.

Ways to compensate for matrix effects include: (1) method of standard addition (GC–MS, LC–MS); (2) use of isotopically labelled internal standards (GC–MS, LC–MS); (3) use of matrix-matched standards (GC–MS, LC–MS); and (4) use of analyte protectants (GC–MS). The latter approach offers the most practical and convenient solution to the problems associated with calibration in routine GC analysis of pesticide residues in diverse food samples. Essentially, analyte protectants are compounds that strongly interact with active sites in the GC system, thus decreasing degradation and/or adsorption of co-injected analytes. Therefore, the application of those compounds can minimise losses of susceptible analytes, thereby significantly improving their peak shapes and lowering detection limits. The analyte protectants are added to sample extracts and matrix-free standards alike to induce response enhancement in both instances, resulting in maximization and equalisation of the matrix-induced response enhancement effect. Various compounds have been evaluated as analyte protectants, and a mixture of 3-ethoxypropane-1,2-diol, L-gulonic acid γ-lactone, and D-glucitol (in MeCN extracts) was found to most effectively cover a wide volatility range of GC-amenable pesticides (47).

7. Quality Assurance

The analysis of POPs/PAHs occurring typically at ultra-trace levels requires an extensive quality assurance/quality control (QA/QC) regime to ensure required data quality objectives can be met. This includes, among others, following major areas:

1. *Calibrants*: Standard solutions (isotopically labelled) internal and syringe standards.
2. *Analytical procedure control*: Chromatographic parameters such as baseline, peak shape, resolution; monitoring the ion intensity ratios; recovery; procedural blanks; estimation of limit of detection (LOD) and quantification (LOQ); matrix effects.
3. *System performance/long-term stability*: Precision (repeatability, reproducibility); method stability and reliability; accuracy (use of spiked samples and certified reference materials, CRM); participation in relevant proficiency testing whenever possible.

The detailed discussion of QA/QC requirements (for regulatory purpose) can be found in various sources such as Commission Decision 2002/657/EC – Performance of Analytical Methods and the Interpretation of Results (48), SANCO/10684/2009 – Method Validation and Quality Control Procedures for Pesticide Residues Analysis in Food and Feed (49), Commission Directive 2002/69/EC – Sampling methods and the methods of analysis for the official control of dioxins and the determination of dioxin-like PCBs in foodstuffs (19), CITAC/Eurachem Guide – Guide to Quality in Analytical Chemistry (50) and in Notes section (see Notes 1–6).

8. Notes

Following recommendations should always be considered within POPs/PAHs analysis (8, 10, 21, 51):

1. *Blank analysis – BFRs*. The use of plastics should be reduced to a minimum in the analysis of BFRs, since they can contain a wide range of these compounds. In addition, higher concentrations of BDE 47 and BDE 99 can originate from the laboratory air; in the case of BDE 209 also the contamination originated from in-house dust has to be taken into account. Therefore, the laboratory glass should be placed in a closed area not allowing the deposition of PBDEs from the air/dust. The correction of the results by applying the analysis of procedural blanks can be applied only if the blank values are relatively constant. In the case of BDE 209 congener the analysis results

can be considerably affected by the degradation under influence of daylight (the use of UV filters at laboratory windows is highly recommended) and poor solubility (this should be checked before preparing stock solution or preparing highly concentrated extracts).

2. *Blank analysis – perfluoroalkylated substances.* The source of the procedural contamination involves a contact with laboratory materials made of, or containing, fluoropolymers (e.g. polytetrafluoroethylene or perfluoroalkoxy compounds). This can be, for example, polypropylene sample bottle, SPE cartridges, purified reagent water, nylon syringe filter, HPLC tubing, autosampler vial septum, the degasser and solvent-selection valves containing fluoropolymer coatings and seals. Therefore, during the validation of the method used for perfluoroalkylated substances, the possible sources of contamination have to be investigated and eliminated (e.g. replacing the type of SPE cartridges, washing the nylon syringe filter prior the filtration, replacing the HPLC tubing constructed from poly(tetrafluoroethylene) (PTFE) by stainless steel and polyetheretherketone (PEEK) tubing). The procedural blank should be run during each sample sequence. Special care should also be taken to blank analysis when replacing product(s) of the manufacturer by the other one(s).

3. *Internal and syringe standards.* The use of internal and syringe standards is highly recommended. A known amount of internal standard (surrogate) added at the beginning of the procedure is used to compensate for the losses throughout the analytical procedure, while the syringe standard is added before the injection for compensation of inter-injection fluctuations. As a general rule, both internal and syringe standards should not be present in the sample, should combine chromatographic and physical properties similar to those of target POPs/PAHs and should not co-elute with target and also non-target analytes (if MS cannot distinguish the co-eluting analytes based on the different mass spectra). During recent years, the number of available internal and syringe standards, in the case of POPs/PAHs, has rapidly increased. This includes the use of (1) unlabelled, (2) ^{13}C-labelled, and (3) D-labelled compounds. Although more expensive, the ^{13}C-labelled analogues are preferred over D-labelled standards due to risk of isotopic exchange process in non-deuterated solvent or with matrix components, but in the case of GC analysis of PBDEs/HBCD this limits the detection to EI-MS only.

4. *GC determinative step – BFRs.* Thermal degradation of BFRs should be checked and minimised by using short and narrow GC columns with thin films of stationary phase. The other aspects involve: (1) temperature during the sample injection and GC separation (should be high enough for high boiling

point POPs), (2) short residence times during injection (this can be achieved by using a pulsed splitless injection).

5. *GC determinative step – OCPs. p,p′*-DDT degrades into *p,p′*-DDE and/or *p,p′*-DDD, endrin into endrin ketone and endrin aldehyde in active or poorly deactivated injection port liners during GC injection that uses liners (splitless, PTV). Degradation products of endrin and *p,p′*-DDT should be checked on a regular basis (by injecting a single analyte). If the breakdown exceeds the 15% level replacement of liner (or even cutting of 10–20 cm of the front part of GC column occupied by active sites) is recommended. Also, various types of commercially available specially deactivated liners can limit the breakdown level.
6. *GC and LC determinative steps.* It is recommended to run a standard with a known amount of target analytes at the beginning and the end of each (longer) sequence. This can provide useful information on the stability of the analyte signal during the analyses and to consider instrument maintenance on the bases of signal decrease and behaviour of analyte peak shape caused by matrix coextracts (matrix effects). This includes, in the case of GC–MS replacement of the liner, cutting of 20–30 cm of the front part of GC column, cleaning of the ion source; in LC–MS, replacing the pre-column (or even the LC column), and cleaning the ion source. The analysis using LC–MS (especially if ultra-high performance LC, UHPLC, with small particles of stationary phase is employed) should always include filtration of the final extract by a syringe filter (0.22 or 0.45 μm for UHPLC or HPLC, respectively). This (simple) procedure significantly prolongs the lifetime of a particular LC column.

Acknowledgments

This chapter was financially supported by the Ministry of Education, Youth and Sports of the Czech Republic (project MSM 6046137305) and the European Commission (project "Contaminants in food and feed: Inexpensive detection for control of exposure" (acronym CON*ff*IDENCE), contract 211326-CP Collaborative Project).

References

1. Stockholm Convention on persistent organic pollutants (POPs). Available: http://chm.pops.int/default.aspx via the Internet. Accessed on Feb 1 2010.
2. Scippo, M.-L., Eppe, G., Saegerman, C., Scholl, G., Pauw, E. D., Maghuin-Rogister, G., and Focant, J.-F.: Persistent organochlorine pollutants, dioxins and polychlorinated biphenyls. In: *Comprehensive Analytical Chemistry* 51, Y. Pico (Ed.), Elsevier, Amsterdam (2008) pp. 457–506.
3. Jensen, A. A. and Leffers, H. (2008) Emerging endocrine disrupters: perfluoroalkylated substances. *Int. J. Androl.* **31**, 161–169.

4. Polycyclic Aromatic Hydrocarbons in Food [1] - Scientific Opinion of the Panel on Contaminants in the Food Chain. Available: http://www.efsa.europa.eu/EFSA/efsa_locale-1178620753812_1211902034842.htm via the Internet. Accessed on Feb 20 2010.
5. Ahmed, F. E. (2003) Analysis of polychlorinated biphenyls in food products. *TrAC-Trend Anal. Chem.* **22**, 170–185.
6. Beyer, A. and Biziuk, M. (2008) Applications of sample preparation techniques in the analysis of pesticides and PCBs in food. *Food Chem.* **108**, 669–680.
7. Ridgway, K., Lalljie, S. P. D., and Smith, R. M. (2007) Sample preparation techniques for the determination of trace residues and contaminants in foods. *J. Chromatogr. A* **1153**, 36–53.
8. Covaci, A., Voorspoels, S., Ramos, L., Neels, H., and Blust, R. (2007) Recent developments in the analysis of brominated flame retardants and brominated natural compounds. *J. Chromatogr. A* **1153**, 145–171.
9. Anastassiades, M., Lehotay, S. J., Stajnbaher, D., and Schenck, F. J. (2003) Fast and easy multiresidue method employing acetonitrile extraction/partitioning and "dispersive solid-phase extraction" for the determination of pesticide residues in produce. *J. AOAC Int.* **86**, 412–431.
10. Covaci, A., Voorspoels, S., and de Boer, J. (2003) Determination of brominated flame retardants, with emphasis on polybrominated diphenyl ethers (PBDEs) in environmental and human samples – a review. *Environ. Int.* **29**, 735–756.
11. de Voogt, P. and Saez, M. (2006) Analytical chemistry of perfluoroalkylated substances. *TrAC-Trend Anal. Chem.* **25**, 326–342.
12. Venter, A., Nefliu, M, and Cooks, R. G. (2008) Ambient desorption ionization mass spectrometry. *TrAC-Trend Anal. Chem.* **27**, 284–290.
13. Hajslova, J. and Cajka, T: Gas chromatography–mass spectrometry (GC–MS). In: *Food Toxicants Analysis*, Y. Picó (Ed.), Elsevier, Oxford (2006) pp. 419–473.
14. Hoh, E. and Mastovska, K.: Large volume injection techniques in capillary gas chromatography. *J. Chromatogr. A* **1186**, 2–15.
15. Cajka, T., Mastovska, K., Lehotay, S. J., and Hajslova, J. (2005) Use of automated direct sample introduction with analyte protectants in the GC–MS analysis of pesticide residues. *J. Sep. Sci.* **28**, 1048–1060.
16. Lehotay, S.J. and Mastovska, K.: Determination of pesticide residues. In: Methods of Analysis of Food Components and Additives, S. Otles (Ed.), Taylor & Francis, Boca Raton, 2005, pp. 329–359.
17. Careri, M., Bianchi, F., and Corradini, C. (2002) Recent advances in the application of mass spectrometry in food-related analysis. *J. Chromatogr. A* **970**, 3–64.
18. Focant, J. F., Eppe, G., Scippo, M. L., Massart, A. C., Pirard, C., Maghuin-Rogister, G., De Pauw, D. E. (2005) Comprehensive two-dimensional gas chromatography with isotope dilution time-of-flight mass spectrometry for the measurement of dioxins and polychlorinated biphenyls in foodstuffs: Comparison with other methods. *J. Chromatogr. A* **1086**, 45–60.
19. Commission Directive 2002/69/EC of 26 July 2002 laying down the sampling methods and the methods of analysis for the official control of dioxins and the determination of dioxin-like PCBs in foodstuffs (http://eur-lex.europa.eu/LexUriServ/LexUriServ.do?uri=OJ:L:2002:209:0005:0014:EN:PDF)
20. Jahnke, A., Ahrens, L., Ebinghaus, R., Berger, U., Barber, J. L., and Temme, C. (2007) An improved method for the analysis of volatile polyfluorinated alkyl substances in environmental air samples. *Anal. Bioanal. Chem.* **387**, 965–975.
21. Villagrasa, M., Lopez de Alda, M., and Barcelo, D. (2006) Environmental analysis of fluorinated alkyl substances by liquid chromatography–(tandem) mass spectrometry: a review. *Anal. Bioanal. Chem.* **386**, 953–972.
22. Poster, D. L., Schantz, M. M., Sander, L. C., and Wise, S. A. (2006) Analysis of polycyclic aromatic hydrocarbons (PAHs) in environmental samples: a critical review of gas chromatographic (GC) methods. *Anal. Bioanal. Chem.* **386**, 859–881.
23. Tamakawa, K.: Polycyclic aromatic hydrocarbons. In: *Comprehensive Analytical Chemistry* 51, Y. Pico (Ed.), Elsevier, Amsterdam (2008) pp. 599–651.
24. Dalluge, J., Beens, J., and Brinkman, U. A. Th. (2003) Comprehensive two-dimensional gas chromatography: a powerful and versatile analytical tool. *J. Chromatogr. A* **1000**, 69–108.
25. Adahchour, M., Beens, J., and Brinkman, U. A. Th. (2008) Recent developments in the application of comprehensive two-dimensional gas chromatography. *J. Chromatogr. A* **1186**, 67–108.
26. Focant, J.-F., Pirard, C., and Pauw, E. D. (2004) Automated sample preparation-fractionation for the measurement of dioxins and related compounds in biological matrices: a review. *Talanta* **63**, 1101–1113.

27. Focant, J.-F., Sjödin, A., and Patterson Jr., D. G. (2004) Improved separation of the 209 polychlorinated biphenyl congeners using comprehensive two-dimensional gas chromatography–time-of-flight mass spectrometry. *J. Chromatogr. A* **1040**, 227–238.

28. Pulkrabova, J., Kalachova, K., Drabova, L., Cajka, T., Poustka, J., and Hajslova, J. (2009) Rapid method for simultaneous determination of PCBs, PBDEs and PAHs in fish samples. *Organohal. Comp.* **71**, 2230–2235.

29. Hoh, E., Lehotay, S. J., Pangallo, K. C., Mastovska, K., Ngo, H. L., Reddy, C. M., and Vetter, W. (2009) Simultaneous quantitation of multiple classes of organohalogen compounds in fish oils with direct sample introduction comprehensive two-dimensional gas chromatography and time-of-flight mass spectrometry. *J. Agric. Food Chem.* **57**, 2653–2660.

30. Hoh, E., Lehotay, S. J., Mastovska, K., and Huwe, J. K. (2008) Evaluation of automated direct sample introduction with comprehensive two-dimensional gas chromatography/time-of-flight mass spectrometry for the screening analysis of dioxins in fish oil. *J. Chromatogr. A* **1201**, 69–77.

31. Focant, J.-F., Reiner, E. J., MacPherson, K., Kolic, T., Sjödin, A, Patterson Jr., D. G., Reese, S. L., Dormand, F. L., and Cochran, J. (2004) Measurement of PCDDs, PCDFs, and non-ortho-PCBs by comprehensive two-dimensional gas chromatography-isotope dilution time-of-flight mass spectrometry (GC × GC-IDTOFMS). *Talanta* **63**, 1231–1240.

32. Korytar, P., Covaci, A., Leonards, P. E. G., de Boer, J., and Brinkman, U. A. Th. (2005) Comprehensive two-dimensional gas chromatography of polybrominated diphenyl ethers. *J. Chromatogr. A* **1100**, 200–207.

33. Schurek, J., Portoles, T., Hajslova, J., Riddellova, K., and Hernandez, F. (2008) Application of head-space solid-phase microextraction coupled to comprehensive two-dimensional gas chromatography–time-of-flight mass spectrometry for the determination of multiple pesticide residues in tea samples. *Anal. Chim. Acta* **611**, 163–172.

34. Zrostlikova, J., Hajslova, J., and Cajka, T. (2003) Evaluation of two-dimensional gas chromatography–time-of-flight mass spectrometry for the determination of multiple pesticide residues in fruit. *J. Chromatogr. A* **1019**, 173–186.

35. Banerjee, K., Patil, S. H., Dasgupta, S., Oulkar, D. P., Patil, S. B., Savant, R., and Adsule, P. G. (2008) Optimization of separation and detection conditions for the multiresidue analysis of pesticides in grapes by comprehensive two-dimensional gas chromatography-time-of-flight mass spectrometry. *J. Chromatogr. A* **1190**, 350–357.

36. Focant, J.-F., Sjodin, A., Turner, W. E., and Patterson Jr., D. G. (2004) Measurement of selected polybrominated diphenyl ethers, polybrominated and polychlorinated biphenyls, and organochlorine pesticides in human serum and milk using comprehensive two-dimensional gas chromatography isotope dilution time-of-flight mass spectrometry. *Anal. Chem.* **76**, 6313–6320.

37. Cajka, T. and Hajslova, J. (2007) Gas chromatography–time-of-flight mass spectrometry in food analysis. *LC GC Eur.* **20**, 25–26, 28–31.

38. Cajka, T., Hajslova, J., and Mastovska, K.: Mass spectrometry and hyphenated instruments in food analysis. In: *Handbook of Food Analysis Instruments*, S. Otles (Ed.), CRC Press, Boca Raton (2008) pp. 197–228.

39. Cajka, T. and Hajslova, J. (2004) Gas chromatography–high-resolution time-of-flight mass spectrometry in pesticide residue analysis: Advantages and limitations. *J. Chromatogr. A* **1058**, 251–261.

40. Patel, K., Fussell, R. J., Hetmanski, M., Goodall, D. M., and Keely, B. J. (2005) Evaluation of gas chromatography–tandem quadrupole mass spectrometry for the determination of organochlorine pesticides in fats and oils. *J. Chromatogr. A* **1068**, 289–296.

41. Santos, F. J. and Galceran, M. T. (2003) Modern developments in gas chromatography–mass spectrometry-based environmental analysis. *J. Chromatogr. A* **1000**, 125–151.

42. Eppe, G., Focant, J. F., Pirard, C., and Pauw, E. D. (2004) PTV-LV-GC/MS/MS as screening and complementary method to HRMS for the monitoring of dioxin levels in food and feed, *Talanta* **63**, 1135–1146.

43. Hoh, E., Mastovska, K., and Lehotay, S. J. (2007) Optimization of separation and detection conditions for comprehensive two-dimensional gas chromatography–time-of-flight mass spectrometry analysis of polychlorinated dibenzo-p-dioxins and dibenzofurans. *J. Chromatogr. A* **1145**, 210–221.

44. Cajka, T., Hajslova, J., Kazda, R., and Poustka, J. (2005) Challenges of gas chromatography–high-resolution time-of-flight mass spectrometry for simultaneous analysis of polybrominated diphenyl ethers and other halogenated persistent organic pollutants in environmental samples. *J. Sep. Sci.* **28**, 601–611.

45. Hajslova, J. and Cajka, T.: Gas chromatography in food analysis. In: *Handbook of Food Analysis*

Instruments, S. Otles (Ed.), CRC Press, Boca Raton (2008) pp. 119–144.

46. Hajslova, J. and Zrostlikova J. (2003) Matrix effects in (ultra)trace analysis of pesticide residues in food and biotic matrices. *J. Chromatogr. A* **1000**, 181–197.
47. Mastovska, K., Lehotay, S. J., and Anastassiades, M. (2005) Combination of analyte protectants to overcome matrix effects in routine GC analysis of pesticide residues in food matrixes. *Anal. Chem.* 77, 8129–8137.
48. Commission Decision 2002/657/EC – Performance of Analytical Methods and the Interpretation of Results. Available: http://eur-lex.europa.eu/LexUriServ/LexUriServ.do?uri=OJ:L:2002:221:0008:0036:EN:PDF via the Internet. Accessed on Feb 1 2010.
49. SANCO/10684/2009 = Method Validation and Quality Control Procedures for Pesticide Residues Analysis in Food and Feed. Available: http://ec.europa.eu/food/plant/protection/resources/qualcontrol_en.pdf via the Internet. Accessed on Feb 1 2010.
50. CITAC/Eurachem Guide – Guide to Quality in Analytical Chemistry. Available: http://www.eurachem.org/guides/CITAC%20EURACHEM%20GUIDE.pdf via the Internet. Accessed on Feb 1 2010.
51. de Boer, J. and Wells, D. E. (2006) Pitfalls in the analysis of brominated flame retardants in environmental, human and food samples – including results of three international interlaboratory studies. *TrAC-Trend Anal. Chem.* **25**, 364–372.

ERRATUM

Multiresidue Pesticide Analysis by Capillary Gas Chromatography-Mass Spectrometry

Jon W. Wong, Kai Zhang, Douglas G. Hayward, and Chin Kai-Meng

Jerry Zweigenbaum (ed.), *Mass Spectrometry in Food Safety: Methods and Protocols*,
Methods in Molecular Biology, vol. 747, DOI 10.1007/978-1-61779-136-9_6, © Springer Science+Business Media, LLC 2011

DOI 10.1007/978-1-61779-136-9_6

The publisher regrets that in the print and online versions of this book, the name of the author of Chapter 6 is listed incorrectly, and should read "Chin-Kai Meng", rather than "Chin Kai-Meng".

The online version of the original chapter can be found at
http://dx.doi.org/10.1007/978-1-61779-136-9_6

Index

Jerry Zweigenbaum (ed.), *Mass Spectrometry in Food Safety: Methods and Protocols*,
Methods in Molecular Biology, vol. 747, DOI 10.1007/978-1-61779-136-9, © Springer Science+Business Media, LLC 2011

D

E

F

G

N

O

P

MIX
Papier aus verantwortungsvollen Quellen
Paper from responsible sources
FSC® C105338
FSC
www.fsc.org

If you have any concerns about our products, you can contact us on
ProductSafety@springernature.com

In case Publisher is established outside the EU, the EU authorized representative is:
Springer Nature Customer Service Center GmbH
Europaplatz 3, 69115 Heidelberg, Germany

Printed by Libri Plureos GmbH
in Hamburg, Germany